Studies in Surface Science and Catalysis 140

OXIDE-BASED SYSTEMS AT THE CROSSROADS OF CHEMISTRY

Studies in Surface Science and Catalysis

Advisory Editors: B. Delmon and J.T. Yates

Vol. 140

OXIDE-BASED SYSTEMS AT THE CROSSROADS OF CHEMISTRY

Second International Workshop, October 8-11, 2000, Como, Italy

Edited by

A. Gamba
Università degli Studi dell'Insubria, Dip. Di Scienze Chimiche, Fisiche e Matematiche, 22100 Como, Italy

C. Colella
Università Federico II, DIMP Chimica Applicata, 80125 Napoli, Italy

S. Coluccia
Università di Torino, Dip. Chimica IFM, 10125 Torino, Italy

2001
ELSEVIER
Amsterdam — London — New York — Oxford — Paris — Shannon — Tokyo

ELSEVIER SCIENCE B.V.
Sara Burgerhartstraat 25
P.O. Box 211, 1000 AE Amsterdam, The Netherlands

First edition 2001

Library of Congress Cataloging in Publication Data
A catalog record from the Library of Congress has been applied for.

ISBN: 0-444-50632-2
ISSN: 0167-2991

The paper used in this publication meets the requirements of ANSI/NISO Z39.48-1992 (Permanence of Paper).
Printed in The Netherlands.

Preface

"Oxide-based Systems at the Crossroads of Chemistry", edited by Aldo Gamba, Carmine Colella and Salvatore Coluccia.

The second International Workshop on Oxide-based Systems at the Crossroads of Chemistry was held at Villa Olmo in Como, October 8-11, 2000, in the framework of A. Volta Celebrations (1996-2000), under the auspices of the Universities of Insubria, Naples Federico II and Turin, Italian National Research Council (CNR) and Italian Zeolite Association. It was attended by nearly 100 scientists, engineers and industrial researchers coming from about 20 countries.

This volume assembles reviews and research papers selected by the editors among those presented in the oral and poster sessions of the workshop. These papers represent a significant picture of the highlights of research in the field of oxides and related oxide-based materials, e.g., zeolites and layer-structured compounds. A wide range of "oxidic" materials is considered and described in terms of preparation methods, structural, physical and chemical characterisation and applications, especially as catalysts.

The application of the most powerful simulation and physical-chemical techniques and their skill in discovering and explain structural and dynamical properties of complex materials is presented. Moreover, the development of sophisticated spectroscopic and analytical techniques are shown to give decisive improvements to the growth of surface oxide science, generating new tools for the knowledge of catalysts structure and reaction mechanisms.

Lastly, a wide selection of papers, focus the most recent achievements in the application of zeolitic compounds, in the use of oxide materials in the fuel cells technology and in the understanding of the complex mechanisms of oxide-based materials in the soil system.

As a consequence to put together different experiences in a global project on oxide materials, we chose to order the papers alphabetically in the scope to furtherly unify different cultural and technological research contributions.

Contents

Studies in Surface Science and Catalysis 140
A. Gamba, C. Colella and S. Coluccia (Editors)

High-Resolution Electron Microscopy, Neutron Diffraction with Isotopic Substitution and X-Ray Absorption Fine Structure for the Characterisation of Active Sites in Oxide Catalysts

John Meurig Thomas

Davy Faraday Research Laboratory, The Royal Institution of Great Britain, 21 Albemarle Street, London W1S 4BS, U.K.; email **jmt@ri.ac.uk**; *and Department of Materials Science, University of Cambridge, Cambridge CB2 1QY*

ABSTRACT

The power of high-resolution electron microscopy (HREM) for both *ex situ* and *in situ* studies of complex oxide catalysts is illustrated with specific reference to La_2CuO_4, zeolite-L and $(VO)_2P_2O_7$. The surfaces of stoichiometric La_2CuO_4 are seen by combined HREM and X-ray emission spectroscopy to be essentially La_2O_3. Unlike the surfaces of simple oxides (e.g. those with corundum structures), those of zeolite-L are unreconstructed, and have essentially the same composition and structure as the bulk zeolite. *In situ* HREM studies (by Gai) shed considerable light on the catalytic properties of $(VO)_2P_2O_7$ in its conversion of butane to maleic anhydride. Isotopic substitution of ^{62}Ni for natural nickel in NaNi-exchanged zeolite Y catalysts (for the cyclotrunerisation of acetylene) offer novel insights into the initial act of bonding (via a π-complex) between C_2H_2 and individual Ni^{2+} active sites in the zeolite. The great advantages attendant on the use of X-ray absorption fine structure (XAFS) for *in situ* studies of active site participation Ti-SiO_2 and FeAlPO-31 catalysts are also illustrated.

1. HISTORICAL INTRODUCTION

"After a long silence, for which I will not try to excuse myself, I have the pleasure of communicating to you, Sir, and through you to the Royal Society, some striking results which I have obtained in following up my experiments ...... the electricity developed by the mere contact of different metals ......"

These words were read by Sir Joseph Banks, President of the Royal Society in London in mid-April, 1800. They constitute the opening sentence of Alessandro Volta's letter, despatched from Como on 20th March, 1800. I chose to open this lecture with these historical comments because, like others here this evening, I rejoice in being in Como, the birthplace of Plinius the Elder and Plinius the Younger, and also the place where Alessandro Volta was born and where he died. In common with millions of scientists world-wide, I have revered the memory of Volta ever since I first heard of his work. That reverence was enhanced when, nearly twenty years ago, my dear friend, the late Massimo Simonetta, with whom I had an exciting collaboration up until he passed away, took me to Pavia, where Volta was Professor. Later, more than a decade ago, my family and I came to this wonderful place, where I purchased (in the Volta Museum) a copy of Bertini's famous painting of the occasion when Volta demonstrated his pile to the Emperor Napoleon in 1801.

My predecessor but seven as Director of The Royal Institution (RI), Michael Faraday, greatly admired Volta, who gave him a special Voltaic pole that Faraday frequently demonstrated at the RI, and which we still have (and exhibit).

In October 1842, Faraday received a letter from W.R. Grove (a Welsh lawyer-scientist working in London), describing the fuel cell, which Grove had invented. The opening of that letter is rather evocative:

"I have just completed a curious voltaic pile, which I think you would like to see ......"

When Faraday died in 1867, his successor John Tyndall (famed for the Tyndall effect, and also as the first person to describe the greenhouse effect and global warming) wrote a brilliant biography entitled *"Faraday as Discoverer"*[1]. Tyndall was a fine stylist and wrote with mellifluous charm. I should like to draw to your attention a passage in that book, which has particular resonance for this occasion when this Symposium honours Volta's name:

"In one of the public areas of the town of Como stands a statue with no inscription on its pedestal, save that of a single name 'Volta' ...... "

Tyndall then proceeds to describe in beautiful English prose what Volta did, and the controversy that raged for some considerable time as to the origin of the electromotive force of a voltaic pile. Tyndall's words merit repetition here, for they echo the essence of many subsequent scientific disputes.

"The objects of scientific thought being the passionless laws and phenomena of external nature, one might suppose that their investigation and discussion would be completely withdrawn from the region of the feelings, and pursued by the cold dry light of the intellect alone. This, however, is not always the case. Man carries his heart with him into all his works. You cannot separate the moral and emotional from the intellectual; and thus it is that the discussion of a point of science may rise to the heat of a battle-field".

2. BACKGROUND

The only major *ex situ* technique that I propose to dwell upon is high-resolution (transmission) electron microscopy (HRTEM) which, nowadays (see later) can also be used for *in situ* studies. Its value as a tool in *post mortem* (as well as *pre natal*) studies of many catalysts has been highlighted on numerous previous occasions — see, for example, refs. [2-5]. I also wish to say a few words about electron crystallography[4].

So far as microporous oxide catalysts are concerned — and here I focus predominantly on zeolitic ones or their aluminophosphate analogues — the great merit of HRTEM[4-7] is that it can be used to "read off", in real-space images, the symmetry elements of the structure, and often yield a reasonably accurate value of the number of tetrahedral sites in a pore aperture. (In general, 5-, 6-, 8-, 10- or 12-membered (T sites) can be straightforwardly discerned from the HRTEM images recorded down the appropriate zone axis). This real-space approach also proves invaluable in identifying intergrowth structures. I have given elsewhere[8] an account of how several of the ostensibly different members of the ZSM family of synthetic zeolites turned out to be no more than intergrowth variants of well defined end-members. HRTEM, twenty years ago, was shown[9] to be capable of "seeing" in real space the intergrowth variants of ZSM-5 and ZSM-11.

With CCD detectors and other instrumental advances[4,6] the age of electron crystallography is well and truly with us. Small, ultrathin specimens of zeolites (or mesoporous silicas) can be processed by a combination of electron diffraction and HRTEM imaging, so as to yield a full, hitherto unknown, structure, thanks to the pioneering work of Terasaki and his colleagues in Tohuku University.

Surface structures of complex oxides are also amenable to direct imaging and determination by HRTEM (which has the advantage over scanning tunnelling microscopy in being able, through the observation of parallel X-ray emission spectra, to yield elemental composition)[10]. There are two examples which I wish to cite.

First, La_2CuO_4, a material that is of great relevance in the field of warm superconductors, has been studied in detail by my former colleague Wuzong Zhou, now of the University of St. Andrews. A strictly stoichiometric powdered sample of La_2CuO_4 was viewed down the [110] direction by HRTEM. The outline of the [001] surface is seen in Figure 1.

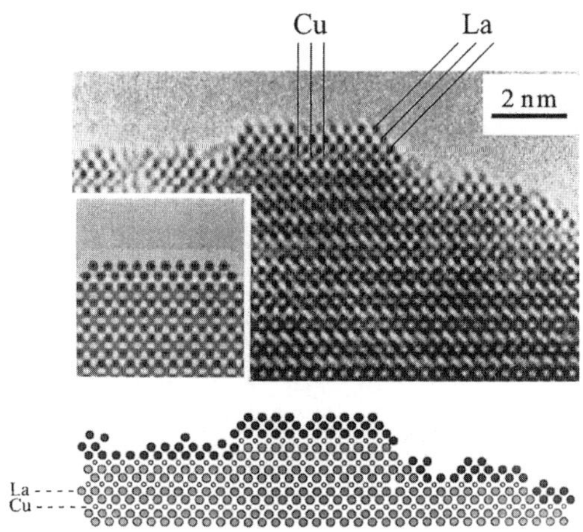

Figure 1: A typical region of the surface of La_2CuO_4 viewed down the [110] direction. The outer layers of the solid have the structure and composition of La_2O_3 (after Zhou *et al*[11])

Significantly, it is found that the surface composition does not correspond to La_2CuO_4. The images leave no doubt (and microanalysis confirms it) that the exterior surfaces (the last three layers or so) have the so-called C-La_2O_3 structure and composition[11]. Zhou and I have recently reported[10] that, in the warm superconductor $HgBa_2CuO_{4+\delta}$, the exterior and immediate sub-surface region is stoichiometrically quite different from the bulk. There are well-known reasons, discussed by Freund[12] recently, which lead us to expect polar surfaces (of oxides) to undergo reconstruction on exposure to vacuum (or an ambient gas). What is surprising here, in this powdered specimen of (nominally) La_2CuO_4, is that there is extensive reconstruction and reconstitution at the surface regions.

Low-energy electron diffraction (LEED) studies conducted by Freund *et al*[13], show that, in simple oxides belonging to the corundum family (e.g. Al_2O_3, Cr_2O_3 and Fe_2O_3), there is quite extensive reconstruction (in a direction perpendicular to the [0001] surface), amounting to between 50 and 60 percent change, compared with the inter-planar distances inside the bulk of the oxide. With the complex oxides that have zeolitic structure, however, the degree of reconstruction is rather minute. Thus, in zeolites L (idealised formula $Na_3K_6[Al_9Si_{27}O_{72}]$) Terasaki *et al* has shown[14] (Figure 2) by HRTEM that the surface structure essentially retains that of the bulk, with the individual cancrinite cages being intact right up to the last surface layer.

In addition to being able to cope with the structure determination of beam-sensitive oxides by the methods alluded to above, many other developments in HRTEM now underway are likely to contribute to further physico-chemical elucidation of oxide surfaces, simple and complex. Thus, at the *ex situ* level, following Bovin's recent classic work[15], one can already see the advantage in recording energy-filtered transmission electron micrographs (EFTEM) for the elemental mapping of 'frozen' particles. This approach of Bovin's promises to be able to follow sub-monolayer structural changes at solid surfaces undergoing reactions with liquids. There is also the (expensive) prospect of doing photo-emission electron microscopy (PEEM) using synchrotron radiation as the primary beam[16].

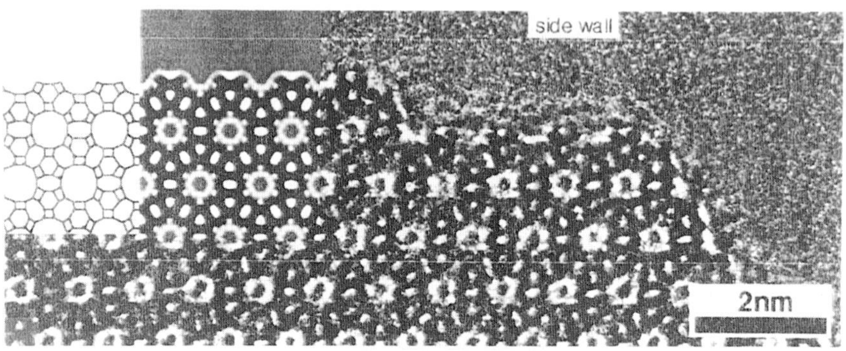

Figure 2: Unlike many simple metal oxides, such as those possessing the corundum structure, there is very little reconstruction at the outer layers of zeolite L

Gai[5,17,18] has used HRTEM successfully as an *in situ* technique in her pioneering work on the dynamic reduction of vanadyl pyrophosphate (also known as VPO) under reaction conditions. She was able to chart the formation of extended defects by a glide shear mechanism. Ordering of the glide shear (so-called GS defects) leads to a new structure by transformation of the orthorhombic vanadyl pyrophosphate [$(VO)_2P_2O_7$] into an anion-deficient tetragonal structure. The GS defects essentially preserve anion vacancies and do not lead to a collapse of the lattice (Figure 3), and are distinct from the well-known crystallographic shear (CS) planes, which eliminate oxygen ion vacancies. These GS defects, as Gai has shown elsewhere[5], have important consequences in the VPO-catalysed conversion of butane to produce maleic anhydride[19].

3. *IN SITU* STUDIES OF OXIDE CATALYSTS USING NEUTRON BEAMS

At first sight, neutron scattering may seem singularly ill-suited to probing the surfaces of heterogeneous catalysts. Compared with electrons or even X-rays, neutrons are only weakly scattered by matter. But neutron scattering studies come into their own with solids of very high areas and for investigating interlamellar species (as in clay intercalates). The study of zeolitic oxide catalysts has benefited enormously from neutron scattering, especially through high-resolution (time-of-flight) neutron diffraction, assisted by Rietveld powder profile refinement. An early example[20] entailed the determination of the precise position of pyridine adsorbed within the channels of a zeolite –L model catalyst. And an ever more relevant application, in the context of catalysis by zeolitic oxides, was the proof by Cheetham *et al*[21], again via the agency of Rietveld analysis, that La^{3+} ions in zeolite Y (idealised formula $Na_{59}Al_{59}Si_{133}O_{384}$) polarise their hydraion shells so much that $(LaOH)^{2+}$ ions, together with "free" protons, loosely attached to the framework, are formed (see Fig. 3.51 of ref. [8]).

Inelastic neutron scattering as a spectroscopic tool is not limited by restrictive selection rules and, despite its relatively low sensitivity, can also reveal a good deal about species attached to oxidic surfaces.

3.1 Neutron Diffraction with Isotropic Substitution (NDIS) for In Situ Studies of Complex Oxides

When the atoms present at the active sites of complex oxide catalysts constitute a small fraction of the total number of atoms in the catalytic system, advantage may be taken of the isotopic substitution method, which then yields (see below) radial distributions of the surrounding species. Take, for example, Ni^{2+} ion-exchanged zeolite Y which is known[22] to be a good catalyst for the cyclotrimerization of benzene. It so happens that there is an enormous difference in neutron scattering lengths between 'natural abundance' nickel atoms (^{nat}Ni) and the isotope ^{62}Ni. The change in scattering length in going from ^{nat}Ni to ^{62}Ni is 18.9fm, even larger than the substantial change (10.4fm) in going from ^{1}H to ^{2}H[23].

By performing neutron diffraction experiments on two samples which are identical in every respect except that a specific atom has been isotopically enriched, it is in principle possible to extract correlations involving just the isotopically enriched species[23-25]. Previously this NDIS technique has been applied to disordered systems such as liquids and glasses, but Turner *et al*[26] have recently shown that it is equally applicable to spatially distributed atoms (as in Ni^{2+}-zeolite Y) with low site occupancies.

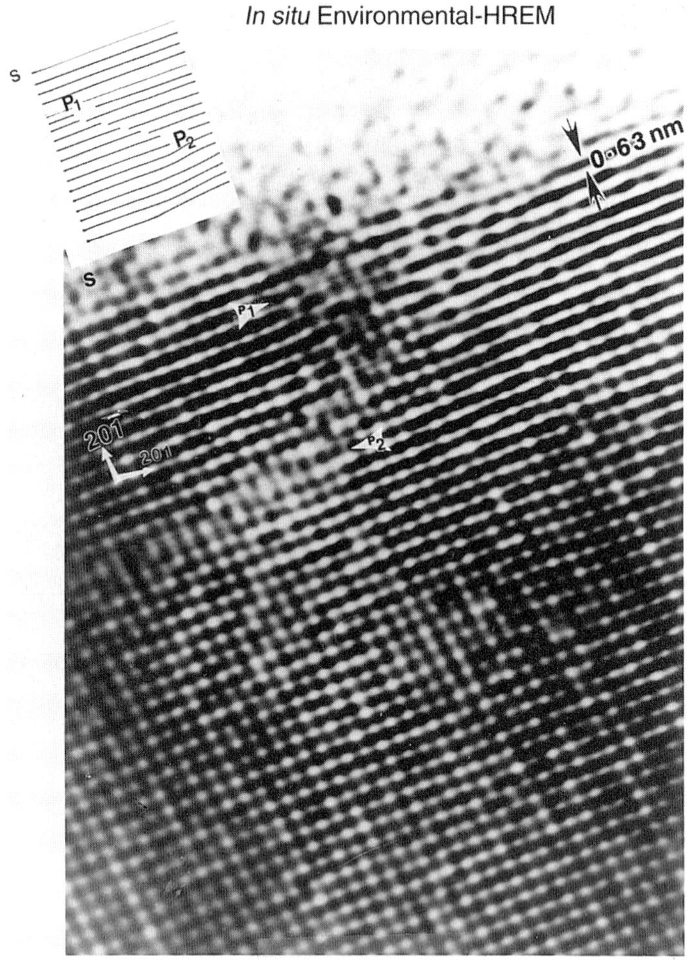

Figure 3: *In situ* studies using HRTEM (See text) reveal that so-called glide shear defects, identified by Gai[19] play an important role in the catalytic action of $(VO_2)P_2O_7$ during the oxidation of butane to maleic anhydride. P_1 and P_2 are partial screw dislocations.

The quantity measured in such an experiment is the differential neutron scattering cross section $\partial\sigma/\partial\Omega$, of which the interference part (or total structure factor) can be expressed as a linear sum of the partial structure factors, S(Q). The structure factor describes the spatial distribution of scattering centres (the atomic nuclei) of the sample in question. Thus, in the total structure factor, all distances between all scatterers are present, weighted according to the concentration of each particular type of atom, *c*, and their scattering length, *b*. The differential neutron scattering cross section can be written as:

$$\frac{\partial\sigma}{\partial\Omega} = N\left\{\sum c_\alpha b_\alpha + \sum c_\alpha b_\alpha c_\beta b_\beta \left[S_{\alpha\beta}(Q) - 1\right]\right\}$$

where the partial structure factor $S_{\alpha\beta}(Q)$ describes interaction between atoms of type α and β. $S_{\alpha\beta}(Q)$ in turn is related directly to its real space radial distribution function $g_{\alpha\beta}(r)$ through Fourier transformations[23,24].

The quintessential points to bear in mind in regard to the NDIS method is that data are collected for the zeolitic (Ni^{2+}-exchanged) catalyst (i) both in its ^{nat}Ni and ^{62}Ni (dehydrated) states, (ii) both in the $^{nat}Ni(C_2D_2)$ and $^{62}Ni(C_2D_2)$ states, at a particular loading (or series of loadings of C_2D_2). As is described in detail elsewhere[26], the real space environment of Ni^{2+} ions both in the absence and presence of the (deuterated) bound acetylene can be measured, if necessary, at a series of temperatures and acetylene pressures. These measurements then yield the precise interatomic distances in the $Ni{:}C_2D_2$ complex, which is the precursor state leading subsequently to trimerisation and benzene formation.

In the experiments of Turner *et al*[26] it transpired that the neutron diffraction data were consistent with the Ni complex of the Dewar-Chatt-Duncanson type involving π-acid coordination (Figure 4).

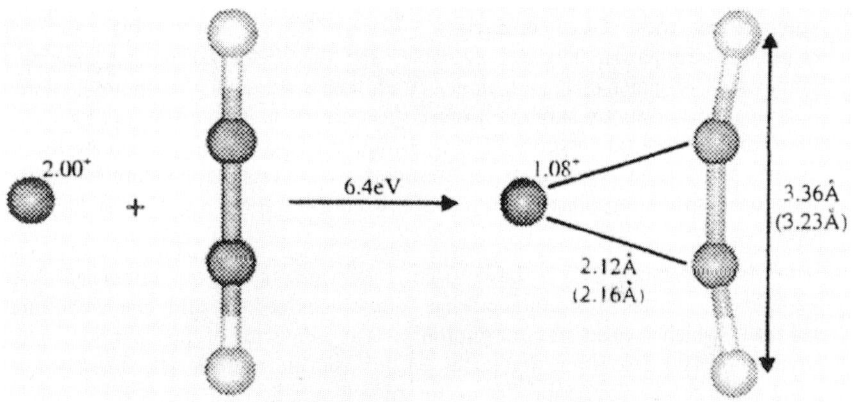

Figure 4: Ni^{2+} ions in zeolite Y form a π-complex of the Dewar-Chatt-Duncanson type. The distances shown here were deduced from neutron diffraction studies[26]. The figure shows the BLYP/DNP optimised geometries, charge on the Ni ion and binding energy for the interaction of a molecule of C_2H_2 with a bare Ni^{2+} ion

What we have outlined above is an experiment that probes the formation of the initial complex between acetylene and the Ni^{2+} ion at the active site. By repeated experiments with progressively increasing doses of (deuterated) acetylene injected into the ^{62}Ni-enriched zeolitic catalyst, one would acquire information that would chart the formation of benzene by trimerisation. There is no doubt that, given adequate neutron-beam time — a not inexpensive factor — the NDIS technique could become a powerful *in situ* technique for exploring active centres in these complex oxide catalysts. What is increasingly apparent, however, is that synchrotron radiation sources at centralised facilities present fewer logistic, and far fewer time-consuming operations, than neutron sources. The experiments that my colleagues and I have been carrying out at the Daresbury Synchrotron Laboratory in the U.K. highlight this point (see refs. [27-30]). In the next section we cite a few illustrative examples of the merits of synchrotron radiation for the *in situ* study of oxide catalysts.

4. X-RAY ABSORPTION FINE STRUCTURE (XAFS) STUDIES FOR *IN SITU* INVESTIGATIONS OF ACTIVE SITES IN OXIDE CATALYSTS

Of the various tools available to the catalytic scientist, few are more versatile and illuminating than XAFS, as has been demonstrated adequately elsewhere[27-34]. Briefly, armed with high fluxes of penetrating and continuously tunable X-rays, XAFS yields atom-specific structural and electronic information concerning bond lengths, coordination numbers and mean-square deviation in bond-distances as well as the oxidation state. And, if an appropriate *in situ* cell is used, this information may be obtained as a function of temperature and under a variety of ambient conditions involving reactant or product species. Such information may be obtained even when there is no long-range (crystallographic) order: XAFS yields strictly local, short-range structural information. When, however, the technique is used under *in situ* conditions in combination with X-ray diffraction (XRD)[28,30,35-37], both short- and long-range structural information becomes available, and this yields far greater insights than the deployment of either technique alone. With microporous or mesoporous oxide catalysts it is vital to know not only the atomic structure of an active site, but also whether the long-range structural integrity of the porous catalysts remains intact during use. The activity and selectivity of micro- or mesoporous catalysts may be completely destroyed if their open structure collapses during use, even if the local structure of the active site is preserved.

My colleagues and I have used two variants of the combined XAFS-XRD experiment. Faster speed in acquiring data is achieved with the dispersion method (see ref [38]), which uses an array of photodiodes as a detector of the absorption spectra. But greater energy resolution (typically $\pm$1.0eV in the positon of absorption edges) is achieved in the scanning mode, schematised in Figure 5.

In each case a position-sensitive detector records the diffractograms over a range of 2θ which typically extends from 10° to 60°. The scanning mode enables so-called QEXAFS (Q for quick) data to be recorded[39]. Often, because of the small concentrations of active sites present in inorganic catalysts — in the Ti/SiO_2 ones discussed below, for example, Ti/Si ratios seldom exceed 1:50 — it is not possible to acquire reliable XAFS data in the transmission mode. It then becomes essential to use a sensitive, multielement fluorescence detector such as the 13-element germanium one that we and our collaborators at Daresbury have routinely used[39-41].

Reaction cells for our studies were constructed to suit the needs of a particular set of experimental conditions. Gas-solid studies have been carried out with cells that permit temperatures up to 900°C to be explored; and also liquid-solid ones for use at lower temperatures. In each instance, the products of catalytic action may be quantitatively monitored using on-line mass-spectrometry and/or gas chromatography[29,35].

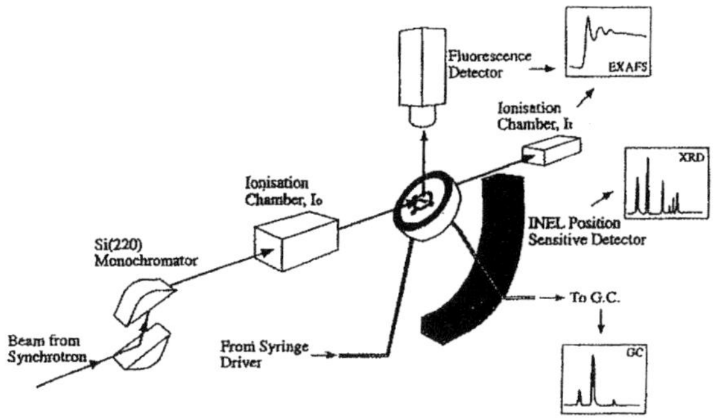

Figure 5: Set-up used for recording combined XAFS and XRD data of a solid catalyst under *in situ* conditions. This scanning arrangement (see [35] and [39] for further details) permits rapid study of traces of catalytic active centres.

4.1 Mesoporous Silica Catalysts with Ti^{IV} Centres

$Ti\text{-}SiO_2$ epoxidation catalysts are important industrially (in, for example, the production of propylene oxide from propylene). In 1995 my colleagues and I set out to determine the nature of the Ti^{IV}-centred sites[42]. Up until that time several contradictory proposals had been made concerning the nature of these centres. Some were based on a 3-coordinated titanyl group; some even on 5- or 6-coordinated centres.

Starting from titanocenedichloride, we grafted on to the inner surfaces of MCM-41 mesoporous silicas, highly active Ti^{IV}-epoxidation centres. Our *in situ* XAFS studies[42-45] left us in no doubt that the active centres are tripodally attached titanol groups: $(Si\text{-}O)_3 - TiOH$ (see Figure 6). Such centres permit facile expansion of the coordination shell (also monitored by *in situ* XAFS) during the catalytic conversion of alkenes (typically cyclohexane) with alkyl hydroperoxides (typically THBP) to the corresponding epoxide.

Armed with such information minute changes in hydrophobicity and local composition (e.g. replacement of one Si by a Ge) can be wrought so as to boost the performance of the catalyst[46].

Recently, Italian workers[47] have shown that our method of producing Ti^{IV}-centred active sites in mesoporous silica yields a catalyst that converts citronellal, in 'one-pot' fashion, into isopulegol, an important fine chemical.

Elsewhere we have summarised[37] how XAFS or combined XAFS-XRD *in situ* studies have elucidated the nature of the catalytically active sites in several oxidic systems:

(i) in FeAlPO-31 for instance, which converts cyclohexane to adipic acid in air[48], the active centres are 4-coordinated (largely tetrahedral) Fe^{III} ions linked to bridging oxygen atoms; and

(ii) in hydrotalcite structures containing copper, magnesium and aluminium in the metal hydroxide layers, we find that, for the decomposition of NO, the active sites are Cu^{I} ions whereas for the reduction of nitrogen oxides, the active sites are Cu° species[49].

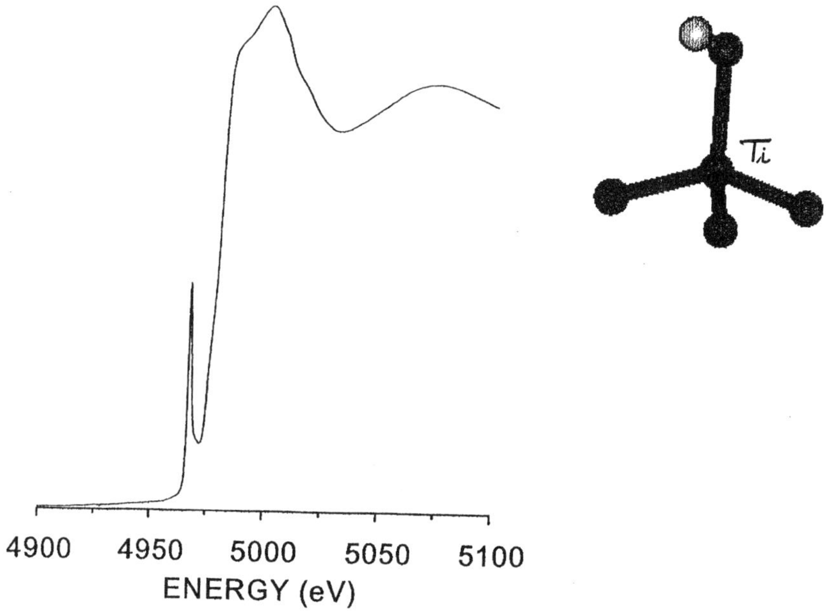

Figure 6: The Ti K-edge XANES spectrum shown here, along with the EXAFS spectrum (not shown) reveals that the catalytically active site in Ti-SiO_2 epoxidation catalysts (see text) are tripodally anchored $Ti^{(IV)}$-OH groups. Only three of the oxygens of the silica support are shown

Many other examples have been described elsewhere; and several new types of catalyst are currently under investigation using either XAFS alone or combined XAFS/XRD (*in situ*) procedures[37,50]/

I acknowledge financial support from the U.K. EPSRC, and continued devoted collaboration with my colleagues, especially G. Sankar, C.R.A. Catlow and T. Maschmeyer.

REFERENCES

1. J. Tyndall, *"Faraday as a Discoverer"*, Longmans, Green & Co., London, 1868
2. J.M. Thomas, *"Proc. 8th Intl. Congr. Catalysis, Berlin 1984"* Vol. 1, p. , VCH Weinheim, 1984
3. J.M. Thomas, D.A. Jefferson and R.F. Egerton, *Chemistry in Britain,* **17**, 514, 1981
4. O. Terasaki, T. Ohsuna, N. Ohnishi and K. Hiraga, *Current Opinion in Solid State and Mater. Sci.,* **2**, 94, 1997
5. P.L. Gai, *Topics in Catalysis,* **8**, 97, 1999
6. J.M. Thomas, O., Terasaki, P.L. Gai, W. Zhou and J.M. Gonzalez-Calbet, submitted
7. O. Terasaki, J.M. Thomas, D. Watanabe and G.R. Millward, *Chem. Mater.,* **1**, 158, 1989
8. J.M. Thomas and W.J. Thomas, *"Heterogeneous Catalysis: Principles and Practice"*, Wiley-VCH, Weinheim, Chap. 3, 1997
9. G.R. Millward, J.M. Thomas, S. Ramdas and M.T. Barlow, *J. Chem. Soc. Farad. Trans. II,* **79**, 1075, 1983
10. W. Zhou and J.M. Thomas, *Current Opinion in Solid State & Mater. Sci.,* 2001 (in press)
11. W. Zhou, D.A. Jefferson and W.Y. Liang, *Surface Sci.,* **310**, 52, 1994
12. H.-J. Freuns, *Angew. Chemie Intl. Ed. Eng.,***36**, 452, 1997
13. H.-J. Freund, M. Baumer, H. Kuhlenbeck, *Adv. Catal.,* **45**, 333, 2000
14. O. Terasaki, T. Oksuna, M. Kaneda, Y. Sakamoto, K. Hiraga, A. Carlson, R. RYoo, M.W. Anderson, *Microscopy and Microanalysis,* **6**, 10, 2000
15. J.-O. Bovin, *Microscopy and Microanalysis,* **6**, 2, 2000
16. B. Gilbert, R. Andres, P. Perfetti, G. Margaritondo, *Ultramicroscopy,* **83**, 129,2000
17. P.L. Gai, K. Kourtakis, *Science,* **267**, 661, 1995
18. P.L. Gai, *Acta. Cryst.,* **B53**, 346, 1997
19. G. Centi, *Catal. Today,* **16**, 1, 1993
20. P.A. Wright, J.M. Thomas, A.K. Cheetham and A.K. Nowak, *Nature,* **318**, 611, 1985
21. A.K. Cheetham, M.M. Eddy and J.M. Thomas, *J. Chem. Soc. Chem. Comm.,* 1337, 1984
22. P.J. Maddox, J. Stahunski and J.M. Thomas, *Catal. Lett.,* **1**, 191, 1988
23. J.E. Enderby, *Chem. Soc. Rev.,* **25**, 159, 1995
24. J.L. Finney, A.K. Soper, *Chem. Soc. Rev.,* **23**, 1, 1994
25. C.J. Benmore, P.S. Salmon, *Phys. Rev. Lett.,* **73**, 264, 1994
26. F.J.C. Turner, C.J. Benmore, C.M. Barker, N. Kaltsoyannis, J.M. Thomas, W.I.F. David and C.R.A. Catlow, *J. Phys. Chem.,* **104**, 7570, 2000

27. E. Dooryhee, G.N. Greaves, A.T. Steel, R.P. Townsend, S.W. Carr, J.M. Thomas and C.R.A. Catlow, *Faraday Discuss. Chem. Soc.*, **89**, 119,1990

28. J.W. Couves, J.M. Thomas, D. Waller, R.H. Jones, A.J. Dent and G.N. Greaves, *Nature,* **354**, 465, 1991

29. J.M. Thomas and G.N. Greaves, *Science,* **265**, 1675, 1994

30. G. Sankar and J.M. Thomas, *Topics in Catalysis,* **8**, 1, 1999

31. H. Kuroda, in *"Dynamic Processes in Solid Surfaces"* (ed. K. Tamaru), Plenum Press, New York, p.51, 1985

32. D.C. Koningsberger and R. Prins (eds.), *"X-ray Absorption: Principles, Applications, Techniques of EXAFS, SEXAFS and XANES",* Wiley, New York, 1986

33. J. Wong, F.W. Lytle, R.P. Messmer and G. Maylott, *Phys. Rev. B,* **5596, 1984**

34. J.H. Sinfelt and G. Meitzner, *Accnts. Chem. Res.,* **26**, 1, 1993

35. J.M. Thomas, *Chem. Europ. J.,* **3**, 1557, 1997

36. G. Sankar, J.M. Thomas and C.R.A. Catlow, *Topics in Catalysis,* **10**, 255, 2000

37. J.M. Thomas and G. Sankar, submitted

38. G. Sankar, J.M. Thomas, D. Waller, J.W. Couves, G.N. Greaves and C.R.A. Catlow, *J. Phys. Chem.,* **96**, 7485, 1992

39. G. Sankar, P. A. Wright, S. Natarajan, J.M. Thomas, G.N. Greaves, A.J. Dent, B.R. Dosbon, C.A. Ramsdale and R.H. Jones, *J. Phys. Chem.*, **97,** 9550, 1993

40. J.M. Thomas, G.N. Greaves and C.R.A. Catlow, *Nucl. Inst. Methods,* **B97**, 1, 1995

41. G. Sankar, J.M. Thomas, G.N. Greaves and A.J. Dent, *J. Phys. IV (France),* 7, 871, 1997

42. T. Maschmeyer, F. Rey, G. Sankar and J.M. Thomas, *Nature,* **378**, 159, 1995

43. G. Sankar, F. Rey, J.M. Thomas, G.N. Greaves, A. Corma and A.J. Dent, *J. Chem. Soc. Chem. Comm.,* 2279, 1994

44. L. Marchese, E. Gianotti, T. Maschmeyer, G. Matra, S. Coluccia, J.M. Thomas, *J. Nuovo Cimento,* **19**, 1707, 1997

45. L. Marchese, T. Maschmeyer, G. Sankar and J.M. Thomas, *Phys. Chem. Chem. Phys.,* **1**, 585, 1999

46. R.D. Oldroyd, G. Sankar and J.M. Thomas, *Chem. Comm.,* 2025, 1997

47. M. Guidotti, G. Moretti, R. Psaro and N. Ravasto, *Chem. Comm.,* 1789, 2000

48. M. Dugal, G. Sankar. R. Raja and J.M. Thomas, *Angew. Chemie Intl. Ed. Eng.,* **39**, 2310, 2000

49. I.J. Shannon, F. Rey, G. Sankar, J.M. Thomas, T. Maschmeyer, A.W. Waller, A.E> Palomanes, A. Corma, A.J. Dent and G.N. Greaves, *J. Chem. Soc. Faraday Trans.,* **92**, 4331, 1996

50. J.M. Thomas and G. Sankar, *J. Synchrotron Radiaton,* 2001 (in press)

Studies in Surface Science and Catalysis 140
A. Gamba, C. Colella and S. Coluccia (Editors)

Location of Brønsted and cation sites in dehydrated zeolites: A comparison

A. Alberti, G. Cruciani, M. C. Dalconi, A. Martucci and S. Caruso

Dipartimento di Scienze della Terra dell'Università, Sezione di Mineralogia, Corso Ercole d'Este 32, 44100 Ferrara, Italy

To date, the location and population of protons in acid forms of zeolites and those of exchangeable cations in dehydrated zeolites have been investigated as two distinct topics. This paper intends to highlight the analogies and differences in the siting of protons and extraframework cations in dehydrated zeolites. The starting point is that a framework oxygen whose bond strength from tetrahedral cations is low and/or whose bond strength from extraframework cations is high presumably has a higher probability of hosting a proton. A comparison between the bond strength on oxygen atoms and the location of hydroxyl groups in mordenite, ferrierite, chabazite, faujasite, and rho showed that this assumption is supported in the majority of cases. Consequently, bond strength information can be utilized to locate Brønsted acid sites even if neutron diffraction data are not at our disposal.

1. INTRODUCTION

Among the factors controlling catalytic activity a fundamental role is played not only by zeolite topology, but also by the type of cation, its location and coordination. Of particular importance is the location and population of proton sites of framework hydroxyls in hydrogen zeolites.

To date, no particular attention has been paid to comparing the location and population of proton and extraframework cation sites. Consequently, it could be of interest to highlight the analogies and differences between proton and extraframework cation siting in dehydrated zeolites. Unfortunately, while a lot of work has been done to locate cations, little information exists on proton location.

First of all, we must consider that peculiar features characterize and differentiate proton and cation sites.

As concerns protons, we accept that:

a) a proton can be bonded to a framework oxygen bridging an Al and a Si atom, but not bridging two Si atoms;
b) only one proton can be allocated at any one time to the oxygens of a tetrahedron.

No experimental or theoretical studies have been carried out, to our knowledge, to confirm or reject the latter assumption; however, the marked increase undergone by the Al-O and Si-O distances when the oxygen is protonated – as found by quantum chemical calculations [1,2] and experimental data [3,4] – makes the presence of two contemporaneous OH groups on the same tetrahedron extremely unlikely.

As concerns cations, we know that:

a) an extraframework cation can be coordinated to framework oxygens bridging two Si atoms;
b) the same cation can be bonded to more than one oxygen of a tetrahedron;
c) normally, in dehydrated phases, cations split over a larger number of sites when compared with hydrated ones [5-7], to improve the charge compensation of as large a number of framework oxygens as possible; in hydrated phases this charge compensation can be obtained by hydrogen bonding.

When comparing the locations of Brønsted and extraframework cation sites, we must take into account that:

1. the distance of a proton from its framework oxygen is around 1Å, whereas the distance of a cation from coordinated framework oxygens can be as much as two Å greater;
2. a proton in a Brønsted site is bonded to one oxygen, whereas the coordination number of a cation in dehydrated phases is usually greater than or equal to three; only in few cases do cations with small ionic radii display a twofold coordination.

In dehydrated zeolites, the Si/Al distribution can play an important role in guiding the location and population of ion sites, no matter if they are occupied by protons or extraframework cations. An oxygen on which bond strength from the tetrahedral cations is low has a higher probability of hosting a proton or bonding cations. This means that Brønsted acid sites are often associated to framework oxygens on which there is a strong bond strength from extraframework cations. Therefore, this strength in dehydrated cation-exchanged forms of zeolites can provide information useful for localizing Brønsted sites. The literature, in fact, abounds with studies on the structure of dehydrated cation-exchanged zeolites, whereas framework hydroxyls have, to date, been recognized only in very few zeolites; it is obvious that other factors can influence the location of protons and cations. Framework topology, for example, can favour the bonding of a proton on an oxygen atom, whereas it can hinder the location of a cation, especially if its ionic radius is large. Nevertheless, the analysis of bond strength on framework oxygens can help to locate Brønsted acid sites. This is the most important goal of the present paper. Sometimes, however, it is not easy to decide if a cation is bonded or not to a framework oxygen. As a general rule, we consider a cation as coordinated to an oxygen when its distance from the oxygen is not greater than (Rc + Ro + 0.8Å), where Rc and Ro are the ionic radii of the cation and oxygen (according to Shannon [8]) respectively.

A limitation which arises in comparing proton and cation sites is that investigations into acidic forms by X-ray or neutron diffraction, in order to obtain the location and population of Brønsted sites, have so far been carried out only for faujasite [3,4,9-12], rho [13,14], mordenite [15], chabazite [16]) and ferrierite [17]). Our work will therefore be concerned with these zeolites. One feature of all these materials, excluding faujasite, is that their acid forms show topological symmetry, whereas their dehydrated cation-exchanged forms undergo a lowering of symmetry, sometimes accompanied by remarkable distortion of the framework only for some cations: this can be an obstacle in the computation of bond strength.

In this paper protons are referred to with the symbol H, followed by the framework oxygen numbers of the Brønsted hydroxyls, which are numbered according to the sequence commonly adopted. As concerns extraframework cations, they are referred to with Roman numbers; the nomenclature commonly adopted in literature to indicate cation sites is followed whenever it is possible, and the nomenclature used by Mortier [18] in his compilation of extraframework sites is moreover reported every time this is useful or possible.

2. RESULTS AND DISCUSSION

2.1 Mordenite

The topological symmetry of mordenite is orthorhombic, space group *Cmcm*. In its framework there are four tetrahedral and eight oxygen sites, symmetrically independent. The topological symmetry is also the real symmetry of dehydrated H-mordenite, and of many of its dehydrated cation-exchanged forms. In Na-, K-, and Ba-mordenites [19-21] the framework is slightly distorted and the real symmetry is lowered to the space group *Pbcn*. In this case framework oxygens O1, O2, and O3 are split over two pseudo-symmetric positions. As this effect does not influence in any remarkable way the data and conclusion reached, their split will be overlooked in the following.

Four Brønsted acid sites have been recognized in mordenite [15], each bonded to one of the four symmetrically independent tetrahedra. One of these sites (H9 in Fig. 1) is bonded to O9, a framework oxygen of the 4-ring, and heads toward the centre of the 8-ring. In monocationic dehydrated mordenites two different sites have been localized in the 8-ring. One, site I in Fig. 1, is at the center of the ring, at z=0 (site A according to Mortier [18]), and is 6-fold coordinated with two O9 oxygens at quite a small distance (around 2.5-2.6Å), and to four O1 oxygens at a rather long distance. Site I is occupied only by quite small cations such as Na [19], Ca [20] and, according to the spectroscopic data of Dedecek and Wichterlová [21], by cobalt.

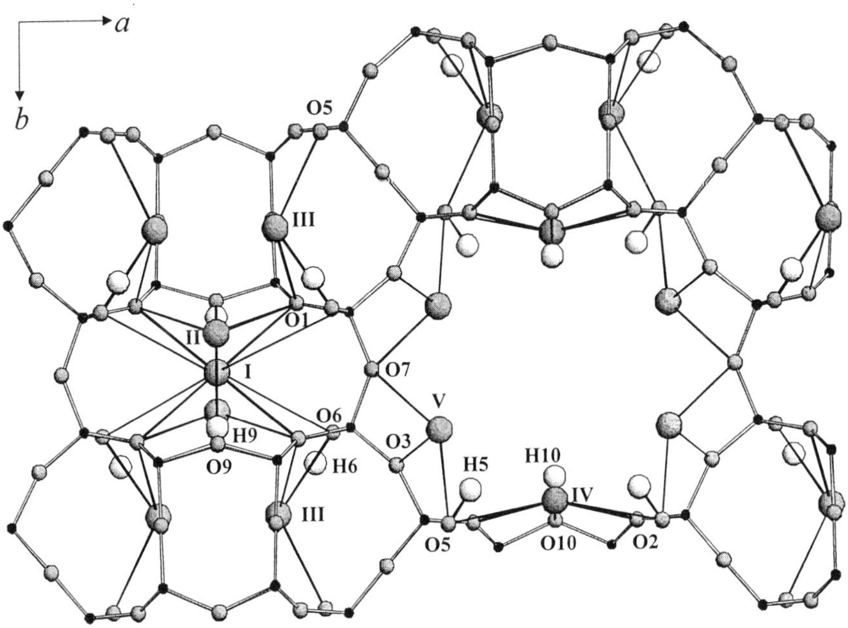

Fig. 1. Projection of the crystal structure of mordenite along the [001] direction.

When mordenite is exchanged with large cations such as K [22], Rb [23], Cs [24] or Ba [25], site I is empty, whereas another site, site II (which corresponds to site B [18]), still in the

8-ring but rather distant from the previous one, hosts these cations and is coordinated to O9, O1 and O6. Both cation sites I and II are quite close to the position of proton H9 (see Fig. 1).

Another proton site is on framework oxygen O6 (H6 in Fig. 1), and heads toward the side pocket; cation site III (site C [18]), which is frequently weakly occupied, is not far from the center of the side pocket and is coordinated, at a short distance, to oxygen O6, like site II, and to O5. Site III is always weakly occupied and is empty in mordenites exchanged with large cations.

Two other Brønsted acid sites have been recognized in the 12-ring channel, and are bonded to framework oxygens O5 (H5 in Fig 1) and O10 (H10 is the proton), respectively. A cation site, site IV of Fig. 1 (site D [18]), is present in the 12-ring in all dehydrated mordenites. This site is normally strongly occupied and is 5-coordinated, twice to O5 and O2, and once to O10. Its position is therefore intermediate to the H5 and H10 positions, coordinating both O5 and O10. Another cation site, site V of Fig. 1 (site E [18]), inside the 12-ring, is often present in dehydrated cation-exchaged mordenites, even if with low occupancy, and is also coordinated to O5.

To sum up, all framework oxygens bonded to protons, *id est* O5, O6, O9, and O10, are bonded to at least one, strongly occupied cation site.

Alberti [26] showed that in mordenite, according to the results of crystal structure refinements of some natural samples, there is a partial order in the Si/Al distribution, and Al is distributed in the four symmetrically independent tetrahedral sites according to the ratio:
T1 : T2 : T3 : T4 = 18 : 10 : 43 : 29.

Therefore, in the calculation of the bond strength on oxygen atoms by tetrahedral cations, the Al content given by the chemical analysis was distributed in the four T sites according to the above ratios.

Table 1 reports the values of bond strength for the 10 symmetrically independent oxygen atoms of the mordenite framework. This Table shows that O9, O1, O5, O10 and O6 receive the lowest bond strength from tetrahedral cations and the highest from extraframework cations. Therefore we can argue that Brønsted acid sites are present on a few of these oxygens. This conclusion is in very good agreement with the results of the neutron structure refinement of dehydrated D-mordenite, which clearly shows that protons are bonded to O9, O5, O10, and O6 [15].

2.2 Ferrierite

In the topological symmetry of ferrierite, orthorhombic, space group *Immm*, there are four tetrahedral and eight oxygen sites, symmetrically independent. Unfortunately the structural information available for this zeolite is not as satisfactory as that of mordenite. To date, only four X-ray single-crystal structure refinements, on natural samples, are available. As reported by Alberti and Sabelli [27], three of these (Mg-rich ferrierites characterized by *Immm* symmetry) seem to indicate an Al enrichment in the T2 tetrahedron, whereas the other (a Mg-poor, K-, and Na-rich ferrierite characterized by a monoclinic symmetry, space group $P2_1/n$) does not show any significant Si/Al ordering in the tetrahedral sites. All the structure refinements of dehydrated cation-exchanged ferrierites have been made, by powder diffraction data, on synthetic Mg-absent materials; therefore we assumed a complete disorder in the Si/Al distribution. Consequently, bond strength by tetrahedral cations cannot provide useful information to locate hydroxyl groups.

Table 1
Bond strength on framework oxygens in mordenites. First line: overall bond strength; second line: bond strength from framework cations; third line: bond strength from extraframework cations.

	Nat.[39]	Ca [20]	Rb [23]	Cs [24]	Na [19]*	K [22]*	Ba [25]*	Average
O1	2.12	2.03	2.00	1.96	2.01	2.00	2.03	2.02
	1.93	1.90	1.89	1.91	1.90	1.90	1.89	1.90
	0.17	0.13	0.11	0.05	0.11	0.10	0.14	0.12
O2	2.04	1.96	2.04	1.99	2.05	2.03	2.03	2.01
	1.98	1.93	1.94	1.93	1.92	1.94	1.95	1.95
	0.06	0.03	0.10	0.06	0.13	0.09	0.08	0.06
O3	1.96	2.00	1.98	2.00	2.00	1.99	1.97	1.99
	1.96	1.97	1.96	1.96	1.96	1.97	1.97	1.97
	0.00	0.03	0.02	0.04	0.04	0.02	0.00	0.02
O4	1.90	1.90	1.90	1.92	1.90	1.90	1.90	1.90
	1.90	1.90	1.90	1.92	1.90	1.90	1.90	1.90
	0.00	0.00	0.00	0.00	0.00	0.00	0.00	0.00
O5	2.01	2.04	2.06	2.04	2.04	2.09	2.03	2.04
	1.87	1.94	1.94	1.94	1.94	1.94	1.92	1.92
	0.14	0.10	0.12	0.10	0.10	0.15	0.11	0.12
O6	1.95	1.96	2.03	1.97	1.94	2.00	2.02	1.98
	1.88	1.93	1.91	1.91	1.94	1.93	1.92	1.92
	0.07	0.03	0.12	0.06	0.00	0.07	0.10	0.06
O7	1.94	2.00	1.98	2.02	2.01	1.99	1.94	1.99
	1.94	1.94	1.94	1.92	1.93	1.94	1.94	1.94
	0.00	0.06	0.04	0.10	0.08	0.05	0.00	0.05
O8	1.98	2.01	1.99	1.98	2.00	1.98	1.97	1.99
	1.98	1.99	1.99	1.98	2.00	1.98	1.97	1.99
	0.00	0.02	0.00	0.00	0.00	0.00	0.00	0.00
O9	2.20	2.12	2.08	2.02	2.16	2.12	2.22	2.13
	1.82	1.83	1.85	1.87	1.84	1.84	1.84	1.84
	0.38	0.29	0.23	0.15	0.32	0.28	0.38	0.29
O10	1.97	1.95	2.08	2.00	1.98	2.02	2.01	2.00
	1.87	1.90	1.91	1.90	1.91	1.90	1.92	1.90
	0.10	0.05	0.17	0.10	0.07	0.12	0.09	0.10

[xx] reference
* Space group *Pbcn*; bond strength values of O1, O2, and O3 averaged over the two pseudo-symmetric positions.
Bond strength calculated according to Ferguson [57].

Two Brønsted acid sites have been recognized in ferrierite [17]. One of them is on framework oxygen O4 (H4 in Fig. 2), and heads toward the centre of the ferrierite cage, while the other is on O6 (H6) and points toward the 10-ring channel, not far from the center of the 8-ring of the ferrierite cage facing towards the channel (see Fig. 2).

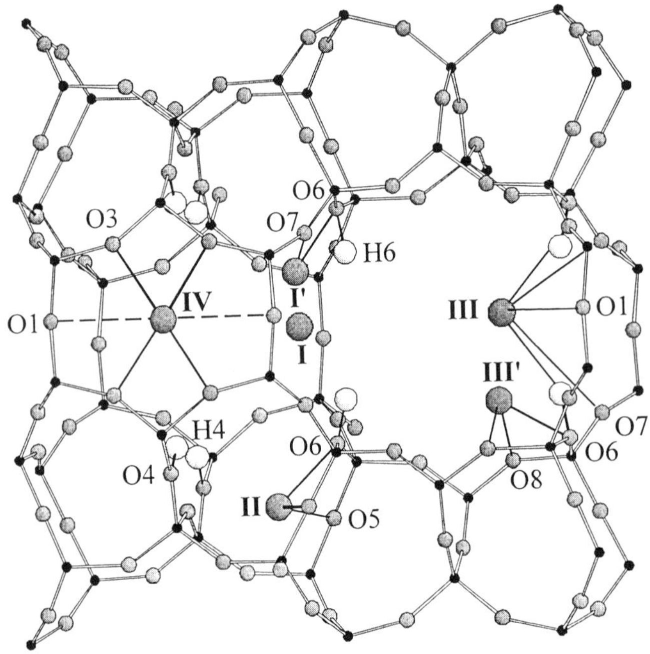

Fig. 2. The crystal structure of ferrierite viewed along [001].

Four cation sites have been recognized in dehydrated ferrierite. In K-ferrierite [28] the most important one, site I in Fig. 2, is at the center of the 8-ring of the ferrierite cage, facing towards the 10-ring channel. This site is coordinated with two O6 framework oxygens at quite a short distance (2.6-2.9Å), and with four O7 at a distance greater than 3Å. In dehydrated Co-, Ni-, and Cu-ferrierite [29-31] this site, called I' in Fig. 2, is shifted from the center of the ring, and is coordinated to only one O6 and one O7; the ionic radii of these cations are too small to remain at the center of the 8-ring. The similarity of location between H6 these sites is evident in Fig. 2.

In Co- and Ni-ferrierite a cation site is located inside the ferrierite cage, in the so called "boat shaped" position (site II), and is bonded to O6 and O5.

A second K site (site III in Fig. 2) has been found in K-ferrierite [28], not far from the center of the 10-ring channel; this site is coordinated to O1 and O7. In Ni- and Co-ferrierites (whose ionic radii are far shorter than that of K) this site is shifted in III' in order to have a better coordination, and is bonded to O6 and O8 (see Fig. 2). Another extraframework site was found in Co- and Ni-ferrierites, and is bonded to O3. Moreover, cobalt exerts a strong interaction on O1 (this oxygen shifts about 0.6Å from its position when Co is present), so that O1 is coordinated to site IV in Co-ferrierite but not in Ni-ferrierite.

Table 2
Bond strength on framework oxygens from extraframework cations in ferrierite.

	O1	O2	O3	O4	O5	O6	O7	O8
Cu[31]	0.00	0.00	0.00	0.00	0.00	0.08	0.03	0.02
Ni[20]	0.00	0.03	0.18	0.00	0.03	0.10	0.03	0.04
Co[29]	0.12	0.02	0.13	0.00	0.02	0.08	0.04	0.05
K[28]	0.07	0.00	0.00	0.00	0.00	0.10	0.14	0.00
Average	0.05	0.01	0.08	0.00	0.01	0.09	0.06	0.03

[xx] reference
Bond strength calculated according to Ferguson [57].

Table 2 reports the values of bond strength from extraframework cations for the 8 symmetrically independent oxygens. Bond strength from tetrahedral cations is omitted because a complete disorder in the Si/Al distribution was assumed.

As pointed out before, protons are attached to O6 and O4. Of these, O6 receives the highest bond strength among all framework oxygens, whereas this value is zero for O4. It is to be noted that the value of bond strength on O1, O3, and O7 strongly depends on the exchangeable cation, whereas this value for O6 does not greatly depend on the cation.

2.3 Chabazite

In the topological symmetry ($R\bar{3}m$) of chabazite there is only one symmetrically independent tetrahedron. Consequently, in all chabazite we can assume as a first approximation that each framework oxygen receive the same bond strength from tetrahedral cations. Nevertheless, the bond strength from extraframework cations is often strongly different, so that this information could be useful for localizing hydroxyl groups. Four oxygen sites are present in the topological symmetry of chabazite. The topological symmetry is also that of the majority of the dehydrated cation-exchanged phases. In Na- and Ag-exchanged chabazites [32,33] the framework is remarkably distorted, and the real symmetry lowers to monoclinic, *C2/m* space group. As a consequence, oxygen sites split over three (O2) or four (O1, O3, O4) symmetrically independent positions. However, this circumstance does not significantly change the bond strength on the pseudo-symmetric oxygen sites, so that their averaged values have been utilized in the following.

In synthetic chabazite two Brønsted sites have been recognized by neutron powder diffraction data [16]. One of these (H1) is on oxygen O1, which is on the 4-ring joining together the 6-rings of the double 6-ring, and the other (H3) is on framework oxygen O3 of the double 6-ring. Both head toward the large cage of the zeolite (see Fig. 3). It is important to note that the latter proton (H3) was found to tilt towards the center of the six-ring at an angle as high as 45° from the plane of the T-O3-T bridge and shifted toward O4, the other oxygen in the six ring, so that the H3-O4 distance was only about 2.6Å.

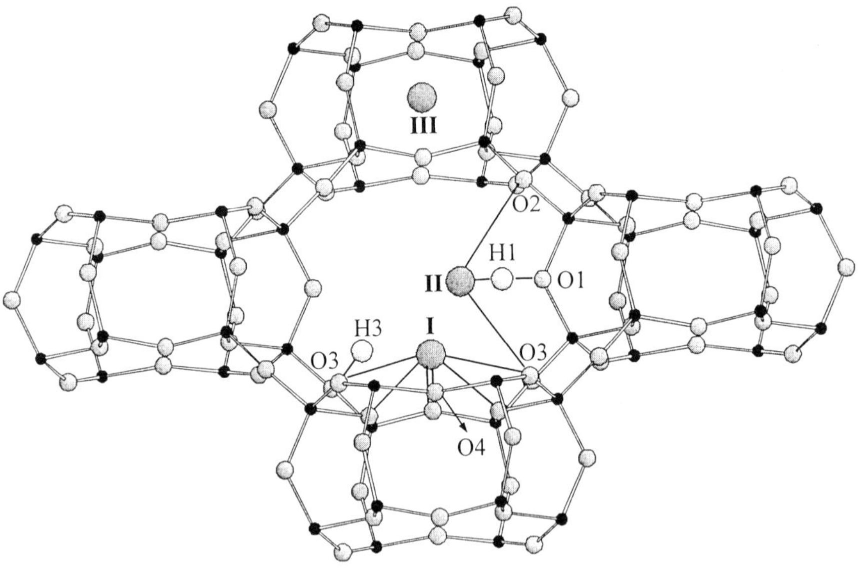

Fig. 3. Perspective view of the crystal structure of chabazite.

Three extraframework cation sites are present in dehydrated chabazites [32-37]. One, site III in Fig. 3 (site A according to Mortier [18]), is at the centre of the double 6-ring; its occupancy is always very low or even zero. The most important, site II in Fig. 3 (site C [18]), is on the threefold axis in the large cage outside the double 6-ring, but near the center of one 6-ring of D6R, and coordinates the 6 oxygens of the ring: three O4 at a short distance (around 2.4Å) and three O3 at about 2.8Å. This site is not far from proton H3. The last cation site is on the 8-ring, coordinating O1 and, at larger distances, O2 and O3. This site is not far from proton H1.

Table 3
Bond strength on framework oxygens from extraframework cations in chabazite.

	Ca [34]	Co[36]	Mn[37]	Cs[38]	Cu[35]*	Na[32]*	Ag[33]*	Average
O1	0.00	0.12	0.05	0.24	0.00	0.03	0.05	0.07
O2	0.00	0.00	0.05	0.00	0.00	0.18	0.17	0.07
O3	0.30	0.24	0.26	0.30	0.25	0.28	0.25	0.27
O4	0.36	0.23	0.29	0.08	0.36	0.14	0.13	0.23

[xx] reference
* Space group *P* $2_1/n$; bond strength values averaged over the pseudo-symmetric positions.
Bond strength calculated according to Ferguson [57].

Table 3 reports the values of bond strength from extraframework cations on the four oxygens. According to these data, O3 appears the most probable location of hydroxyl groups, not only because the highest bond strength is on this oxygen, but also because this value is about the same independently of the exchanged cation (and indeed a proton is bonded to O3). As concerns the other oxygen sites, there is a remarkable dishomogeneity in the bond strength depending on the exchangeable cation. O1 is usually characterized by low or even zero bond strength, but in Cs-chabazite this value is high; O2 receives a remarkable bond strength only in the Na- and Ag-forms, whereas the value of the bond strength on O4 is high every time this value is low on O1 or O2. Experimental data has shown that another proton is bonded to O1; from the data of Table 3 we must conclude that, for this Brønsted site, the information given by the bond strength is unsatisfactory, whereas O3 is clearly the best candidate to host a hydroxyl group.

2.4 Faujasite

The topological symmetry of faujasite-type zeolites is *Fd3m* with only one tetrahedral and four oxygen sites, symmetrically independent. This is also the real symmetry of Al-poor faujasite (zeolite X), whereas Al-rich faujasite (zeolite X), as a consequence of ordering in the Si/Al distribution, lowers its symmetry to the $Fd\bar{3}$ space group with two symmetrically non equivalent tetrahedral sites (and four oxygen sites). However, also in this case each oxygen receives the same bond strength as is bonded to an Al tetrahedron and to a Si tetrahedron. Consequently, only the bond strength from extraframework cations is taken into account.

The location of Brønsted acid sites in dehydrated faujasite has been studied by many authors, who agree that two proton sites are present in this zeolite [3,9-12]. One has been recognized on framework oxygen O1 (H1), which is the oxygen joining the two 6-rings of the D6R, and points toward the large cavity, while the other is bonded to O3 (H3), an oxygen of the 6-ring. Proton H3 heads toward the center of the 6R, but outside the D6R, i.e. in the sodalite cage. According to Czjzek and coworkers [4], a third proton site (H2) with lower occupancy is bonded with O2, the other oxygen on the 6R and, like H3, points toward the center of the 6R. It is evident that the position of H1 in faujasite resembles that of H1 in chabazite, whereas H3 (and H2) can be compared with H3 in chabazite. Protons have not been found in either structure on the oxygen of the 4-ring which joins the D6R.

There is a wealth of literature on the dehydrated forms of faujasite, which is difficult to schematise. In fact, the Si/Al ratio strongly varies in this phase, and significant differences have often been found by different authors in materials exchanged with the same cation and having comparable Si/Al ratio.

Basically, in dehydrated faujasite cations can occupy four sites on the triad, normally called sites I, I', II and II' (according to Mortier [18], sites A, C, G, E, respectively), whereas two other sites, usually called III and III', very near each other and normally weakly occupied or even empty, are often present in the large cage. It is unusual that all these sites are occupied in a single phase, whereas cations are always present in site I or I' and in site II or II'

Site I is 6-coordinated to oxygen O3, while is bonded to proton H3, and I' is 3-coordinated to the same oxygen at quite a short distance (around 2.5Å) and to O2 at a greater distance (around 3Å); sites II and II' are 3-coordinated (at very short distances, 2.3-2.5Å) with O2, bonded to proton H2, and with O4, whereas sites III and III' are bonded to O4 and O1 (see Fig. 4). When protons begin to exchange extraframework cations, they tend to occupy Brønsted site O1, and only at a higher cation exchange do they occupy site O3. It may seem strange that the first Brønsted site involves framework oxygen O1, which is weakly bonded to cation sites (see Table 4), whereas O3, O2 and also O4 are strongly bonded to cations; an explanation can be found in the fact that site II is almost fully occupied in dehydrated Y, so that O2 and O4 cannot host a proton, and the fact that O3 cannot host a proton until both I and I' sites are empty.

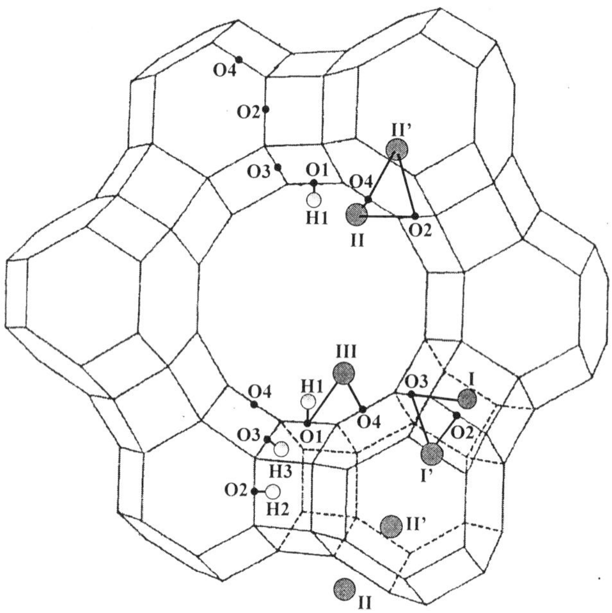

Fig. 4. Schematic representation of faujasite structure viewed along [111].

Table 4
Bond strength on framework oxygens from extraframework cations in faujasite.

	NaX [40]	NaX [41]	NaX [42]	CdX [43]	TlX [47]	AgXY [46]	AgY [46]	CaXY [45]	CaY [44]	LaY [49]	CuY [48]	Average
O1	0.18	0.06	0.20	0.00	0.15	0.14	0.00	0.00	0.00	0.00	0.03	0.07
O2	0.31	0.31	0.31	0.42	0.23	0.27	0.20	0.33	0.16	0.06	0.20	0.25
O3	0.21	0.17	0.21	0.21	0.20	0.27	0.20	0.34	0.33	0.41	0.20	0.25
O4	0.28	0.19	0.26	0.27	0.28	0.25	0.11	0.23	0.11	0.02	0.10	0.19

[xx] reference
Bond strength calculated according to Ferguson [57].

According to Bosacek *et al.* [12] the affinity of Brønsted sites with the framework oxygens is:

$$O3 > O1 > O2 \approx O4.$$

According to the data of Table 4, O3 is the best candidate to host a hydroxyl group due to its high bond strength with extraframework cations, whatever these are. Also O2 seems a feasible candidate to coordinate a Brønsted acid site, much more so than O4 and O1.

2.5 Zeolite Rho

The topological symmetry of zeolite rho is cubic, space group *Im3m*, with only one tetrahedral and two oxygen sites symmetrically independent.
It is interesting to note that the H-form of zeolite rho maintains its topological symmetry, whereas all dehydrated cation-forms of this zeolite are strongly distorted and assume an $I\bar{4}3m$ symmetry. In this space group there is again only one tetrahedral site, but there are three symmetrically independent oxygens, due to the splitting of site O2 over two sites, O2 and O3.

Only one Brønsted acid site has been recognized in synthetic zeolite rho [13,14]: proton H1 is bonded to framework oxygen O1 and heads toward the 8-ring (Fig. 5, on the right).

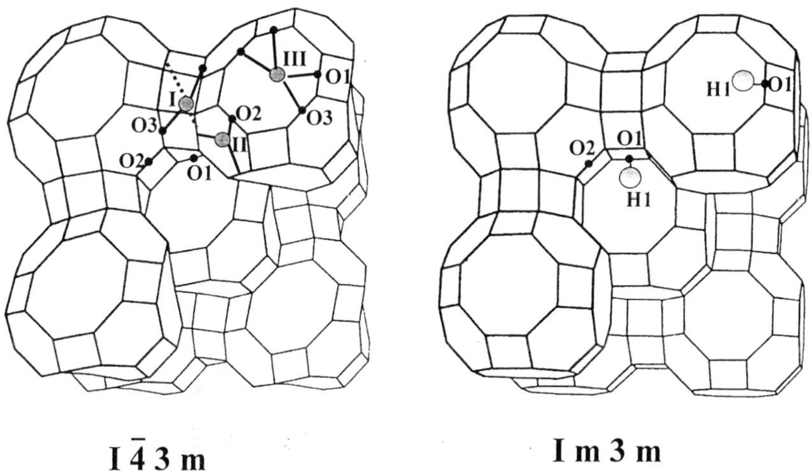

Fig. 5. Schematic representation of the structure of zeolite rho in its acid form with topological symmetry (on the right), and in its cation-exchanged form, with real symmetry (left), viewed along [001].

Three cation sites are present in this zeolite: one, site I, is at or near the center of the double 6-ring (see Fig. 5); a second, site II, is at the center of the S6R; and the last, site III, approximately on the plane of the S8R, but shifted with respect to the center of the S8R in order to have better coordination distances. Site I is 4-fold coordinated to O3 and, at a greater distance, to O1; this site is normally fully occupied, whereas it is empty when zeolite is exchanged with small cations. Site II is 3-fold coordinated, at short distances, to oxygen O2. Site III is 4-fold coordinated, twice to O1 and twice to O3.

To sum up, both sites I and III are located in positions which resemble that of proton H1. It is to be noted that when site I is weakly occupied, or empty, site III is largely occupied, and if site III is empty site I is fully occupied.

Table 5
Bond strength on framework oxygens from extraframework cations in zeolite Rho.

	NaCs[50]	SrCs[53]HT	SrCs[53]LT	Ca[51]	Cs[56]c	Rb[52]*	Tl[55]*	Ba[54]	Average
O1	0.10	0.11	0.13	0.16	0.08	0.10	0.07	0.10	0.11
O2	0.08	0.00	0.00	0.00	0.00	0.06	0.17	0.00	0.04
O3	0.23	0.21	0.16	0.18	0.09	0.13	0.19	0.10	0.16

[xx] reference
HT: heated at 573 K and collected at 573 K; LT: heated at 473 K and collected at 293 K
Bond strength calculated according to Ferguson [57].

However, from Table 5 we do not have a clear picture of the most probable location of hydroxyl groups. Bond strengths seem to suggest a preference of protons in the order O3 > O1 > O2 but, if we consider that in the H-form (space group *Im3m*) site O2 is not split over O2 and O3 (as happens in $I\bar{4}3m$ symmetry), no indication can be obtained from the location of extraframework sites in dehydrated zeolite rho.

3. CONCLUSIONS

It has been seen that Brønsted acid sites are frequently associated to framework oxygens on which there is a strong bond strength. Therefore, the information given by the location and occupancy of cation sites in dehydrated cation-exchanged forms can help to localize Brønsted sites. The literature, in fact, abounds with studies on the structure of dehydrated cation-exchanged zeolites, whereas framework hydroxyls have, to date, been recognized only in the zeolites discussed above. Also the Si/Al distribution could provide useful information. An oxygen on which bond strength from tetrahedral cations is low has a higher probability of hosting a proton. Mordenite is a good example of this statement. Unfortunately, zeolite rho, chabazite, and faujasite, having only one symmetrically independent tetrahedral site, cannot provide information in this way, while the data from the literature on Si/Al distribution in ferrierite, is insufficient to utilize this information for our aim.

Spectroscopic data (IR and NMR in particular), as well as the information given by quantum-chemical computations, can be extremely useful, when supported by structural data, for recognizing Brønsted sites. In our opinion, if all this information is taken into account together, it is possible in many cases to locate framework hydroxyls, even if neutron diffraction data are not at our disposal.

4. ACKNOWLEDGEMENTS

The MURST is thanked for financial support of the research programme "Transformations, reactions, and ordering in minerals" (Cofin 99). The CNR of Italy is acknowledged for financial support.

REFERENCES

1. J. Sauer, C.M. Kölmel, J.R. Hill and L. Ahlrichs, Chem. Phys. Lett., 164 (1989) 193.
2. L.A. Curtis, H. Brand, J.B. Nicholas and L.E. Iton, Chem. Phys. Lett., 184 (1991) 215.
3. D.H. Olson and E. Dempsey, J. Catal., 13 (1969) 221.
4. M. Czjzek, H. Jobic, A.N. Fitch and T. Vogt, J. Phys. Chem., 96 (1992) 1535.
5. W.H. Baur and W. Joswig, N. Jb. Miner. Mh., 1996 (1996) 171.
6. G. Vezzalini, A. Alberti, A. Sani and M. Triscari, Micr. Mes. Mat., 31 (1999) 253.
7. A. Alberti, M. Sacerdoti, G. Vezzalini and S. Quartieri, Phys. Chem. Minerals, 26 (1999) 181.
8. R.D. Shannon, Acta Cryst., A32 (1976) 751.
9. W.J. Mortier, J.J. Pluth and J.V. Smith, J. Catal., 45 (1976) 367.
10. Z. Jirák, S. Vratislav, J. Zajícek and V. Bosacek, J. Catal., 49 (1977) 112.
11. Z. Jirák, S. Vratislav, V. Bosacek, J. Phys. Chem. Solids, 41 (1980) 1089.
12. V., Bosacek, S., Beran and Z. Jirak, J. Phys. Chem., 85 (1981) 3856.
13. R.X. Fischer, W.H. Baur, R.D. Shannon and R.H. Staley, J. Phys. Chem., 91 (1987) 2227.
14. R.X. Fischer, W.H. Baur, R.D. Shannon, R.H. Staley, L. Abrams, A.J. Vega and J.D. Jorgensen, Acta Cryst., B44 (1988) 321.
15. A. Martucci, G. Cruciani, A. Alberti, C. Ritter, P. Ciambelli and M. Rapacciuolo, Micr. Mes. Mat., 35-36 (2000) 405.
16. L.J. Smith, A. Davidson and A.K. Cheetham, Catalysis Letters, 49 (1997) 143
17. A. Martucci, A. Alberti, G. Cruciani, P. Radaelli, P. Ciambelli and M. Rapacciuolo, Micr. Mes. Mat., 30 (1999) 95 .
18. W.J. Mortier, Compilation of Extra Framework Sites in Zeolites, Butterworth, 1982.
19. J.L. Schlenker, J.J. Pluth and J.V. Smith, Mat. Res. Bull., 14 (1979) 751.
20. W.J. Mortier, J.J. Pluth and J.V. Smith, Mat. Res. Bull., 10 (1975) 1037.
21. J. Dedecek and B. Wichterlová, J. Phys. Chem., B 103 (1999) 1462.
22. W.J. Mortier, J.J. Pluth and J.V. Smith, In "Natural Zeolites", L.B. Sand and F.A. Mumpton (eds.), Pergamon Press, Elmsford, USA, 1978, p. 53.
23. J. L. Schlenker, J.J. Pluth and J.V. Smith, Mat. Res. Bull., 13 (1978) 77.
24. J. L. Schlenker, J.J. Pluth and J.V. Smith, Mat. Res. Bull., 13 (1978) 901.
25. J. L. Schlenker, J.J. Pluth and J.V. Smith, Mat. Res. Bull., 13 (1978) 169.
26. A. Alberti, Zeolites, 19 (1997) 441.
27. A. Alberti and C. Sabelli, Z. Kristallogr., 178 (1987) 249
28. I.J. Pickering, P.J. Maddox, J.M. Thomas and A.K. Cheetham, J. Catal., 119(1989) 261.
29. M.C. Dalconi, pers. comm.
30. M.C. Dalconi, G. Cruciani, A. Alberti, P. Ciambelli and M.T. Rapacciuolo, Micr. Mes. Mat., 39 (2000) 423.
31. M.P. Attfield, S.J. Weigel and A.K. Cheetham, J. Catal., 172 (1997) 274.
32. W.J. Mortier, J.J. Pluth and J.V. Smith, Mat. Res. Bull., 12 (1977) 241.
33. M. Calligaris, A. Mezzetti, G. Nardin and L. Randaccio, Zeolites, 4 (1984) 323.

34. W.J. Mortier, J.J. Pluth and J.V. Smith, Mat. Res. Bull., 12 (1977) 97.
35. J.J. Pluth, J.V. Smith. and W.J. Mortier, Mat. Res. Bull., 12 (1977) 1001.
36. M. Calligaris, G. Nardin and L. Randaccio, Zeolites, 4 (1984) 251.
37. M. Calligaris, A. Mezzetti, G. Nardin and L. Randaccio, Zeolites, 5 (1985) 317.
38. M. Calligaris, A. Mezzetti, G. Nardin and L. Randaccio, Zeolites, 6 (1986) 137.
39. A. Martucci, pers. comm.
40. F. Porcher, M. Sonhasson, Y. Dusansoy and C. Lecomte, Eur. J. Miner., 11 (1999) 333.
41. Yu. I. Smolin, Yu. F. Shepelev, I.K. Butikova and V.P. Petranovskii, Sov. Phys. Cryst., 28 (1983) 36.
42. D.H. Olson, Zeolites, 15 (1995) 439.
43. Yu. I. Smolin, Yu. F. Shepelev, A. E. Lapshin and E.A. Vasil'eva, Cryst. Reports, 43 (1998) 383.
44. J. M. Bennet and J.V. Smith, Mat. Res. Bull., 3 (1968) 633.
45. J.J. Pluth and J.V. Smith, Mat. Res. Bull., 7(1972) 1311.
46. L.R. Gellens, W.J. Mortier and J.B. Uytterhoeven, Zeolites, 1 (19981) 11.
47. Y. Kim, W.H. Young and K. Seff, Zeolites, 18 (1997) 325.
48. I.E. Maxwell and J.J. de Boer, J. Phys. Chem., 79 (1975) 1874.
49. J. M. Bennet and J.V. Smith, Mat. Res. Bull., 3 (1968) 865.
50. L.B. McCusker, C. Baerlocher, In: Proc. Sixth Int. Zeolite Conference, D. Olson and A. Bisio (eds.), Butterworths, Guildford, UK, 1984, p. 812.
51. D.R. Corbin, L. Abrams, G.A. Jones, M.M. Eddy, W.T.A. Harrison, G.D. Stucky and D.E. Cox, J. Am. Chem. Soc., 112 (1990) 4821.
52. J.B. Parise, D.R. Corbin, T.E. Gier, R.L. Harlow, L. Abrams and R.B. Von Dreele, Zeolites, 12 (1992) 360.
53. A. Bieniok and W.H. Baur, Mat. Sci. Forum, 79-82 (1991) 721.
54. A. Bieniok, Ph. D. Thesis, 1992.
55. J.B. Parise, D.R. Corbin, L. Abrams, P. Northrup, J. Rakovan, T.M. Nenoff and G.D. Stucky, Zeolites, 14 (1994) 25.
56. J.B. Parise and E. Prince, Mat. Res. Bull., 18 (1983) 841.
57. R.B. Ferguson, Acta Cryst., B30 (1974) 2527.

Studies in Surface Science and Catalysis 140
A. Gamba, C. Colella and S. Coluccia (Editors)

Characteristics in the Photocatalytic Reactivity of the Tetrahedrally Coordinated Ti-Oxide Species Designed within various Types of Zeolites and on Support Surfaces

Masakazu Anpo[a]*, Shinya Higashimoto[a], Yasushi Shioya[a], Keita Ikeue[a], Masaru Harada[b], and Mamoru Watanabe[b]
[a]Department of Applied Chemistry, Graduate School of Engineering
Osaka Prefecture University
1-1 Gakuen-cho, Sakai, Osaka 599-8531, Japan
E-mail: anpo@ok.chem.osakafu-u.ac.jp
[b]National Institute for Research in Inorganic Materials
Tsukuba, Ibaraki 305-0044, Japan

In situ characterizations of Ti-oxides included within various types of zeolites or anchored on support surfaces were carried out by means of UV-Vis, ESR, photoluminescence and XAFS (XANES and EXAFS) investigations, providing important insights into their local structure and their photocatalytic reactivity for the decomposition of NO into N_2 and O_2 as well as the reduction of CO_2 with H_2O into CH_3OH and CH_4. It was found that Ti-oxides prepared within zeolites or on support surfaces exhibit quite different and high photocatalytic reactivity as compared to bulk powdered TiO_2 and also that the high reactivity of the charge transfer excited triplet state of the tetrahedrally coordinated Ti-oxides, (Ti^{3+} - O^-)*, plays a significant role in the unique and high photocatalytic reactivity of these catalysts. The reactivity of such a charge transfer excited state was also found to strongly depend on differences in the molecular environment of the Ti-oxides such as the rigidity or flexibility of the zeolite framework and the local structures of the TiO_4 unit, *i. e.*, $Ti(OSi)_4$, $Ti(OH)(OSi)_3$ or $Ti(OH)_2(OSi)_2$.

1. INTRODUCTION

The design of highly efficient and selective photocatalytic systems that work without any loss in the use of solar energy is of vital interest for future energy needs. In particular, the development of efficient photocatalytic systems which are able to decompose NO directly into N_2 and O_2 or to reduce CO_2 with H_2O into chemically valuable compounds such as CH_3OH or CH_4 are among the most desirable and challenging goals in the research of enviromentally-friendly catalysts [1]. From this point of view, much attention has been focused on the useful application of Ti-oxide photocatalysts, mainly as TiO_2 semiconductors, TiO_2 thin films as well as isolated tetrahedrally coordinated Ti-oxides [1-3]. Recently, we have reported that the highly dispersed tetrahedrally coordinated Ti-oxides, when compared with bulk TiO_2 powder, exhibit high and unique photocatalytic reactivity for the NO decomposition reaction as well as the reduction of CO_2 with H_2O [2-5]. Thus, it is of special importance to understand the relationship between the local structure of the Ti-oxides and their photocatalytic reactivity, especially their selectivity among various types of Ti-oxide catalysts.

The present study deals with the distinct characteristics of Ti-oxides incorporated within zeolite frameworks (TS-1, TS-2, Ti-MCM-41, Ti-MCM-48) or on the surface of supports (Ti/FSM-16, Ti/Vycor glass) and their photocatalytic reactivity including their selectivity for the NO decomposition reaction as well as for the reduction of CO_2 with H_2O at the molecular level using various in situ spectroscopic techniques such as photoluminescence, UV-Vis, XAFS (XANES and EXAFS) and ESR spectroscopy.

2. EXPERIMENTAL

TS-1 and TS-2 (Si/Ti = 85), Ti-MCM-41 (Si/Ti = 100) and Ti-MCM-48 (Si/Ti = 80) were hydrothermally synthesized according to procedures previously reported [6-8]. Ti/FSM-16 and Ti/Vycor glass were prepared by a chemical vapor deposition method (CVD) of the Ti-oxides onto FSM-16 or Vycor glass, respectively, through a facile reaction of $TiCl_4$ with the surface OH groups of these supports in the gas phase at 453-473 K, followed by treatment with water vapor to replace the chlorine atoms with the OH groups [2,9]. A TiO_2 powdered catalyst, JRC-TIO-4 (anatase 92 %, rutile 8 %), supplied by the Catalysis Society of Japan was used. Prior to photocatalytic reactions and spectroscopic measurements, the catalysts were heated in O_2 at 723 K and then evacuated at 475 K. The photoluminescence and lifetimes of the catalysts were measured at 77 K with a Shimadzu RF-501 spectrofluorophotometer and an apparatus for lifetime measurements. The XAFS (XANES and EXAFS) spectra were obtained at the BL-7C facility of the Photon Factory at the National Laboratory for High Energy Physics (KEK-PF) in Tsukuba. The Ti K-edge absorption spectra were recorded in the fluorescence mode at 295 K with a ring energy of 2.5 GeV. UV-Vis spectra were recorded at 295 K with a Shimadzu UV-2200A spectrophotometer. The ESR spectra were recorded with a JEOL-2X spectrometer (X-band) at 77 K. The photocatalytic decomposition reactions of NO and the reduction of CO_2 with H_2O were carried out using the catalysts (150 mg for each) in a quartz cell with a flat bottom (ca. 100 ml) under UV irradiation using a high pressure Hg lamp (λ > 240 nm) at 295 K or 323 K, respectively.

3. RESULTS AND DISCUSSION

3.1 Characterization of the Local Structure of the Ti-oxides within various Types of Ti-oxide Catalysts

The results of the XRD patterns and UV-Vis absorption spectra indicated that TS-1, TS-2, Ti-MCM-41, Ti-MCM-48 and Ti/FSM-16 have original zeolitic structures with high crystallinity while no aggregated Ti-oxides are formed [6-9]. The FT-IR spectra of TS-1 and TS-2 exhibit significant bands at 960-970 cm^{-1}, attributed to the stretching vibration mode of the Si-OH groups bonded to the metal ions, indicating the successful incorporation of the Ti ions within the zeolite framework [6]. The ESR signals of all Ti-containing zeolites as well as Ti/Vycor glass show that the Ti^{3+} ions are produced by photoreduction in the presence of H_2 at 77 K, indicating that they involve the tetrahedrally coordinated Ti-oxides within the framework or on the surfaces of the supports.

Figure 1 shows the XAFS (XANES and EXAFS) spectra of Ti-MCM-41, Ti-MCM-48 and the reference TiO_2 powdered sample. The shape of the XANES spectra provides useful and detailed information on the local structure and coordination geometry of the central Ti atoms. One of the characteristic features of the XANES spectra of these Ti-containing zeolites was the appearance of a single pre-edge peak and its shape was similar to that of $Ti(O\text{-}iPr)_4$ having tetrahedral coordination, indicating that the Ti-oxides in these catalysts form a tetrahedral coordination, while the bulk TiO_2 exhibited three peaks with lower intensities, characterizing an octahedral coordination as in a rutile or anatase structure. However, the pre-edge peak in the XANES spectra reported in Fig. 1 (A, B) for Ti-MCM-41 and Ti-MCM-48 exhibit about 30 % of the edge jump, suggesting that these samples include a rather large fraction of Ti in square pyramidal or octahedral coordination, which was probably caused by pre-treatment problems in the samples, i.e., some water may have adsorbed on the samples before and/or during the XAFS measurements. Moreover, in the Fourier transforms of the EXAFS spectra of these catalysts, there is only one peak at around 1.6 Å due to the neighboring oxygen atoms (Ti-O), indicating the presence of isolated Ti-oxides in these catalysts. The curve-fitting analysis of the EXAFS spectra of Ti-MCM-41 and Ti-MCM-

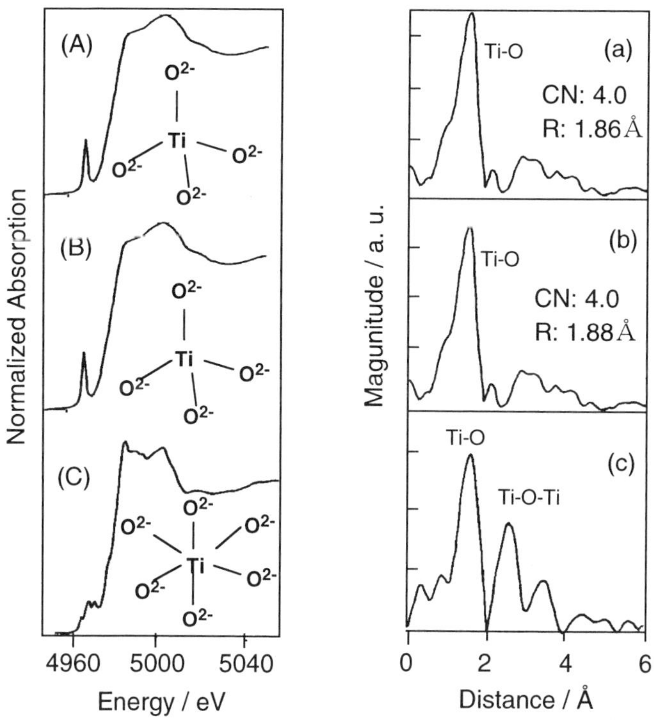

Fig. 1. XANES (left) and Fourier transforms of normalized k^3 weighted EXAFS (right) without phase-shift correction of Ti-MCM-41 (A, a), Ti-MCM-48 (B, b), and the bulk powdered TiO_2 (C, c) as the reference sample.

48 showed the tetrahedrally coordinated Ti-oxides to have four oxygen atoms in Ti-O distances of 1.86 Å and 1.88 Å, respectively. Analyses of the XAFS (XANES and EXAFS) spectra of the dehydrated TS-1 and TS-2 showed that they involve tetrahedrally coordinated Ti-oxides within zeolite framework having four oxygen atoms in Ti-O distances of 1.83 Å and 1.84 Å, respectively, as shown in Table 1. In TS-1 and TS-2, Ti-O distances of 1.83 Å and 1.84 Å, respectively, are rather large when compared those reported in the previous literature. This would probably be caused by the average bonding distance involving the presence not only of tetrahedral but also Ti in square pyramidal or octahedral coordination [8, 10-14].

Table 1
Comparison of the various chemical parameters of the Ti-oxide species formed within zeolites and on FSM-16 and SiO_2 surfaces and their photocatalytic reactivity.

	TS-1	TS-2	Ti-MCM -41	Ti-MCM -48	Ti/FSM -16	Ti/Vycor glass	Bulk TiO_2
Coordination	Tetra	Tetra	Tetra	Tetra	Tetra	Tetra	Octa
Ti-O bond length, Å	1.83	1.84	1.86	1.88	---	---	---
Photoluminescence wavelength, nm	480	480	480	480	465	445	---
Selectivity for the CH_3OH formation, %	20.6	26.5	30.8	28.8	41.2	35.0	$\cong 0$
Selectivity for the N_2 formation, %	82.0	80.0	---	89.0	---	88.0	> 18

Figure 2 shows that Ti-MCM-48 exhibits a photoluminescence spectrum at around 490 nm by excitation at around 270 nm at 77 K. The observed photoluminescence and absorption bands are in good agreement with those previously observed with highly dispersed tetrahedrally coordinated Ti-oxides prepared in silica matrices except for the band position. We can, therefore, conclude that the observed photoluminescence spectrum is attributed to the following radiative decay process (eq. 1) from the charge transfer excited state to the ground state of the highly dispersed Ti-oxides in tetrahedral coordination [3,5].

$$(Ti^{4+} - O^{2-}) \underset{h\nu'}{\overset{h\nu}{\rightleftharpoons}} (Ti^{3+} - O^{-})^* \qquad \text{(eq. 1)}$$

The maximum positions of the photoluminescence spectra on the various types of Ti-oxide catalysts are shown in Table 1. It can be seen that the Ti-containing zeolites (TS-1, TS-2, Ti-MCM-41 and Ti-MCM-48) exhibit photoluminescence at a maximum position of 480 nm, while Ti/FSM-16 and Ti/Vycor glass exhibit photoluminescence at 465 and 445 nm, respectively. Marchese et al. proposed that the various band positions in the spectra are due to the relaxation of the different oxygen molecules to tetrahedral Ti (IV) charge transfer processes in the different environments [11]. Moreover, it is supposed that Ti-oxides included within Ti/FSM-16 or Ti/Vycor glass prepared by the CVD method have an open structure with one or two terminal OH ligands, $Ti(OH)(OSi)_3$ or $Ti(OH)_2(OSi)_2$, while Ti-oxides incorporated within the zeolite framework have a closed structure of $Ti(OSi)_4$, as suggested by Coluccia et al. [15] Moreover, it is likely that the molecular environment of the Ti-oxides such as the rigidity or flexibility of the zeolite framework causes a significant and pronounced effect on the chemical properties of the TiO_4 unit. Taking these results into consideration, the different band positions of the observed photoluminescence of the various Ti-oxide catalysts may be attributed to the differences in the molecular environment of the Ti-oxides such as the rigidity or flexibility of the zeolite framework and the local structures of the TiO_4 unit, *i. e.*, $Ti(OSi)_4$, $Ti(OH)(OSi)_3$ or $Ti(OH)_2(OSi)_2$, as shown in Fig. 3 [14,15].

Figure 2 also shows that the addition of NO, H_2O or CO_2 molecules onto the Ti-MCM-48 catalyst leads to an efficient quenching of the photoluminescence as well as a shortening of the lifetime of the charge transfer excited state, its extent depending on the amount of gasses added. Such an efficient quenching of the photoluminescence with NO, CO_2 or H_2O molecules

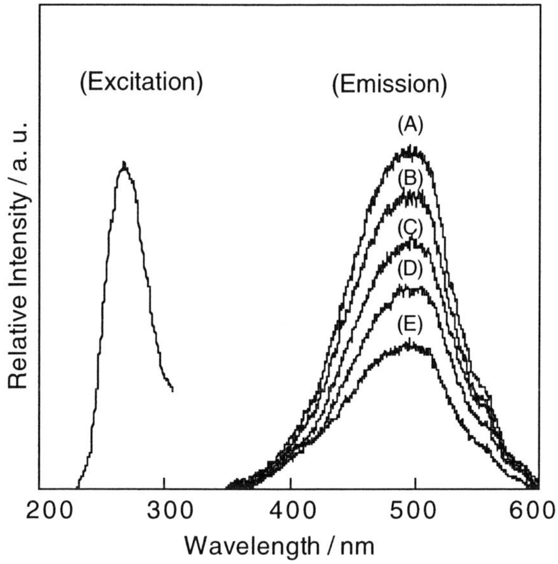

Fig. 2. Effect of the addition of CO_2 on the photoluminescence spectra of the Ti-MCM-48 catalyst measured at 77 K. Pressure of added CO_2: 0 (A), 0.1 (B), 0.5 (C), 2.0 (D) and (E) 10.0 Torr

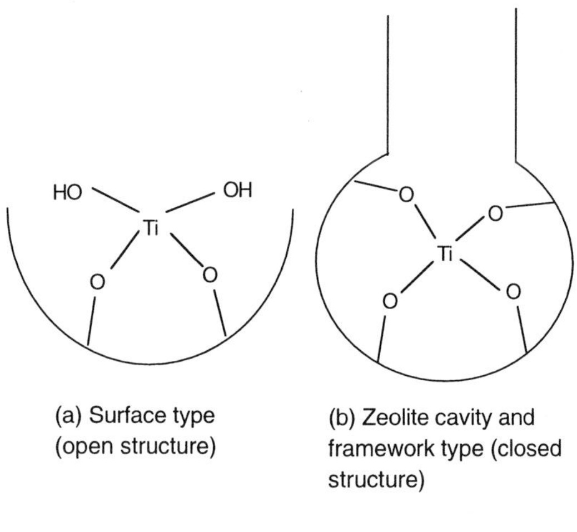

Fig. 3. Local structure models of the tetrahedral Ti-oxides between the open structure type (a) and the closed structure type (b) in different environments.

suggests not only that tetrahedrally coordinated Ti-oxides locate at positions accessible to these small molecules but that they also interact with the Ti-oxides in both its ground and excited states.

3.2 Interaction of the Ti-oxide Catalysts with the Guest Reactant Molecules and their Photocatalytic Reactivity

3.2.1 Photocatalytic Reactivity for the Reduction of CO_2 with H_2O and NO Decomposition on the Ti-oxide Catalysts

UV-irradiation of the Ti-oxide catalysts in the presence of a mixture of CO_2 and H_2O led to the formation of CH_3OH and CH_4 as the main products as well as CO, C_2H_4 and C_2H_6 as the minor products in the gas phase at 323 K, while the yields of these photo-formed products increased in proportion to the irradiation time [4]. The yields of CH_4 and CH_3OH as well as the selectivity for CH_3OH formation, which have been normalized for the unit amount of TiO_2 in the catalysts, are shown in Fig. 4 and Table 1, respectively. It can be seen that Ti-oxide catalysts exhibit high photocatalytic reactivity for CO_2 reduction with H_2O as compared bulk TiO_2 in the following order: TiO_2 << TS-1 < Ti-MCM-41 < Ti-MCM-48 << Ti/FSM-16 ≒ Ti/Vycor glass. Thus, it can be seen that highly dispersed tetrahedrally coordinated Ti-oxides play a significant role in the reactivity and selectivity for CH_3OH formation in this reaction. In particular, Ti/FSM-16 shows the highest reactivity among these Ti-oxide catalysts, which may be attributed to the high reactivity of the charge transfer excited triplet state of $(Ti^{3+} - O^{-})^*$. It seems possible that the reactivity strongly depend on the molecular environment of the Ti-oxide species such as the rigidity or flexibility of the zeolite framework, including the differences in the local structures of the TiO_4 unit, *i. e.*, $Ti(OH)(OSi)_3$ or $Ti(OH)_2(OSi)_2$. These results clearly indicate that a high photocatalytic efficiency and selectivity for the formation of CH_3OH in

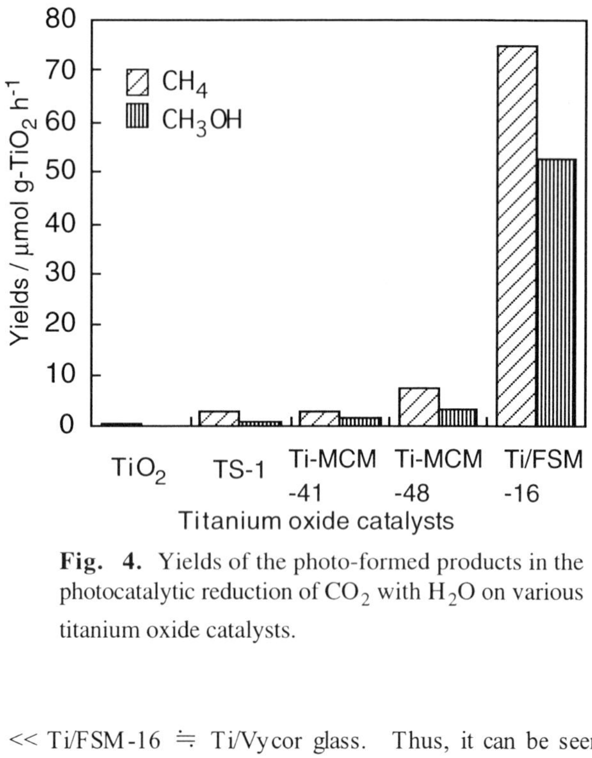

Fig. 4. Yields of the photo-formed products in the photocatalytic reduction of CO_2 with H_2O on various titanium oxide catalysts.

the reduction of CO_2 with H_2O could be achieved by the design of catalysts having highly dispersed tetrahedrally coordinated Ti-oxides within the mesoporous zeolite cavities.

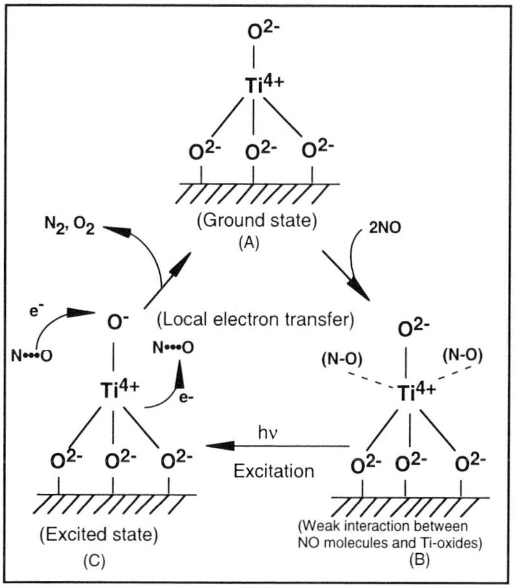

Fig. 5. Reaction scheme of the photocatalytic decomposition of NO into N_2 and O_2 on the Ti-oxide catalyst with tetrahedral coordination.

Furthermore, the photocatalytic decomposition reaction of NO was seen to proceed on these Ti-oxide catalysts, leading to the formation of N_2, O_2 and N_2O with a good linearity against the UV-irradiation time, while under dark conditions, no products could be observed. The selectivity for N_2 formation in the photocatalytic decomposition of NO on various Ti-oxide catalysts are shown in Table 1. It can be seen that the specific photocatalytic reactivity and selectivity for N_2 formation on the Ti-oxide catalysts which have been normalized for the unit amount of titanium oxide in the catalysts are much higher than the bulk TiO_2 catalysts. These results indicate that highly dispersed tetrahedrally coordinated Ti-oxides exhibit high selectivity as well as efficiency for N_2 formation in the photocatalytic decomposition of NO into N_2 and O_2 as compared with bulk TiO_2 having an octahedral coordination. From the above results, the reaction mechanism for the photocatalytic decomposition of NO on the isolated tetrahedrally coordinated Ti-oxides can be proposed, as shown in Fig. 5. The NO molecules are able to adsorb onto these Ti-oxides as weak ligands to form the reaction precursors. Under UV-irradiation, the charge transfer excited complexes of the Ti-oxides, $(Ti^{3+} - O^{-})^*$, are formed. Within their lifetimes the decomposition of NO into N_2 and O_2 as compared with bulk TiO_2 having an octahedral coordination. These electron transfers lead to the direct decomposition of two sets of NO on (Ti-O) into N_2 and O_2 under UV-irradiation. On the other hand, with the bulk TiO_2 catalyst, the photo-formed holes and electrons rapidly separate from each other with large space distances between the holes and electrons, thus preventing the simultaneous activation of two NO molecules on the same active sites while the decomposed N and O species react with NO on different sites to form N_2O and NO_2, respectively [5].

3.2.2 Relationship between the Coordination Number of Ti-oxides and the Selectivity for Photo-formed Products on the various Types of Ti-oxide Catalysts

As mentioned above, XAFS (XANES and EXAFS) investigations of Ti-oxide catalysts at the Ti K-edge were performed and the results revealed the Ti-oxides to have a tetrahedral coordination within the Ti-oxide catalysts, while the Ti-oxides have an octahedral coordination in the case of the bulk TiO_2 photocatalyst. Figure 6 shows the relationship between the coordination number of the Ti-oxides and the selectivity for N_2 formation in the photocatalytic decomposition reaction of NO on the various Ti-oxide photocatalysts. The clear dependence of the N_2 selectivity on the coordination number of the Ti-oxides can be seen, *i. e.*, the lower the coordination number of Ti-oxides, the higher the N_2 selectivity [5]. Moreover, the clear dependence of the CH_3OH selectivity in the photocatalytic reduction of CO_2 with H_2O on the coordination number of the Ti-oxides could also be observed, *i. e.*, the lower the coordination number of Ti-oxides, the higher the CH_3OH selectivity. From these results, it can be proposed that the highly efficient and selective photocatalytic decomposition of NO into N_2 and O_2 as well as reduction of CO_2 with H_2O into CH_3OH can be achieved using Ti-containing zeolites as a photocatalyst involving highly dispersed tetrahedrally coordinated Ti-oxides as the active species.

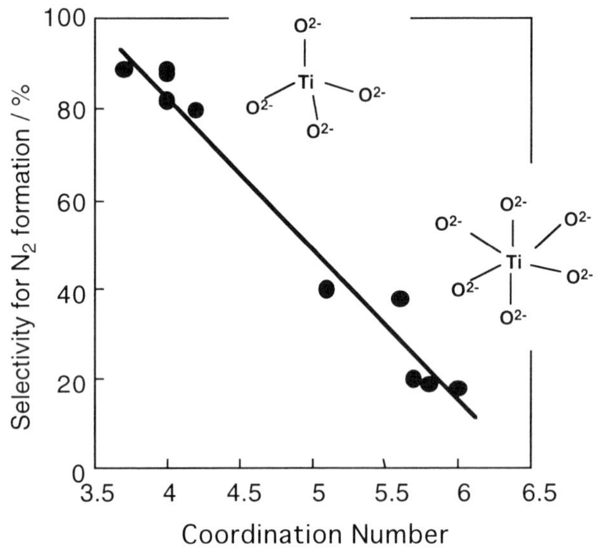

Fig. 6. Relationship between the coordination number of Ti-oxides and the selectivity for the N_2 formation in the photocatalytic decomposition of NO into N_2 and O_2 on the various titanium oxide catalysts.

CONCLUSIONS

We have demonstrated that a highly efficient and selective photocatalytic decomposition reaction of NO into N_2 and O_2 as well as the photocatalytic reduction of CO_2 with H_2O into CH_3OH can be achieved using Ti-containing zeolites as a photocatalyst which includes highly dispersed tetrahedrally coordinated Ti-oxides. Furthermore, it was found that a lower coordination number of Ti-oxides (from octahedral to tetrahedral coordination) caused a higher selectivity for photo-formed N_2 and CH_3OH formation in the photocatalytic decomposition of NO and reduction of CO_2 with H_2O, respectively. Moreover, it was found that the Ti-oxides anchored onto the surface of FSM-16 and porous Vycor glass exhibited much higher photocatalytic reactivity for the reduction of CO_2 with H_2O into CH_3OH and CH_4 than Ti-oxides incorporated within the micropores of the zeolite framework such as in the TS-1 and TS-2, the reactivity strongly depending on the molecular environment of the Ti-oxides such as the rigidity or flexibility of the zeolite framework as well as the local structures of the TiO_4 units, *i.e.*, $Ti(OSi)_4$, $Ti(OH)(OSi)_3$ or $Ti(OH)_2(OSi)_2$.

REFERENCES

1. M. Anpo, in Stud. Surf. Sci. Catal., 130 A, Part A (2000) 157.
2. M. Anpo, N. Aika, Y. Kubokawa, M. Che, C. Louis and E. Giamello, J. Phys. Chem., 89 (1985) 5017.
3. M. Anpo and M. Che, Adv. Catal., 44 (1999) 119, and references therein.
4. M. Anpo and K. Chiba, J. Mol. Catal., 74 (1992) 207.
5. M. Anpo (ed.), in Photofunctional Zeolites, NOVA Publishers Inc. (2000), and references therein.
6. T. Tatsumi, M. Nakamura, S. Negishi and H. Tominaga, J. Chem. Soc., Chem. Commun., (1990) 476.
7. W. Zhang, M. Froba, J. Wang, P. T. Tanev, J. Wong, and T. J. Pinnavaia, J. Am. Chem. Soc., 118 (1996) 9164.
8. A. Corma, Chem. Rev., 97 (1997) 2373.
9. K. Ikeue, H. Yamashita, M. Anpo, Chem. Lett. (1999) 1135.
10. T. Maschmeyer, F. Rey, G. Sankar, and J. M. Thomas, Nature, 378 (1995) 159.
11. K. I. Zamaraev, J. M. Thomas, Adv. Catal., 41 (1996) 335.
12. S. Bordiga, S. Coluccia, C. Lamberti, L. Marchese, F. Boscherini, F. Buffa, F. Genoni, G. Leofanti, G. Petrini, and G. Vlaic, J. Phys. Chem. 98 (1994) 4125.
13. R. D. Oldroyd, G. Sankar, J. M. Thomas, and D. Özkaya, J. Phys. Chem. B 102 (1998) 1849.
14. C. Lamberti, S. Bordiga, D. Arduino, A. Zecchina, F. Geobaldo, G. Spano, F. Genoni, G. Petrini, A. Carati, F. Villain, G. Vlaic, J. Phys. Chem. B 102 (1998) 6382.
15. L. Marchese, T. Maschmeyer, E. Gianotti, S. Coluccia, and J. M. Thomas, J. Phys. Chem. B, 101 (1997) 8836.

Studies in Surface Science and Catalysis 140
A. Gamba, C. Colella and S. Coluccia (Editors)

Polymer-silica composite membranes for Direct Methanol Fuel Cells

V. Antonucci, A.S. Aricò, E. Modica, P. Cretì, P. Staiti and P.L. Antonucci*

Institute CNR-TAE, Salita S. Lucia sopra Contesse 5, 98126 S. Lucia, Messina, Italy

* Dipartimento di Meccanica e Materiali, Università di Reggio Calabria, Feo di Vito, 89100 Reggio Calabria, Italy

Heteropolyacid-modified polymer silica membranes for Direct Methanol Fuel Cells have been prepared and tested under high temperature operation conditions (145°C) in single cell configuration. A maximum power density of 0.4 W/cm^2 in oxygen with 2 M methanol has been obtained; with air at the cathode, this value decreased to 0.25 W/cm^2. The higher performance of the heteropolyacid-Nafion-silica membrane, with respect to Nafion-silica, is attributed to its better ion transport properties, since the measured cell resistance value is similar for both membranes.

1. INTRODUCTION

Solid polymer electrolyte direct methanol fuel cells (SPE-DMFCs) are emerging as candidates for power sources in transportation applications, due to their very low emission levels. A SPE-DMFC comprises a membrane electrolyte coated on one side with a platinum-based cathode and the other side with a platinum ruthenium-based cathode. At present state-of-the-art perfluorosulphonic polymer electrolyte membranes suffer two main drawbacks, i.e. dehydration (with consequent lower conductivity and performance losses), and methanol cross-over. The first drawback severely hinders the operation at temperatures higher than 100°C which is a prerequisite for the suitable electro-oxidation of small organic molecules involving formation of strongly adsorbed reaction intermediates such as CO-like species. Thus, a breakthrough would be needed to increase the working temperature and, consequently, improve the performance of the methanol-fed devices; the development of of high temperature resistant proton exchange electrolytes is an ambitious goal to be pursued in order to favour the acceptability of DMFC worldwide.

Recently, Dahr [1], Stonehart [2] and Watanabe [3] have made an attempt to reduce the humidification constraints in solid polymer electrolyte fuel cells (SPEFCs) by using modified perfluorosulfonic membranes. A recast Nafion film sandwiched between the two electrodes was first proposed by Dahr [1] for the realization of an internally humidified SPEFC. Stonehart [2] suggested the inclusion of small amounts of silica powder into the recast film in order to retain the electrochemically produced water inside the membrane. Watanabe et al [3] have tried to exploit the H_2/O_2 crossover through the membrane to produce a chemical recombination to water on small Pt clusters inside the membrane. All of these membranes were operated with H_2/O_2 at 80°C and allowed the development of systems without assisted humidification or with near ambient humidification.

The basic idea of the present work was to use a composite silica – recast Nafion membrane in a SPE-DMFC operating at a temperature (about 150°C) well above that of a H_2/O_2 fuel cell, in order to enhance the kinetics of methanol oxidation. In a liquid-feed fuel cell water is supplied together with the fuel at the anode compartment and is produced at the cathode side during the electrochemical operation. Thus, we have sought a first advantage from the water-uptake properties of both recast Nafion and silica in order to allow fuel cell operation at 145°C.

In a further series of experiments, phosphotungstic acid (PWA) was impregnated onto silica particles and the resulting material was loaded in the recast Nafion. As is well known, heteropolyacids (HPAs) have demonstrated suitable characteristics to be used as proton conductive materials in fuel cells [4-6]. Due to dissolution phenomena in water, however, previous experiments using solid heteropolyacid did not result in stable fuel cell performance [5]. To overcome this problem, resulting in a short lifetime of the fuel cell, experiments of blocking the HPA in a host material were carried out [7-9]. Thus, in this work the phosphotungstic acid-modified membrane was compared, in terms of performance, with the bare silica-recast Nafion membrane in direct methanol fuel cell at 145°C.

1.1. The heteropolyacids

Heteropolyacids (HPAs) have suitable characteristics to be used as solid electrolytes or as aqueous solutions in fuel cells [4-6]. They are known to show very strong Bronsted acidity as solid electrolytes [10], exceeding that of such conventional solid acids such as SiO_2-Al_2O_3, H_3PO_4/SiO_2, and HX and HY zeolites. The acid strength of concentrated heteropolyacid aqueous solutions in terms of the Hammett acidity function is higher than that of the constituent oxo-acids and ordinary inorganic acids [10-12]. In fact, the three protons of $H_3PW_{12}O_{40}$ are completely dissociated in dilute aqueous solutions. It has been recently confirmed by calorimetric titration that anhydrous $H_3PW_{12}O_{40}$ in acetonitrile is a superacid [13]. The strong acidity is caused by two main factors: (1) dispersion of the negative charge over many atoms of the polyanion and (2) the fact that the negative charge is less distributed over the outer surface of the polyanion owing to the double-bond character of the M=O bond, which polarizes the negative charge of O to M [14]. Keggin was the first to determine the structure of $[PW_{12}O_{40}]^{3-}$ using X-ray analysis [15]. The acidity of the solutions of heteropolyacids is reflected on their proton conductivity which is of the same order of magnitude as that of ordinary mineral acids. Heteropolyacids also show high proton conductivity when they are in crystalline forms with determined numbers of water molecules in their structure [16-18]. Proton conductivities of 0.17 and 018 S cm^{-1} were measured by Nakamura et al. [16], respectively, for phosphomolybdic and phosphotungstic acids containing in their structures 29 water molecules of crystallization. The proton conductivities of the heteropolyacids in this solid form are higher than those of other known solid conductors.

2. EXPERIMENTAL

An appropriate amount of Nafion® ionomer (5% wt/wt, Aldrich) was mixed with 3% wt/wt SiO_2 (Aerosil 200, Degussa) in an ultrasonic bath for 30 min. This solution was cast [19,20] in a Petri dish and heated at 80°C for 30 min. The recast composite Nafion® film was detached from the Petri dish by addition of distilled water and allowed to dry for 15 hrs at room temperature. Afterwards, it was cut to obtain a regular shape and then hot pressed between two PTFE foils at a few bars and increasing temperatures. The final treatment was 160°C for 10 min. The latter step allowed to increase the crystalline fraction inside the composite membrane with consequent improvement of the mechanical properties. The thickness of the dry membrane was 80 μm and contained 3% wt/wt of silica. The membrane was purified in the usual way [21].

The same general procedure was adopted for the preparation of the PWA-silica-Nafion membrane. The PWA-containing silica powder was composed of 30% wt/wt phosphotungstic acid; this amount was dictated by the uptake characteristics of silica.

2.1. Membrane-electrode assembly (MEA)

The catalysts for methanol oxidation and oxygen reduction were 40% Pt-20% Ru/Vulcan XC 72 (E-TEK Inc.) and 20% Pt/Vulcan XC 72 (Electrochem Inc.), respectively. The reaction layers were prepared by directly mixing in an ultrasonic bath a suspension of Nafion® ionomer in water with the catalyst powder; the obtained paste was spread on carbon cloth backings. Pt loading was 2 mg cm^{-2} in both electrodes. The MEA was manufactured by pressing the electrodes onto the composite membrane at 130° and 50 atm [22].

2.2. Single Cell

The MEA was loaded into a single cell test fixture (Globe Tech Inc.) which was connected to an HP 6060B electronic load. Aqueous solutions of methanol, as well as humidified oxygen, were preheated at 85°C and fed to the cell. The operating temperature of the cell was 145°C. Oxygen humidification at significantly lower temperatures produced thermal gradients which resulted in some fluctuations during the electrochemical polarization. The outlet fluxes from anode and cathode compartments were measured on-line before being condensed in two separate reservoirs. The solutions were analysed by a Carlo Erba VEGA Series 2 GC 6000 chromatograph equipped with Carbopack 3% SP 1500 column and flame ionization detector[23].

An appropriate experiment was devised to analyze the gas compositions at both anode and cathode outlets. The outlet stream composition was chromatographically analyzed by drawing a dried gas sample with a syringe. Water and methanol from the stream were removed by a magnesium perchlorate bed while the gas sample was being drawn from a glass sampling bulb provided with a plug-type system. The gaseous products were analyzed with two columns in series: (i) a 2.5 m Porapak QS 80/100 and (ii) a 2.5 m molecular sieve 5 A kept at 70°C and connected to a thermal conductivity detector (TCD).

2.3. Physico-chemical characterization

X-ray diffraction (XRD) analysis was carried out with a Philips X-Pert diffractometer using a CuKα source. Instrumental broadening was determined by a standard Pt sample under the same analysis conditions.

The morphology of the MEA was investigated with a Philips XL20 scanning electron microscope (SEM), equipped with a LaB_6 filament.

3. RESULTS AND DISCUSSION

X-ray diffraction analysis showed that both anode and cathode catalysts exhibit the diffraction peaks of Pt fcc structure (Fig. 1). A significant shift to higher 2θ values is observed for the diffraction peaks in the Pt-Ru/C sample with respect to the Pt/C catalyst. The lattice parameters, a_{fcc}, were 3.921Å in Pt/C and 3.87 Å in Pt-Ru/C catalysts, respectively, as determined by the peak profile fitting of the (220) reflection (Fig. 2).

The observed lattice parameter for the Pt-Ru/C sample follows from the presence of a solid solution of Pt and Ru. According to the variation of a_{fcc} with composition for Pt-Ru bulk alloys, an atomic fraction of 45% Ru should be present in the carbon supported alloy. An average particle size for the metal crystallites of 23 Å and 20 Å in the Pt/C and Pt/Ru/C catalysts, respectively, was determined from the broadening of the (220) diffraction peak by using the Debye-Sherrer equation.

X-ray diffraction patterns of the powder consisting of 30% wt/wt PWA revealed an overlapping of the typical crystalline structure of heteropolyacid and the amorphous structure of silica (Fig. 3); the PWA-modified membrane shows even larger crystalline characteristics, indicating a higher degree of backbone chain packing.

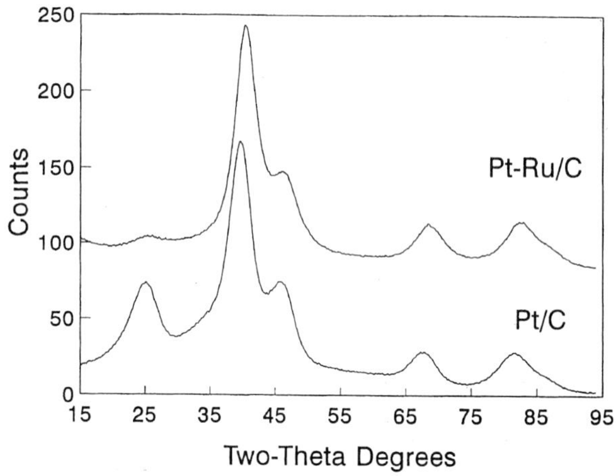

Fig. 1. X-ray diffractogram of Pt-Ru/C and Pt/C catalysts.

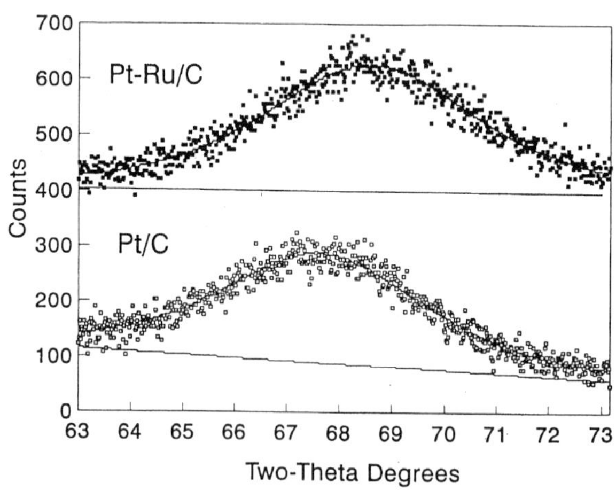

Fig. 2. Peak profile fitting of the (220) reflection in Pt-Ru/C and Pt/C catalysts.

Accordingly, Fig. 4 shows the X-ray diffraction profiles of Nafion® 117 and the composite membranes, prepared as described in the experimental section, in the protonic form and dry state. The broad diffraction peaks at $2\theta = 12\text{-}20°$ result from a convolution of amorphous ($2\theta = 16$) and crystalline ($2\theta = 17.50$) scattering from the polyfluorocarbon chains of Nafion®.

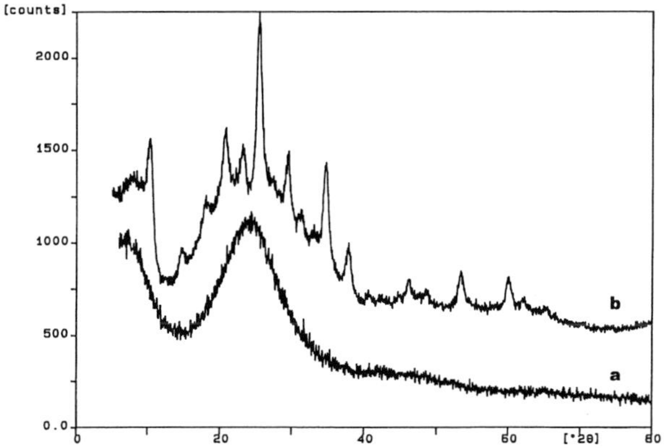

Fig. 3. X-ray diffraction patterns of bare (a) and PWA-impregnated silica powder

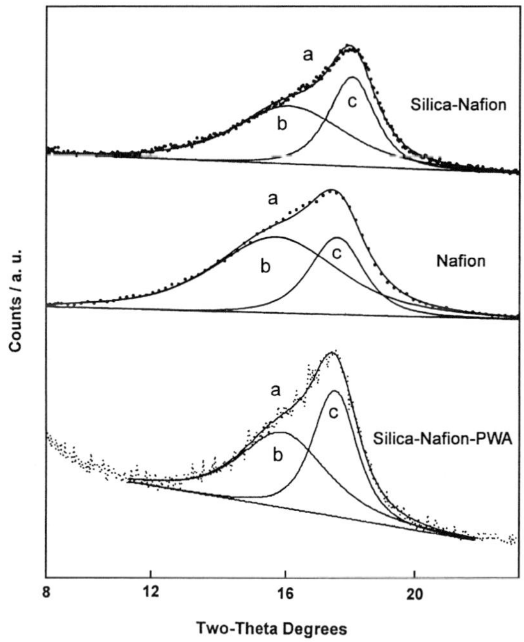

Fig. 4 X-ray diffraction profiles of Nafion 117 and composite membranes in protonic form and dry state. Curve a is the convolution of curves b and c. Measured data are shown by dots.

By comparison of the diffractograms, an increase of the ratio between the crystalline peak and the amorphous scattering, as well a shift of the crystalline peak to higher Bragg angles, is apparent in the Nafion-silica membrane with respect to Nafion® 117.

Thus, the addition of silica and heteropolyacid modifies the relative ratio between crystalline and amorphous structure of cast Nafion with respect to bare Nafion. Yet, due to the low loading of silica and heteropolyacid in the membrane (3% silica, less than 1.5% PWA), it appears unlikely that the inorganic components are responsible of such significant modification in the Nafion structure. More probably, the observed changes have to be attributed to the final thermal treatment (160°C).

A secondary electron micrograph of the complete MEA is shown in Fig. 5 . At the bottom and top sides of the micrograph, diffusion layers including carbon cloth fibers are easily recognized.

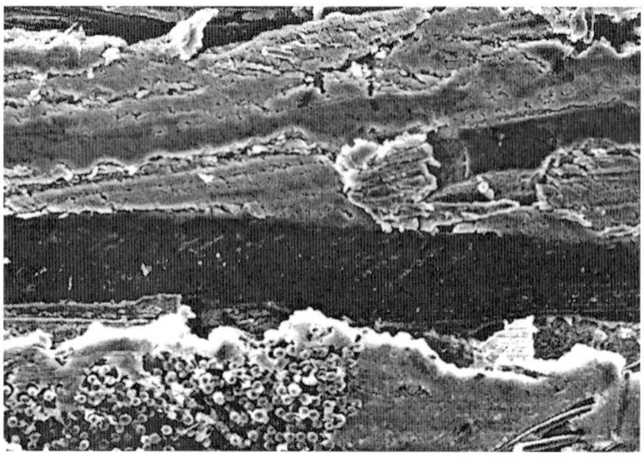

Fig. 5. SEM image of a cross section of the membrane-electrode assembly. Anode, bottom; cathode, top.

The electrode-electrolyte assembly was investigated in a single cell test station. After installing the MEA in the fuel cell housing, water was supplied to the anode and cathode backing layers and the cell was warmed-up step-wise from room temperature to 145°C. The polarization curves obtained for the fuel cells equipped with the Nafion-silica and Nafion-silica-PWA membranes, under same conditions in presence of oxygen feed at cathode and 2M methanol solution at anode, are reported in Fig. 6.

Similar electrochemical characteristics are observed up to 0.5 A/cm^2. The PWA-based membrane shows lower mass transport limitation with consequent higher voltage at higher current densities, giving a maximum power density of 0.4 W/cm^2 at current density of about 1.4 A/cm^2 under oxygen feed operation at 145°C; maximum power density of 0.34 W/cm^2 was obtained with the silica-modified membrane. It is clear that a positive effect on cell performance is given by the addition of phosphotungstic acid to the silica-Nafion membrane.

Since the enhancement was observed at high current densities, one may ascribe this behaviour to a better ion transport through the membrane under such conditions. Yet, this conjecture is not supported by the analysis of the cell resistance-temperature relationship (Fig. 7).

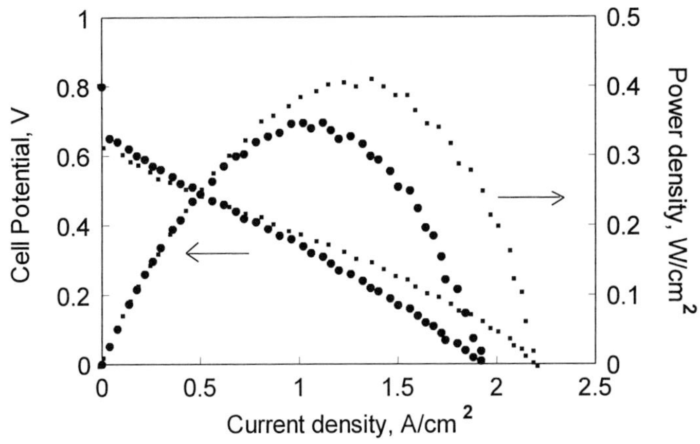

Fig. 6. Polarization and current density curves for MEAs equipped with Nafion-silica (●) and Nafion-silica-PWA (▪) membranes. Conditions: 145°C, oxygen, 2M methanol.

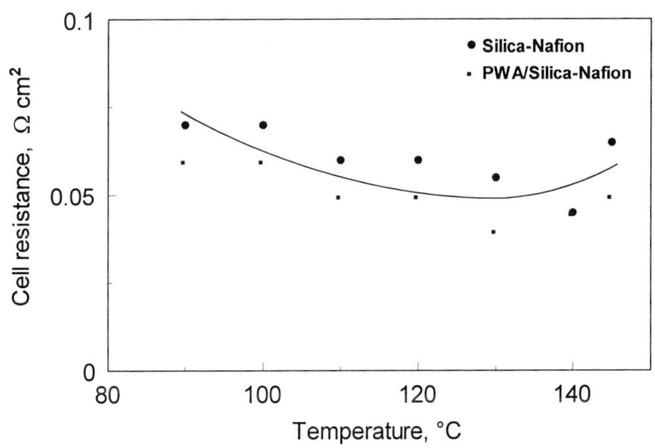

Fig. 7. Cell resistance vs. temperature during DMFC operation (current interrupter method).

In a previous paper [24] we have demonstrated a promoting effect of phosphotungstic acid on the kinetics of electrochemical reduction of oxygen. This was mainly attributed to the high oxygen solubility in the PWA electrolyte. Thus, the increase in limiting current density here obtained with the PWA-based membrane-electrode assembly could be more properly related to such high oxygen solubility. For what concerns a possible promoting effect on methanol oxidation, it is known that CO-like products, which form at the anode during methanol oxidation reaction giving rise to a strong chemisorption on the catalyst surface, are better removed in the presence of heteropolyacid [25]. Yet, the expected improvement in the activation- controlled region is not observed. Thus, the hypothesis of lower voltage losses at high currents due to enhanced oxygen solubility at the interface appears to be the most probable. In the presence of oxygen, the best fuel cell performance was obtained with 2M methanol solution. With air at the cathode, the best performance (0.25 W/cm^2 at 145°C, PWA membrane) was obtained by feeding 1M methanol solution at the anode, on account of the more significant effect of methanol cross-over on cathode polarization in the presence of air as oxidant; 0.21 W/cm^2 were obtained with the silica-Nafion membrane at the same temperature (Fig. 8).

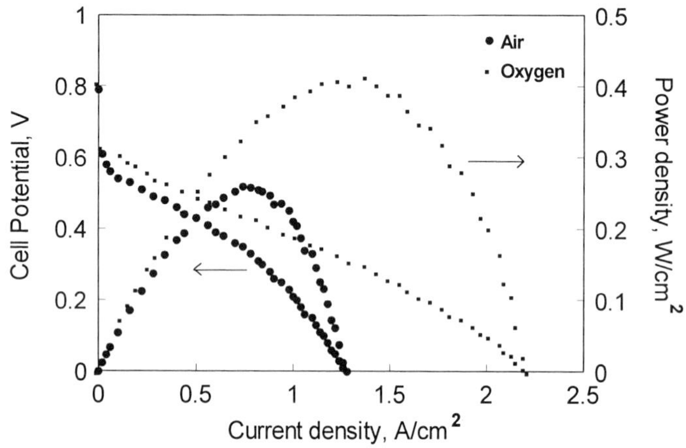

Fig. 8. Comparison of the polarization and power density curves with air and oxygen for the MEA equipped with the PWA-modified membrane at 145°C. Operating conditions: oxygen feed, 2M methanol; air feed, 1M methanol.

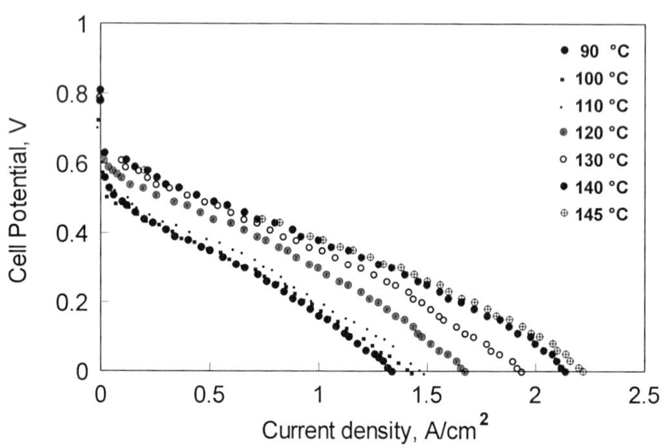

Fig. 9. Influence of operation temperature on the DMFC polarization characteristics for the MEA equipped with the PWA-modified membrane (oxygen, 2M methanol).

The influence of temperature on the polarization characteristics of the fuel cell working with the PWA membrane is reported in Fig. 9. The voltage losses progressively decrease with temperature in the range 90-145°C for both activation- and diffusion-controlled regions; experiments carried out at 150°C gave rise to an unstable electrochemical behaviour. It is pointed out that the voltage gain at very low current densities (50 mA/cm^2), passing from 90 to 145°C, is about 100 mV, denoting the strong activation nature of the methanol electro-oxidation process.

4. CONCLUSIONS

Heteropolyacid-modified silica-Nafion membranes showed suitable properties for operation at 145°C in direct methanol fuel cell. Since the cell resistance is similar to that of Nafion-silica membrane and the main improvement in polarization is observed at high current density, it is thought that the excellent oxygen solubility characteristics at the electrode/PWA interface are responsible for the significantly higher limiting current density.

REFERENCES

1. H.P. Dahr, Proc. 1994 Fuel Cell Seminar, San Diego ,USA.
2. P. Stonehart, Proc. 1996 Fuel Cell Seminar, Orlando, USA.
3. M. Watanabe, H. Uchida, Y. Seki, M. Emori and P. Stonehart, J. Electrochem. Soc., 143 (1996) 3847.
4. N. Giordano, P. Staiti, S. Hocevar and A.S. Aricò, Electrochim. Acta, 41 (1996) 397.
5. P. Staiti, S. Hocevar and N. Giordano, Int. J. Hydrogen Energy, 22 (1997) 809.
6. P. Staiti, A.S. Aricò, S. Hocevar and V. Antonucci, J. New. Mat. Electrochem. Systems, 1 (1998) 1.
7. A.J. Appleby, Proc. 9th World Hydrogen Energy Conference, Paris, 1992.
8. S. Hocevar, P. Staiti and N. Giordano, Proc. 4th Grove Fuel Cell Symp., London, 1995.
9. P. Staiti, S. Freni and S. Hocevar, J. Power Sources, 79 (1999) 250.
10. M. Misono, Catal. Rev. Sci. Eng., 29 (1987) 269.
11. I.V. Kozhevnikov, Chem. Rev., 98 (1998) 171.
12. N. Mizuno and M. Misono, Chem. Rev., 98 (1998) 199.
13. R.S. Drago, J.A. Dias and T.O. Maier, J. Am. Chem. Soc., 119 (1997) 7702.
14. M. T. Pope, Heteropoly and Isopoly Oxometallates, Springer, Berlin, 1983.
15. J.F. Keggin, Nature, 131 (1933) 908.
16. O. Nakamura, T. Kodama, I. Ogino and Y. Miyake, Chem. Lett., (1979) 17.
17. Poulsen, RISO Report M-2244, 1980.
18. D.E. Katsoulis, Chem. Rev., 98 (1998) 359.
19. R.B. Moore and C.R. Martin, Analyt. Chem., 58 (1986) 2569.
20. G.Gebel, P. Aldebert and M. Pineri, Macromolecules, 20 (1987) 1428.
21. A. Ticianelli, C.R. Derouin, A. Redondo and S. Srinivasan, J. Electrochem. Soc., 135 (1988) 2209.
22. P.L. Antonucci, A.S. Aricò, P. Cretì, E. Ramunni and V. Antonucci, Solid State Ionics, 125 (1999) 431.
23. A.S. Aricò, P. Cretì, P.L. Antonucci and V. Antonucci, Electrochem. and Solid State Lett., 1 (1998) 66.
24. N. Giordano, A.S. Aricò, S. Hocevar, P. Staiti, P.L. Antonucci and V. Antonucci, Electrochim. Acta, 38 (1993) 1733.
25. A.S. Aricò, E. Modica, I. Ferrara and V. Antonucci, J. Appl. Electrochem., 28 (1998) 881.

Studies in Surface Science and Catalysis 140
A. Gamba, C. Colella and S. Coluccia (Editors)

Mixed vanadium oxides for application as electrode materials in the Li-ion cells

M. Arrabito[a], S. Bodoardo[a], S. Ronchetti[a], N. Penazzi[a], G. Martra[b], D. Sidoti[b], Y. Wang[c], X. Guo[c], S. G. Greenbaum[c]

[a] Dip. SMIC, Politecnico di Torino, C.so Duca Abruzzi 24 10129 Torino ITALY

[b]Dipartimento di Chimica IFM - Universita' di Torino, via P. Giuria 7 Torino ITALY

[c]Department of Physics, Hunter College of CUNY - New York, NY 10021 USA

ABSTRACT

The results obtained from the investigation of the crystallochemical properties of a family of mixed Li-vanadates of general formula $LiCo_yNi_{(1-y)}VO_4$ are reported. This kind of material is believed to be a valid alternative to the traditional positive electrode materials in the Li-ion cell, an electrochemical power source of outstanding importance in the portable electronics field. The powders, prepared via two different methods (a wet chemistry route and a solid state method), have been characterised by XRPD analysis, NMR and diffuse reflectance Vis-NIR spectroscopy. The findings allowed to correlate the electrochemical performance of the vanadates, which is higher for the samples prepared via wet chemistry, to the crystallinity of the powders and to the transition metal cations distribution.

1. INTRODUCTION

The successful results of the Lithium – ion battery, a cell based on the rocking chair concept [1], as an optimized power source system for portable electronics [2], has stimulated research on new types of lithium insertion materials. This kind of accumulator needs, in fact, that Li^+ ions can be released during discharge from the negative electrode (anode) in correspondence to the oxidation of some species in the electrodic material. Conversely, the positive electrode (cathode) must intercalate Li^+ ions, during discharge, in correspondence to the reduction of some component. Being this cell an accumulator, it can be recharged, at the end of discharge, via a process opposite to that just described.

As far as the positive electrode material is concerned, promising alternatives appear to be the mixed lithium vanadates of general formula $LiCo_yNi_{(1-y)}VO_4$ [3,4]. These compounds adopt an inverse spinel structure, where Li^+ ions reside in octahedral sites, Ni^{2+} and Co^{2+} ions and V^{5+} ions can be distributed both in octahedrally and tetrahedrally coordinated sites [5,6]. Such structure should allow easy removal of lithium at potentials of the order of 4.2V to 4.8 V vs. Li, depending upon the Ni/Co ratio.

We have therefore investigated the family of the mixed Li transition metal vanadates trying to get an insight as complete as possible of the crystallochemical properties of these compounds. Our goal for the future is to unravel the correlations between the structural characteristics of these materials and their electrochemical behaviour.

Different preparation routes lead to powders with different electrochemical performance, so, in order to investigate this aspect, we have prepared, by two different methods, series of $LiCo_yNi_{(1-y)}VO_4$ (y=0, 0.2, 0.5, 0.8, 1) compounds. These powders have been characterized by X-ray powder diffraction (XRPD), diffuse reflectance Vis-NIR and NMR spectroscopies.

The XRPD technique, while an indispensable tool in the structural characterization of battery cathode materials, provides information by averaging over hundreds of lattice sites. Thus it is relatively insensitive to the local structural variations that occur when one kind of transition metal ion is sequentially substituted for another (although it does give the average change in lattice parameter according to Vegard's Law[rr]). On the other hand, NMR is particularly sensitive to short range interactions, and then it yields information on short-range spatial correlations, especially when the ions being substituted into the lattice are paramagnetic.Diffuse reflectance Vis-NIR spectroscopy appeared, and proved to be, a useful tool to investigate the features of nanometric thickness surface layers of the materials.

2. EXPERIMENTAL

A series of $LiCo_yNi(_{1-y})VO_4$ materials, namely $LiCoVO_4$, $LiCo_{0.2}Ni_{0.8}VO_4$, $LiCo_{0.5}Ni_{0.5}VO_4$, $LiCo_{0.8}Ni_{0.2}VO_4$ and $LiNiVO_4$ have been considered in this work. These compounds have been synthesised starting from reagent grade chemicals (ALDRICH chemicals) by two different preparation methods: a wet chemistry and a solid state route. The wet chemistry preparation procedure (hereafter referred to as WCh) has been described elsewhere [5]. In the solid state route (hereafter referred to as SS) the precursors, $LiCO_3$, CoO, NiO, V_2O_5, were finely ground in a ball mill and then annealed at 530 °C for about 200 to 230 hours. During this treatment the powders were re-ground each 24 hours to reduce the size of the grains.

The structure of the compounds was probed by XRPD (Philips PW Diffractometer with graphite monochromator). The measured integrated intensities were used for the refinement of the unit cell dimensions, the atomic coordinates, the site occupancies and the isotopic thermal factor related to a crystallographic model developed by XRD123, a publically available software (internet address: http://www.polito.it/Ulisse/Corsi/TLC/F0620/materiali).

The sample (500 mg) was packed into 5 mm OD pyrex tubes for the NMR measurements, which were performed with a Chemagnetics CMX300 Spectrometer, operating at ^{7}Li and ^{51}V resonance frequencies of 116.8 and 79.2 MHz respectively. Single pulse excitation sequence was employed. A typical B/2 pulse width was 2.5 μs. For ^{7}Li, typically 1200 spectral acquisitions were required. Chemical shift references were aqueous solutions of LiCl.

Diffuse reflectance Vis-NIR spectra were obtained using a Perkin-Elmer Lambda 19 spectrophotometer equipped with an integrating sphere and using $BaSO_4$ as reference powder.

3. RESULTS AND DISCUSSION

3.1. XRPD analysis

The analysis of XRPD data indicates that all prepared materials have an inverse spinel structure belonging to the cubic Fd3m space group. In the case of SS samples, as the duration of annealing increased, diffraction peaks progressively narrower were recorded, indicating a a corresponding increase of the cristallite size. After the annealing step, the diffraction peaks obtained for the SS appeared narrower than those for the corresponding WCh samples, indicating that this second set of samples is constituted by smaller size crystallites. SEM micrographs [5,7] confirm this trend: SS powders are composed by about 5 μm particles, while WCh samples contain grains smaller than 1 μm.

Figure 1 reports the variation of the lattice parameter a_0 for the materials prepared via the two preparation procedures. It can be observed that for both sample sets the lattice parameter increases linearly with the Co^{2+} content, mainly because of the larger radius of this ion with respect to Ni^{2+} and that for each composition the SS and WCh samples exhibit the same value of a_0, within the limits of the experimental error.

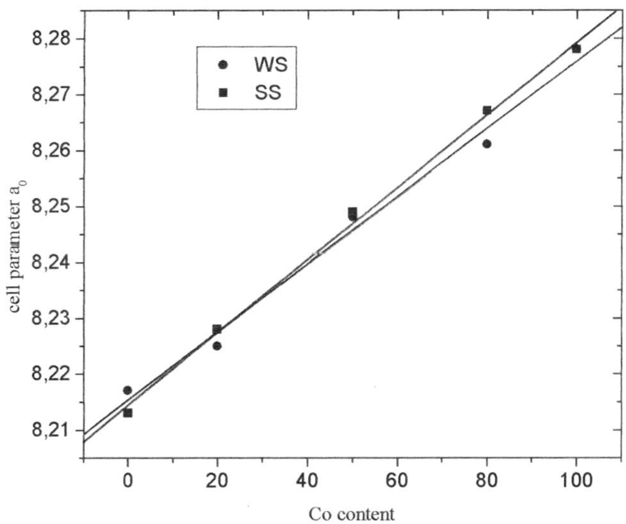

Figure 1. Variation of the lattice parameter a_0 [Å] *vs* Co content for all studied samples in dependence on the preparation method.

As the crystallographic site occupancy is concerned, in inverse spinels all Li^+ ions reside in octahedral coordination sites, while the transition metal ions (TMI) can occupy both octahedral and tetrahedral sites. The results obtained are reported in Table 1; they suffer a limitation due to the fact that the XRPD technique can discriminate between V^{5+} ion and Co^{2+} or Ni^{2+} ions, while it fails in the distinction between Co^{2+} or Ni^{2+} cations. The data reported clearly indicate, anyway, that the occupancy of octahedral and tetrahedral sites by Co^{2+}- Ni^{2+} and V^{5+} ions depends on both composition and preparation method.

Table 1:
Number of Co^{2+} and/or Ni^{2+} and V^{5+} ions in octahedral (Oh) and tetrahedral (Td) sites per unit cell for the $LiCo_yNi_{(1-y)}VO_4$ materials prepared *via* solid state (SS) or wet chemical (WCh) method.

Entry	Sample	Preparation method	Co^{2+}_{Oh}- Ni^{2+}_{Oh}	Co^{2+}_{Td}- Ni^{2+}_{Td}	V^{5+}_{Oh}	V^{5+}_{Td}
1	$LiCoVO_4$	SS	8	0	0	8
2	$LiCoVO_4$	WCh	7	1	1	7
3	$LiCo_{0.8}Ni_{0.2}VO_4$	SS	8	0	0	8
4	$LiCo_{0.8}Ni_{0.2}VO_4$	WCh	7	1	1	7
5	$LiCo_{0.5}Ni_{0.5}VO_4$	SS	5	3	3	5
6	$LiCo_{0.5}Ni_{0.5}VO_4$	WCh	7	1	1	7
7	$LiCo_{0.2}Ni_{0.8}VO_4$	SS	5	3	3	5
8	$LiCo_{0.2}Ni_{0.8}VO_4$	WCh	6	2	2	6
9	$LiNiVO_4$	SS	5	3	3	5
10	$LiNiVO_4$	WCh	6	2	2	6

considering SS samples it can be seen that, in $LiCoVO_4$ and $LiCo_{0.8}Ni_{0.2}VO_4$, Co^{2+} - Ni^{2+} ions are confined in octahedral and V^{5+} ions, consequently, occupy tetrahedral sites (Table 1, entries 1 and 3). From $LiCo_{0.5}Ni_{0.5}VO_4$ down to $LiNiVO_4$ every kind of ion appears distributed almost equally between both tetrahedral and octahedral sites (Table 1, entries 5,7 and 9) with some preference of the Co^{2+} - Ni^{2+} cations for the octahedral sites and, conversely, of the tethraedral ones for V^{5+}.

These findings are quite interesting since they are different with respect to what has been reported for the pure spinels where Co^{2+} exhibits a tendency (of the order of 13 kJ/mole, difference between the octahedral and tetrahedral coordination lattice energies) to assume a tetrahedral coordination and Ni^{2+} even a more marked tendency (of the order of 50kJ/mole, difference between the tetrahedral and octahedral coordination lattice energies) to assume an octahedral coordination [8].

As for the WCh materials, in $LiCoVO_4$ already some sharing of the Co^{2+} ions between the tetrahedral and octahedral sites appears, the same happening, in a complementary way, for the V^{5+} ions (Table 1, entry 2). With the increase of the y value, in $LiCo_yNi_{(1-y)}VO_4$ the extent of the tetrahedral - octahedral distribution of the same kind of ion becomes more marked. This takes place for the highest y values and it is not so extended as in the SS samples (Table 1, entries 8 and 10).

A possible explanation of the differences in the TMI site occupancy between the SS and WCh samples comes from the fact that during the long annealing treatment of the SS materials the cations can have enough time to migrate so to achieve the most thermodynamically stable coordination, whereas only metastable conditions are reached in the mild conditions used for the WCh route.

3.2. NMR spectra

Lithium-7 NMR spectra of all samples are displayed in Fig. 2 (Section A and B).

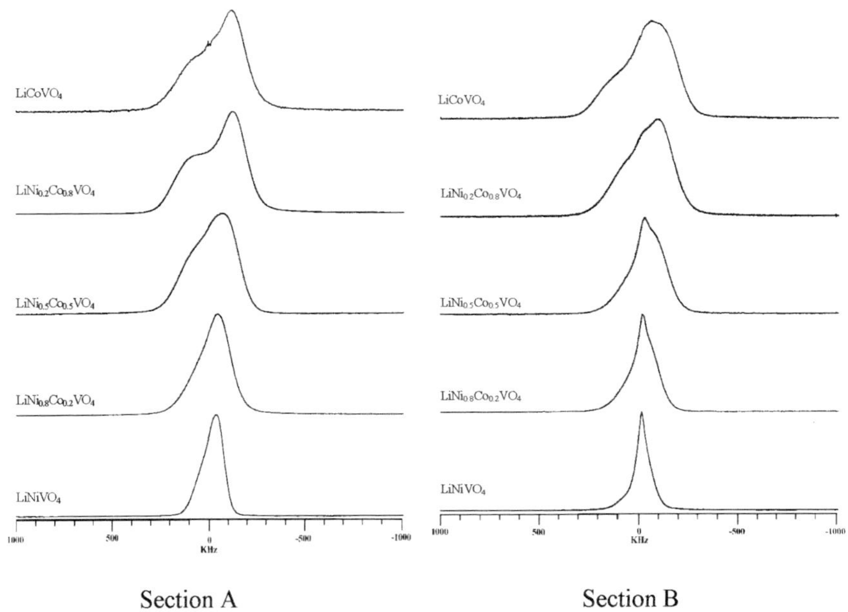

Figure 2. ^{7}Li NMR signal at room temperature. Section A: WCh samples. Section B: SS samples.

The spectrum of $LiNiVO_4$ (bottom) is a relatively narrow line with no evidence of quadrupole satellite transitions, consistent with Li in an octahedral environment. Substitution of Co^{2+} for Ni^{2+} broadens the ^{7}Li resonance line significantly due to the strong magnetic dipolar interaction with spins localized on the Co site. Comparison of samples with the same composition but different preparation methods shows that the ^{7}Li NMR line shapes are substantially different, especially when the Co-substitution is in the range of 50%, reflecting a different spatial distribution of paramagnetic Co^{2+} ions with respect to the Li^{+} ions. The line width of the solid synthesis (SS) sample at the 1:1 Ni:Co composition is narrower than that of the corresponding wet synthesis (WCh) material, and the peak positions of the two lines differ, with the SS peak appearing between the main peak and observable shoulder of the WCh sample. This is consistent with a statistically random distribution of Li-Co spatial correlations in SS as opposed to a non-random distribution in WCh.

3.3. Diffuse reflectance Vis-NIR spectra

Figure 3A reports the diffuse reflectance Vis-NIR spectra of the $LiCoVO_4$ materials prepared through the two different routes. The spectrum of the $LiCoVO_4$ WCh sample exhibits an intense edge absorption at 21300 cm^{-1} and two weaker signals at 16500 and 7500 cm^{-1} (with a shoulder at 6800 cm^{-1}). (Fig.3A, a). The edge at high frequency is attributed to the lowest-energy charge transfer transition (LCT) associated with electron transfer from O^{2-} ligands to V^{5+} ions in octahedral coordination [9]. As for absorptions at lower frequency, by comparison with the spectra of Co^{2+} ions in oxidic matrices [10,11), they can be assigned to the $^4A_2 \rightarrow ^4T_1(P)$ (component at 16500 cm^{-1}) and $^4A_2 \rightarrow ^4T_1(F)$ (band at 7500 and shoulder at 6800 cm^{-1}) *d-d* transitions of Co^{2+} in tetrahedral coordination, their broadness and complexity resulting from the lack of resolution of the transition to the various states of the excited terms. Apparently, Co^{2+} ions in octahedral sites, which represent the overwhelming fraction of Co^{2+} in the material (see Table 1, entry 2), do not contribute significantly to the spectrum. This is not surprising, because the absorption coefficients of their *d-d* transitions are more than one order of magnitude weaker than those of ions in tetrahedral one [11]. It can be then supposed that their absorptions, expected to occur in the 18000-21000 and 8000-8700 cm^{-1} ranges [11], contribute as weak, unresolved features to the low frequency side of the edge absorption at 21300 cm^{-1} and to the high frequency side of the band at 7500 cm^{-1}, respectively.

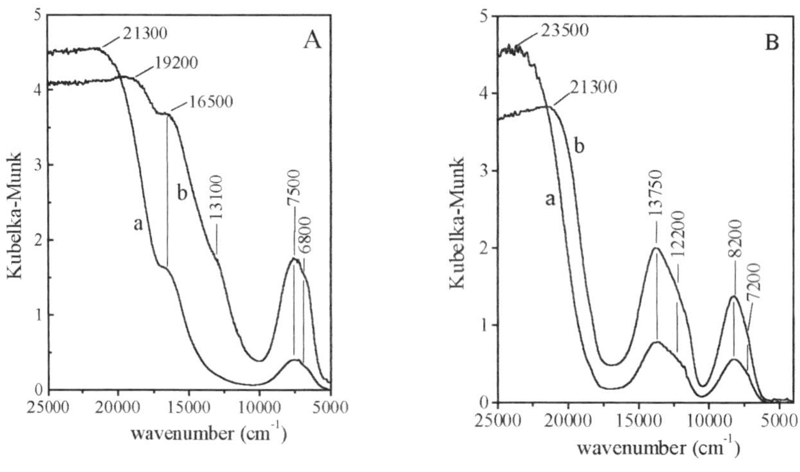

Figure 3. Diffuse reflectance Vis-NIR spectra of $LiCoVO_4$ and $LiNiVO_4$ materials prepared *via* both wet chemistry (WCh) and solid state (SS) methods. Section A: a) $LiCoVO_4$ WCh, b) $LiCoVO_4$ SS; Section B: $LiNiVO_4$ WCh, b) $LiNiVO_4$ SS.

On this basis, only very weak *d-d* bands due to octahedral Co^{2+} should be expected in the spectrum of the $LiCoVO_4$ SS material, as XRPD data indicate that in this case all Co^{2+} ions are in octahedral sites (see Table 1, entry 1). By contrast, the absorptions at 16500 and 7500 cm^{-1} due to tetrahedral Co^{2+} ions are present also in the spectrum of this materials, and exhibit an intensity significantly higher than for the previous sample, while new components appear at 19200 and 13100 cm^{-1} (Fig. 3A, b). Furthermore, the LCT edge due to the V^{5+} ions appears less intense than for the $LiCoVO_4$ WCh sample.

The electronic spectroscopic pattern is then really different from that expected on the basis of the structural insights provided by the XRPD data. However, it must be considered that the depth of penetration of a radiation in a solid where it can be efficiently absorbed is confined to a layer close to the surface [12], whereas highly energetic radiation that are simply diffracted can travel through the solid for larger distances. It can be considered that XRPD data provide information on the structural features of thick portions (at least a few μm) of the grains of the materials, whereas the DR Vis-NIR spectra monitor the status of a thin (about 50 nm) surface layer, the contribution of which to the XRPD pattern is negligible. Apparently, in the case of the $LiCoVO_4$ SS material such thin surface layer contains tetrahedral Co^{2+} ions in larger amount than the corresponding WCh sample. The assignment of the new components at 19200 and 13100 cm^{-1} is not so evident, and further studies and experiments devoted to clarify their nature are in progress.

Finally, it can be noticed that the lower intensity of the edge due to V^{5+} species might indicate that the surface layer of the $LiCoVO_4$ SS grains monitored by the Vis-NIR radiation is poorer in such type of ions in octahedral coordination.

The DR Vis-NIR analysis of the $LiNiVO_4$ WCh and SS materials was also performed, and the results are reported in figure 3B.

In the spectrum of the $LiNiVO_4$ WCh sample, an edge absorption at 23500 cm^{-1} and two bands at 13750 (with a shoulder at 12200 cm^{-1}) and 8200 cm^{-1} (with a shoulder at 7200 cm^{-1}) are present. Also in this case, the edge signal is assigned to the LCT transition from O^{2-} to octahedral V^{5+} ions, its position at higher frequency with respect to that observed in the $LiCoVO_4$ samples probably resulting from a decreased number of structural units containing octahedral V^{5+} ions sharing their electronic structure.

The absorptions at lower frequency can be attributed to the $^3T_1(F) \rightarrow {}^3T_1(P)$ (band with maximum at 13750 cm^{-1}) and $^3T_1(F) \rightarrow {}^3A_2(F)$ (band with maximum at 8200 cm^{-1}) *d-d* transitions of Ni^{2+} ions in tetrahedral coordination [10,13]. The position of the $^3T_1(F) \rightarrow {}^3T_1(P)$ absorption suggests that such ions experience a ligand field weaker than in other oxidic matrices (e.g. ZnO) [10,13].

The same *d-d* bands observed for the $LiNiVO_4$ WCh material are present also in the spectrum of the corresponding SS sample (Fig. 3B, b), but *ca.* 2.5 times more intense. XRPD data indicate that in the bulk of this material tetrahedral Ni^{2+} ions are only 1.5 times more abundant than in the bulk of the $LiNiVO_4$ WCh grains. It can be then proposed that such higher relative content in tetrahedral Ni^{2+} ions is a characteristic of the surface layer monitored by the Vis-NIR radiation.

For both samples no *d-d* bands due to octahedral Ni^{2+}ions are observed, because of the very low value of their extinction coefficient with respect the tetrahedral species (13), as in the case of Co^{2+} ions.Finally, it can be noticed that for the $LiNiVO_4$ SS sample the edge associated with the LCT transition from O^{2-} to octahedral V^{5+} ions appear less intense and located again at 21300 cm^{-1}, indicating that the surface region of the grains of such material are less rich in octahedral V^{5+} ions, and that these species are located in larger domains of structural units sharing their electronic structure.

DR Vis-NIR spectra of the WCh and SS materials containing both Co^{2+} and Ni^{2+} ions were recorded also. Owing to the heavy overlapping of components due to Co^{2+} and Ni^{2+} they appeared very complex, and do not allow a detailed recognition of specific features of each material.

4. CONCLUSIONS

All the experimental techniques used have put in evidence well detectable differences between materials of the same chemical composition prepared via different routes. Diffraction data showed differences in the kind of distribution of the MTI between the tetrahedral and octahedral crystallographic sites. As y increases, the MTI ditribution for SS $LiCo_yNi_{(1-y)}VO_4$ samples markedly varies tending to an almost even occupation of the tetrahedral and octaedral sites. The WCh samples show only a slight MTI distribution variation.

As expected, some of the results obtained from different experiments seem at a first sight in contrast. This is not true because, as reported in the introduction, every experimental technique gives informations on a particular aspect only of the tested system. So, while X ray diffraction describes the overall situation of the TMI distribution, NMR data have shown that one can expect that WCh samples present locally a preferred distribution of the ions, different from the general one. The DR Vis-NIR spectra have made clear that the ions distribution varies with the distance from the surface.

Previously reported electrochemical data [14,15] showed that WCh materials are in general characterized by superior performance with respect to SS samples. This result has been related to the higher crystallinity of the SS powders, that is the larger crystallites size. Now we now that the differences in the structure, the TMI distribution between the octahedral and tetrahedral sites and the composition in the surface layers of the materials prepared via the two different methods could play a role in determining their electrochemical behaviour.

ACKNOWLEDGEMENTS

The authors wish to thank Dott. Priscilla Reale of the University La Sapienza of Rome for the preparation of the WS samples and the electrochemical characterization.

REFERENCES

1. M. Lazzari, B. Scrosati, J. Electrochem. Soc., 127 (1980) 773.
2. B. Scrosati, Nature, 373 (1995)557.
3. G. T. K. Fey, K. S. Wang, S. M. Wang, J. Power Sources, 68 (1997) 159.
4. G. T. K. Fey, D. L. Huang, Electrochim. Acta, 45 (1999) 295.
5. S.Panero, P.Reale, F.Bonino, B.Scrosati, M.Arrabito, S.Bodoardo, D.Mazza, N.Penazzi, Solid State Ionics, 128 (2000) 43.
6. G. Ting-Kuo Fey, Wu Li and J. R. Dahn, J. Electrochem. Soc., 141 (1994) 2279.
7. M. Arrabito, PhD Thesis, Torino 2000.
8. U. Müller, in " Inorganic Structure Chemistry", J.Willey & Son (eds.), Chapter. 8.
9. W. Hanke, R. Beinert and H.-G. Jerschkewitz, Z, Anor. Allg. Chem., 414 (1975) 109.
10. H. A. Weakliem, J. Chem. Phys., 36 (1962) 2117.
11. R. Pappalardo, D.L. Wood and R.C. Linares, J. Chem. Phys., 35 (1961) 2041.
12. R.I. Bickley, in "Photoelectrochemistry, Photocatalysis and Photoreactors. Fundamental and applications, M. Schiavello (ed.), Reidel Publishing Company, Dordrecht, 1985; NATO ASI Series C: Mathematical and Physical Sciences, Vol. 146, p. 379.
13. R. Pappalardo, D.L. Wood and R.C. Linares, J. Chem. Phys., 35 (1961) 1460.
14. M.Arrabito, S.Bodoardo, N.Penazzi S.Panero, P.Reale, B.Scrosati, Y. Wang, X. Guo, S. G. Greenbaum, J. Power sources, in press.
15. Priscilla Reale, degree thesis, Rome 1999.

Studies in Surface Science and Catalysis 140
A. Gamba, C. Colella and S. Coluccia (Editors)

Search and Optimization of Multi-Metal-Oxide Catalysts for the Oxidative Dehydrogenation of Propane - A Combinatorial and Fundamental Approach -

M. Baerns[1], O. V. Buyevskaya, A. Brückner, R. Jentzsch, E. Kondratenko, M. Langpape and D. Wolf

Institute for Applied Chemistry Berlin-Adlershof
Richard-Willstätter-Straße 12, D-12489 Berlin, Germany

This study presents a strategy for elaborating multi-metal-oxide materials for the oxidative dehydrogenation of propane, which was chosen as a model reaction. This strategy is based on an evolutionary approach, which uses fundamental knowledge on catalysts operation for supporting design of a first generation of catalytic materials to start with. The additional knowledge derived from further characterisation of catalytic materials of subsequent generations was shown to be the important factor for further development of catalytic materials with improved performance.

Key words: combinatorial catalysis, isolated VO_x sites, MCM-41 and -48 support for highly dispersed VO_x, oxidative dehydrogenation of propane

1. INTRODUCTION

In spite of very significant progress in the fundamental understanding of catalysis there is nevertheless still a need for an empirical approach when searching and optimizing catalytic materials. This is illustrated by the revival of high-throughput experimentation including synthesis and catalyst testing which has been already applied in a somehow different manner by A. Mittasch in the development of an iron-based catalyst for ammonia synthesis [1]. Present-day high-throughput experimentation aims at diminishing the time required for the above mentioned steps. Usually only activity and yield of tested materials is screened for. Only very recently also physico-chemical characterization of such tested materials has become known [2]; further development will certainly aim also at a rapid and parallel characterization of catalysts which are being tested. Another characterization approach for the catalytic materials besides activity or yield respectively would be a complete kinetic assessment of the catalytic performance. The kinetics would give the opportunity to identify the optimum operating conditions for achieving the maximum selectivity or yield respectively of each material which may differ significantly on changing tem-

[1] *Corresponding author:* Manfred Baerns, e-mail: baerns@aca-berlin.de

perature and partial pressures. In the present contribution the combinatorial process described recently by our group [3] is illustrated by a case study in which the search and optimization procedure for catalytic materials applied to the oxidative dehydrogenation of propane (ODP) is explained for illustration. ODP was chosen as a model reaction since some catalysts have been already suggested [4] and we were interested to learn whether the chosen combinatorial "grass-roots" procedure which was only based on some general knowledge on partial hydrocarbon oxidation would lead to similar results.

2. COMBINATORIAL METHODOLOGY

Combinatorial approach in heterogeneous catalysis means search for optimal combinations between chemical composition and structure of solid catalytic materials which are related to a diversity of preparation parameters (type of precursors, calcination temperature, calcination media, etc.) and operating conditions (temperature, contact time, reactant concentrations).

In the present work, a genetic algorithm was applied for searching new catalytic materials with high efficiency which was described in detail in [3]. The algorithm implies:

(i) stochastic creation of a library of catalytic materials of the 1st generation based on the pool of primary components,

(ii) preparation and testing of catalytic materials of the first generation,

(iii) selection of a certain proportion of those materials from the 1st generation which resulted in the best performance,

(iv) creation of a new library (2nd generation) based on the selected materials of the 1st generation using mutation and cross-over methods,

(v) preparation and testing the 2nd generation of catalytic materials,

(vi) repetition of (iii) – (v) for the next generations.

These steps were performed in the following way:

The selection of primary redox compounds was based on the simplifying assumption that a medium O-Me binding energy is required to ascertain the participation of lattice oxygen in ODP. From the metal oxides having a free energy of oxide formation $\Delta_f G$ per oxygen atom in the range between -400 and -200 kJ/mol similar to V_2O_5 ($\Delta_f G$ = -311 kJ/mol), the following 13 compounds were selected which were expected to contribute to the ODP catalysis: V_2O_5, MoO_3, MnO_2, Fe_2O_3, ZnO, Ga_2O_3, GeO_2, Nb_2O_5, WO_3, Co_3O_4, CdO, In_2O_3, NiO.

In an early stage of our work (procedure (A)), the different materials of the first generation (Table 1) consisted of only three primary components chosen from the pool of 13 oxides mentioned above; the total number of combinations of the 1st generation amounted to 20. In the subsequent generations only 10 new compositions were applied based on the information of the 5 best catalysts; among these 10 new catalytic materials 7 had been generated by mutation and 3 by cross-over. This approach had the advantage that fast improvement of the catalyst performance from generation to generation can be achieved. However, this might also lead to a local convergence and a rapid loss of potentially active components from the population. Due to this experience, a modified procedure (B) was derived and applied to in an advanced step of combinatorial search.

Procedure (B) was performed based on 5 different redox metal oxides which had proven to be catalytically active towards propene formation as a result of procedure A and 3 additional metal oxides of strong metal oxygen bond strength; i. e., MgO of basic nature, B_2O_3 of acidic nature and La_2O_3 on which dissociative oxygen adsorption takes place.

According to the advanced procedure, five generations were prepared and tested using always a population size of 56. Contrary to procedure A, catalytic materials with low performance were not completely removed from the population when the new generation was formed. Catalytic materials for preparing the next generation were chosen from the preceding generation with a certain probability which was proportional to their ranking in propene yield. Thus, a proportion of less active and selective materials was maintained in the population during the whole search process. Although this procedure leads to a slower overall convergence of the properties of the population, the danger of a rapid local convergence can be suppressed.

Table 1
Compositions of catalytic materials of the first generation (Procedure A)

No.	Content												
	V	Mo	Mn	Fe	Zn	Ga	Ge	Nb	W	Co	Ni	Cd	In
1				0.79		0.02		0.19					
2		0.44									0.23		0.33
3					0.7		0.08			0.22			
4	0.33			0.43								0.24	
5						0.009		0.33			0.66		
6		0.33			0.42								0.25
7							0.11		0.33			0.56	
8	0.26		0.33			0.41							
9								0.16		0.33			0.5
10		0.42		0.53			0.44						
11	0.47								0.19		0.33		
12			0.41			0.51		0.08					
13		0.45								0.22			0.33
14				0.75			0.05		0.2				
15	0.33		0.43								0.23		
16					0.67			0.1		0.23			
17		0.33		0.42								0.25	
18						0.065			0.33		0.6		
19			0.33		0.41								0.25
20							0.14			0.33		0.53	

Both procedures (A) and (B) were subsequently applied. The variation of the chemical composition was accomplished according to [3] by qualitative mutation and cross-over. By quantitative mutation the stoichiometry of the best samples was varied.

The best materials were partly characterized by their ability to adsorb oxygen by applying pulse experiments and by the state of the active material by applying XRD, XPS, EPR and UV/vis.

For comparison and further optimization with respect to the materials obtained by the combinatorial procedure, V-Mg-O, V-Ga-O and V-MCM catalysts were also prepared and tested.

3. EXPERIMENTAL

3.1. Catalyst preparation

The catalytic materials used in procedure A and B were prepared by dilution or dispersion of the starting compounds in water in the presence of granulated α-Al_2O_3 (CONDEA, particle size = 1mm, S_{BET} = 5 m^2g^{-1}) served as inert support for the active mass amounting to 30 wt.%. The mixture was stirred at 353 K for 30 min before evaporating the water at 373 K under stirring. The solid material obtained was dried at 393 K for 10h and calcined at 873 K for 3h in air. A more detailed description of the preparation can be found in [1].

The catalytic materials of the series V-Ga-O have been prepared by depositing VO_x on Ga_2O_3 [2]: Vanadium(V)oxide was dissolved at 353 K in oxalic acid. The gallium oxide was added to the solution; after stirring for 30 min the water was evaporated at 353 K. For the preparation of V-Mg oxide with Mg/V atomic ratios of 9/1, 4/1, 1/1 and 1/4, freshly prepared $Mg(OH)_2$ was added to a aqueous solution of NH_4VO_3 (Merck, p.a.) followed by water evaporation [3]. Drying and calcination of the samples were performed as described above.

Two samples with a vanadium loading of 2.8 % on MCM were prepared by impregnation of conventionally synthetisized siliceous MCM-41 and MCM-48 with $(NH_4)VO_3$ [4].

3.2. Catalyst testing

The set-up for catalytic tests contained a parallel arrangement of 6 U-form quartz reactors ($\varnothing_{int}$ = 6 mm), kept at a constant temperature by means of a fluidized bed of sand. The catalyst (m_{cat} = 0.3 g, d_p = 250 - 355µm) was packed between two layers of quartz particles of the same size. The reactors were operated at ambient pressure and the reaction mixture was sequentially passed over the catalysts. The analysis of the products is achieved by on-line gas chromatography. A more detailed description of the set-up can be found in [5].

The catalytic materials were tested for their catalytic performance at a reaction temperature of 773 K. The reaction mixture consisting of C_3H_8, O_2 and N_2 ($C_3H_8/O_2/N_2$ = 30/10/60 or 40/20/40) was passed over the catalyst at a total flow rate between 30 and 150 ml/min, depending on the activity of the catalytic material, which corresponds to a contact time of 0.6 to 0.12 g s·cm^{-3}.

3.3. Catalyst characterization

Specific surface areas of the catalytic materials were determined by the one-point BET method (Gemini III Micromeritics).

XRD powder analyses were carried out using a STADIP transmission powder diffractometer (Stoe) with $CuK_{\alpha 1}$ radiation.

The near-surface composition of the catalytic materials was derived from XPS spectra (ESCALAB 220i-XL, Fisons Instruments) using AlK_α radiation (1486.6 eV).

For analysing oxygen adsorption by transient experiments the TAP-2 reactor system was used which has been described in detail elsewhere [6]. The catalytic material (30-140 mg; d_p = 250-355 μm) was packed between two layers of quartz of the same particle size in the reactor. Before each experiment the oxide material was treated in a flow of O_2 (30 ml/min) at 823 K for ca 1 hour at 100 kPa. Then, the reactor was evacuated at 773 K to 10^{-4} Pa. Subsequently a reaction mixture of oxygen and neon (O_2:Ne = 1:1) was pulsed over the catalytic material. During a pulse experiment the response to a single pulse is monitored at a fixed AMU value. Pulses are repeated regularly and the responses were averaged to improve the signal-to-noise ratio. The repetition time was 1.3 sec to avoid accumulation of adsorbed species at the oxide surface.

EPR spectra were acquired with an ELEXSYS 500-10/12 (Bruker) spectrometer in X band at both 298 and 77 K. The magnetic field was measured with respect to the standard 2,2-diphenyl-1-picrylhydrazyl hydrate (DPPH).

UV/VIS-DRS measurements were performed using a Cary 400 UV/VIS spectrometer (Varian) equipped with a diffuse reflectance accessory (praying mantis, Harrick). The reflectance spectra were converted into the Kubelka-Munk function F(R) which is proportional to the absorption coefficient for low values of F(R). Spectra were deconvoluted into Gaussian bands using the GRAMS/386 program (Galactic Industries Corporation).

4. RESULTS

4.1. Catalytic performance

The detailed results of the catalytic tests of the different catalytic materials have been reported earlier [7]. The catalytic results for the best catalysts of each generation of procedures A and B are presented in Figure 1a and 1b, respectively. In both cases an improvement of the catalytic performance was observed. It became obvious that V_2O_5, MgO and Ga_2O_3 are the essential compounds of the best materials; however, even the best materials contained Fe, Mn, and Mo in a minor amount. It could not yet be decided whether minor amounts of these elements were of any significant effect.

In order to evaluate separately the effect of the major components 36 V-Mg-Ga mixed metal oxides of different composition supported on α-Al_2O_3 were investigated. The propene yields obtained were slightly lower than the results described above which may be due to the absence of Fe, Mn or Mo.

These results were compared to those obtained for vanadium oxide supported on Ga_2O_3 and MgO with comparable vanadium content. It was found that V-Mg-O-materials performed catalytically better and the V-Ga-O-materials performed worse than the V-Mg-Ga-O mixed metal oxides supported on Al_2O_3. Table 2 presents the catalytic performance of the best catalytic materials from V-Mg-Ga, V-Mg-O and V-Ga-O solid material libraries.

Based on the characterization results (see below) meso-porous siliceous MCM-48 and MCM-41 was used as support for VO_x.

The catalytic performance of these materials was the best of all materials investigated, maximal propene yields of 17.4 % were reached (Table 2).

4.2. Catalyst characterization

Bulk and surface compositions as well as surface areas of the best catalytic materials are also summarised in Table 2. The V-Ga-O materials only consisted of phases of pure

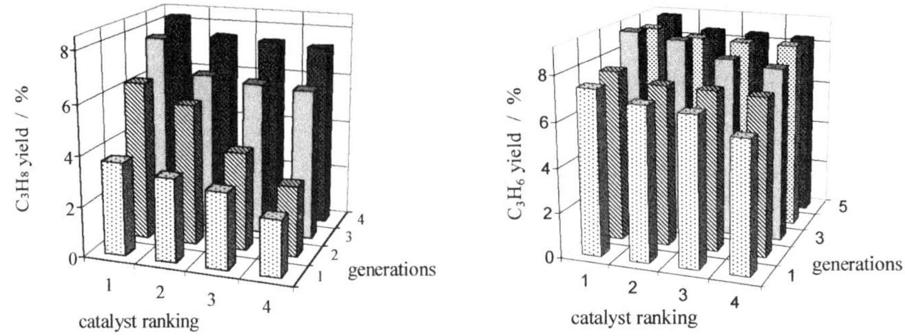

Fig. 1. Best propene yields achieved in each generation on the following catalytic materials (in the order of decreasing catalyst quality)
Case A: 1st generation: $V_{0.33}Fe_{0.43}Cd_{0.24}O_x$, $V_{0.33}Mn_{0.43}Ni_{0.23}O_x$, $V_{0.26}Mn_{0.33}Ga_{0.41}O_x$, $Mo_{0.42}Ge_{0.44}Fe_{0.53}O_x$; 2nd generation: $V_{0.47}Mn_{0.22}Ga_{0.31}O_x$, $V_{0.43}Mn_{0.54}Fe_{0.03}O_x$, $V_{0.26}Mo_{0.33}Fe_{0.41}O_x$, $V_{0.30}Fe_{0.48}Cd_{0.22}O_x$, 3rd generation: $V_{0.2}Mn_{0.17}Fe_{0.32}Ga_{0.32}O_x$, $V_{0.47}Mn_{0.13}Ga_{0.38}O_x$, $V_{0.26}Mn_{0.33}Ga_{0.41}O_x$, $V_{0.37}Mn_{0.22}Fe_{0.41}O_x$,
4th generation: $V_{0.19}Mn_{0.24}Fe_{0.32}Ga_{0.25}O_x$, $V_{0.06}Mn_{0.02}Ga_{0.92}O_x$, $V_{0.49}Mn_{0.13}Ga_{0.38}O_x$, $V_{0.47}Fe_{0.19}Ga_{0.33}O_{x\cdot}$;
Case B: 1st generation: $V_{0.24}La_{0.33}Fe_{0.15}Ga_{0.28,}$ $V_{0.42}Mo_{0.11}Mn_{0.26}Ga_{0.19}$, $V_{0.44}Mg_{0.4}Fe_{0.04}Ga_{0.14}$, $V_{0.27}Mg_{0.34}Mo_{0.21}Fe_{0.18}$; 2nd generation: $V_{0.47}Mo_{0.05}Mn_{0.27}Ga_{0.20}$, $V_{0.39}Mo_{0.22}Mn_{0.23}Ga_{0.17}$, $Mg_{0.26}Mo_{0.31}Mn_{0.44}$; 3rd generation: $V_{0.22}Mg_{0.47}Mo_{0.11}Ga_{0.20}$, $V_{0.14}Mg_{0.2}Mn_{0.15}Fe_{0.08}Ga_{0.42}$, $V_{0.4}Mg_{0.42}Mn_{0.09}Ga_{0.09}$, $V_{0.39}Mg_{0.24}Mo_{0.1}Mn_{0.1}Ga_{0.17}$; 4th generation: $V_{0.27}Mg_{0.37}Mo_{0.12}Fe_{0.13}Ga_{0.11}$, $V_{0.29}Mg_{0.31}Mo_{0.14}Ga_{0.26}$, $V_{0.19}Mg_{0.39}Mo_{0.09}Ga_{0.33}$, $V_{0.41}Mg_{0.25}Mo_{0.1}Fe_{0.05}Ga_{0.18}$;
5th generation: $V_{0.32}Mg_{0.18}Mo_{0.04}Mn_{0.09}Ga_{0.33}$, $V_{0.16}Mg_{0.11}Mo_{0.17}Fe_{0.09}Ga_{0.47}$, $V_{0.38}Mg_{0.4}Mo_{0.18}Ga_{0.04}$, $V_{0.46}Mg_{0.28}Mn_{0.06}Ga_{0.2}$

oxides (V_2O_5 or Ga_2O_3) and they had the lowest selectivity and yields of propene among all the materials listed. All the active and selective catalytic materials stabilise different $Mg_xV_yO_z$ phases and differ in Mg/V ratio on the surface as derived from XPS.

The catalysts $V_{0.8}Mg_{0.2}O_x$ and $V_{0.1}Mg_{0.9}O_x$ which are associated with the highest and lowest propene yields, respectively, have been also studied by EPR and UV/VIS diffuse reflectance spectroscopy to gain information on the degree of vanadium-site isolation [8]. The UV/VIS spectrum of sample $V_{0.8}Mg_{0.2}O_x$ (Figure 2b) shows the typical charge-transfer (CT) bands of pentavalent vanadyl species in crystalline V_2O_5 [9]. This is in agreement with XRD results which confirm the presence of V_2O_5 as the main component in $V_{0.8}Mg_{0.2}O_x$. The mean vanadium-valence state determined by potentiometric titration [10] amounts to 4.91 indicating that 8.2 % of all vanadium ions are tetravalent. This percentage is too small to be detected besides V^{5+} in the UV/VIS spectra. However, the EPR spectrum, in which only tetravalent vanadium ions can be detected under the conditions applied, shows a narrow isotropic singlet (Figure 3) which is characteristic of closely neighbouring VO^{2+} ions coupled by effective spin-spin exchange interactions [11]. These species are most likely formed by partial reduction of V_2O_5. Thus, as a rough approximation, the distribution of tetravalent vanadium sites reflected by EPR can be taken as a monitor for the distribution of VO^{3+}, too.

In the UV/VIS spectrum of sample $V_{0.1}Mg_{0.9}O_x$ CT bands of isolated tetrahedral V^{5+} [12] can be identified by spectra deconvolution at 224, 268 and 326 nm (Figure 2a). They arise most likely from magnesium orthovanadate detected, besides MgO, as a minor component in this catalyst by XRD. Additionally, a weak band at 380 nm points to the presence of some V^{5+} in isolated octahedral or square-pyramidal coordination [13]. The EPR spectrum of $V_{0.1}Mg_{0.9}O_x$ shows the characteristic hyperfine splitting multiplet (hfs) of well-isolated octahedrally coordinated VO^{2+} species [14] superimposed on a broad isotropic singlet of weakly interacting VO^{2+} species (Figure 3). By spectra simulation it has been derived that the latter comprise about 97 % of the overall EPR active V^{4+} content. Those species are probably dispersed on the surface of MgO being the main component in $V_{0.1}Mg_{0.9}O_x$. In contrast to sample $V_{0.8}Mg_{0.2}O_x$ in which a narrow EPR singlet arises from effective spin-spin exchange interactions between closely neighbouring VO^{2+} sites, the mutual distance of these species in sample $V_{0.1}Mg_{0.9}O_x$ should be markedly longer since the the large line width of the isotropic EPR singlet suggests weak dipolar interactions only. This is confirmed, too, by the UV/VIS spectrum (Figure 2a) in which no bands of oligomeric $V^{5+}O_x$ units at wavelength higher than 390 nm are observed. Thus, all vanadium species in sample $V_{0.1}Mg_{0.9}O_x$ including those giving rise to the broad EPR singlet can be regarded as more or less isolated insofar as they do most probably not form V-O-V bonds.

Table 2
Results of bulk and surface characterisation and catalytic performance (T=773 K, $C_3H_8/O_2/N_2$=40/20/40)

Oxide materials	S_{BET} m^2/g	Crystalline phases XRD	Mg/V, Ga/V* XPS	τ $s{\cdot}g{\cdot}cm^{-3}$	$S(C_3H_6)$ %	$X(C_3H_8)$ %
		Catalytic materials prepared by a combinatorial approach				
$V_{0.3}Mg_{0.63}Ga_{0.07}$	10	$Mg_2V_2O_7$	2	0.18	49	22.4
$V_{0.22}Mg_{0.47}Mo_{0.11}Ga_{0.2}$	17	$Mg_2V_2O_7$	3.33	0.30	49.7	22.3
$V_{0.32}Mg_{0.18}Mo_{0.04}Mn_{0.19}Ga_{0.27}$	13	MgV_2O_6, $Mn(VO_3)_2$, $Mn(V_2O_7)$	0.47	0.08	45.2	21.1
$V_{0.16}Mg_{0.11}Mo_{0.17}Fe_{0.09}Ga_{0.47}$	14	Ga_2O_3, $FeVO_4$, MoO_3, $FeGa_2O_4$	0.21	0.12	40.3	19.5
		V-Mg-Ga-O/α-Al_2O_3 catalytic system				
$V_{0.2}Mg_{0.2}Ga_{0.6}$	13.5	MgV_2O_6, Ga_2O_3, V_2O_4	0.34	0.12	42.1	19.5
$V_{0.2}Mg_{0.5}Ga_{0.3}$	14.1	$Mg_3(VO_4)_2$, V_2O_4, $Mg_xV_yO_z$	0.5	0.06	45.4	20.8
		V-Mg-O catalytic system				
$V_{0.8}Mg_{0.2}$	1.9	V_2O_5, MgV_2O_6	0.2	0.80	32.8	16.3
$V_{0.5}Mg_{0.5}$	7.9	$Mg_2V_2O_7$	1.2	1.20	46	18.9
$V_{0.2}Mg_{0.8}$	46.9	$Mg_3V_2O_8$, MgO	4.5	0.24	55.2	23.6
$V_{0.1}Mg_{0.9}$	87.4	$Mg_3V_2O_8$, MgO	9.3	0.24	57.5	24
		V-Ga-O/α-Al_2O_3 catalytic system				
*$V_{0.5}Ga_{0.5}$	7.2	V_2O_5, Ga_2O_3	2.1	0.18	31	24.2
*$V_{0.33}Ga_{0.67}$	4.9	V_2O_5, Ga_2O_3	0.91	0.18	36.4	22
		VO_x-MCM catalytic system				
VO_x(2.8 wt.%)/MCM-41	682			0.04	49.1	30.3
VO_x(2.8 wt.%)/MCM-48	878			0.08	32.8	53.0

The EPR spectra of catalytic materials with vanadium contents between those of $V_{0.1}Mg_{0.9}O_x$ and $V_{0.8}Mg_{0.2}O_x$ (3.3 < V content / wt.-% < 15) show also a superposition of a hyperfine splitting multiplet (hfs) of isolated VO^{2+} species and a more or less broad isotropic singlet of interacting VO^{2+} species. With increasing vanadium concentration the fraction of interacting vanadyl species increases.

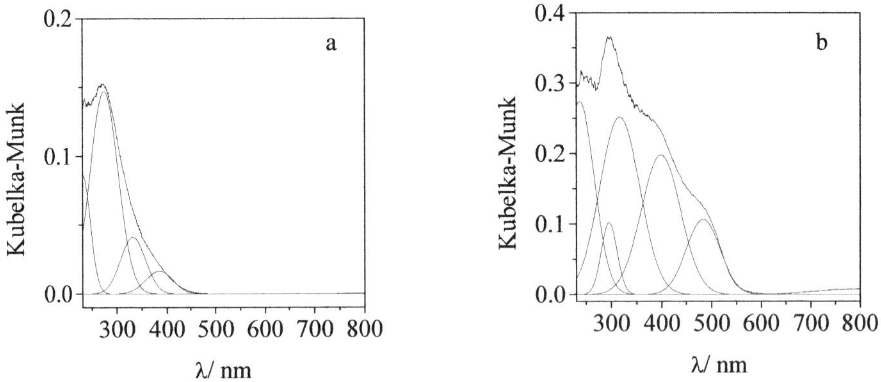

Fig. 2. UV/VIS-DRS spectra of the catalysts $V_{0.1}Mg_{0.9}O_x$ (a) and $V_{0.8}Mg_{0.2}O_x$ (b)

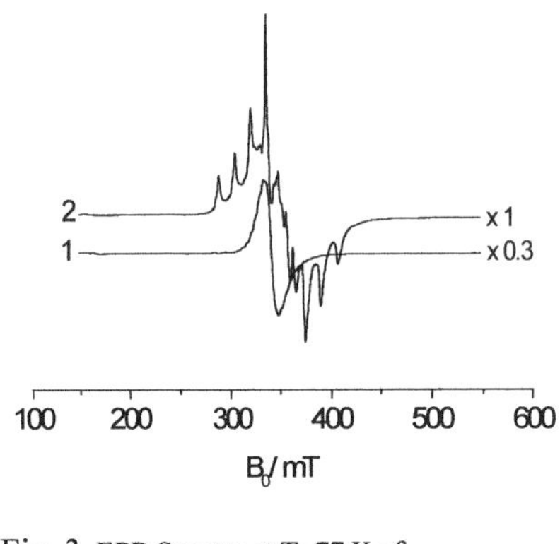

Fig. 3. EPR Spectra at T=77 K of $V_{0.8}Mg_{0.2}O_x$ (1), $V_{0.1}Mg_{0.9}O_x$ (2).

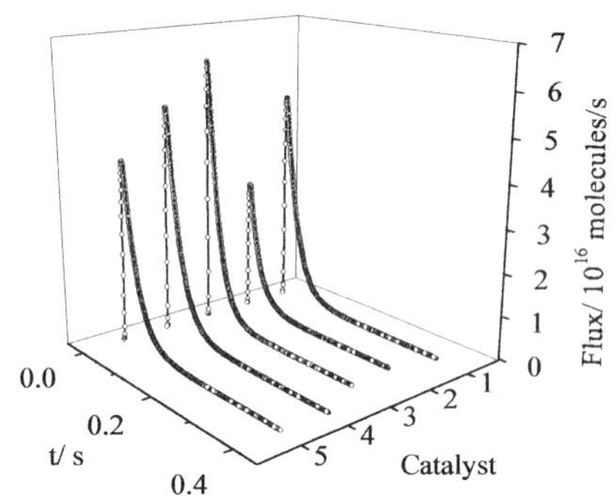

Fig. 4. Comparison between simulated (line) and and experimental (circle) responses of oxygen pulses over different catalytic materials at 773 K: (1)$V_{0.16}Mg_{0.11}Mo_{0.17}Fe_{0.09}Ga_{0.47}$, (2) $V_{0.32}Mg_{0.18}Mo_{0.04}Mn_{0.19}Ga_{0.27}$, (3) $V_{0.8}Mg_{0.2}$, (4) $V_{0.2}Mg_{0.8}$, (5) $V_{0.22}Mg_{0.47}Mo_{0.11}Ga_{0.2}$

The finding that in all well performing catalysts vanadium existed as isolated VO_x sites lead to the use of MCM-41 and -48 as already mentioned above.

The analysis of oxygen adsorption kinetics led to further information on the mechanistic background of catalyst activity and selectivity. The oxygen transient response were fitted by simultaneous numerical integration applying first a genetic algorithm for preliminary parameter estimation and a Simplex procedure for final optimisation assuming that irreversible dissociative adsorption of oxygen occurs on two catalytically active sites. The mechanistic model and the oxygen-mass balances for data evaluation can be written as following:

Oxygen adsorption on catalyst surface sites denoted z

$$O_2 + 2z \xrightarrow{k_{ads}} 2z - O \quad (1)$$

mass balance of gas-phase oxygen

$$\frac{\partial C_{O_2}}{\partial t} = D_{eff} \cdot \frac{\partial^2 C_{O_2}}{\partial x^2} - k_{ads} \cdot C_{tot}^2 \cdot (1-\Theta_O)^2 \cdot C_{O_2} \cdot \quad (2)$$

mass balance of surface species

$$\frac{\partial \Theta_O}{\partial t} = 2 \cdot k_{ads} \cdot C_{tot} \cdot C_{O_2} \cdot (1-\Theta_O)^2 \quad (3)$$

notations

$1=\Theta_O + \Theta_z$, $\Theta_z = C_z/C_{tot}$, $\Theta_O = C_{z\text{-}O}/C_{tot}$ and D_{eff} : effective Knudsen diffusion coefficient

The search for the kinetic parameters was performed using a reactor model for the pulse reactor and a numerical optimisation procedure as outlined above (see also [15]). A comparison of experimental oxygen responses and the simulated ones is presented in Figure 4. It is obvious that the model describes the experimental data very well. This observation is in agreement with experimental data on isotopic exchange of oxygen [16] and kinetic modelling of ODP [17]. From these earlier studies O_2 chemisorption was assumed to be irreversible and dissociative which was confirmed by the present work. The optimised values of parameters obtained from the modelling are summarised in Table 3.

Table 3
Transient kinetic parameters for the elementary steps of oxygen interaction with different catalysts

	$C_{tot}^2 \cdot k_{ads}$ s^{-1}	$C_{tot} \cdot k_{ads}$ $m^2 \cdot sites^{-1} \cdot s^{-1}$	C_{tot} $sites \cdot m^{-2}$
$V_{0.22}Mg_{0.47}Mo_{0.11}Ga_{0.2}$	14	$3.2 \cdot 10^{-14}$	$4.5 \cdot 10^{14}$
$V_{0.32}Mg_{0.18}Mo_{0.04}Mn_{0.19}Ga_{0.27}$	36	$1.6 \cdot 10^{-13}$	$2.3 \cdot 10^{14}$
$V_{0.16}Mg_{0.11}Mo_{0.17}Fe_{0.09}Ga_{0.47}$	50	$6.0 \cdot 10^{-14}$	$8.3 \cdot 10^{14}$
$V_{0.2}Mg_{0.8}$	6.3	$6.2 \cdot 10^{-14}$	$1.0 \cdot 10^{14}$
$V_{0.8}Mg_{0.2}$	190	$2.5 \cdot 10^{-15}$	$7.6 \cdot 10^{16}$

The effective parameter $k_{eff}=C_{tot}^2 \cdot k_{ads}$ has units of s^{-1} and it is an apparent first-order rate constant of oxygen activation. The composition of the catalytic materials influences strongly the apparent first-order rate constant of oxygen activation (Table 3). This constant determines how high is the rate of oxygen activation and, therefore, how high is the surface coverage by active oxygen species. As catalytic results show, the lowest selective catalytic material ($V_{0.8}Mg_{0.2}$) has the highest surface oxygen coverage, while the best performing catalytic material ($V_{0.2}Mg_{0.8}$) has the lowest surface oxygen coverage.

Thus, kinetic and mechanistic information on oxygen activation and catalyst characterisation data permit to conclude that a good catalytic performance for ODP is achieved by an optimal concentration of the active VO_x species which have to be well dispersed on the catalyst surface

CONCLUSIONS

In order to illustrate that the combination of fundamental knowledge on catalyst operation and an evolutionary approach is a valuable method to improve existing catalytic effective compositions or to develop new catalytic materials, the oxidative dehydrogenation of propane was studied as a model reaction over catalytic materials, which were prepared based on such methodology. Fundamental knowledge supports the design of a first generation of catalytic materials. Based on the further characterisation of best performing materials identified in subsequent materials generations, fundamental insights on catalyst operation and designing were derived which give additional hints for improvement of the catalytic material.

Acknowledgement The authors thank Dr. M. Schneider for XRD analysis, Dr. J. Radnik for XPS measurements and Dr. H. Berndt for providing the VO_x(2.8 wt.%)/MCM-48 and VO_x(2.8 at.%)/MCM-41 materials. The work was supported by the German Federal Ministry for Education and Research and the Department of Science of the State of Berlin.

REFERENCES

1. D. Wolf, O. V. Buyevskaya, M. Baerns; Appl. Catal. A: 200 (2000) 63.
2. O. V. Buyevskaya, A. Brückner, E.V. Kondratenko, D. Wolf and M. Baerns, Catal. Today, submitted for publication.
3. Siew Hew Sam, D., Soenen, V. and Volta, J. C., J. Catal., 123 (1990) 417.
4. H. Berndt, A. Martin, A. Brückner, E. Schreier, D. Müller, H. Kosslick, G.-U. Wolf, and B. Lücke, J. Catal. 191 (2000) 384.
5. O. V. Buyevskaya and M. Baerns, Catal. Today 42 (1998) 315.
6. J. T. Gleaves, G.S. Yablonskii, P. Phanawadee, Y. Schuurman, Appl. Catal., 160 (1997) 55.
7. O. V. Buyevskaya, A. Brückner, E. V. Kondratenko, D. Wolf and M. Baerns, Catal. Today, accepted/D. Wolf, O. V. Buyevskaya and M. Baerns, Appl. Catal. A: Chemical, 200 (2000) 63.
8. P. Rybarczyk, H. Berndt, J. Radnik, M.-M. Pohl, O. Buyevskaya, M. Baerns and A. Brückner, J. Catal., submitted
9. G. Centi, S. Perathoner, and F. Trifitó, J. Phys. Chem., 96 (1992) 2617 .
10. M. Niwa and Y. Murakami, J. Catal., 76 (1982) 9.
11. A. Brückner, A. Martin, B. Kubias, B. Lücke, J. Chem. Soc. Faraday Trans., 94 (1998) 2221.

12. G. Centi, S. Perathoner, and F. Trifiró, J. Phys. Chem., 96 (1992) 2617.
13. M. Morey, A. Davidson, H. Eckert, and G. Stucky, Chem. Mater., 8 (1996) 486.
14. M. Wark, A. Brückner, T. Liese and W. Grünert, J. Catal. 175 (1998) 48.
15. M. Rothaemel and M. Baerns, Ind. Eng. Chem. Res., 35 (1996) 1556/M. Soick, D. Wolf and M. Baerns, Chem. Eng. Sci. 55 (2000) 2875.
16. K. Chen, A. Khodakov, J. Yang, A. T. Bell and E. Iglesia, J. Catal., 186 (1999) 325.
17. D. Creaser and B. Anderson, Appl. Catal. A: 141 (1996) 131.

Studies in Surface Science and Catalysis 140
A. Gamba, C. Colella and S. Coluccia (Editors)

FT-IR study of Pt, Cu and Pt-Cu phases supported on hydrotalcite-derived mixed oxides.

L. Balduzzi [a], F. Prinetto [a], G. Ghiotti [a], A. Bianchini [b], M. Livi [b], A. Vaccari [b].

[a] Dip. di Chimica IFM, Università di Torino, Via Pietro Giuria 7, 10125 Torino, Italy.
[b] Dip. di Chimica Industriale e dei Materiali, Università di Bologna, Viale Risorgimento 4, 40136 Bologna, Italy.

Pt, Cu and Pt-Cu dispersed catalysts were prepared by impregnation of mixed oxides obtained by thermal decomposition of hydrotalcite materials. In this preliminary work, the nature of the metallic Pt, Cu or Pt-Cu phases was investigated by CO and NO adsorption, followed by FT-IR spectroscopy. On catalysts containing Pt and Cu the formation of a bimetallic Pt-Cu phase was evidenced.

1. INTRODUCTION.

Synthetic hydrotalcite-type (HT) compounds, having general formula $[M^{z+}_{1-x} M^{3+}_{x} (OH_2)]^{b+} [A^{n-}_{b/n}] \cdot mH_2O$ (M= metal, A= anion and B= x or 2x-1, respectively, for z= 2 or 1) are very attractive precursors of catalysts or catalyst supports [1, 2]. In particular, controlled thermal decomposition of HT materials allows to obtain mixed oxides with high surface areas, homogeneous and thermally stable interdispersion of the elements, synergetic effects and peculiar acid-base properties. These features make the HT-derived mixed oxides interesting supports for highly dispersed metal catalysts to be used in a wide variety of reactions of industrial relevance. New applications of HT-based catalysts concern the field of environmental catalysis, including SO_x, NO_x and N_2O removal [2, 3]. Very good stability and activity in the reduction of nitrogen oxides has been reported for mixed oxides obtained from Cu/Mg/Al HT precursors, which are able to operate in O_2 excess in the simultaneous presence of SO_x and NO_x [4]. In the frame of the development of novel catalysts active and selective in NO_x storage-reduction cycles and stable towards poisoning by SO_x, our interest has been focused on Pt, Cu and Pt-Cu catalysts supported on HT-derived mixed oxides. Preliminary tests have evidenced that these catalysts show activity comparable or higher than that of a reference $Pt/BaO/Al_2O_3$ catalyst and a better resistance towards the deactivation by SO_x [5]. A careful characterisation of the metal phases plays a fundamental role to shed light on the catalytic behaviour as well as to optimise the chemical composition and activation conditions of the catalysts.

2. EXPERIMENTAL.

2.1. Sample preparation.

Mg/Al mixed oxide supports were obtained by calcination at 923 or 1173 K of a commercial HT precursor [Mg^{2+}/Al^{3+} = 60:40 as atomic ratio, by CONDEA (D)] and named HT_{923} and HT_{1173}, respectively. Pt, Cu and Pt-Cu supported catalysts (metal loading: Pt 1 wt-%, Cu 4 wt-%) were obtained by incipient wetness impregnation of the calcined HT supports, using $Pt(NH_3)_2(NO_2)_2$ and/or $Cu(NO_3)_2{\cdot}3H_2O$ as precursors, then dried at 363 K and calcined at 823 K for 12 h.

2.2. Characterisation techniques and procedures.

XRD patterns were recorded on a Philips PW 1710 instrument using Cu $K_{\alpha 1}$ radiation (λ = 0.15418 Å, 40 kW, 25 mA). N_2 sorption experiments at 77 K were carried out with a Sorpty 1750 instrument (Carlo Erba Strumentazione) and specific surface areas (SS) calculated using the BET method (Table 1). Absorption/transmission IR spectra were run at RT on a Perkin-Elmer FT-IR 1760-X spectrophotometer equipped with a Hg-Cd-Te cryodetector. For IR analysis powders were pelletted in self-supporting discs (10-15 mg cm^{-2}), activated in vacuo at increasing temperature up to 823 K, heated in dry O_2 at the same temperature and cooled down in oxygen or subsequently reduced in H_2 at 623 K and outgassed at the same temperature. CO and NO (Matheson C.P.) were used as probes.

Table 1.
Some characteristics of the samples.

Sample	Calcin. temp. of the support (K)	SS ($m^2\ g^{-1}$)	Metal loading (wt-%) Pt	Cu	Cristallographic phases
HT	-	208	-	-	Hydrotalcite
HT_{923}	923	161	-	-	Mg(Al)O
$1Pt/HT_{923}$	923	147	1	-	Mg(Al)O
$4Cu/HT_{923}$	923	128	-	4	Mg(Al)O
$1Pt\text{-}4Cu/HT_{923}$	923	125	1	4	Mg(Al)O
HT_{1173}	1173	82	-	-	$MgO + MgAl_2O_4$
$1Pt/HT_{1173}$	1173	99	1	-	$MgO + MgAl_2O_4$
$4Cu/HT_{1173}$	1173		-	4	$MgO + MgAl_2O_4$
$1Pt\text{-}4Cu/HT_{1173}$	1173	94	1	4	$MgO + MgAl_2O_4$

3. RESULTS AND DISCUSSION.

As seen in Table 1, XRD patterns of HT_{923} samples were characteristic of Mg(Al)O mixed oxides with rock-salt-type structure, while for HT_{1173} they evidenced the formation of MgO and of the $MgAl_2O_4$ spinel phase. In the impregnated samples, Pt and/or Cu massive oxide phases were never detected. Upon impregnation with Pt and/or Cu the specific surface area of the HT_{923} support (161 m^2 g^{-1}) decreased of 10-20 %, while that of the HT_{1173} support (82 m^2 g^{-1}) slightly increased.

The nature of the metal phases was investigated by interaction with CO and NO at RT followed by IR spectroscopy. The study was carried out on the catalysts both pre-oxidised and pre-reduced in H_2, since the NO_x storage-reduction cycle involves an alternation of lean and rich steps, where the catalysts work under oxidising or reducing atmospheres, respectively.

Interaction with CO is first examined. No surface species were detected upon admission of CO on the pure HT supports. At variance, on oxidised samples containing Pt and/or Cu intense bands due to bridging and chelating bidentate carbonates (1680-1550 cm^{-1}, $\nu_{C=O}$, 1400-1250, 1060-1040 cm^{-1}, ν_{OCO} asym and sym, respectively) and monodentate carbonates (1550-1500 and 1450-1350 cm^{-1}, ν_{OCO} asym and sym, respectively) were formed [6]. Carbonate species arose from the interaction of the basic oxygens of the support with CO_2 produced from the reduction of Pt and/or Cu ions by CO at RT. CO co-ordinated to the reduced metal sites thus formed gave monocarbonylic species. Noticeably, the relative amounts of carbonate and carbonyl species was higher on Pt/HT than on Cu/HT catalysts (Fig. 1a and b, respectively), evidencing the higher reducing effect of CO at RT on Pt than on Cu ions (*vide infra*).

On oxidised Pt/HT samples (Fig. 1a), linear Pt^0 carbonyls were mainly detected, exhibiting a peak at 2095 cm^{-1} (this value corresponds to the highest CO coverage studied), stable upon heating under vacuum up to 573 K. CO bridged species were formed in negligible amounts. Weak components at 2185 and 2160 cm^{-1}, the former reversible upon evacuation at RT, were ascribed to Al^{3+} and Pt^{2+} carbonyls, respectively [7]. It is noted that the frequency of the Pt^0 carbonyl band was quite high with respect to those generally found for $Pt^0(CO)$ species [8]. This accounts for an incomplete reduction of PtO particles by CO, so that electron donation from the surface Pt^0 atoms to bulk Pt^{2+} ions occurs in competition with the π back-donation from surface Pt^0 to chemisorbed CO. Accordingly, after reduction in H_2 at 623 K (Fig. 1c) the band due to $Pt^0(CO)$ increased in intensity and shifted to 2065 cm^{-1}, while carbonate species were formed in very weak amounts, revealing the presence of a small fraction of Pt not completely reduced during the sample pre-treatment.

Upon CO admission on oxidised Cu/HT samples (Fig. 1b) Cu^{2+} ions were mainly reduced to Cu^+ which chemisorbed CO giving linear carbonyls at 2118 cm^{-1} [9,10]. Anyway, Cu^{2+} ions could not be detected by CO adsorption at RT [9]. $Cu^+(CO)$ species were quite stable, only

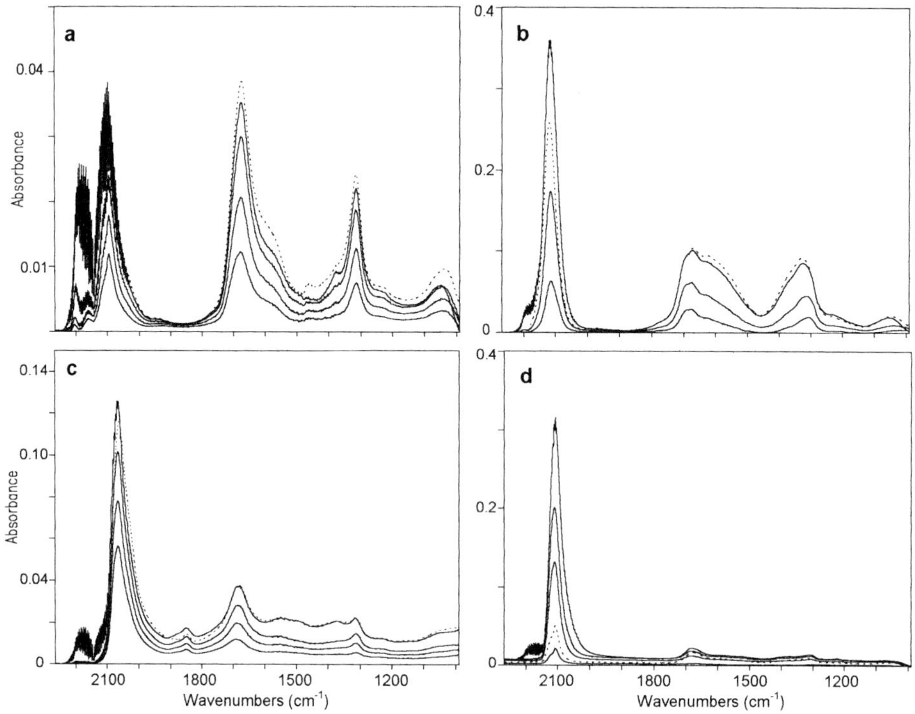

Figure 1. IR difference spectra upon CO adsorption at increasing pressure up to 50 mbar and subsequent evacuation at RT (dotted lines) on Pt/HT_{923} (sections a, c) and Cu/HT_{923} (sections b, d) oxidised at 823 K (sections a, b) or reduced in H_2 at 623 K (sections c, d).

20-30% being removed by evacuation at RT. After reduction in H_2 at 623 K (Fig. 1d), carbonates were not formed, while Cu^0 linear carbonyls were detected at 2108 cm^{-1}. In agreement with the literature [11], Cu^0(CO) species showed lower stability than Cu^+(CO), over 85% being removed by evacuation at RT.

Going from the Cu/HT_{923} to the Cu/HT_{1173} sample, the bands due to chemisorbed CO exhibited a smaller half-width, evidencing a lower dispersion of the metal phase, in agreement with the decreased specific surface area of the support. At variance, the effect of the pre-calcination temperature of the support could not be evidenced in the case of Pt/HT samples, possibly due to the low Pt loading (spectra not reported).

The spectra of oxidised Pt-Cu/HT catalysts upon CO adsorption (not reported) were very similar to those of Cu/HT samples, while peculiar features were observed for Pt-Cu/HT catalysts reduced in H_2 at 623 K. In particular, on Pt-Cu/HT_{923} (Fig. 2 curve 3) a band at 1996 cm^{-1} (at the maximum CO coverage) was noted and ascribed to CO adsorbed on negatively charged Pt atoms, thus evidencing the formation of a bimetallic Pt-Cu phase, where electron transfer from Cu to Pt atoms occurs [12]. As a consequence, the σ donation from CO to Cu and the π back-donation from Pt to CO increase. For the same reason a blue shift of the band due to $Cu^0(CO)$ with respect to Cu/HT samples should take place. On the contrary it was not observed, possibly accounting for the presence of a part of copper not involved in the alloying, owing to the high amount of Cu relatively to Pt (Cu:Pt = 12:1 atomic ratio).

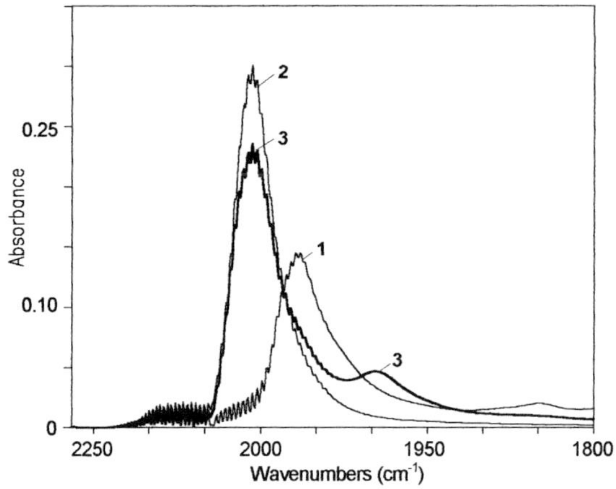

Figure 2. IR difference spectra upon CO adsorption (50 mbar) on Pt/HT_{923} (curve 1), Cu/HT_{923} (curve 2), and Pt-Cu/HT_{923} (curve 3) reduced in H_2 at 623 K.

By increasing the CO pressure and hence the CO coverage over Cu, the band due to $Pt^{\delta-}$ carbonyls progressively shifted towards the lower frequencies (-10 cm^{-1}), evidencing a further increase of the electron density over Pt (Fig. 3a). Upon the subsequent evacuation at RT (dotted line in Fig. 3a), CO adsorbed on Cu^0 was partially removed while the band due to CO on negatively charged Pt atoms shifted towards the higher frequencies and was enhanced, due to an increase in the extinction [12]. Further components were observed at 2057 and 2065 cm^{-1}, *i.e.* in the CO stretching region observed for Pt^0 carbonyls on Pt/HT samples, revealing the presence of a fraction of Pt not interacting with Cu.

A similar behaviour was observed for Pt-Cu/HT_{1173} catalyst (Fig. 3b), although some significant differences could be evidenced. Indeed, the components at 2057 and 2065 cm^{-1} were absent while the band around 2000 cm^{-1} showed a higher intensity, showing that in this case all the Pt atoms were involved in the bimetallic phase. Furthermore, the carbonyl bands exhibited a smaller half-width, accounting for a lower dispersion of the metal phase over the HT_{1173} support.

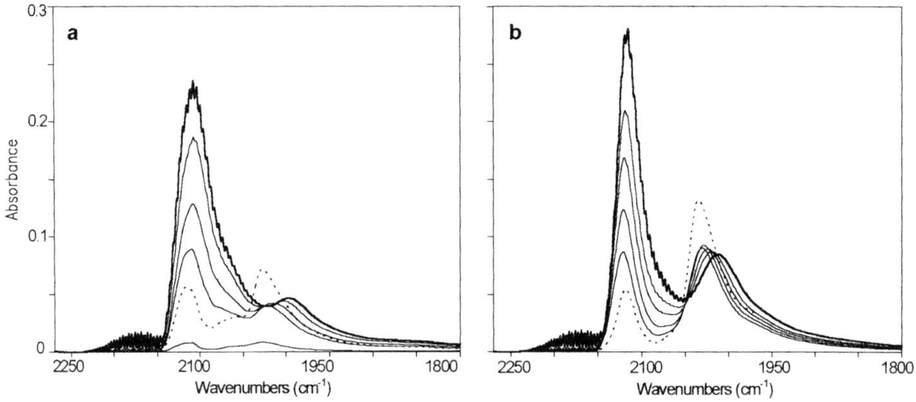

Figure 3. IR difference spectra upon CO adsorption at RT at increasing pressure up to 50 mbar and subsequent evacuation at RT (dotted lines) on Pt-Cu/HT_{923} (section a) and Pt-Cu/HT_{1173} (section b) reduced in H_2 at 623 K.

NO molecule was then used in complement with CO as a probe of the metal sites. This became particularly necessary in the case of Cu^{2+} sites which could be not revealed by CO, as shown above. No surface species were formed upon NO admission on the pure HT supports (Fig. 4 curves 1). On oxidised Pt/HT catalysts, NO adsorption caused a partial reduction of Pt ions with the consequent formation of small amounts of Pt^0 mononitrosyls (1770 cm^{-1}) and nitrites (1250 cm^{-1}). Weak amounts of NO chemisorbed on $Pt^{\delta+}$ sites (~ 1900 cm^{-1}) were also observed (Fig. 4a curve 2). After reduction in H_2, linear, bent and bridged Pt^0 nitrosyls (1780, 1700, 1595 cm^{-1}, respectively, [13]) were formed in high amounts (Fig. 4b curve 2). Conversely, on oxidised Cu/HT samples NO was mainly chemisorbed on Cu^{2+} sites (band at 1855 cm^{-1} with a shoulder at 1885 cm^{-1}), while only a small fraction of copper was reduced by NO, giving then Cu^0(NO) species (1747 cm^{-1}), nitrites (1250 cm^{-1}) and small amounts of nitrates (1600-1300 cm^{-1}) (Fig. 4a curve 4). The formation of nitrites has been suggested to involve the reduction of Cu^{2+} to Cu^+, as in eqn. (1) [14]. The nitrites thus formed are strongly co-ordinated on Cu^+ sites, which therefore are not able to give nitrosyl species.

$$Cu^{2+}\text{-} O^{2-} + NO_{gas} \rightarrow Cu^+ \text{-} ONO^- \qquad (1)$$

After reduction (Fig. 4b curve 4), the absorption due to Cu^0(NO) species was enhanced and shifted at 1740 cm^{-1}. Parallely, the bands due to Cu^{2+} mononitrosyls, nitrites and N_2O (2250 cm^{-1}) developed, accounting for the dissociative chemisorption of NO on metallic Cu^0, which was at least partially reoxidised. Nitrites and N_2O arose therefore from the reaction of gaseous NO with N and O ad-atoms, respectively, formed upon NO dissociation.

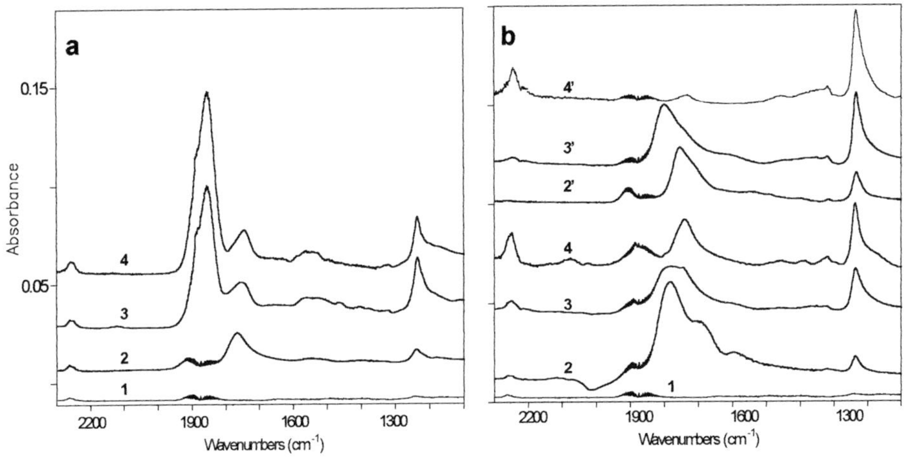

Figure 4. IR difference spectra upon NO adsorption (5 mbar) at RT on HT_{923} (curves 1), Pt/HT_{923} (curves 2), Pt-Cu/HT_{923} (curves 3), Cu/HT_{923} (curves 4), Pt/HT_{1173} (curve 2'), Pt-Cu/HT_{1173} (curve 3') and Cu/HT_{1173} (curve 4') oxidised at 823 K (section a) or reduced in H_2 at 623 K (section b).

The spectra of oxidised Pt-Cu/HT catalysts upon NO adsorption (Fig. 4a curve 3) were very similar to those of Cu/HT samples. After reduction in H_2 (Fig. 4b curve 3), they represented an intermediate situation between those of Pt/HT and Cu/HT samples, showing a broad absorption with maximum at 1780 cm^{-1} and shoulders at 1750 and 1610 cm^{-1} where the contributions due to Pt^0 and Cu^0 mononitrosyls could be hardly distinguished. Noticeably, the component due to bridged NO on Pt sites (1595 cm^{-1}) was markedly decreased in intensity in comparison to Pt/HT sample, pointing out to the diluition of Pt by copper. Moreover, Cu^{2+} mononitrosyls and N_2O were not detected and nitrites were formed in minor amounts in comparison to Cu/HT sample, evidencing the lower ability of NO to reoxidise the copper involved in the bimetallic Pt-Cu phase.

The effect of the pre-calcination temperature of the HT support was then examined. Very similar features were observed upon NO adsorption on pre-oxidised Pt and/or Cu containing catalysts supported on HT_{923} or HT_{1173} (spectra not reported). Conversely, on pre-reduced Pt/HT and Cu/HT samples the amounts of nitrosyl species decreased going from the HT_{923} to the HT_{1173} support (Fig. 4b curves 2, 2' and 4, 4'). These findings evidenced a lower dispersion of the metal phases over the HT_{1173} supports and therefore a lower accessibility of the metallic sites, not revealed for Pt/HT samples using CO as a probe. At variance, the spectra of the bimetallic Pt-Cu/HT_{923} and Pt-Cu/HT_{1173} samples upon NO adsorption (Fig. 4b curves 3, 3') were similar and characterised by the strong superposition of the bands due to Pt and Cu nitrosyls. Therefore, in this case NO molecule did not allow us to obtain information on the effect of the pre-calcination temperature of the support, which was instead provided by CO.

4. CONCLUSIONS.

The complementary use of CO and NO as probe molecules allowed us to obtain information on the nature and dispersion of the Pt and/or Cu supported phases and on the effect of the pre-calcination temperature of the hydrotalcite support. The influence of the sample pre-treatment was also taken into account in view of the catalytic applications of these materials in the NO_x storage-reduction process, which involves an alternation of oxidising or reducing steps. We would just underline that the formation of a bimetallic phase evidenced in the case of the catalysts containing Pt and Cu can be particularly interesting from a catalytic point of view. Indeed, the modification observed in the metal properties due to the alloying can induce favourable changes in the specific activity and selectivity of the catalysts.

ACKNOWLEDGEMENTS.

Financial support was provided by the Ministero dell'Università e della Ricerca Scientifica e Tecnologica (MURST). The authors gratefully thank CONDEA Chemie (D) for providing the Mg/Al HT precursor.

REFERENCES.

1. F. Cavani, F. Trifirò and A. Vaccari, Catal. Today 11 (1991) 173.
2. A. Vaccari, Appl. Clay Sci. 14 (1999) 161.
3. A. Corma, A.E. Palomares, F. Rey and F. Marquéz, J. Catal. 170 (1997) 140.
4. J.N. Armor, T.A. Braymer, T.S. Farris, Y. Li, F.P. Petrocelli, E.L. Weist, S. Kannan, C.S. Swammy, Appl. Catal. B: Environmental 7 (1996) 397.
5. A. Calciolari, Graduation Thesis, Bologna, 2000.
6. G. Busca, V. Lorenzelli, Mater. Chem. Phys. 7 (1982) 89.
7. N. Sheppard, T.T. Nguyen, in Advances in Infrared Spectroscopy, R.J.H. Clark, R.E. Hester (Eds.), Heyden, London, 1978, p.67.
8. Yu. A. Lokhov, A. A. Davydov, Kinet. Katal. 21 (1980) 1523.
9. G. Ghiotti, F. Boccuzzi, A. Chiorino, in Adsorption and Catalysis on Oxide Surfaces, M. Che and G.C. Bond (Eds.), Elsevier Science B.V., Amsterdam, 1985, p.235.

10. D. Scarano, S. Bordiga, C. Lamberti, G. Spoto, G. Ricchiardi, A. Zecchina, C. Otero Areán, Surf. Sci. 411 (1998) 272.
11. A. A. Davydov, Infrared Spectroscopy of Adsorbed Species on the Surface of Transition Metal Oxides, C.H. Rochester (Ed.), John Wiley & sons, Chichester, 1990.
12. N.A. Solokova, A.P. Barkova, D.B. Furman, V.Yu. Borovkov and V.B. Kazansky, Kinet. Katal. 36 (1995) 434.
13. F. Boccuzzi, E. Guglielminotti, Surf. Sci. 271 (1992) 149.
14. R. Hierl, H.P. Urbach, H. Knözinger, J. Chem. Soc. Faraday Trans. 88(3) (1992) 355.

Studies in Surface Science and Catalysis 140
A. Gamba, C. Colella and S. Coluccia (Editors)

Chemisorption and catalytic properties of gold nanoparticles on different oxides: electronic or structural effects?

Flora Boccuzzi and Anna Chiorino

Dipartimento di Chimica I. F. M. Università di Torino, via P. Giuria 7, 10125 Torino, Italy

Infrared spectra of CO adsorbed on nanometric gold particles supported on different supports, both oxidised and reduced, are reported. Depending on the nature of the supports, on the pretreatments and on the size of the gold nanoparticles, different gold carbonylic species have been evidenced. The electronic and structural properties of gold nanoparticles and the electron-donating effects of reducible supports are shown to be relevant in determining the chemisorption and catalytic properties of the samples.

1. INTRODUCTION

The great interest shown in recent years in gold catalysts, due to the high catalytic activity at low temperatures in a number of different important reactions as CO oxidation, WGS reaction, propylene epoxidation, produced a large number of papers devoted to analyse the different factors controlling the activity of gold catalysts. Among them the most relevant appear to be the gold particle size and the nature of the support. However it is not completely clear if the observed differences are simply related to the concentration of special, uncoordinated surface sites, or also to electronic and chemical modifications of the adsorption sites by interaction with the different supports.

No clear explanation has been found to date for the different catalytic properties observed in samples containing gold particles of quite similar size, supported on a variety of supports [1] or for the deactivation observed on some supports [2,3]. The problem is probably that, up to now, there is not enough information at the atomic level on the adsorption sites and on their electronic properties and on the influence of the chemical treatments on the nanostructure of the particles. In order to get a deeper insight on these points, the adsorption and the reactivity of different molecules, as CO, O_2 and H_2, will be reported, studied by FTIR spectroscopy, at 90 K and at 300 K on gold catalysts supported on different oxides (TiO_2, ZrO_2, Fe_2O_3, ZnO) differently pretreated.

2. EXPERIMENTAL

2.1 Materials

In table 1 are summarised the main characteristics of the examined catalysts that in the following will be indicated by the capital letters reported in the first column of the table. All the samples have been prepared by the deposition-precipitation (dp) method, but the sample A, prepared by coprecipitation (cp). Before the experiments the samples have been submitted to the following pretreatments: heated up to 673 K in dry oxygen and cooled down in the same atmosphere, indicated as (ox) or reduced in hydrogen up to 523 K, indicated in the following as (red). As for sample C we will report also experiments on a called “used” catalyst, that is a catalyst pretreated some times in hydrogen and successively in oxygen at 353 K and finally quickly treated in oxygen again at 673 K.

Table 1

Sample	Supporting oxide	Au loading (wt%)	Au size (nm)	Preparation method	Calcination temperature (K)
A	ZnO (a)	11	5 ±0.8	cp	673
B	Fe_2O_3 (b)	3	3.5±0.8	dp	673
C	ZrO_2 (c)	1.7	10.2±4.7	dp	673
D	TiO_2 (b)	3	2.5±0.7	dp	673
E	TiO_2 (a)	1	2.5±0.5	dp	573
F	TiO_2 (a)	1	10.6±3.0	dp	873

(a) prepared in the Haruta laboratory [1]
(b) prepared in the Andreeva laboratory [14]
(c) prepared in the Pinna laboratory [3]

2.2 Methods

The FTIR spectra have been taken on a 1760 Perkin-Elmer spectrometer, with the samples in self-supporting pellets pressed at 4 x 10^3 kg/cm^2 in a cell allowing to run the spectra in controlled atmospheres at 90 K and at room temperature.

3. RESULTS AND DISCUSSION

3.1 Comparison of CO chemisorption at room temperature between calcined samples with gold particle size lower than 5 nm and larger than 10 nm

In Figure 1a and 1b the spectra produced by CO interaction at different equilibrium pressures on the samples D and A (ox) are reported. The strong bands observed at full coverage exhibit an asymmetric broadening from the low frequency side. By reducing the equilibrium pressure the bands blue shift and the intensities gradually reduce without significant changes in the shape. The main absorption band at about 2110 cm^{-1}, on the basis of many previous literature data, summarised in Table 10 of a recent review [4], is assigned to CO adsorbed on Au° sites.

The blue shift observed by decreasing the CO coverage is a clear evidence of lateral interactions between the adsorbed molecules. The main absorption band and the asymmetric broadening from the low frequency side can be ascribed to CO adsorbed on step sites and on step sites at the borderline of the metallic particles, respectively. These last sites interact with a reduced number of vicinal CO and may be in some extent modified by an electron transfer from the support. It must be stressed that, although the two samples have been prepared by different methods on two different supports and with a quite different Au loading, the spectroscopic features are quite similar. Looking at the data reported in Table 1, the only characteristic similar in the two samples is the size of the metallic particles, ≤5 nm. It appears evident that on these two samples, at the end of the pretreatment and after cooling in oxygen, CO is adsorbed on clean gold sites. Very different features are observed in the spectra reported in Fig. 2, as for the samples exhibiting particles of size larger than 10 nm.

In Figure 2a we show a comparison of the spectra produced by CO interaction at full coverage on sample E (fine line) and on sample F (bold line), while in figure 2b the spectra of CO adsorption on "fresh" sample C (fine line) and on "used" sample C (bold line) are reported.

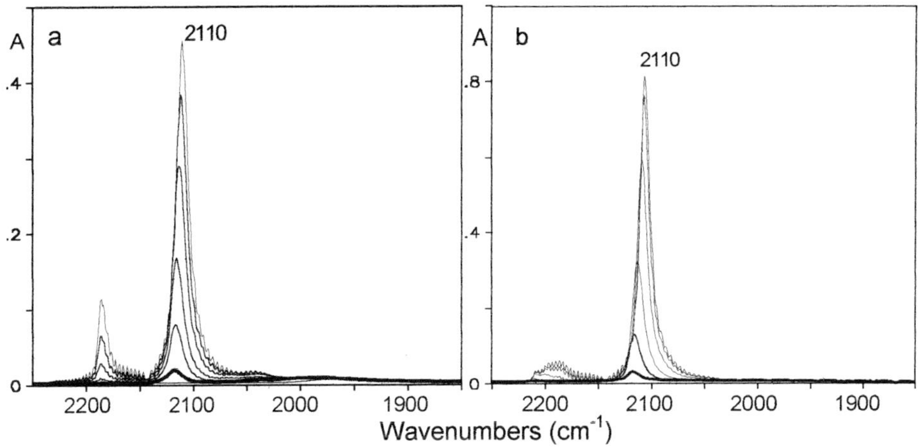

Figure 1. FTIR absorption spectra obtained at RT by CO interaction on sample D (ox) (section a) and on sample A (ox) (section b). CO pressure decreases from 40 mbar to $1*10^{-2}$ mbar going from the most intense spectrum (thin line) to the less intense one (thick line).

It appears immediately evident that, on sample F and on "used" sample C, the CO absorption band is significantly blue shifted, indicating that on these samples the adsorption gold sites of CO are more electron-deficient than in the previously illustrated ones, while on sample E and on "fresh" sample C a band at about 2110 cm^{-1}, similar to that already shown on the samples A and D is observed. The samples F and C are characterised by the largest particle size, i.e. more similar to bulk gold, therefore the effect cannot be related to a direct positivization of the gold sites by interaction with the oxidic support.

It can be postulated that on these large particles some oxygen remains adsorbed in a molecular peroxidic form on the step sites. On the "fresh" sample C, CO is adsorbed on the fraction of small particles present in the sample, on the "used" sample C, a roughness due to

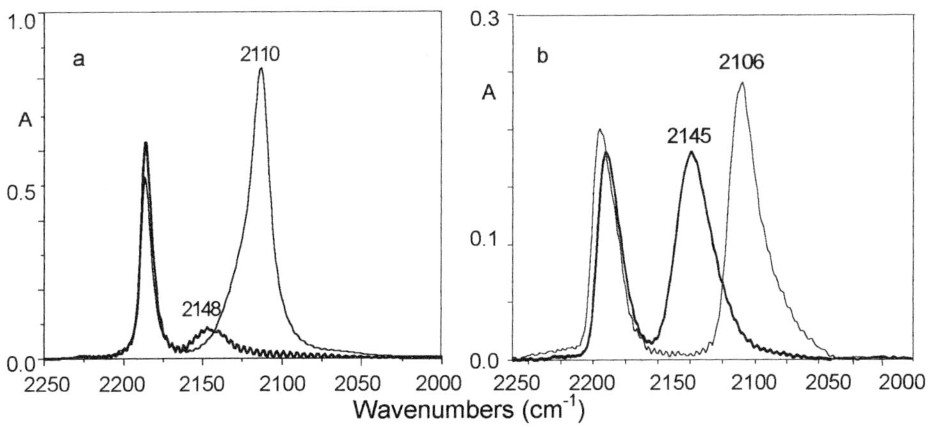

Figure 2. FTIR absorption spectra obtained at RT by 20 mbar CO interaction on: a) sample E (thin line) and sample F (thick line); b) sample C "fresh" (thin line) and sample C "used" (thick line).

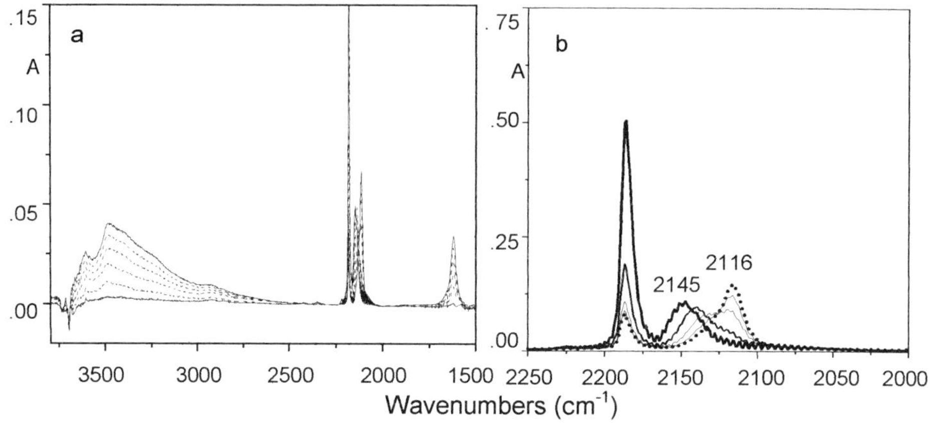

Figure 3. FTIR absorption spectra obtained at RT by 10 mbar H_2 admission on sample F already covered by 20 mbar CO. Spectra are reported from pure CO (0 min) to 30 min mixture contact in the 3800-1500 cm^{-1} range (section a) and in the carbonylic range (section b). Solid thick line in section b: 0 min, dotted line in section b: 30 min.

an increased concentration of step sites has been probably produced by the repeated treatments in hydrogen and oxygen in mild conditions. A confirm to this hypothesis comes from the experiment shown in Fig. 3. The inlet of molecular hydrogen at RT gradually reduces the intensity of the 2145 cm^{-1} band and starts the growth of a new component at 2116 cm^{-1} (Fig. 3b). Moreover new absorptions are observed in the regions typical of adsorbed water (Fig. 3a), in the 3700 – 3200 cm^{-1} and at 1620 cm^{-1}. It can be proposed that the electron-deficient adsorption sites related to the 2145 cm^{-1} band are gold step sites exposed at the surface of the large particles where CO and molecular oxygen are coadsorbed and that the broadening towards high frequencies of the 2110 band is related to CO coadsorbed with atomic oxygen. The smaller blue shift, in respect to the frequency of CO adsorbed on Au° clean sites, 2110 cm^{-1}, is a consequence of a smaller electron transfer from metallic gold to the atomic oxygen.

How can be explained that the steps of the particles larger than 10 nm do not dissociate molecular oxygen while step sites of particles smaller than 5 nm can dissociate oxygen producing atomic oxygen? In agreement with Bondzie et al [5] analysis and considerations it can be proposed that on the largest and thickest particles the activation energy for oxygen dissociation is larger and the adsorption energy of atomic oxygen is smaller than on the smallest ones. Therefore on the large particles oxygen remains adsorbed in a molecular form while on the small ones the dissociation and the atomic adsorption can occur more readily.

3.2 CO chemisorption at room temperature on samples reduced in hydrogen

Figures 4a and 4b show the spectra at different coverages of freshly reduced samples B and D, Figure 4c shows a comparison at full coverage between samples E and F, reduced and already used.

In Figure 4a and 4b a band reversible to the outgassing at 2110 cm^{-1} and some more strongly bonded ones in the 2050 -1950 cm^{-1} range have been observed. Looking at figure 4c it appears

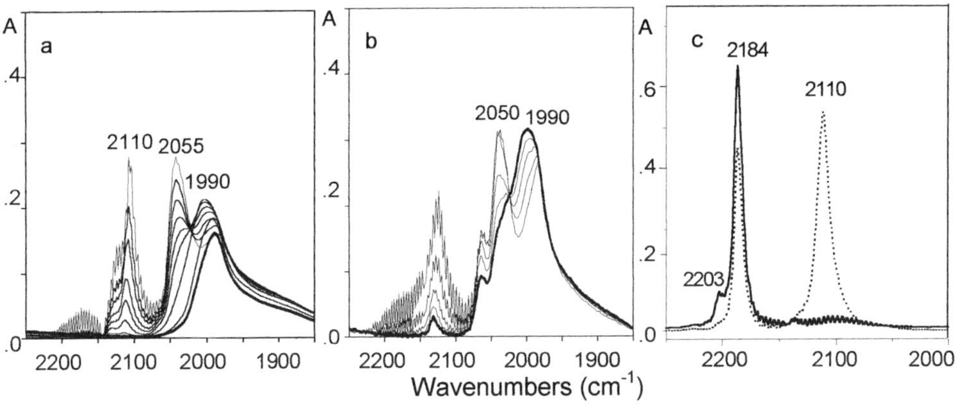

Figure 4. FTIR absorption spectra obtained at RT by CO interaction on reduced samples. Section a: sample B (red), CO pressure decreases from 90 mbar to 1 mbar going from the thinnest to the thickest line. Section b: sample D (red), CO pressure decreases from 20 mbar to $1*10^{-3}$ mbar going from the thinnest to the thickest line. Section c: 20 mbar CO on samples E (red) (dotted line) and F (red) (solid line), already used.

evident that bands in the 2050 – 1950 cm^{-1} range are not present, while a band at 2184 cm^{-1} and two bands, at 2186 and 2203 cm^{-1}, ascribed to CO adsorbed on terrace and step Ti^{4+} sites respectively, are observed. It can be hypothesized that on the freshly prepared and oxidized samples (Fig. 1) a fraction of the gold is highly dispersed in very small clusters, covered by adsorbed oxygen. By reduction in H_2 at 523 K the oxygen adsorbed on the gold clusters and some of the surface oxygen atoms of the titania react, giving rise to the formation of water and to oxygen vacancies and/or to Ti^{3+} defects on titania. The presence of these defects allows an electron transfer from the support to the gold particles and to the gold clusters. In particular, on the very small clusters the negative charge can produce a significant effect on the chemical properties of the gold sites, giving rise to bands red-shifted and more resistant to the outgassing than on the larger particles. The frequency shift between CO adsorbed on top on gold particles, ≈ 2110 cm^{-1}, and CO adsorbed on top on gold small clusters, 2055 cm^{-1}, can be related either to a Stark electrostatic effect or to an enhanced metal back-donation into the $2\pi^*$ CO orbitals. The absence of the bands in the 2050 – 1950 cm^{-1} range in the spectra of already used samples E and F (Fig. 4c) appears as an evidence that these absorptions are related to unstable gold clusters. By decreasing the CO pressure the 2110 cm^{-1} band, assigned to the CO chemisorption on Au° step sites and the bands at 2170 - 2190 cm^{-1}, attributed to CO adsorbed on the support cations, smoothly reduce their intensities in the way already discussed in previous papers, while the bands at 2050 - 1950 cm^{-1} exhibit on both the supports a different and interesting behaviour: the band at 2055 cm^{-1} reduces its intensity while the band at 1990 cm^{-1} grows up and an isosbestic point is evident at about 2015 cm^{-1}. It can be proposed that, by decreasing the CO coverage, a decrease of the CO adsorbed linearly on top on the small clusters occurs (band at 2055 cm^{-1}) and CO bridge bonded grows up (band at 1990 cm^{-1}). It appears that a bridge-bonded configuration is stabilised at low coverages and that CO adsorbed on these small clusters is more strongly bonded than CO adsorbed on the other, larger gold particles. Bands in the 2050 - 1950 cm^{-1} range are reported in literature on gold electrodes during the electrooxidation of CO at negative potentials [6].

Moreover, looking at Fig. 4c, it appears evident that the intensity of the absorption bands in the 2200 – 2180 cm^{-1} range is sample E< sample F, while the intensity of absorption related to CO adsorbed on gold sites is maximum on the sample E and almost nihil on the sample F. The gradual increase in the intensities of the bands assigned to CO adsorbed on the support is in qualitative agreement with the increase observed by TEM of the metallic particle size with the temperature of calcination of the samples. Taking in account that the gold loading of the two samples is the same, it appears reasonable that the increase in the particle size would be related to an increase in the free support surface. However, the increase of the mean gold particle sizes from 2.5 nm up to 10 nm cannot justify by itself the intensity close to zero of the absorption at about 2100 cm^{-1} on the sample F, and it appears as a strong indication that CO is only adsorbed on gold step sites and not on terrace sites. We will discuss this point in the next section.

3.3 CO chemisorption at 90 K on two Au/TiO_2 samples with different gold sizes

Figure 5a shows the FTIR spectra of CO adsorption on the two samples, E and F, calcined in air at 573 and 873 K, simply outgassed at room temperature and run after interaction with 4 mbar of CO at 90 K, normalised on the intensity of the absorption band at 2150 cm^{-1}. This band, assigned to CO interaction by hydrogen bond with the surface OH groups, can be taken as a measure of the exposed hydroxylated titania area. In fact, by outgassing at the same temperature, the band is completely depleted (Fig. 5b) and at the same time the strong perturbation observed in the OH stretching region of the titania hydroxyls completely disappears (not shown for the sake of brevity).

Looking at the spectra reported in Fig. 5a, relevant differences are evident, in particular the band at about 2100 cm^{-1} is completely absent in the sample F and it is strong on sample E. The *complete absence* of the band on sample F, characterised by metallic particles quite large, 10 nm, with a ratio edge Au atoms/total Au atoms < 0.01, appears as a strong indication that already at 90 K *only the step sites* are able to adsorb CO while gold terrace sites, that are the large majority of the gold exposed sites on this sample, do not adsorb at all CO at 90 K. While the estimated step concentration on particles of 10 nm is less than 1% of the total gold

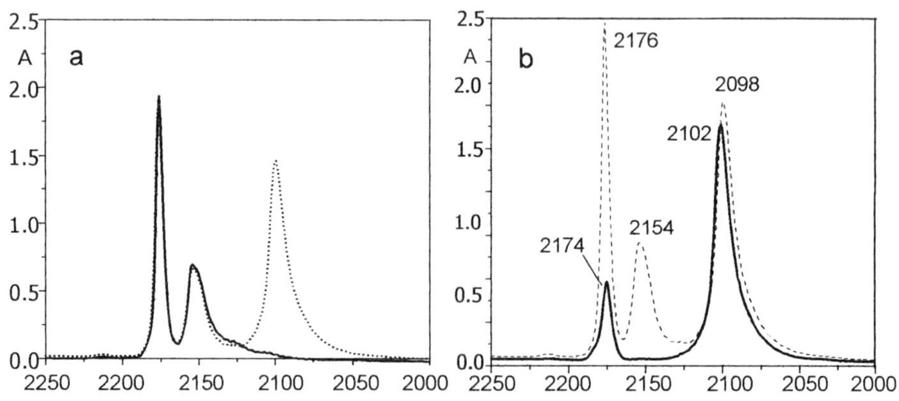

Figure 5. FTIR absorption spectra obtained at 90 K by CO interaction on calcined samples. a): 4 mbar CO on sample E (dotted line) and on sample F (solid line). b): sample E, CO pressure decreases from 6 mbar to $1*10^{-2}$ mbar going from dashed to solid thick line.

atoms, on samples with particle size of 2.5 nm the step site density is close to 30%. As shown by Mavrikakis et al. [7] by self-consistent density functional calculations, «free» step sites are found to bind CO considerably stronger than terraces: the adsorption energy on these sites is similar to that of CO adsorbed on flat Cu(111) copper surface [8], in excellent agreement with the experiments by Ruff et al. [9], where it is found that Au steps give rise to CO adsorption with a thermal desorption temperature close to that of CO on Cu surfaces.

As for the 2176 cm^{-1} band, the frequency region is typical of CO adsorbed on cationic sites and the frequency and intensity are almost identical on the two samples. Different hypotheses can be proposed for the assignment of this band: i) CO adsorbed on Ti^{4+} ions; ii) CO on cationic gold; iii) CO on gold atoms bonded to a peroxo species. The first hypothesis can be ruled out taking in account that the surface is simply outgassed at room temperature, that is completely hydrated and hydroxylated and therefore it does not expose coordinatively unsaturated Ti^{4+} ions. In fact it was shown that the spectra of CO adsorbed at 77 K on pure Degussa P25 exhibit a strong absorption in this region only after activation and dehydration at 873 K [10]. Another hypothesis, i.e. CO adsorbed on cationic gold can be considered. A band at about 2175 cm^{-1}, assigned to CO on Au^{x+} sites, is reported on Au/SiO_2 samples [11]. In the case of small particle sizes the fraction of the metal in ionic form can be quite high and it can act as a chemical glue at the interface between the support and the particle. However, the intensity of the band is almost the same on the two samples, therefore this hypothesis does not appear very likely. The adsorption on isolated gold atoms bonded to a peroxo species can be postulated, looking at the literature. It was shown by Huber et al. [12] that, by cocondensing Au atoms with CO and oxygen in a inert matrix, a band at 2176 cm^{-1} assigned to the CO stretching of a monoatomic carbonyl peroxo gold species was identified. The comparison of the spectrum produced by the inlet of 4 mbar of CO at 90 K on a sample F with the spectrum produced by the interaction with CO at room temperature on the same sample and by successive cooling at 90 K (not shown for sake of brevity) shows that at the end of this procedure a weak 2100 cm^{-1} band is detected; at the same time, the intensity of a 2176 cm^{-1} peak is significantly reduced. [13].

Figure 5b illustrates the behaviour of the absorption bands of the E sample by decreasing the CO pressure from 4 mbar to 10^{-2} mbar at 90 K. The 2150 cm^{-1} band is completely depleted and the 2176 cm^{-1} one strongly reduces its intensity. At the same time the 2100 cm^{-1} one shows an erosion from the low frequency side, producing a blue shift of the band up to 2102 cm^{-1}: this is an evidence that at 90 K a component centred at 2098 cm^{-1} is reversible to the outgassing. Two different hypotheses can be made, as for the assignment of these two species: the first hypothesis is that the two species are adsorbed on two kinds of step sites, on top of the particles, the 2102 cm^{-1} band and on step sites at the contact line with the support, the 2098 cm^{-1} one; the other hypothesis, as already proposed previously [14], is that the more labile species may be assigned to CO adsorbed on step sites, possibly contaminated by atomic hydrogen. Hydrogen in our experimental conditions can be produced by reaction of CO with water and hydroxyls present on the surface of the catalysts, as it was shown previously on Au/TiO_2 catalysts in RT experiments [15].

3.4 Two CO – $^{18}O_2$ interactions at 90 K on sample E

In Fig. 6a and 6b two different CO – $^{18}O_2$ interactions at 90 K on sample E are shown. The admission of O_2 at 90 K (solid line) on preadsorbed CO (dotted line) on the sample E (Fig. 6a) produces a very small blue shift of the 2098 cm^{-1} band without changes in the intensity, a small erosion of the absorption from the low frequency side, the appearance of a shoulder from the high frequency side, at 2116 cm^{-1} (see inset), already assigned to CO coadsorbed with atomic oxygen and a band at 2323 cm^{-1}. The shift and the small erosion can be taken as an indication that molecular oxygen is activated and reacts with the CO species weakly adsorbed on step sites at the borderline of the gold particles. The band at 2323 cm^{-1} indicates the formation of $C^{16}O^{18}O$ already at 90 K, its intensity in respect to the 2340 cm^{-1} one indicates that at this temperature only the oxygen present in the gas phase and activated on gold step sites at the borderline of the particles participates to the reaction. On the contrary, the preadsorbed $^{18}O_2$ (Fig. 6b) causes a strong decrease of the 2103 cm^{-1} band, a new and well evident component at 2125 cm^{-1} and almost no formation of $C^{16}O^{18}O$. As for the origin of the asymmetric behaviour of the two CO-O_2 interactions illustrated in Fig. 6a and 6b, it can be hypothesised that the asymmetry is generated by the intrinsic differences of the two ligands, a two electron donor the CO, and a one electron acceptor the oxygen. The first one makes easier the successive adsorption and dissociation of the oxygen, while the second one will saturated the most uncoordinated exposed sites, the corners, inhibiting the successive chemisorption, and will modify the vicinal step sites. It has been recently well established by Wallace and Whetten [16,17] that oxygen on gold clusters of size between 2-22 atoms acts exclusively as a one electron acceptor: the process exhibits a binary-like behaviour (zero or one molecule adsorbed) and appears correlated to the electron affinity of the cluster; the CO acts as a two electron donor and up to seven molecules are adsorbed on the largest clusters. Sanchez et al. [18], on the basis of studies of CO oxidation and ab initio simulations for gold clusters ($n < 20$) supported on either a defect-free or defect-rich MgO, proposed different models for the CO oxidation and, among them, a Langmuir-Hinshelwood model (LH_p) where CO is adsorbed on top facet of the gold cluster and the peroxo species is bonded at the perimeter between the cluster and the support. This model appears related to the cluster size and to the defect concentration of the support and it can be taken as a quite good model for the behavior of the small gold particles supported on titania.

It can be relevant to observe that the temperature of CO combustion on gold clusters supported on defect-rich MgO, 240 K, is very close to the temperature of the rapid increase in

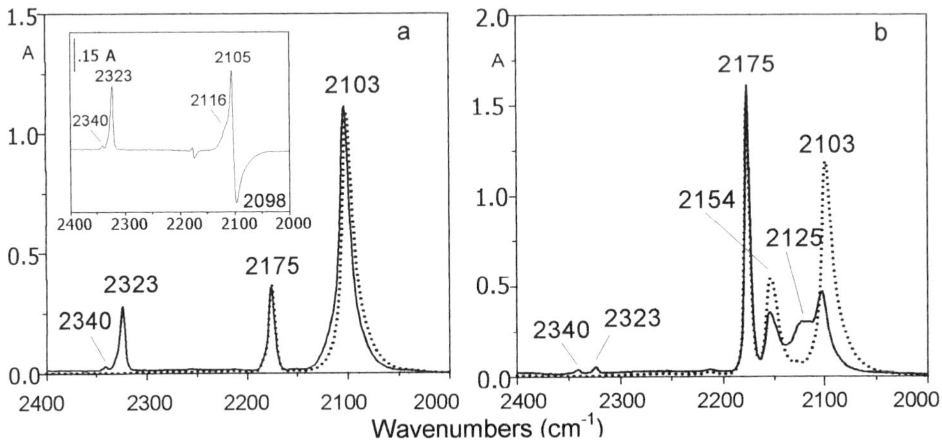

Figure 6. FTIR absorption spectra obtained by CO-$^{18}O_2$ interaction at 90K on calcined sample E. Section a: admission of 1 mbar $^{18}O_2$ (solid line) on preadsorbed 0.5 mbar CO (dotted line); in the inset: difference between solid and dotted lines. Section b: 3 mbar CO on $^{18}O_2$ preadsorbed (solid line) compared with 3 mbar CO (dotted line).

the conversion of CO on the E sample [19]. Therefore it can be proposed that the high activity of gold supported on titania can be explained by the LH_p mechanism.

In conclusion, on gold catalysts at low temperatures the reaction occurs according to a mechanism like that of Langmuir-Hinshelwood, generally accepted for the reaction on more traditional metallic catalysts (Pt, Pd and Rh) at higher temperatures between adsorbed CO and adsorbed oxygen; however on the gold supported catalysts, oxygen can be reactively activated in presence of CO on sites at the borderline of the metallic particles, probably at oxygen-vacancy defects.

3.5 CO - O_2 interactions at RT on sample E, F and C

The presence of a multiplet of bands in the CO_2 stretching region after interaction with $^{18}O_2$ on sample E, unlike at low temperature, clearly shows that the strong increase in the reaction rate at room temperature is mainly due to the participation of oxygen activated on the support (Fig. 7): in CO oxidation in fact, an *extensive exchange reaction with the oxygen atoms of the support occurs,* as evidenced by the absorption bands related to the different CO_2 isotopomers. Moreover on sample E, by oxygen preadsorption, no competition or inhibition effects on the CO oxidation have been put in evidence: on the contrary, unlike at low temperature, before the growth of adsorbed CO bands, very strong CO_2 bands are detected in the first two minutes. After two minutes the growth of the three bands stops, the CO absorption grows up and a weak band at 2070 cm^{-1}, assigned to $C^{18}O$ is observed (Fig. 7b). From these data it appears that the $C^{18}O$ formation is not a direct scrambling reaction, as previously proposed [14], but could be the consequence of secondary reactions, i.e. the CO_2 readsorption in a labile and mobile molecular form on the support, that can exchange oxygen atoms with the oxide, giving rise to $C^{18}O$, $C^{16}O^{18}O$ and also to the doubly marked products $C^{18}O^{18}O$ and $C^{16}O^{16}O$, not observed in the low temperature experiments. Anyway, carbonate-like species are not involved in these reactions as already shown previously. It can be proposed that at room temperature, the first step of the reaction is the adsorption of a mobile, molecular oxygen species on the support, its dissociation, probably at the interface between the oxide and the metallic step sites, a spill-over of O_{ad} to the Au particles and finally there the reaction with the incoming CO. On sample F and on the used sample C, molecular CO_2 is

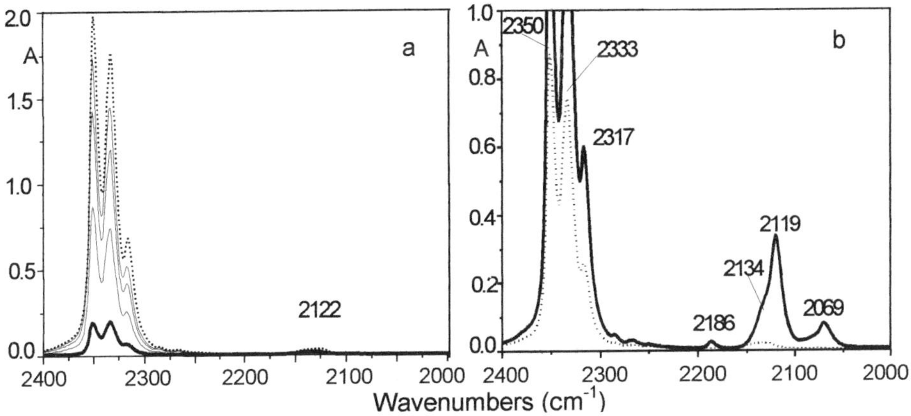

Figure 7. FTIR absorption spectra obtained by CO-$^{18}O_2$ interaction at 300K on calcined sample E. Spectra run at different times during diffusion of 10 mbar of CO on 5 mbar of preadsorbed $^{18}O_2$: every 15 sec in section a, after 2 and 3 min (dotted and full lines) in b.

not produced by $CO-O_2$ interactions. As already discussed previously both these two samples are characterised by particles of size larger than 10 nm, however, as a consequence of the pretreatment, on the used sample C the number of exposed step sites is quite high, as indicated by the high relative intensity of the 2145 cm^{-1} band in Fig. 2b. On the basis of the experiment presented in Fig. 3, relative to the interaction with hydrogen on sample F, the absorption at 2145-48 cm^{-1} has been assigned, differently from the first assignment to two dimensional metallic particles reported in [3], to CO coadsorbed with molecular oxygen on highly stepped, large particles.

Therefore, the inactivity in the CO oxidation both of sample F, characterised by a low step concentration on regular large metallic particles, and of used sample C, where as a consequence of the pretreatments an high step concentration on large particles is produced, indicates that electronic and quantum size effects are important in determining the activity of gold catalysts.

It can be concluded that the high activity of gold nanoparticles in the CO oxidation is not simply related to the high density of step sites but also to a correlated electronic effect of the particles with d< 2.5 nm in the oxygen activation and dissociation. An high concentration of step sites is needed but their presence is not sufficient in order to get an high activity.

REFERENCES

1. M. Haruta, Catal. Today , 36 (1997) 153.
2. A. Baiker, M. Kilo, M. Maciejewski, S. Menzi, A. Wakaun, in L. Guczi et al.(eds.), New Frontiers in Catalysis, Elsevier Science B. V. (1992), pp 1257-1266.
3. F. Boccuzzi, E. Guglielminotti, F. Pinna, G. Strukul, Surf. Sci., 728 (1997) 377.
4. G. C. Bond, D. T. Thompson, Catal. Rev.- Sci. Eng., 41 (1999) 319.
5. A. V. Bondzie, S. C. Parker, C. T. Campbell, Catal. Lett., 63 (1999) 143.
6. S. C. Chang, A. Hamelin and , M. J. Weaver, Surf. Sci., 239 (1990) L543.
7. M. Mavrikakis, P. Stoltze and J. K. Norskov, Catal. Lett., 64 (2000) 101.
8. R. A. van Santen and M. Neurock, Catal. Rev.- Sci. Eng., 37 (1995) 557.
9. M. Ruff, S. Frey, B. Gleich and R. J. Behm, Appl. Phys. A, 66 (1998) S513.
10. G. Cerrato, L. Marchese and C. Morterra, Appl. Surf. Sci., 70/71 (1993) 200.
11. S. Galvagno and G. Parravano, J. Catal., 55 (1978) 178.
12. H. Huber, D. McIntosh, G. A. Ozin, Inorg. Chem., 16 (1977) 975.
13. N. Sheppard, N. N. Nguyen, in R. J. H. Clark and R. E. Hester (eds.), Advances in Infrared and Raman Spectroscopy, Heyden, London, 1978.
14. F. Boccuzzi, A. Chiorino, S. Tsubota and M. Haruta, J. Phys. Chem. 100 (1996) 3625.
15. F. Boccuzzi, A. Chiorino, M. Manzoli, D. Andreeva, T. Tabakova, J. Catal., 188 (1999) 176.
16. W. T. Wallace and R. L. Whetten, J. Phys. Chem. B, 104 (2000)10964.
17. Salisbury, B. E., Wallace, W. T. and Whetten, R. L., Chem. Phys., 262 (2000) 119.
18. A. Sanchez, S. Abbet, U. Heiz, W. D. Schneider, H. Hakkinen, R. N. Barnett, U. Landman, J. Phys. Chem. A, 103(1999) 9573.
19. F. Boccuzzi, A. Chiorino, M. Manzoli, P. Lu, T. Akita, S. Ichikawa and M. Haruta, J. Catal., submitted.

Studies in Surface Science and Catalysis 140
A. Gamba, C. Colella and S. Coluccia (Editors)

Modeling soil ped formation: properties of aggregates formed by montmorillonitic clay, Al or Fe poorly-ordered oxides and polyphenol in acidic *milieu*

A. Buondonno and E. Coppola

Dipartimento di Scienze Ambientali, Polo Scientifico di Caserta, Seconda Università di Napoli, via Vivaldi 43, 81100 Caserta, Italy.

A modeling approach was applied to understand and clarify the composite phenomena of the organo-mineral complexing and the ped formation in soil. The properties of conglomerates formed at pH 5.5 by montmorillonitic clay (*C*), synthetic *Al*- or *Fe*-oxy-hydroxides, and polyphenol (*pp*) as humic material were studied. Investigational methods included SEM, FT-IR spectroscopy, X-ray diffraction, particle-size distribution, mechanical stability against ultrasonic dispersion and chemical stability to mild and strong reactants and to oxidative treatments. The models showed distinct morphological, mineralogical, mechanical, chemical and physico-chemical features. By SEM studies, the *C-Al-pp* model exhibited an irregularly rugged surface, while the *C-Fe-pp* showed evidence of spheroidal outgrowths. FT-IR spectroscopy provided clues for the effective interaction among the organic and inorganic components of conglomerates. X-ray diffraction patterns showed that *Al*- and, by a lesser extent, *Fe*- hydrolytic products partly entered the clay layers, thus expanding the basal period. The presence of broad and weak diffusion bands also suggested the formation of poorly-ordered *Al*- and *Fe*-oxy-hydroxides. The presence of *Al*- led to the prevailingly formation of small massive packets (<20 μm) and microaggregates (20-50 μm), quite stable to mechanical and oxidative breakdown; on the contrary, in the presence of *Fe*-, larger micro- and mesoaggregates (50-130 μm), well resistant to acidic etching, were formed. The experimental data highlighted the different role of *Al*- and *Fe*-oxy-hydroxides in determining the peculiar properties of the organo-mineral complexes, also validating the suitability and reliability of the adopted modeling approach to improve the understanding of factor and processes of soil ped formation.

1. INTRODUCTION

Soils are complex, dynamic systems, in which an almost countless number of processes are taking place; by and large, they can be classified as chemical, physical or biological, but there are no sharp distinction between these three groups. The *aggregation* is the primary physical process whereby a number of particles are held or brought together to form soil structure units known as *peds* [1].

The ped formation is an essential component of soil development, *i.e.*, of the pedogenesis, and results from the substitution of soil structure for the geologic structure of parent material [2]. While geologic structure is inherited from jointing cracks and bedding planes, soil structure forms and develops as a function of several mechanisms involving the movement of soil solution and the translocation of the soil colloidal phase. This last generally consists of soil colloid associations, also referred to as *conglomerates* or *organo-mineral complexes,* formed by mutual interaction of soil organic matter (SOM), clay minerals (and interstratified clays), oxy-hydroxides, and polyvalent bridge cations [3,4,5]. Such conglomerates play a fundamental role in soil structure evolution as the main cementing agent of discrete mineral particles, thus inducing the formation of larger and more heterogeneous and developed aggregates [6].

It is evident that both the nature of the conglomerates, as well as the relative prevalence of the single components widely varies in soil, according to the different pedoclimatic conditions and soil use. The SOM is usually considered as the prevailing responsible for the maintenance of the structural stability of most of soil types, particularly Mollisol, Alfisol, Ultisol and Inceptisol [5,7]. Organic compounds that may be involved in organo-mineral complexing in soils have several origins. Some are released by living plants, others are formed at various stages of decomposition of plants and animals and still others are products of microbial synthesis or abiotic reactions. Many of the most active complexing substances in soils are low-molecular-weight organic acids and phenolic compounds at various degree of polymerisation [4,7]. The polysaccharide component of soil has received the greatest attention in producing stable aggregates, but other organic substances bonded to clay through association with Al or Fe, such as polyphenols, may also be involved.

Natural polyphenols (*syn.* vegetable tannins) constitute an important compartment of the soil organic matter, both as plant metabolites and vegetal residues [8], as well as precursory molecules in the humification process [4,5,7,9]. They are effective chelators of metal ions and colloids [10], and can easily interact with expansible phyllosilicates and polyvalent cations, differently involved in the evolution of the organo-mineral complexes in the natural soils [4,5,7,11,12]. In particular, polyphenols likely play a fundamental role in the so-called *"cheluviation"* process in Spodosols, involving the complexing, transport and re-arranging of Al/Fe organo-mineral complexes downward the soil profile [12]. Laboratory studies also showed that polyphenols can form discrete conglomerates with a variety of chemical, physical, and morphological properties [13,14,15].

The involvement of SOM as cementing agent tends to be less relevant in soils such as Oxisols and Andisols, where amorphous Al- or Fe-oxides could play a more important role as inorganic cementing materials, or in self-mulching soils (*e.g.*, some Vertisols) that contain clays with a high shrink/swell potential [5,7]. The inorganic cementing materials increase bonding between mineral particles, having the greatest influence on structural characteristics by increasing the strength of failure zones. These materials exist as coatings on the surface of mineral particles and are usually amorphous or poorly crystallized [6]. Examples include Al- and Fe-oxides, precipitated aluminosilicates or combinations of these materials complexed with SOM. Amorphous inorganic materials have dominantly a variable charge nature, and can alter the surface charge characteristics of clay minerals when adsorbed, or existing as a surface film. Changes in surface charge characteristics results in changes in particle-particle interaction, but these changes are not directly predictable.

From literature, Fe oxides appear to be the major responsibles for soil aggregation and strength development in hardsetting soils [6,16,17,18]. However, in soils with high content of both SOM and amorphous materials, Al- and Fe-oxy-hydroxides and Ca or Al-complexes with high molecular weight organic materials were found to be the prevailing cementing agents [20]. Differently, the nature and content of clay minerals becomes the main factor influencing the ped evolution in fine-textured soils with smectitic mineralogy. These clays, such as the montmorillonite group, exhibit striking swelling/shrinking properties under wetting/drying cycles, with subsequent bulk soil volume changes which develop strongly expressed structural units [2].

Another relevant factor for organo-mineral complexing and ped formation is the presence and nature of bridge polyvalent cations. As it is known, organic anions are normally repelled from negatively clay surfaces, so that stable SOM/clay interactions occur only when polyvalent cations are available in the medium as bonding bridges. Unlike monovalent cations typically found in soils (Na^+, K^+), polyvalent cations such as Ca^{2+}, Fe^{3+}, and Al^{3+} are effective in partly neutralize both the charge on the clay and the acidic functional groups on SOM [5]. The divalent Ca^{2+}, prevailing in neutral-alkaline soil environments, does not form strong coordination complexes with organic molecules; on the contrary, laboratory studies showed that Ca tends to segregate as crystalline calcite even in the presence of excess of humus precursors [15]. Differently, Fe^{3+} and Al^{3+}, more abundant in neutral-acidic soils, easily form stable coordination complexes with organic compounds, so that SOM and clay can be strongly stabilized in organo-mineral complexes. Furthermore, both Fe and Al can cover the clay as surface film of amorphous materials, becoming even more effective in holding inorganic and organic anions [21].

From the considerations above, it appears that the formation and the development of peds strictly reflects the pedogenic history of a soil and its relationships with the environment. Thus, studies on the mechanisms for ped aggregation and expression represent a fundamental tool not only to elucidate the pedogenic processes, but also to better understand the phenomena of soil degradation related to the loss of structure stability.

A good approach attempting to understand and clarify the complex phenomena of ped evolution is to model conglomerates from clay minerals, synthetic oxy-hydroxides, and humus components [3]. This approach would be relevant since detailed knowledge is available on the nature and amount of each component, thus allowing to discriminate their peculiar effects on the final chemical, physico-chemical and morphological properties of the produced models. In a previous research, the properties of montmorillonite-Ca-polyphenol aggregates produced at pH 8.5 were illustrated [15]. This paper deals with the properties of organo-mineral aggregates formed by interaction at pH 5.5 of a montmorillonitic clay, polyphenol, and Al or Fe poorly-ordered oxides.

2. MATERIAL AND METHODS

2.1. Model preparation

Laboratory models were prepared following the procedures described elsewhere [13,14], here modified in accordance with the specific aim of the investigation. Starting materials were:

- 500 ml of aqueous suspension containing 1 g of the <2-μm fraction of a montmorillonitic clay *(C)* as bentonite (Prolabo Rhone-Poulenc), previously purified from organic matter and alkaline-earth carbonates, dialyzed, and then freeze-dried;

- 6 mmol of $AlCl_3$ or $FeCl_3$ (Baker Chemicals); hereafter, the aluminium or iron oxy-hydroxides component of the aggregate models will be indicated as *Al-* or *Fe-*, respectively;
- 0.6 mmol of vegetable tannin polyphenol *(pp)* as gallotannin ($C_{76}H_{52}O_{46}$, mw 1.702 kD, Fluka); because of its low solubility in pure water, the *pp* was previously dissolved in 100 ml H_2O/CH_3OH (v:v = 5:1).

The models *C-Al-pp* and *C-Fe-pp* were obtained by thoroughly mixing the components in a suspension with a final volume of 1 litre at pH 5.5, and then allowed to react in stoppered polyethylene bottles at 40°C. After ageing 90 days, the materials were dialyzed, air-dried at 40°C, and then ground to pass a 0.25 mm sieve. Reference standards by *C* and *pp* with or without *Al-* or *Fe-* were also prepared at the same conditions.

Nine replicates of each sample were prepared.

2.2. Morphological and mineralogical properties

Scanning electron microscopy *(SEM)* studies were carried out by a Cambridge S4 Stereoscan. X-ray patterns were obtained by a Rigaku-DmaxB apparatus, using a Fe-filtered Co$k\alpha$ radiation. FT-IR spectroscopy patterns were obtained by a Perkin Elmer 1720 apparatus.

2.3. Aggregate size distribution and mechanical stability

The Particle-Size Distribution *(PSD)* was determined by the laser light scattering technique [22], using a Malvern M6.10 apparatus. Analyses were performed on water-suspended material after 1 min stirring.

The mechanical stability was evaluated on the basis of the change of the *PSD* of each sample, determined either before and after 3-min sonication treatment by ultra-sound. Sonication was performed by a Hielscher Ultraschallprozessor UP 200s apparatus.

2.4. Aggregate chemical stability

The chemical stability of the aggregate models was evaluated on the basis of Al and Fe extractions and organic C oxidative treatments, each performed by different conventional or experimental procedures.

Al and Fe were extracted by various mild to strong reactants:

- H_2O, w:v = 100 mg:25 ml, 25 °C, 5 min, before or after sonication treatment (*H_2O-b.s.* or *H_2O-a.s.*, respectively);
- 1 M NaClO, w:v = 50 mg:25 ml, 25 °C, 2h;
- oxalic acid-ammonium oxalate mixture at pH 3 *(Oxa)* [23], w:v = 100 mg:10 ml, 25 °C, 4h in the dark;
- Mehlich No. 3 reactant *(M3)*, 0.2 M CH_3CCOH, 0.001 M EDTA, 0.013 M HNO_3, 0.015 M NH_4F, 0.25 M NH_4NO_3, [24], w:v = 100 mg:2 5 ml, 25 °C, 5 min.

The organic C was determined according to the standard Walkley-Black procedure, by 30-min oxidative treatment with $K_2Cr_2O_7/H_2SO_4$ [23], as well as by the modified procedure with 120-min oxidation at 80 °C [13].

All data were expressed on the 105° oven-dried weight basis.

3. RESULTS

3.1. Morphological and mineralogical properties

Figure 1a shows the micrograph of *C-pp* model. Typical montmorillonitic clay flakes are well visible and clearly distinguishable from the irregular blocky mass of consolidated polyphenol. This suggests that the *C-pp* system behaves as a mixture of separate discrete phases rather than a homogeneous aggregate structured by complex physicochemical interactions between the two components. Differently, both *C-Al-pp* and *C-Fe-pp* show more dense and consistent *habitus*, even at high magnification (Figures 1b and 1c, respectively). In particular, the *C-Al-pp* exhibits an irregularly rugged surface, while the *C-Fe-pp* shows evidence of spheroidal outgrowths, whose morphology is strictly consistent with that of globular ferrihydrite/organic matter aggregates described by Schwertmann and Taylor [25]. In any case, there is no evidence for the formation of any crystalline structure.

Table 1 lists the main peaks detected by FT-IR analyses in the reference models *C*, *pp*, and *C-pp*, together with the associated functional groups.

Table 1
Main FT-IR peaks (wavenumber, cm^{-1}) and associated functional groups detected in the reference models *C*, *pp*, and *C-pp*

	C	*pp*	*C-pp*
Al-OH	3627	nd	3627
OH(s)	3420	3415	3382
COOH	nd	1708	1715
OH	1636	nd	nd
C=C(Ar)/C=O	nd	1615	1616
C=C(Ar)	nd	1520	1520
C-H	nd	1449	1455
C-O(b)	nd	1287	1283
OH(b)	nd	1206	1206
Si-O-Si	1044	nd	1039

(s) = stretching; (Ar) = aromatic C; (b) = bending; nd = not detected

The *C-pp* model exhibits the main peaks of both *C* and *pp*, but the –OH bound signal at 1636 cm^{-1} is masked and the –OH stretching effect is shifted to 3382 cm^{-1}.

Figure 2 depicts the FT-IR spectra of *C-Al-pp* and *C-Fe-pp* models, contrasted with *C*. In the *C-Al-pp* and *C-Fe-pp* aggregates, the peculiar *C* reference peak at 3627 cm^{-1} disappears, and the peak at 3420 cm^{-1} is left-shifted toward larger wavenumber values. The strong carbonilic group effects at 1708-1715 cm^{-1}, detected in the *pp* model, also fade away, whereas the peak at 1520 cm^{-1}, associated to the aromatic C=C in *pp* and *C-pp*, are left-shifted to 1505 cm^{-1}, with decreasing intensity from *C-Al-pp* to *C-Fe-pp*.

The X-ray diffraction patterns of *Al-pp* and *Fe-pp* (not shown) did not give any significant regular reflection, thus suggesting that the presence of polyphenol inhibits the crystallization of Al or Fe polymers, likely by effective chelation reactions. However, two contiguous, broad diffusion bands, with minimum at 8.72, 23.92 and 40.00 °2θ, were observed in the *Fe-pp* model. This could be attributed to the higher charge density of Fe^{3+} ion, which promotes the

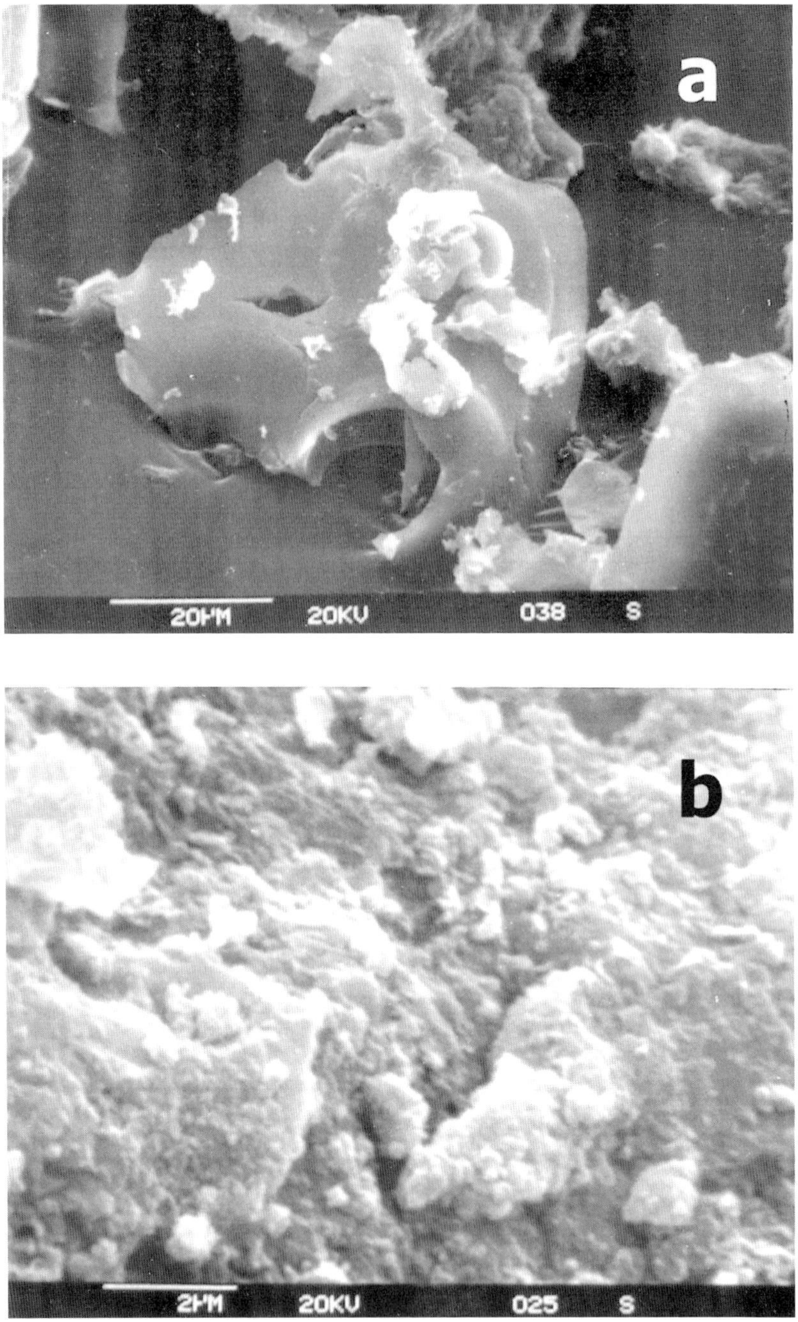

Fig. 1a, b. SEM micrograph of representative *C-pp* aggregate (a) and *C-Al-pp* (b).

Fig. 1 c. SEM micrograph of representative *C-Fe-pp* (c) aggregates.

formation of large, stable polycations [25], whose regular crystallization is nevertheless hindered by the presence of organic matter.

Figure 3 represents the X-ray diffraction patterns of *C-Al-pp* and *C-Fe-pp* models, compared with *C*. For this last model, the very strong first-order effect at 1.491 nm, and the secondary effects with lower intensity at 0.499, 0.446, 0.424, 0.406, 0.321, and 0.298, are detected. The vast majority of such reflections undergo substantial change as a consequence of the different interaction between clay, polyphenol and *Al-* or *Fe-* polycations. In the *C-Fe-pp* model, the peak at 1.491 nm exhibits a moderate and symmetrical expansion to 1.540 nm, suggesting a fair and almost regular intercalation of *Fe-* polymers into clay layers. Differently, in the *C-Al-pp* model, the 1.491 peak is extensively modified as a wide, low-intensity diffraction band, diffuse from 4.74 to 7.80 °2θ, and centered at 6.04 °2θ, corresponding to a basal period of 1.698 nm. For the same model, another large, asymmetrical band is detectable from 7.80 to 23.00 °2θ, with maximum at 11.44 °2θ, corresponding to a basal period of 0.897 nm. Such evidences suggest two concomitant phenomena on the development of *C-Al-pp* aggregate:

- the intercalation of large, poorly-ordered pillars of *Al*-polymers or *Al-pp* complexes between clay layers [13], with subsequent expansion of basal period as a wide, diffuse and asymmetrical band at 1.698 nm;
- the "template effect" of outer clay surface [26], which favors the segregation of *Al*-polymers and their re-arrangement as discrete phases of Al-oxy-hydroxides at early

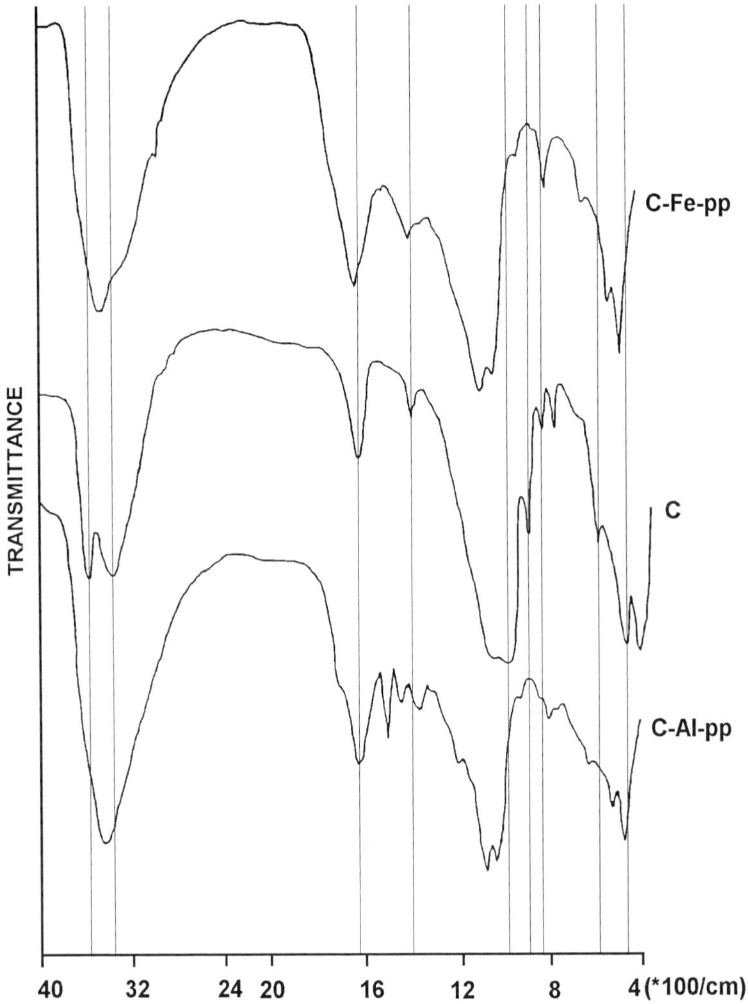

Fig. 2. FT-IR spectra of *C*, *C-Al-pp*, and *C-Fe-pp* models. Vertical bars indicate standard reference peaks of montmorillonite clay.

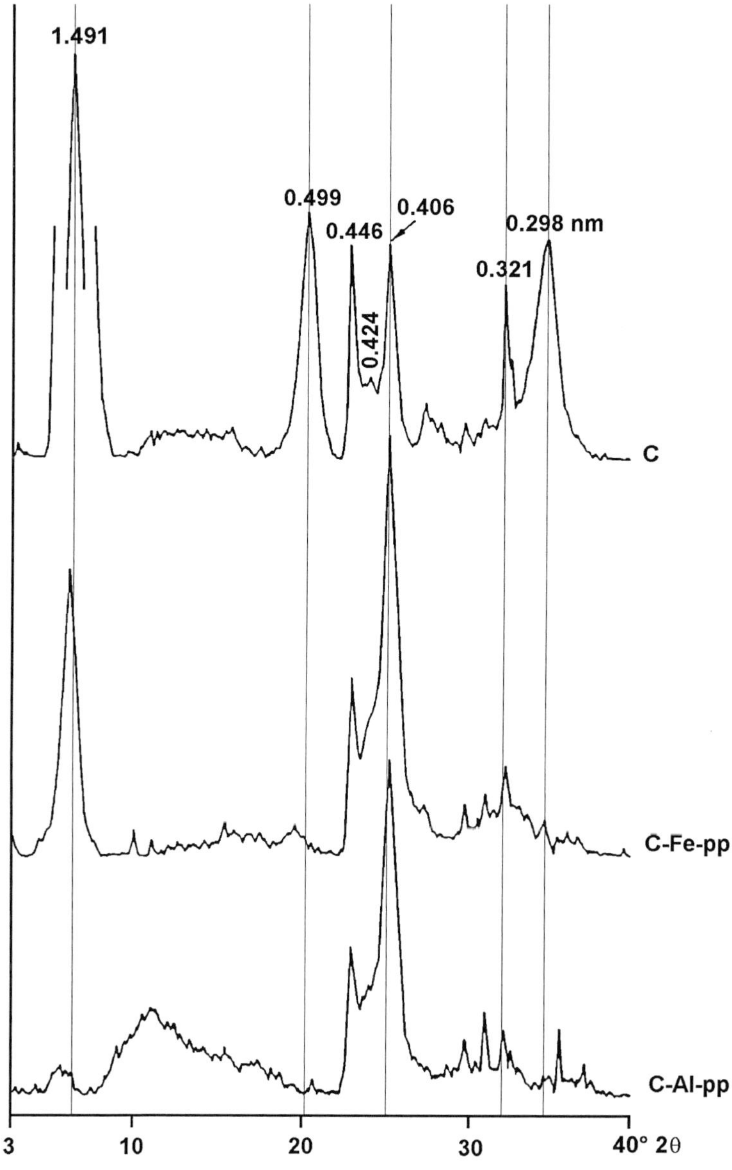

Fig. 3. X-ray diffraction patterns of *C*, *C-Al-pp*, and *C-Fe-pp* models. Vertical bars indicate standard reference peaks of montmorillonite clay.

crystallization stage (peak centered at 0.897 nm), thus limiting the perturbing effect of the organic matter.

The weak reflections between 10 and 20 °2θ observed for the *C-Fe-pp* aggregate also suggest the incipient formation of *Fe*-polymer para-crystalline phases.

The *SEM* studies, as well as the comparative analysis of X-ray and FT-IR patterns, altogether highlight that clay, polyphenol, and *Al-* or *Fe-* polymers effectively interacted among them, producing aggregates with distinct and peculiar features as a function of the bridge cation.

3.2. Aggregate size distribution and mechanical stability

As postulated by Golchin et al. [27], three levels of soil particles aggregation exist: (1) the binding together of clay plates by organo-mineral cements into *packets* < 20 μm, (2) the binding of *packets* into stable *microaggregates* 20-250 μm, and (3) the binding of stable *microaggregates* into *macroaggregates* > 250 μm. For each aggregation level, organo-mineral cements are involved as binding agents. From the observations of various authors [7], we can further on classify as *mesoaggregates* the soil particles conglomerates having a dimension 50-250 μm.

In our study, no macroaggregates were found after experimental models were suspended in water. However, the various clay/*Al-* or *Fe-*/polyphenol interactions produced aggregates characterized by widely different texture and mechanical stability, as evaluated by the Particle-Size Distribution (*PSD*) of the solid phases determined before and after sonication treatments.

Table 2 presents the *PSD* of the binary models obtained by reaction of clay or polyphenol with *Al-* or *Fe-* polymers. Both *Al-* and *Fe-* were able to cement clay particles with initial size < 2 μm, into small packets with particle size prevailingly ranging from 5 to 15 μm. However, the *C-Al* packets are characterized by a noticeable dimensional homogeneity, since most of particles (70.5%) lies in the size range of 5-10 μm. Differently, the size of the *C-Fe* packets is more widely distributed between 5 (50.1%) and 15 (41.6%) μm.

Both models exhibit a moderate physical stability, since about 30% of *C-Al* and 40% of *C-Fe* materials keep a particle size larger than 5 μm after the disruptive ultra-sound treatment.

Table 2
Particle-Size Distribution (%) determined before (b.s.) and after (a.s.) 3-min sonication

	C-Al		*C-Fe*		*Al-pp*		*Fe-pp*	
Size μm	b.s.	a.s.	b.s.	a.s.	b.s.	a.s.	b.s.	a.s.
<2	0.6	4.0	0.4	3.4	2.0	3.9	0.0	2.6
2-5	11.4	65.5	5.4	56.7	40.9	66.1	0.0	48.4
5-10	70.5	30.5	50.1	39.8	56.1	30.0	0.0	48.6
10-15	17.0	0.0	41.6	0.1	1.0	0.0	0.0	0.4
15-20	0.0	0.0	2.4	0.0	0.0	0.0	0.0	0.0
20-50	0.0	0.0	0.1	0.0	0.0	0.0	86.6	0.0
50-100	0.0	0.0	0.0	0.0	0.0	0.0	13.3	0.0
>100	0.0	0.0	0.0	0.0	0.0	0.0	0.1	0.0

The interaction of polyphenol with *Al-* and *Fe-* led to the formation of complexes with very different physical features. The *Al-pp* model formed small microaggregates, with particle size ranging from 2 to 10 μm, and moderate stability (30% of material with particle size 2-5 μm after sonication). Differently, *Fe-pp* formed larger aggregates, up to 100 μm, which were easily disrupted by ultra-sound treatment. However, about 49% of particles still keep a dimension of 5-10 μm after sonication. This should confirm the overall tendency of *Fe-* to form polymers larger than *Al-*.

Figure 3 shows the *PSD* of *C-Al-pp* and *C-Fe-pp* aggregates determined before and after sonication treatment. Even in this case, the largest aggregates were formed in the presence of iron oxy-hydroxides. Before sonication, the particle-size of *C-Fe-pp* was widely distributed from 10 to 120 μm, with 66.5% of material forming mesoaggregates > 50 μm. Most of these mesoaggregates exhibits a dimension of 55-75 μm. The sonication treatment had some disruptive effect on mesoaggregates, thus leading to the formation of finer material which was prevailingly represented by microaggregates 20-50 μm (85% of total particles), with a peak for particles ≈ 40 μm, but still including a significant proportion of mesoaggregates (14% of total particles). The *C-Al-pp* model was formed by aggregates with dimension smaller than *C-Fe-pp*, almost equally distributed between microaggregates (42%) and mesoaggregates (53%). After sonication, the *C-Al-pp* particles were pull down to microaggregates (82%) and packets (14%), with only a small residue (4%) of particles forming mesoaggregates.

3.3 Aggregate chemical stability

The data in Tables 3a and 3b show the different stability of the samples after chemical extractions and organic C oxidative treatments, respectively. To make the comparison easier, the values are also expressed as the percentage of the respective amounts of Al and Fe, or of C, initially added to the models.

Table 3a
Aggregate chemical stability: Al or Fe extracted by various extractants from aggregate models and reference binary materials with polyphenol*

	Al-pp		*Fe-pp*		*C-Al-pp*		*C-Fe-pp*	
Extractant	mg/g	*% i.a.*	mg/g	*% i.a.*	mg/g	*% i.a.*	mg/g	*% i.a.*
H_2O-b.s. [1]	0.22	0.16	3.10	1.86	0.02	0.03	2.33	1.63
H_2O-a.s. [1]	2.04	1.49	5.90	2.39	1.67	2.25	5.24	3.68
M3 [2]	43.25	31.57	30.69	14.61	20.30	27.36	26.21	18.42
NaClO	14.24	10.39	46.44	18.88	17.39	23.44	41.72	29.32
Oxa [3]	61.58	44.95	58.87	23.83	47.10	63.48	26.08	18.33

* Values are expressed as mg of Al or Fe per g material, and as the percentage of the amount of Al or Fe initially added (*% i.a.*); [1] H_2O before (*b.s.*) or after (*a.s.*) sonication treatment; [2] Melich No. 3 reactant; [3] oxalic acid-ammonium oxalate mixture at pH 3.

All the compared models were sparingly soluble in water, with particular reference to the *Al*-containing materials, albeit the sonication treatment significantly increased the amounts of both *Al-* and *Fe-* released in solution. Considerably higher amounts of *Al-* and *Fe-* were extracted by the other tested reactants.

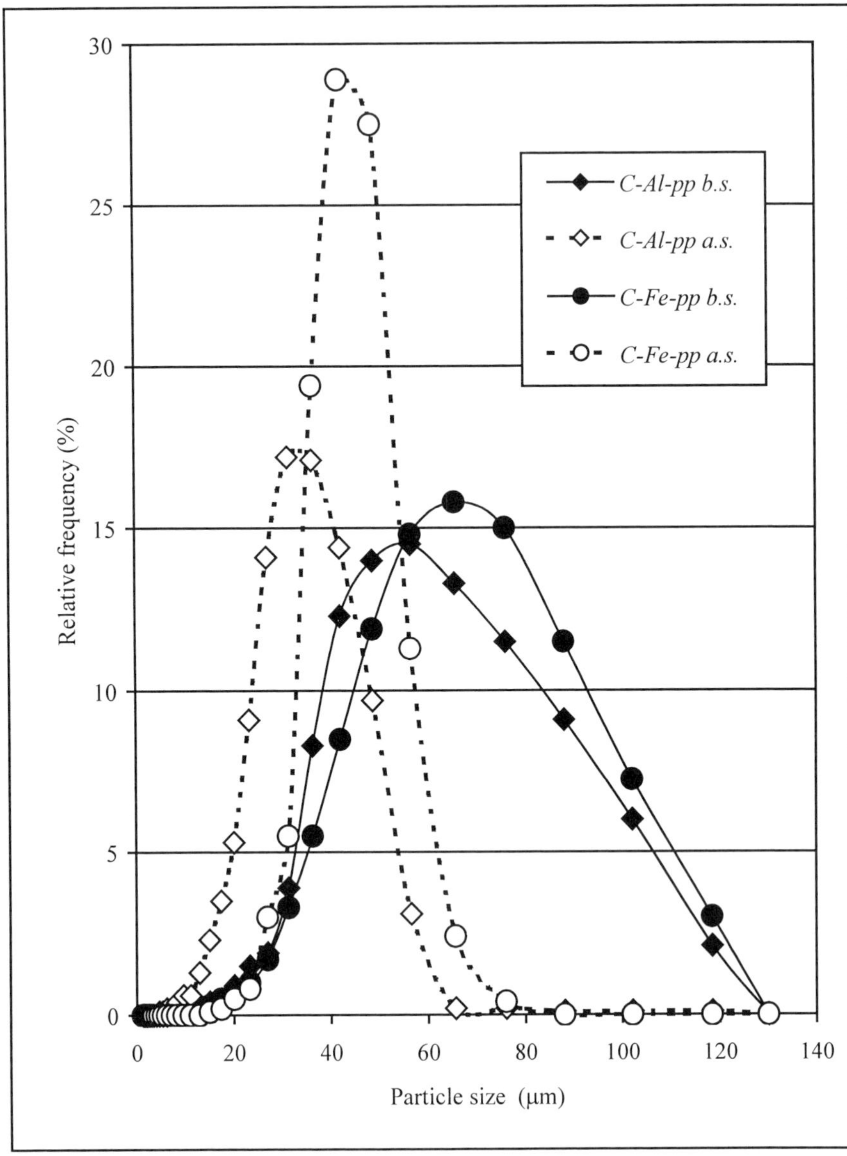

Fig. 3. Particle-size distribution of *C-Al-pp* and *C-Fe-pp* aggregates determined before (b.s.) and after (a.s.) sonication treatment.

However, larger proportions of *Al-* than *Fe-* were removed by the acidic reactants *M3* and *Oxa*, while the alkaline-oxidizing *NaClO* was more effective in destabilizing the *Fe-*containing materials. This suggests that the organic component of the latter materials would be less stable to the oxidative etching. Such a conjecture could be confirmed by comparing data in Table 3b. The amounts of C oxidized after 30-min treatment in the binary and ternary *Fe*-containing models were always larger than the respective determined for the *Al*-containing ones, with percentage values of nearly 80% for the former, and 60% for the latter.

Table 3b
Aggregate chemical stability: organic C oxidized after 30-min or 120-min treatment with $K_2Cr_2O_7/H_2SO_4$ at 80 °C*

	after 30-min		*after 120-min*	
Model	mg/g	*% i.a.*	mg/g	*% i.a.*
pp	398.2	74.29	536.8	100.15
C-pp	250.7	92.62	270.2	99.80
Al-pp	274.8	59.40	282.6	61.09
Fe-pp	325.5	80.65	337.7	83.67
C-Al-pp	156.4	62.39	208.8	83.29
C-Fe-pp	183.6	79.10	193.5	83.37

* Values are expressed as mg of C per g material, and as the percentage of the amount of C initially added (*% i.a.*).

By extending the hot oxidative treatment to 120 minutes, the amount of oxidized C did not exceed 84% for *Fe-pp*, *C-Al-pp*, and *C-Fe-pp* models, but it was still nearly 60% for *Al-pp*. This clearly proves that *Al*-oxy-hydroxides were more effective than *Fe*-oxy-hydroxides in stabilizing and protecting the organic matter.

As reference values, Table 3b also presents the amounts of C oxidized after 30 or 120 min for pure tannic acid *pp* and for the *C-pp* model. It is evident that, in the absence of an efficient complexing cation, the whole amount of organic C is definitively oxidized within 120 min. Interestingly, in the *C-pp* model, the large amount of C already oxidized after 30-min could be ascribed to a possible priming catalytic effect by the clay mineral; differently, the flakes of pure *pp* probably clotted, thus needing more time to be completely oxidized [15].

All together, these data agree with those obtained for models prepared in alkaline conditions [15], and confirm that the interactions with clay minerals and polyvalent cations substantially protect the organic component of the aggregates even under strong oxidative conditions.

4. CONCLUSIONS

Soil ped models with peculiar features were obtained in our study. The nature of the bridge-cation binding clay and organic matter considerably affected the distinct morphological, mineralogical, mechanical, chemical and physico-chemical properties of the produced aggregates. The polyphenol component substantially complexed *Al-* or *Fe-*, thus hampering the formation and the separation of well-crystallized discrete phases as Al- or Fe-oxides. In the interaction with clay, a part of *Al-* or *Fe-*, likely binding organic matter, were more or less regularly intercalated into clay layers, contributing to structural distortion. In any case, almost chemically and mechanically stable conglomerates of different size were formed by interaction of clay and poliphenol *via Al-* or *Fe-* bridge-cations. The presence of *Al-* led to the prevailingly formation of small massive packets and microaggregates with rugged surface, quite stable to mechanical and oxidative breakdown; on the contrary, in the presence of *Fe-*, larger micro- and mesoaggregates, with globular *habitus*, well resistant to acidic etching, were formed. On the whole, the congruity among the experimental data, and their consistency with the properties of soil aggregates and organo-mineral complexes in soil environment should validate the suitability and reliability of model systems studies to improve the understanding of natural processes of soil genesis and evolution.

REFERENCES

1 E.A. FitzPatrick, Soils. Their formation, classification and distribution, Longman, New York, 1983.
2 O.A. Chadwick and R.C. Graham, Pedogenic processes. In: M. E. Sumner (ed.), Handbook of Soil Science, CRC press, Boca Raton, FL, USA, 2000, p. A41.
3 M.H.B. Hayes and F.L. Himes, Nature and properties of humus-mineral complexes. In: P.M. Huang and M. Schnitzer (eds.), Interactions of soil minerals with natural organics and microbes, SSSA Sp. Publ. No. 17, Soil Sci. Soc. Am. Ed., Madison, WI, USA, 1986, p. 103.
4 J.A. McKeague, M.V. Cheshire, F. Andreux and J. Berthelin, Organo-mineral complexes in relation to pedogenesis. In: P.M. Huang and M. Schnitzer (eds.), Interactions of soil minerals with natural organics and microbes, SSSA Sp. Publ. No. 17, Soil Sci. Soc. Am. Ed., Madison, WI, USA, 1986, p. 549.
5 F.J. Stevenson, Humus chemistry. Genesis, composition, reactions, J. Wiley and Sons, New York, 1982.
6 B.D. Kay and D.A. Angers, Soil Structure. In: M. E. Sumner (ed.), Handbook of Soil Science, CRC press, Boca Raton, FL, USA, 2000, p. A229.
7 J.A. Baldock and P.N. Nelson, Soil Organic Matter. In: M. E. Sumner (ed.), Handbook of Soil Science, CRC Press, Boca Raton, FL, USA, 2000, p. B25.
8 E. Haslam, J. Chem. Ecology, 14(10) (1988) 1789.
9 F.H. Frimmel and R.F. Christman (eds.), Humic substances and their role in the environment, Chichester, UK, 1988.
10 F.J. Stevenson and A. Fitch, Chemistry of complexation of metal ions with soil solution organics. In: P.M. Huang and M. Schnitzer (eds.) Interactions of soil minerals with natural organics and microbes, SSSA Sp. Publ. No. 17, Soil Sci. Soc. Am. Ed., Madison, WI, USA, 1986, p. 29.
11 W.W. Emerson, R.C. Foster and J.M. Oades, Organo-mineral complexes in relation to soil aggregation and structure. In: P.M. Huang and M. Schnitzer (eds.), Interactions of soil minerals with natural organics and microbes, SSSA Sp. Publ. No. 17, Soil Sci. Soc. Am. Ed., Madison, WI, USA, 1986, p. 521.

12 N. van Breemen and P. Buurman, Soil Formation, Kluwer Academic Pub., Dordrecht, The Netherlands, 1998.
13 A. Buondonno, D. Felleca and A. Violante, Clays and Clay Minerals, 37(3) (1989) 235.
14 A. Buondonno and A. Violante, Can. J. Soil Sci., 71 (1991) 285.
15 A. Buondonno, M.L. Ambrosino, E. Coppola, D. Felleca, F. Palmieri, P. Piazzolla and A. De Stradis, Models of organo-mineral aggregates with polyphenols. 2. Stability of bentonite-Ca-gallotannin interaction products. In: N. Senesi e T.M. Miano (eds.), Humic substances in the global environment and implications on human health, Elsevier, Amsterdam, 1994, p. 1101.
16 E. Barberis, F. Aimone Marsan, V. Boero and E. Arduino, J. Soil Sci., 42 (1991) 535.
17 W.W. Emerson, Inter-particle bonding. In: Soils: an Australian viewpoint, Academic Press, London, 1983.
18 D.P. Franzmeier, C.J. Chartres. and J.T. Wood, Soil Sci. Soc. Am. J., 60 (1996) 1178.
19 W.D. Kemper and E.J. Koch, USDA Tech. Bull. 1355 (1966).
20 F. Bartoli, R. Philippy and G. Burtin, J. Soil Sci., 42 (1988) 535.
21 D.J. Greenland, Soil Sci., 111 (1971) 34.
22 R. Stern, B.E. Eisenberg and M.C. Laker, S. Afr. J. Plant and Soil, 8(3) (1991) 136.
23 P. Buurman, B. van Lagen and E.J. Velthorst (eds.), Manual for soil and water analysis, Backhuys Publishers, Leiden, The Netherlands, 1996.
24 A. Mehlich, Commun. in Soil Sci. Plant Anal., 15(12) (1984) 1409.
25 U. Schwertmann and R.M. Taylor, Iron oxides. In: J.B. Dixon and S.B. Weed (eds.), Minerals in Soil Environments, SSSA Publ. No. 1, Madison, WI, USA, 1989, p. 379.
26 P.H. Hsu, Aluminium Oxides and Oxyhydroxides. In: J.B. Dixon and S.B. Weed (eds.), Minerals in Soil Environments, SSSA Publ. No. 1, Madison, WI, USA, 1989, p. 331.
27 A. Golchin, J.A. Baldock and J.M. Oades, A model linking organic matter decomposition, chemistry and aggregate dynamics. In: R. Lal, J.M. Kimble, R.F. Follett and B.A. Stewart (eds.), Soil processes and the carbon cycle, CRC Press, Boca Raton, FL, USA, 1998, p. 245.

Studies in Surface Science and Catalysis 140
A. Gamba, C. Colella and S. Coluccia (Editors)

The vibrational spectra of phosphorous oxynitride at high pressures

M. Capecchi, R. Bini, G. Cardini and V. Schettino

Dipartimento di Chimica, Università di Firenze
Via G. Capponi 9, 50121 Firenze, Italia
and
European Laboratory for Non-linear Spectroscopy
Largo Enrico Fermi 2, 50100 Firenze, Italy

The infrared and Raman spectra of phosphorous oxynitride PON, a homologue of silica, have been measured at pressures up to 23 Gpa. The moganite and the α-quartz type crystalline forms of PON have been studied. A correlation has been established between the spectra of PON and those of silica. In the moganite type PON a phase transition has been observed at ~ 4 Gpa. Some preliminary results of an ab initio molecular dynamics simulation are discussed.

1. INTRODUCTION

Phosphorous oxynitride PON has been obtained [1,2] by reacting ammonium dihydrogen phosphate with urea or melamine and crystallized by heating at appropriate temperatures and pressures. PON is isoelectronic with silica and X-ray diffraction studies [2-4] have shown that PON is a chemical and structural analogue of silica being made of networks of $P(O_{0.5}N_{0.5})_4$ tetrahedra. Phosphorous oxynitride is intrinsically disordered as a consequence of the random distribution of the oxygen and nitrogen atoms in the anionic sites of the crystal. Depending on the temperature and pressure treatment PON can be crystallized in a β-cristobalite, in a α-quartz and in a moganite type structures. Interest in phosphorous oxynitride arises from the similarities with silica and from the possibility, studying the phase transformations in PON, of indirectly exploring regions of the phase diagram of silica not easily accessible experimentally. For instance, it has been found that the cristobalite type structure, stable in silica at high temperature, can be isolated at room temperature in PON. With these perspectives in mind we have studied the vibrational spectra of PON measuring the infrared and microRaman spectra of the α-quartz type crystal as a function of temperature and the microRaman spectrum of the moganite type crystal. This latter has been obtained as a function of pressure up to 23 Gpa in a diamond anvil cell (DAC) and at ~4.5 GPa a phase transformation has been observed that has not previously been reported for silica. A correlation of the spectra of PON with those of silica has been discussed with the help of results of DFT calculations of simple PON clusters and of preliminary Car-Parrinello ab initio molecular dynamics simulations.

2. EXPERIMENTAL

Phosphorous oxynitride was prepared in the moganite and quartz type structures according to the procedure reported in ref. [4]. Infrared spectra were measured on KBr pellets by means of an FTIR Bruker IFS 120 HR spectrometer equipped with a closed cycle cryostat allowing measurements between 10 and 300 K. Raman spectra were measured using the 514.5 nm emission line of an Ar ion laser and a double monochromator (Jobin-Yvon U-1000) equipped with a CCD detector. The same source (0.1 mW) was used for the microRaman experiment whose set-up consisted of a BH2 Olympus microscope coupled to a single monochromator and a CCD detector. High pressure measurements were performed by using a membrane diamond anvil cell (MDAC). Few grains of a finely ground sample were placed, joint to a couple of 5 μm ruby crystals necessary for the pressure calibration, in the 100 μm diameter hole of a steel gasket. The cell was then filled with liquid argon, the pressure transmitting medium, by means of a cryogenic loading apparatus. Local pressure was always measured , according to the shift of the R_1 ruby fluorescence line [5], before and after each Raman measurement.

2. RESULTS AND DISCUSSION

2.1. Infrared and Raman spectra

The infrared spectrum at room temperature of the α-quartz type PON is shown in fig.1 where it is compared with the spectrum of α-quartz and of silica glass [6].In the low frequency region the resemblance with the spectrum of silica glass is considerable. Some broadening of the spectral features in comparison with α-quartz can also be noticed, as expected for a disordered system. The broadening of the spectral bands is much more

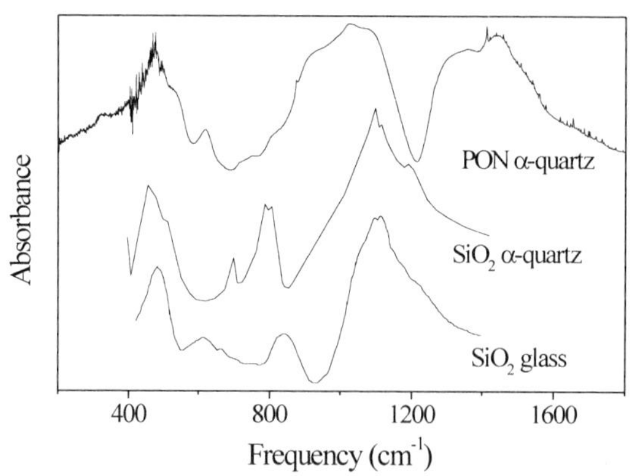

Fig. 1 – Room temperature infrared spectrum of moganite PON.

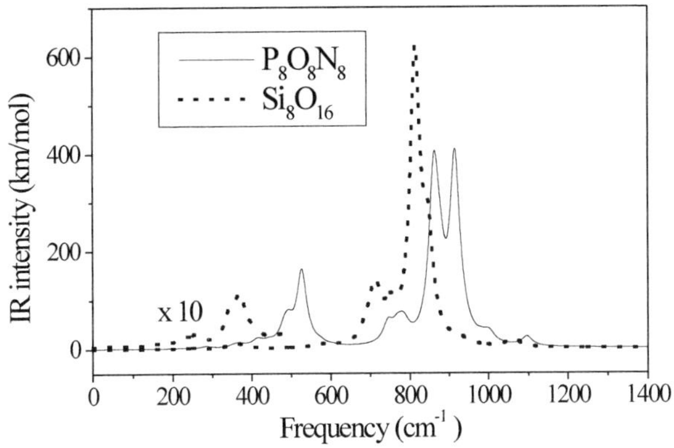

Fig. 2 – Calculated DFT infrared spectra of PON and silica clusters.

evident in the higher frequency region where, in addition, the stretching modes of silica split into two well separated components. This arises from the differences in the P-O and P-N stretching force constants in PON. To check that this is indeed the source of the observed splitting in the high frequency region, model calculations were carried on $(PON)_n$ and $(SiO_2)_n$ clusters for n up to 8. The calculations were performed using the density functional theory approach with the B3-LYP exchange-correlation functional [7] and the 6-31++G(d) basis set. The calculated infrared spectra of the $(PON)_8$ and $(SiO_2)_8$ clusters are shown in fig. 2. It can be seen that in agreement with experiments, there is a general blue shift of the vibrational frequencies in PON and that the major spectral differences are observed in the high frequency region where a splitting of the stretching modes absorption occurs. The normal vibrations calculation in the model clusters also shows that the spectrum can be divided in three regions corresponding, respectively, to bending vibrations (0 – 500 cm^{-1}), mixed bending – stretching vibrations (500 – 800 cm^{-1}) and stretching vibrations (above 800 cm^{-1}).

The micro Raman spectra of the α-quartz and of the moganite type PON are shown in fig. 3 where they are compared with the corresponding spectra of silica [8]. Again the blue shift of the vibrational frequencies in PON is evident. Also the Raman bands show a broadening due the intrinsic disorder in the PON structure. The Raman spectrum is a quite significant structural probe in these systems. It has been reported [8] that the Raman modes frequencies in the 400-550 cm^{-1} region can be correlated with the number of corner sharing tetrahedra forming rings in the silicate structures. In particular, it has been found that four membered tetrahedra rings (like in moganite) give rise to Raman bands at frequencies higher than 500 cm^{-1} while six membered (or larger) rings (like in quartz) are associated with Raman bands below 500 cm^{-1} . It can be seen from fig. 3 that the relative band positions in the α-quartz and in the moganite type structures follow tha same pattern in PON and in silica.

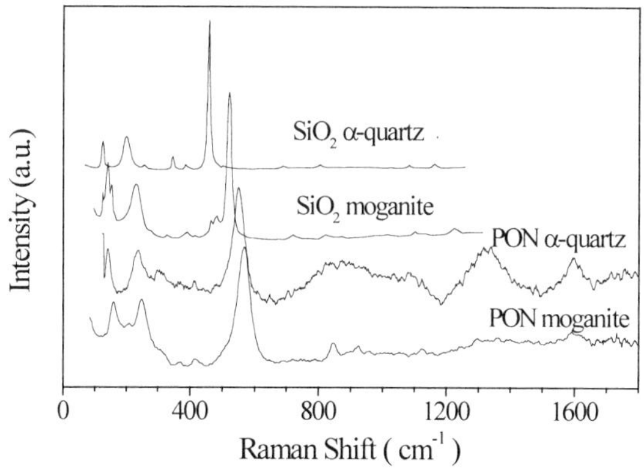

Fig. 3 – Raman spectrum of moganite and α-quartz PON

A preliminary ab initio molecular dynamics simulation has been carried using the Car-Parrinello approach [9]. The volume of the simulation cell was kept constant at the experimental value of the moganite type PON. The cell contained 72 atoms ($P_{24}O_{24}N_{24}$). The weighted density of states of PON obtained from the ab intio molecular dynamics simulation is compared with the infrared spectrum in Fig. 4. It can be seen that the gross features of the experimental spectrum are reasonably reproduced in the simulation.

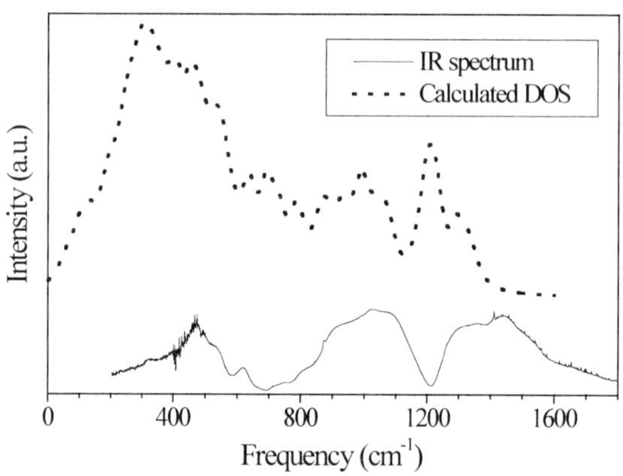

Fig. 4 – Infrared spectrum and density of states of PON

3.2. High pressure experiments

The micro Raman spectrum of the moganite type PON as a function of pressure is shown in fig. 5. The spectral changes are completely reversible in compression and

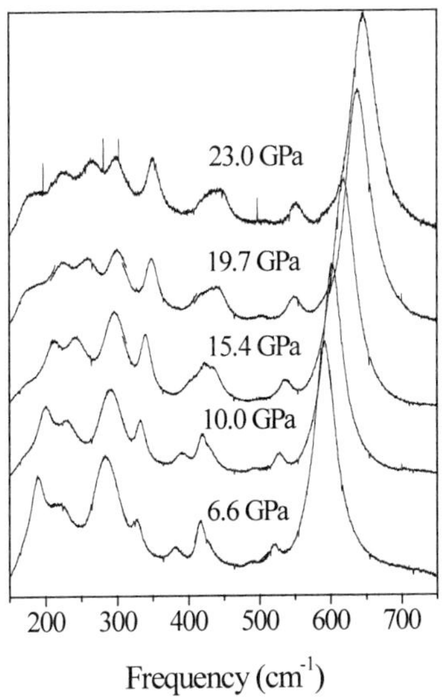

Fig 5 – Raman spectrum of moganite PON as a function of pressure

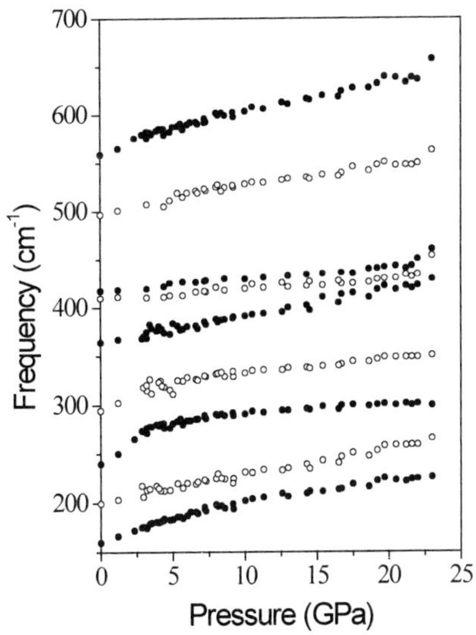

Fig 6 – Raman frequencies of moganite PON as a function of pressure

decompression of the sample. The Raman bands of Fig 5 have been deconvoluted and fitted with pseudoVoigt lineshapes and the resulting peak positions as a function of pressure are reported in fig 6. It can be seen that the frequency variation as a function of pressure changes appreciably from mode to mode. It can also be seen that an abrupt change of the slope is observed at a pressure of ~ 4.5 Gpa. This is particularly evident for some of the modes. The change in slope is an evidence that a phase transition may occur at this pressure in the moganite type PON. The relative Raman intensities of the deconvoluted peaks have been found to change appreciably with pressure. The results for some of the Raman bands are shown in fig. 7 where it can be seen that positive or negative slopes are found to occur for different bands. A change of slope seems to occur for some of the bands at ~ 4.5 Gpa, although in the present case the results are not very accurate.

In a high pressure X-ray diffraction experiment [4] it has been found that the diagram of the cell parameters of the moganite type PON exhibits a change in slope at ~ 8 GPa . This is also an evidence for the occurrence of a phase transformation in the crystal. The pressure of the phase transformations observed in the Raman and in the X-ray experimerimerimeriments are
the same, a possible explanation for the difference in the transition pressure could arise from the different pressure transmitting medium used in the Raman and X-ray measurements.

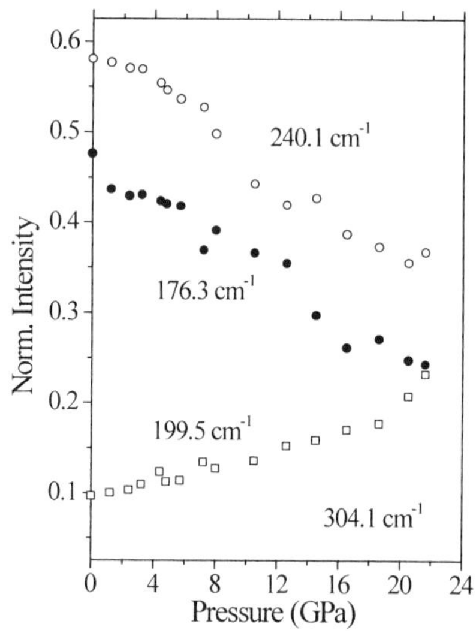

Fig. 7 - Raman intensities of moganite PON as a function of pressure

Aknowledgements.

This work was supported by the Italian Ministry for the University (MURST) and by the Italian Consiglio Nazionale delle Ricerche (Progetto Finalizzato Materiali Speciali). The authors wish to thank Julian Haine for kindly supplying the samples of PON and for helpful discussions.

REFERENCES

1. S.Boukbir, L. Marchand, Y. Laurent, P.Bacher and G. Roult, Ann. Chim. Fr. 14 (1989) 475
2. J.M. Léger, J.Haines, L.S. de Oliveira, C.Chateau, A. Le Sauze and R. Marchand, J. Phys. : Condens. Matter, 8 (1996) L773
3. J.M. Léger, J.Haines, L.S. de Oliveira, C.Chateau, A. Le Sauze, R. Marchand and S. Hull, J. Phys. Chem. Solids, 60 (1999) 145
4. J.M. Léger, J.Haines, C.Chateau and R. Marchand, J. Phys. Chem. Solids, 61 (2000) 1447
5. H.K. Mao, P.M. Bell, J.V. Shaner and D.J. Steinberg, J. Appl. Phys., 49 (1978) 3276
6. Q. Williams, R.J. Hemley, M.B. Kruger and R. Jeanloz, J. Geophys. Res., 98 (1993) 22,157
7. A.D. Becke, J. Chem. Phys., 98 (1993) 98
   C. Lee, W. Yang and R.G. Parr, Phys. Rev. B 37 (1988) 785
8. K.J. Kingma and R.J. Hemley, Am. Miner., 79 (1994) 269
9. R. Car and M. Parrinello, Phys. Rev. Lett., 55 (1985) 2471

Studies in Surface Science and Catalysis 140
A. Gamba, C. Colella and S. Coluccia (Editors)

Fixed-bed ion-exchange process performance of Pb^{2+} removal from a simulated ceramic wastewater by Neapolitan yellow tuff

D. Caputo[a], B. de Gennaro[a], B. Liguori[a], M. Pansini[b], C. Colella[a]

[a] Dipartimento di Ingegneria dei Materiali e della Produzione, Università Federico II, Piazzale V. Tecchio 80, 80125 Napoli, Italy

[b] Dipartimento di Meccanica, Strutture, Ambiente e Territorio, Università di Cassino, Via G. Di Biasio 43, 03043 Cassino (Fr), Italy

A phillipsite-rich Neapolitan yellow tuff bed, pre-exchanged in Na^+ form, was used to remove Pb^{2+} from a simulated wastewater of a ceramic manufacture through cation-exchange in dynamic conditions. The composition of the simulated wastewater was the following: Pb^{2+}, 22 mg/l; Na^+, $1{\cdot}10^{-3}$ M; K^+, $1{\cdot}10^{-4}$ M; Ca^{2+}, $1.5{\cdot}10^{-3}$ M; Mg^{2+}, $1{\cdot}10^{-3}$ M.

A complete Pb^{2+} breakthrough curve, obtained flowing the fixed bed with the simulated wastewater up to exhaustion, showed that the interfering cations present in the water did not result in a substantial reduction of Pb^{2+} exchange capacity of the tuff, which means that Na-phillipsite shows a very strong selectivity for Pb^{2+} in the reported experimental conditions.

The optimisation of the regeneration step suggested to flow the bed with about 12 bed volumes of regenerant solution (1 M $NaNO_3$) which allowed to remove about 53% of Pb^{2+} present in the bed. Such partial regeneration allowed to keep outlet Pb^{2+} concentration below the limit allowed by law for about 1700 bed volumes of eluate, thus giving rise to a yield of the cation-exchange operations, i.e., the ratio between the volumes collected in the ion-exchange step and in regeneration step, respectively, of about 140.

1. INTRODUCTION

Pb release in the environment is a very serious environmental problem on account of its toxic effects on living organisms, either animal or vegetal [1]. Pb removal from water is usually performed by precipitation as a hydroxide by direct addition of lime or soda to the waste [2]. This procedure gives rise to a series of problems such as the following:

(a) possible failure to attain the concentration limit allowed by law (0.2 mg/l, Law No. 152, G.U. - Official Gazette - No. 124, 29th of May 1999);
(b) long settling and/or filtration times of the gelatinous particles of $Pb(OH)_2$, even in the presence of proper coagulants and flocculants;
(c) difficulty in disposal of the resulting sludge;
(d) increased alkalinity of the wastewater.

Ion-exchange is a valid alternative to procedures based on precipitation inasmuch as it allows Pb recovery. Moreover it gives rise, as a rule, to stoichiometric reactions which is very useful when the Pb concentration in the wastewater is not constant over time.

The use of natural zeolites of sedimentary origin (namely zeolitized tuffs) as cation-exchangers in the treatment of waters polluted by Pb has been recommended on the basis of previous studies [e.g., 3,4]. Actually natural zeolites display in general cation selectivity sequences [5-7] which may turn out favourable in selective Pb removal, also in the presence of interfering cations. Moreover the large availability of sedimentary zeolites in many countries of the world [8] contributes to lower the cost of the process.

In a previous paper [9] Na-phillipsite was reported to display a great affinity for Pb and the exchange of Pb for Na on the same zeolite was found to be moderately fast. Equilibrium constant, free energy of exchange and rate constant were determined to be 22.8, -3.9 kJ/equiv and $1.3 \cdot 10^{-2}$ l/mequiv·min, respectively. These positive features gave rise to very favourable performances in Pb removal from water by using a fixed bed of granular phillipsite tuff even in the presence of a considerable amount of Na as interfering cation (500-1000 mg/l) [9]. Such presence, in fact, resulted only in a moderate reduction of Pb cation-exchange capacity and of both dynamic selectivity and process efficiency [9].

Pb is found in wastewaters of various industries, especially those manufacturing storage batteries and antiknock compounds [2]. In addition, it is present in some process waters of the ceramic manufacture, where its concentration is at level of a few tens mg/l in a typical interfering cation matrix of a tap water [10,11].

In view of a possible utilization of Na-exchanged phillipsite-rich tuff in Pb removal from ceramic manufacture wastewaters, this paper aims at optimising this process. Such an optimisation is carried out by subjecting a phillipsite-rich Neapolitan yellow tuff bed to several service-regeneration cycles, in which different volumes of regenerant solution are used. Tuff selected belongs to the huge formation of Neapolitan Yellow Tuff, which normally contains also subordinate amounts of chabazite [12], but the results can be considered generally valid, because of the good performances exhibited also by chabazite in the removal of the same cation [3].

2. EXPERIMENTAL

2.1 Materials and their characterisation

The phillipsite-rich tuff sample came from a quarry in Chiaiano (Naples, Italy), belonging to the huge formation of Neapolitan Yellow Tuff (NYT) spread over a vast area NW of Naples, called *Campi Flegrei* [12].

The X-ray diffraction analysis of the tuff (Philips PW 1730 diffractometer) showed the presence, among the crystalline phases, of phillipsite and smaller amounts of K-feldspar, augite and biotite. The chemical analyses of the tuff sample and the pure phillipsite obtained from the tuff by the usual enrichment techniques employed in mineralogy [13], are reported in Table 1. A suitable sample of this tuff was ground and screened, and the grain size fraction ranging from 0.25 to 0.71mm was used for preparing the fixed bed. The phillipsite content of the tuff grains, determined by the water vapour desorption technique [14] was 61%.

The cation-exchange capacity of the tuff was determined to be 2.01 mequiv/g. This value was calculated by multiplying the cation-exchange capacity of the pure phillipsite (3.30 mequiv/g), determined by the cross exchange method [15] and in good agreement with the value calculated from the composition reported in Table 1 (= 3.41 mequiv/g), by its phillipsite content.

Table 1
Chemical analysis of the phillipsite-rich tuff and phillipsite (wt. %)

	Tuff	Phillipsite
SiO_2	51.60	52.35
Al_2O_3	17.03	17.99
Fe_2O_3	2.55	
MgO	1.29	0.12
CaO	3.42	3.11
SrO	0.04	
BaO	0.09	
Na_2O	3.33	1.35
K_2O	7.23	8.51
H_2O	13.50[a]	16.50[a]

[a] Ignition loss

2.2 Ion-exchange runs

Dynamic runs were carried out with a fixed bed, prepared by allowing wet tuff grains of NYT to settle into a glass column, partially filled with water. The column was then rapped until bed depth showed no further reduction. Before starting the percolation of the solution simulating the wastewater of a ceramic manufacture, the bed was exhaustively Na^+-exchanged by passing through a 1M $NaNO_3$ solution until the concentration of the most strongly bound cation (K^+) in the effluent was below 1 mg/l. The exchanging solution was prepared by dissolving Carlo Erba reagent grade $NaNO_3$ in distilled water.

The model wastewater simulating the wastewater of a ceramic manufacture was prepared by dissolving in distilled water the relevant Carlo Erba reagent grade nitrates so as to obtain the following cation concentrations: Pb^{2+}, 22 mg/l ; Na^+, $1{\cdot}10^{-3}$ M; K^+, $1{\cdot}10^{-4}$ M; Ca^{2+}, $1.5{\cdot}10^{-3}$ M; Mg^{2+}, $1{\cdot}10^{-3}$ M. This solution was percolated through the bed at a constant flow rate, regulated by a peristaltic pump (Pharmacia Mod. P-3). The Pb concentration in the effluent solutions, collected with a fraction collector (Pharmacia Mod. PF-3), was determined by AAS using a Perkin-Elmer AA100 spectrophotometer. Table 2 summarises the experimental conditions for the column runs.

Table 2
Column operating conditions

Equipment specification		Service conditions	
Column diameter	1.43 cm	Feed composition (Pb)	22 mg/l
Bed depth	37 cm	Feed flow rate	0.420 l/h
Weight of tuff	48 g	Holdup time	8.5 min

In the first cation-exchange run the NYT bed was eluted up to exhaustion to obtain the complete S-shaped Pb breakthrough curve. The breakthrough curve was processed to calculate the cation-exchange capacity (CEC) and the dynamic parameters. The cation-exchange capacity for Pb^{2+} (Pb-CEC) was estimated through a graphical integration of the area delimited by the two coordinate axes, the horizontal line denoting the value of the original Pb^{2+} concentration and the S-shaped curve.

After this run the bed was subjected to alternate service-regeneration cycles. During the service step the bed was eluted with the simulated ceramic manufacture wastewater, whose composition has been reported above, up to the breakthrough point. This point was considered to have been attained when outlet Pb concentration steadily exceeded 0.2 mg/l, which is Pb concentration allowed in wastewaters by the Italian law (Law No. 152, G.U. - Official Gazette - No. 124, 29th of May 1999). The regeneration step was performed by percolating downflow the bed with prefixed volumes of 1 M $NaNO_3$ with a flow rate of 0.42 l/h. The volumes of regenerant solution percolated through the bed were: 5.50, 1.40 and 0.70 l, which corresponded to about 92, 24 and 12 bed volumes (BV), respectively, being 1 BV equal to about 60 ml.

3. RESULTS AND DISCUSSION

The breakthrough curve, obtained by eluting the fixed bed of NYT with the simulated ceramic manufacture wastewater is reported in Fig. 1. The elongated shape of the breakthrough curve and the sluggish attainment of Pb^{2+} inlet concentration appear related to the high concentration of the interfering cations compared with that of Pb ($36 \cdot 10^{-4}$ mole/l and $1.1 \cdot 10^{-4}$ mole/l, respectively). In spite of this, the most striking feature of this run is that the bed, made of only 48 g of phillipsite-rich tuff in its Na form, is able to keep Pb^{2+} outlet concentration below the law limit (0.2 mg/l) up to about 320 l of eluate, which is more than 5300 BV. It must be said that some occasional Pb^{2+} leakage exceeding 0.2 mg/l was recorded before reaching the breakthrough point, likely because of the formation of preferential pathways through the bed.

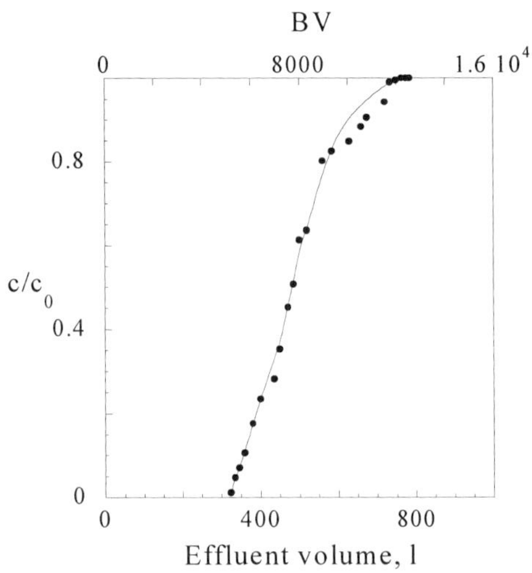

Fig. 1. Pb^{2+} breakthrough curve obtained by eluting the fixed bed of granular NYT with a simulated ceramic manufacture wastewater (composition in the text). Pb conc. in the feed, 22 mg/l; feed flow rate, 420 ml/h; c, actual Pb^{2+} conc. in the effluent; c_0, original Pb^{2+} conc. in the influent; BV= bed volume.

The Pb^{2+} breakthrough curve was subjected to the treatment reported by Micheals [16] and Colella and Nastro [17], in order to determine the length, L_z, of the mass transfer zone (MTZ), namely the length of the bed in which the cation-exchange reaction is supposed to occur. This parameter represents a sort of dynamic selectivity of Na-phillipsite for Pb^{2+} in that the shorter the MTZ, the higher the selectivity is. According to this theory the value of MTZ is given by the following expression [16,17]:

$$L_z = \frac{L_b(V_t - V_B)}{FV_t + (1-F)V_B}, \tag{1}$$

where L_b is the length of the bed, V_t the eluted volume at the exhaustion, V_B the breakthrough volume and F the symmetry index of the breakthrough curve, defined by the relation:

$$F = \frac{\int_{V_B}^{V_t}(c_0 - c)dV}{c_0(V_t - V_B)}, \tag{2}$$

in which c_0 and c represent the original and the actual Pb^{2+} concentrations, respectively. Given the uncertainty in evaluating the real values of V_B and V_t, fictitious values were considered, according to a consolidated practice, at $0.05c_0$ and $0.95c_0$ [16,17].

Table 3
Dynamic data from the lead breakthrough curve in Fig. 1

$[Pb^{2+}]$ mg/l	Interferent matrix	L_Z cm	Pb-CEC meq/g	WEC meq/g	S	E
22	*	32.4	2.06	1.33	~1.00	0.65

* Na^+, $1{\cdot}10^{-3}$ M; K^+, $1{\cdot}10^{-4}$ M; Ca^{2+}, $1.5{\cdot}10^{-3}$ M; Mg^{2+}, $1{\cdot}10^{-3}$ M.

L_Z value is reported in Table 3 together with other dynamic data calculated from the breakthrough curve in Fig. 1. In this Table Pb-CEC represents the cation-exchange capacity that Na-phillipsite displays for Pb^{2+} in the presence of the interfering cation matrix previously reported, whereas WEC represents the cation-exchange capacity for Pb^{2+} at the breakthrough point. The parameters S (selectivity) and E (efficiency) are calculated by dividing Pb-CEC and WEC, respectively, by the total cation-exchange capacity of NYT (CEC = 2.01 mequiv/g).

L_Z value is shorter than L_b which is a proof of the reliability of column exchange operations [16,17]. Its reasonably low value even in the presence of the reported interfering cation matrix confirms the previously ascertained good selectivity of Na-phillipsite for Pb^{2+} [9].

The Pb-CEC value reported in Table 3 practically coincides, within the limits of accuracy of the experimental determinations, with the total cation-exchange capacity, which means that Na-phillipsite practically exchanges only Pb^{2+} even in the presence of the interfering cation matrix previously reported ($S \approx 1$). Also the value calculated for the parameter E (0.65) appears sufficiently high, as it allows about 65% of CEC to be exploited for Pb^{2+} exchange up to the breakthrough point in the presence of the reported cation interfering matrix.

Figure 2 reports the regeneration curve obtained by partially eluting the exhausted NYT bed with the regenerant solution. The amount of Pb^{2+} released was calculated by multiplying the Pb^{2+} concentration of each collected sample by its volume. The regeneration was deliberately stopped when about 70% Pb^{2+} stored in the bed was released, which occurred after it was flowed with about 5.50 l of regenerant solution.

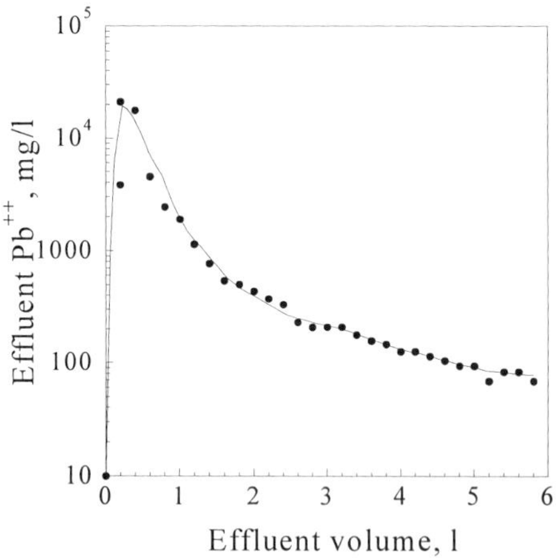

Fig. 2. Regeneration curve of the exhausted NYT fixed-bed with a 1 M $NaNO_3$ regenerant solution. Flow rate, 0.42 l/h.

In the next service step the bed was flowed again with the simulated ceramic manufacture wastewater until the breakthrough point was reached and the breakthrough volume (V_B) recorded. The above service-regeneration cycle was repeated six times. The results of these runs, as well as the results of two other groups of service-regeneration cycles, carried out similarly as the previous cycle in the former step, but different in that the bed was regenerated with only 1.40 or 0.70 l of solution, are summarised in Table 4.

Table 4
Pb^{2+} breakthrough volumes (V_B) after partial regenerations

Group of cycles performed with 5.50 l of regenerant volume (group A)

Cycle	1A	2A	3 A	*	4 A	5 A	*	6 A
V_B, l	240	184	142		249	170		223

Group of cycles performed with 1.40 l of regenerant volume (group B)

Cycle	1B	*	2B	*	3B
V_B, l	126		121		124

Group of cycles performed with 0.70 l of regenerant volume (group C)

Cycle	1C	*	2C	*	3C
V_B, l	94		101		100

*After the regeneration step of previous service-regeneration cycle, distilled water was pumped upflow for about one hour at 0.20 l/h feed flow rate.

As far as the group of cycles A is concerned (see Table 4), it can be observed that Pb^{2+} breakthrough volume dramatically decreases in the second and the third cycle with respect to the first cycle. Two explanations appear, at least in principle, possible:

(a) the increasing reduction of Pb-CEC recorded in the second and the third cycle might be related to the progressive storage of K^+ in the phillipsite-rich tuff bed. Actually phillipsite is known to display a great affinity for this cation [18].

(b) during the service-regeneration cycles, the flow of the liquid stream through the NYT bed results in head losses which make it more compact. Moreover, Neapolitan yellow tuff is reported to be moderately friable [4] and, thus, such head losses give even rise to a progressive production of fines which clog a part of the paths through the grains in which water flows. These phenomena, usually known as *channelling*, give rise to the creation of preferential paths of flow, which exclude considerable portions of the bed from exchange reactions and considerably decrease the Pb-CEC. Head losses are also known to increase as time goes by, and, thus, the recorded drawbacks created by channelling are supposed to become more and more serious.

The fact that the calculated value of selectivity S practically corresponds to unity (Table 3) makes unlike the hypothesis (a) of the progressive storage of K^+ in the phillipsite-rich tuff bed and, thus, gives strong indications of the occurrence of channelling (hypothesis b). To verify this supposition, after the regeneration step of the third service-regeneration cycle (Table 4, group A) distilled water was pumped upflow through the bed for about one hour at 0.20 l/h feed flow rate. This treatment, which is indicated in Table 4 by an asterisk, results in strongly tackling the detrimental effects of channelling and in bringing back the bed to the original efficiency (V_B = 249 l in the fourth cycle, compared to 240 l in the first cycle). This finding is further confirmed by the data of following two cycles in Table 4 (group A): when the treatment of pumping upflow distilled water through the bed was not performed (between the 4th and 5th cycles) V_B decreased to 170 l, whereas, repeating this treatment (stage between the 5th and 6th cycles) V_B increased again to 223 l.

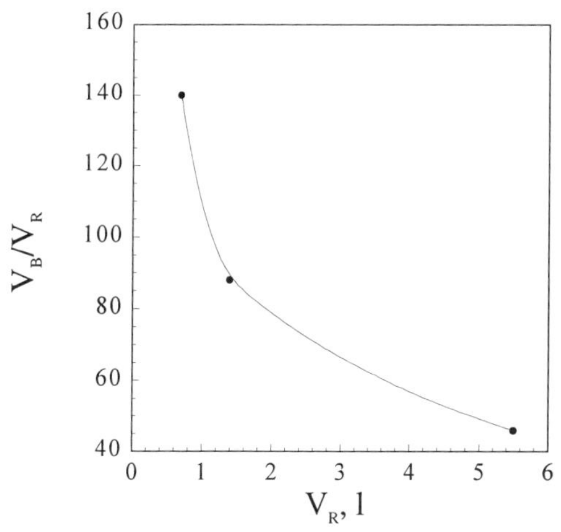

Fig. 3. Yield of the cation-exchange process (V_B = breakthrough volume; V_R = regenerant volume) as a function of the regenerant volume.

Table 4 reports also the results of two other group of service-regeneration cycles in which the regenerant volumes (V_R) were 1.40 and 0.70 l and the above described treatment aiming at tackling the detrimental effects of channelling was performed after the regeneration step of every cycle (group of cycles B and C, respectively). The selected volumes of the regenerant solution, which enabled to remove from the bed about 60 and 53% Pb^{2+} originally stored in the bed, respectively, were chosen inasmuch as they represent typical regenerant volumes used in the industrial processes. The V_B values collected, which demonstrate a substantial constancy over the various cycles, obviously decrease with decreasing V_R.

The average V_B values for the three groups of service-regeneration cycles A-C, which were calculated leaving out the data affected by channelling, divided by the relevant V_R values, are plotted, as a function of V_R, in Fig. 3. The adimensional numbers in ordinates, that represent a sort of yield of the cation-exchange process appear to dramatically increase with decreasing V_R. The optimum volume of regenerant solution appears therefore to be 0.70 l, namely about 12 BV.

An attempt to further reduce V_R, in order to obtain a further increase in the yield of the cation-exchange process, was not successful, because of an increased frequency of the occasional Pb^{2+} leakage during the service step of the various cycles.

4. CONCLUSIONS

This study fully confirms the excellent selectivity of Na-phillipsite for Pb^{2+}, already ascertained previously [9]. Na-phillipsite is in fact able to exchange quantitatively Pb^{2+} even in the presence of an interfering cation matrix of a typical ceramic manufacture wastewater.

The optimisation of this process, performed in the present study, evidences that a sufficiently small regenerant volume, equal to only 12 BV, is able to remove more than 50% of the lead stored in the bed during the service step. Such a partial regeneration, however, allows the fixed bed of NYT to keep outlet Pb^{2+} concentration below the limit allowed by law (0.2 mg/l) for about 1700 BV of eluate, thus giving rise to a yield of the cation-exchange operation of about 140, which is a very satisfying performance.

It must be pointed out that the efficiency of this process is greatly decreased by channelling. Thus, if a more welded and more attrition-resistant sample of phillipsite-rich tuff is not available, a minimal amount of water, which does not affect at all the process performance, should be pumped upflow after the service step of each cycle in order to bring every time the bed to its original efficiency.

REFERENCES

1. E.A. Pfitzer, in Industrial Environmental Health, L.V. Cralley, L.J. Cralley, G.D. Clayton, J.A. Jugiel (eds.), Academic Press, New York, 1972, p. 84.
2. J.W. Patterson, Wastewater treatment technology, Ann Arbor Science Publishers, Ann Arbor, Michigan, 1975, p. 129.
3. C. Colella and M. Pansini, in Perspectives in Molecular Sieve Science, ACS Symp. Series No. 368, W.H. Flank and T.E. Whyte, Jr., (eds.), Amer. Chem. Soc., Washington, D.C., 1988, p. 500.
4. M. Pansini, Mineral. Deposita, 31 (1996) 563.
5. G. Eisenmann, Biophys. J., 2 (1962) 259.
6. H.S. Sherry, in Ion Exchange. A Series of Advances, J. A. Marinsky (ed.), Marcel Dekker, New York, 1969, Vol. 2, p. 89.

7. C. Colella, Mineral. Deposita, 31 (1996) 554.
8. C. Colella, in Natural Microporous Materials in Environmental Technology, P. Misaelides, F. Macasek, T.J. Pinnavaia and C. Colella (eds.), NATO Sciences Series, Series E: Applied Sciences, Vol. 362, Kluwer A.P., Dordrecht, The Netherlands, 1999, p. 207.
9. M. Pansini, C. Colella, D. Caputo, M. de' Gennaro and A. Langella, Microporous Mater., 5 (1996) 357.
10. F. Mazzocco, Inquinamento, 19(9) (1977) 129.
11. I. Mariani and L. Tenti, La Ceramica, 26(7-8) (1973) 22.
12. M. de' Gennaro and A. Langella, Mineral. Deposita, 31 (1996) 452.
13. M. de' Gennaro and E. Franco, L'Industria Mineraria, 30 (1979) 329.
14. M. de' Gennaro and C. Colella, Thermochim. Acta, 154 (1989) 345.
15. C. Colella, M. de' Gennaro, E. Franco and R. Aiello, Rend. Soc. Ital. Min. Petr., 38 (1982-83) 1423.
16. A.S. Michaels, Ind. Eng. Chem., 44 (1952) 1922.
17. A. Nastro and C. Colella, Ing. Chim. Ital., 19(5-6) (1983) 41.
18. R.M. Barrer and B.M. Munday, J. Chem. Soc. A, (1971) 2904.

Studies in Surface Science and Catalysis 140
A. Gamba, C. Colella and S. Coluccia (Editors)

Adsorption properties of clinoptilolite-rich tuff from Thrace, NE Greece

D. Caputo[a], B. de Gennaro[a], B. Liguori[a], M. Pansini[b] and C. Colella[a]

[a] Dipartimento di Ingegneria dei Materiali e della Produzione, Università Federico II, Piazzale V. Tecchio 80, 80125 Napoli, Italy

[b] Dipartimento di Meccanica, Strutture, Ambiente e Territorio, Università di Cassino, Via G. Di Biasio 43, 03043 Cassino (FR), Italy

Clinoptilolite tuff from Pentalofos (Greece) was mineralogically and chemically characterized and its adsorption properties towards H_2O, CO_2, SO_2 and NH_3 were investigated. Adsorption isotherms at 25°C were determined and analyzed with Langmuir, Freundlich and Dubinin-Astakhov models. The Langmuir equation was found to provide excellent fits in the case of sulphur dioxide and ammonia and a good fit in the case of carbon dioxide. Dubinin-Astakhov equation, based on a pore filling-type mechanism, appears to be a suitable model for water adsorption, which is characterized by condensation phenomena close to saturation pressure. The high affinity of clinoptilolite for strongly polar molecules is confirmed. The use of Pentalofos clinoptilolite-rich tuff in separation of gaseous mixtures of polar and non polar substances is envisaged and evaluated.

1. INTRODUCTION

The use of naturally occurring zeolite-bearing materials in separation processes fits into a wiser management of natural resources and in the increasing demand of environmentally friendly technologies. Moreover these naturally-occurring materials are usually very cheap on account of the occurrence of huge outcrops of zeolite-rich deposits in many countries of the world (Hawkins, 1984; Colella, 1999).

A long term study on the use of phillipsite and chabazite bearing volcanic tuff coming from the huge outcrops of central-southern Italy in separation processes was undertaken in this laboratory. Interesting results were obtained as far as the separation of gaseous water-ethanol mixtures (Colella *et al.*, 1991; 1994) and selective removal of heavy metal cations from water by ion exchange (e.g., Colella and Pansini, 1988; Colella *et al.*, 1995; 1998; Pansini *et al.*, 1996) are concerned.

In the Orestia basin of Evros county (Thrace, north-eastern Greece), near the Pentalofos village, a clinoptilolite-rich tuff deposit was found to occur (Kassoli-Fournaraki *et al.,* 2000). The field observations and detailed investigations of the collected samples revealed the deposit to have a large economic potential. Actually, clinoptilolite grade was found to reach values up to 90% and reserves of the rock were evaluated to be as high as $5 \cdot 10^9$ kg.

Considering that clinoptilolite is a HEU-type zeolite, characterized by a two-dimensional pore system with three different channels having pore openings of 2.6×4.7, 3.0×7.6 and 3.3×4.6 Å, respectively (Baerlocher *et al.*, 2001), it was thought worthwhile to promote a study aiming at evaluating the technological potential of this material in application connected to adsorption. Accordingly, this paper reports the adsorption characterization of the above material at 25.0 °C towards four gaseous substances displaying different shape and polarity, i.e., H_2O, CO_2, SO_2 and NH_3.

2. MODELING OF THE ADSORPTION ISOTHERMS

The adsorption isotherms were analyzed using three different models, which have been previously described in detail (Caputo *et al.*, 1999).

In the model proposed by Langmuir the adsorbate mass percent on dry adsorbent base, Q, is related to the partial pressure P of the adsorbate, according to the following equation:

$$\frac{Q}{Q_m} = \frac{kP}{1+kP}, \tag{1}$$

where Q_m and k are two constants. Q_m represents the adsorbate mass percent, referred to dry adsorbent, necessary to cover the whole surface of the adsorbent with a monomolecular layer and $k=k_a/k_d$ is the adsorption equilibrium constant, namely the ratio between the kinetic constants of adsorption (k_a) and desorption (k_d) processes. The Q/Q_m ratio is usually defined as coverage θ, i.e. the fraction of the adsorbent surface covered by the monolayer of adsorbate molecules. The above equation may be written in the following linear form:

$$\frac{P}{Q} = \frac{1}{kQ_m} + \frac{P}{Q_m}. \tag{2}$$

Plotting P/Q against P gives a straight line, whose slope represents the reciprocal of Q_m. It is also possible to compute k value from the intercept on the ordinate axis.

The Freundlich equation is an empirical relationship which has been found best suited for correlating experimental values of Q and P if a multilayer of adsorbed molecules is considered:

$$Q = \beta P^{1/n}, \tag{3}$$

which in the linearized form becomes:

$$\log Q = \log\beta + \frac{1}{n}\log P, \tag{4}$$

where β and n are specific constants for the system adsorbate-adsorbent.

Langmuir and Freundlich models assume an adsorption mechanism of surface coverage-type. A useful model for correlating experimental data with adsorption phenomena based on a pore filling-type mechanism, is the Dubinin-Astakhov equation:

$$\frac{Q}{Q_0} = \exp\left[-\left(\frac{RT}{E}\right)^n \log^n\left(P^0/P\right)\right], \tag{5}$$

which can be linearized as:

$$\log Q = \log Q_0 - 2.303^{n-1}\left(\frac{RT}{E}\right)^n \log^n(P^0/P), \tag{6}$$

where Q_0 is the adsorbate mass percent on dry adsorbent base at saturation pressure P^0, T the absolute temperature, R the gas constant, E the free energy of adsorption and n a small integer (frequently equal to 1 or 2). If the temperature of the adsorption isotherm is well below the critical point, i.e., near the boiling point of the adsorbate, dividing Q_0 by density of the liquid condensed in the pores of the adsorbent, the total volume of the microporous system can be evaluated (Gregg and Sing, 1982).

3. EXPERIMENTAL

3.1. Materials

The X-ray diffraction analysis of the Pentalofos tuff showed that the phase present to the greatest extent was clinoptilolite together with lower amounts of quartz, opal-C, K-feldspar, mica and smectite (A. Hall, Royal Holloway, University of London, 1997, private communication). Its clinoptilolite content was determined to be 62% by a properly modified water vapour desorption technique which had previously been set up for evaluating phillipsite and chabazite content in Italian tuffs (de' Gennaro and Colella, 1989).

The zeolite sample was ground by a ball mill to a grain size lower than 150 μm. After grinding it was dried at 80°C for 12 hours and then stored in conditions of constant humidity (R.H. 50%) before chemical and adsorption characterization.

The chemical analysis of Pentalofos tuff gave the following results (A. Hall, Royal Holloway, University of London, 1997, private communication): SiO_2 = 65.13% TiO_2 = 0.15%, Al_2O_3 = 11.57%, Fe_2O_3 = 1.00%, MgO = 1.11%, MnO = 0.02%, CaO = 3.43%, Na_2O = 0.60%, K_2O = 1.47%, P_2O_5 = 0.08%, H_2O = 15.50%, total = 100.06%.

3.2. Adsorption runs

Adsorption properties were investigated by a gravimetric technique using a McBain-type adsorption balance. A detailed description of the apparatus and methods utilized in the adsorption runs can be found in a previous paper (Caputo *et al.,* 1999).

Before adsorption runs the zeolite-bearing material was activated *in situ* heating it at 350°C for 4 hours under high vacuum ($P < 10^{-5}$ mbar). After cooling to 25.0 °C, the adsorbate gas was allowed into the balance and sufficient time (30 to 60 minutes, depending on the specific adsorbate) was awaited to attain equilibrium. Tuff samples were used only once, i.e., the adsorbent was renewed before any new thermal activation and adsorption run.

Water vapour, sulphur dioxide, carbon dioxide and ammonia were used to disclose the adsorption features of the tuff. Gases were supplied by SON (Società Ossigeno Napoli) with the following specifications: CO_2 was analytical grade (99.99%), whereas SO_2 and NH_3 contained nitrogen at percentages of 15.84% and 15.76%, respectively.

4. RESULTS AND DISCUSSION

In Figs. 1-4 the adsorption isotherms of H_2O, SO_2, NH_3 and CO_2 by Pentalofos tuff at 25.0 °C are presented. In these diagrams the adsorbate percent mass, referred to dry adsorbent, Q, is plotted as a function of partial pressure of the adsorbate, P.

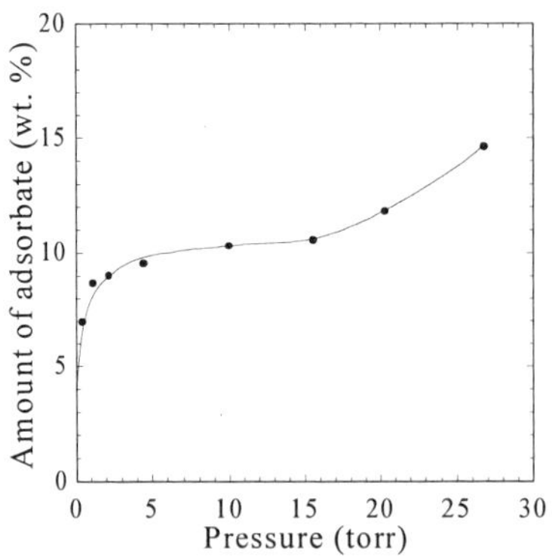

Fig. 1 – Adsorption isotherm of H_2O by Pentalofos tuff at 25 °C

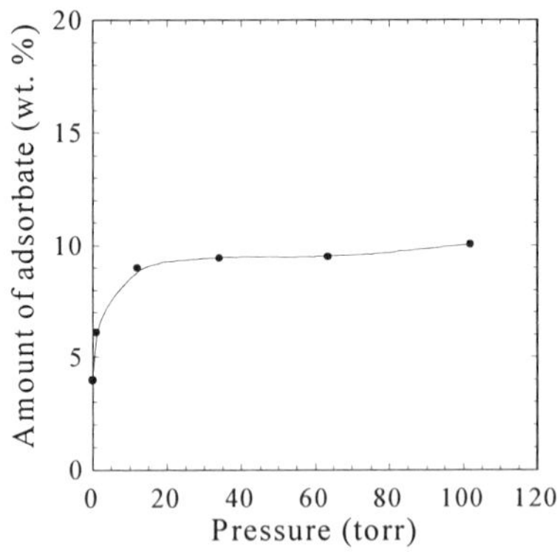

Fig. 2 – Adsorption isotherm of SO_2 by Pentalofos tuff at 25 °C

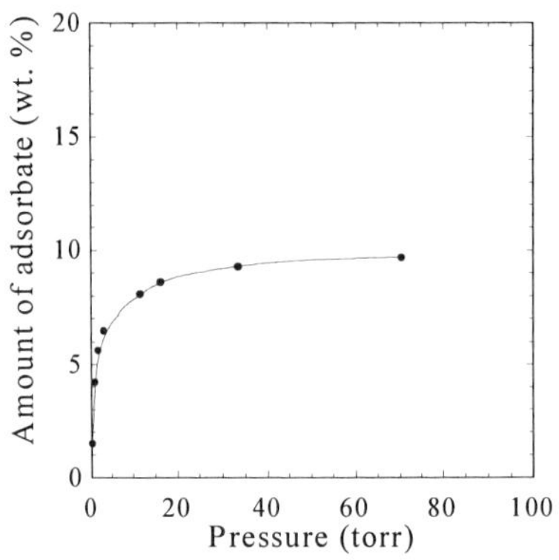

Fig. 3 – Adsorption isotherm of NH_3 by Pentalofos tuff at 25 °C

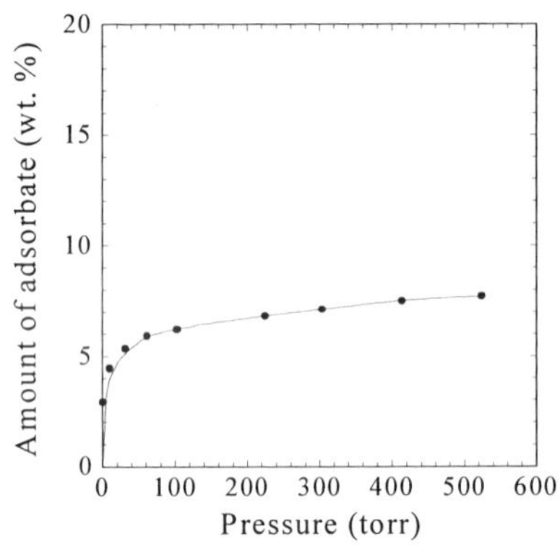

Fig. 4 – Adsorption isotherm of CO_2 by Pentalofos tuff at 25 °C

Adsorbate mass percent at every value of the partial pressure of the adsorbate was found to decrease according to the order $H_2O > SO_2 \approx NH_3 > CO_2$. The polarity of adsorbed molecules appears to play a crucial role in adsorption properties of Pentalofos clinoptilolite-rich tuff. The high permanent dipole moment displayed by H_2O, SO_2, NH_3 molecules, equal to 1.85, 1.63 and 1.47 debye, respectively (Weast *at al.,* 1985), results in their high affinity for the electrically charged adsorption sites present on the surface of clinoptilolite channels, thus giving rise to the typical rectangular shape of adsorption isotherms. In the final part of the curve concerning water vapor (Fig. 1, p >15 torr) a sudden rise of amount of adsorbate is recorded. This must be ascribed to the incipient capillary condensation of water (the saturation pressure of water vapor at 25°C is 23.76 torr) arising from the small dimensions of clinoptilolite channels (see Introduction) (Ruthven, 1984). The apolarity of CO_2 molecule results in a lower affinity denoted by the lower amount of adsorbate at every partial pressure and by the monotonously increasing trend displayed by the adsorption isotherm in the investigated range of partial pressures (up to 520 torr).

Adsorption equilibrium data of SO_2, NH_3 and CO_2 were analyzed by the linearized forms of the Langmuir and Freundlich equations (see Eqns. 2 and 4). The water vapour isotherm was analyzed by the linearized forms of the Dubinin-Astakhov equation (6), because of the above mentioned condensation phenomena.

Table 1 collects the equilibrium parameters computed through a linearization of the adsorption isotherms according to the above mentioned models.

Table 1
Model parameters and coefficients of r^2 to obtain straight lines

Parameters	H_2O	SO_2	NH_3	CO_2
Langmuir				
Q_m	13.7363	10.0806	9.8328	7.7459
k	0.5801	3.8450	0.6368	0.06476
r^2	0.8500	0.9994	0.9997	0.9985
Freundlich				
β	7.9488	7.7535	3.7966	3.0591
n	7.1378	9.5511	3.5689	6.6050
r^2	0.8723	0.9300	0.8513	0.9888
Dubinin-Astakhov				
Q_0	12.3112	-	-	-
n	1	-	-	-
r^2	0.9812	-	-	-

Comparison between model fittings and experimental equilibrium data of Figs. 1-4 are shown in Figs. 5-8, for H_2O, SO_2, NH_3 and CO_2, respectively.

Inspection of the figures and/or the values of the coefficient of r^2 to obtain a straight line (Table 1) shows that the Langmuir model is an excellent representation for the adsorption of SO_2 and NH_3 on Pentalofos tuff and a fairly good representation in the case of CO_2. In the latter case, however, the Freundlich model appears to be more suitable. On the contrary, as expected, in the fitting of the adsorption isotherm of water vapor best results were obtained using the Dubinin-Astakhov equation. In fact, condensation phenomena occurring in proximity of saturation pressure are best described by a pore filling mechanism (Gregg and Sing, 1982)

Concerning the validity of Q_m and k values computed by Langmuir equation, more careful considerations are required. Actually, the physical significance of the constants derived by matching experimental points to the model is strongly conditioned by the extension and distribution of the concentration range of the experimental points themselves.

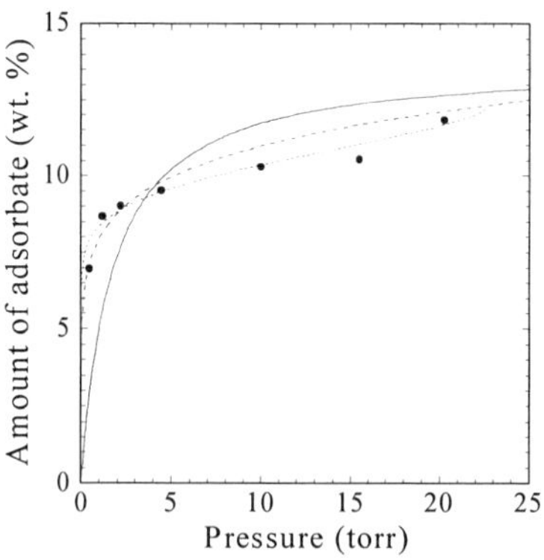

Fig. 5 – Comparison between Langmuir (—), Freundlich (--), Dubinin-Astakhov (...) model fittings and the experimental equilibrium data (points) of Fig. 1

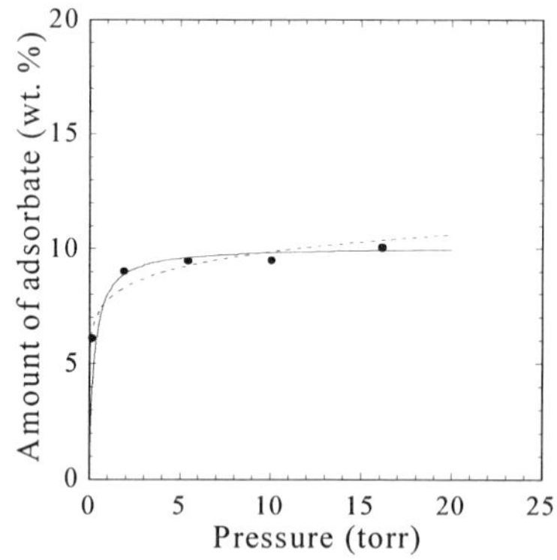

Fig. 6 – Comparison between Langmuir (—), Freundlich (---) model fittings and the experimental equilibrium data (points) of Fig. 2

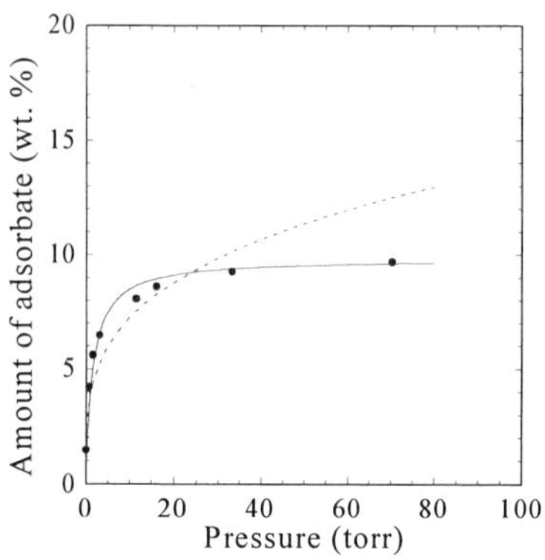

Fig. 7 – Comparison between Langmuir (—), Freundlich (--) model fittings and the experimental equilibrium data (points) of Fig. 3

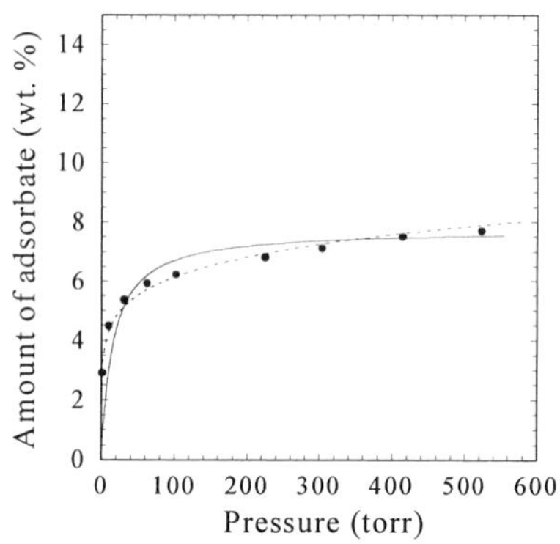

Fig. 8 – Comparison between Langmuir (—), Freundlich (--) model fittings and the experimental equilibrium data (points) of Fig. 4

Obviously Q_m value, which is the adsorbate percent mass when saturation of the adsorbent is attained, is essentially correct when the experimental data cover the high concentration region. As far as k is concerned it must be borne in mind that the Henry's law:

$$\lim_{P\to 0}\left(\frac{Q}{P}\right) = kQ_m = H, \tag{7}$$

where H is the Henry's constant, is valid at low sorbate concentration (Ruthven, 1984). Thus the computed k value is correct provided that the low concentration region is covered by a sufficiently high number of experimental points. Data hereby reported seem to allow a reliable computation only of Q_m values, whereas the computed k values do not appear to have physical significance. Q_m values decrease, in fact, in the same order $H_2O > SO_2 \approx NH_3 > CO_2$ as the adsorbate mass percent does, whereas no logical explanation can be found for k variation.

As the Freundlich equation is an empirical relationship, the parameters β and n are not connected to quantities of physical significance. On the contrary, from the parameter Q_0 of the Dubinin-Astakhov equation an estimation of the volume V_t of microporous system may be obtained. In fact, the adsorbed amount at the saturation pressure P^0, Q_0, can be converted in microporous volumes V_t, dividing it by the density of water at 25 °C (see Section 2). The obtained value of 0.116 cm^3/g of dry tuff is in reasonable agreement with analogous values (0.101-0.136 cm^3/g), estimated for clinoptilolite-rich rocks from Cuba (Roque-Malherbe *et al.,* 1995).

Adsorption properties of several naturally-occurring zeolite-bearing materials were previously investigated by Hayhurst (1980). He reported even data concerning the adsorption of SO_2, NH_3 and CO_2 by some clinoptilolites coming from sedimentary deposits in the United States. Comparing Hayhurst's data with our data, several discrepancies are observed. In particular:

i) Hayhurst reported that times ranging from 1 to 3 hours were needed to attain equilibrium, whereas in our case times not higher than 1 hour appeared sufficient for every adsorbate;
ii) Hayhurst found Freundlich model to fit experimental data better than Langmuir model did. However, the coefficients to obtain straight lines from Freundlich model were scattered over a wide range going from 0.86 to 1.00.

These discrepancies could be justified by the fact that clinoptilolite displays different thermal behaviours depending on its Si/(Si+Al) ratio and cation composition (Gottardi and Galli, 1985). Actually, these two authors found that:

(1) high aluminium and alkaline-earth metal content in zeolites of the clinoptilolite type resulted in the so-called type 1 thermal behaviour, i.e., clinoptilolite undergoes an irreversible phase transition in the temperature range 230-260°C which implies a 15% volume contraction of the elementary cell and a partial loss of its rehydration ability;
(2) high silicon and alkaline metal content in zeolites of the clinoptilolite type resulted in the so-called type 3 thermal behaviour, i.e., clinoptilolite undergoes continuous reversible dehydration with only a very small lattice contraction, lattice being undamaged up to 750°C.

The chemical composition of the clinoptilolite used in this study might result in a thermal behaviour of type 3 and, thus, the clinoptilolite should not have undergone any phase transition during activation which was performed at 350°C. Hayhurst did not report chemical analysis of the clinoptilolites used in his work. Those clinoptilolites might have suffered some structural damage upon heating, with consequent modification, at least to some extent, of their adsorption properties.

5. CONCLUSIONS

The data collected confirm the good affinity of Pentalofos clinoptilolite-rich tuff for strongly polar adsorbate molecules such as H_2O, SO_2 and NH_3, opening the way to its possible utilization to separate gaseous streams in which these substances are present together with other non polar molecules.

Langmuir model is an excellent fit for the adsorption of SO_2 and NH_3 on Pentalofos tuff and a fairly good representation in the case of CO_2. In the latter case, the Freundlich model appears, however, to be more suitable. On the contrary, as expected, in the fitting of the water vapour adsorption isotherm, best results are obtained using the Dubinin-Astakhov equation.

As the adsorption capacity of the tuff is related to its clinoptilolite content, a better performance of Pentalofos clinoptilolite-rich tuff as adsorbent should be obtained if samples bearing a higher clinoptilolite content were used.

6. ACKNOWLEDGMENTS

The authors wish to thank the European Union for the financial support which enabled to carry out this investigation (EC Contract No. BRE-CT94-0954).

REFERENCES

Baerlocher Ch., Meier W.M., and Olson D.H. (2001) Atlas of Zeolite Framework Types, Elsevier, Amsterdam, p. 146.

Caputo, D., de' Gennaro, B., Pansini, M. and Colella, C. (1999) In "Natural Microporous Materials in Environmental Technology", P. Misaelides, F. Macasek, T. Pinnavaia and C. Colella (eds.), NATO Science Series, Series E: Applied Sciences, Vol. 362, Kluwer Academic Publishers, Dordrecht, The Netherlands, p. 225.

Colella C. and Pansini M. (1988) In "Perspectives in Molecular Sieve Science", ACS Symposium Series 368, W.H. Flank and T.E. Whyte (eds.), Am. Chem. Soc., Washington, DC, p. 500.

Colella C., Gallifuoco A., Alfani F., Annetta A., Pansini M. and Cantarella M. (1991) In Proc. I Convegno Nazionale di Scienza e Tecnologia delle Zeoliti, C. Colella (ed.), De Frede, Napoli, Italy, p. 181.

Colella C., Pansini M., Alfani F., Cantarella M. and Gallifuoco A. (1994) Microporous Materials, 3, 219.

Colella C., de' Gennaro M., Langella A. and Pansini M. (1995) In "Natural Zeolites '93", D.W. Ming and F.A. Mumpton (eds.), I.C.N.Z., Brockport, New York, p. 377.

Colella C., de' Gennaro M., Langella A. and Pansini M. (1998) Separation Science and Technology, 33(4), p. 467.

Colella, C. (1999) In "Porous Materials in Environmental Friendly Processes", I. Kiricsi, G. Pál-Borbély, J.B. Nagy and H.G. Karge (eds.), Studies in Surface Science and Catalysis, Vol. 125, Elsevier, Amsterdam, p. 641.

de' Gennaro M. and Colella C. (1989) Thermochimica Acta, 154, p. 345.

Gottardi, G. and Galli, E. (1985) Natural Zeolites, Springer-Verlag, Berlin, pp. 267-270.

Gregg, S.J. and Sing, K.S.W. (1982) Adsorption, Surface Area and Porosity, Academic Press, London.

Hawkins, D.B. (1984) In "Zeoagriculture. Use of Natural Zeolites in Agriculture and Aquaculture", W.G. Pond and F.A. Mumpton (eds.), Westview Press, Boulder, Colorado, p. 69.

Hayhurst, D.T. (1980) Chem. Eng. Comm, 4, 729.

Kassoli-Fournaraki A., Stamatakis M., Hall A., Filippidis A., Michailidis K., Tsirambides A. and Koutles Th. (2000). In "Natural Zeolites for the Third Millennium", C. Colella and F.A. Mumpton (eds.), De Frede, Napoli, Italy, p. 193.

Pansini M., Colella C., Caputo D., de' Gennaro M. and Langella A. (1996) Microporous Materials, 5, 357.

Roque-Malherbe R., Lemes-Fernandez L., Lopez-Colado L., de las Pozas C. and Montes-Carballa A. (1995). In "Natural Zeolites '93", D.W. Ming and F.A. Mumpton (eds.), I.C.N.Z., Brockport, New York, p. 299.

Ruthven D. M. (1984) Principles of Adsorption and Adsorption Processes, Wiley and Sons, New York.

Weast, R.C., Astle, M.J. and Beyer, W.H. (1985) CRC Handbook of Chemistry and Physics, 66th edition, CRC Press Inc., Boca Raton, Florida, p. E-58.

Studies in Surface Science and Catalysis 140
A. Gamba, C. Colella and S. Coluccia (Editors)

Dissociative Adsorption of H_2 on Defect Sites of MgO: A Combined IR Spectroscopic and Quantum Chemical Study

Matteo Cavalleri,[a,c,*] Alexander Pelmenschikov,[a,c] Gabriele Morosi,[a] Aldo Gamba,[a] Salvatore Coluccia,[b] and Gianmario Martra[b]

[a]Dipartimento di Scienze Chimiche, Fisiche e Matematiche, Universita' dell' Insubria, Via Lucini 3, 22100 Como, Italy

[b]Dipartimento di Chimica IFM, Universita' di Torino, Via P. Giuria 7, 10125 Torino, Italy

[c]Department of Physics, Stockholm University, Box 6730, S-113 85 Stockholm, Sweden

The dissociative adsorption of H_2 on defect sites of MgO is studied by means of FTIR spectroscopy and ab-initio HF cluster modelling. IR data indicate irreversible and reversible heterolytic splittings of H_2. Calculations allow us to assign a band of the reversible species at 1325 cm^{-1} to the $Mg_{3C}H$ hydride group, and a band of the irreversible species at 3712 cm^{-1} to the $O_{3C}H$ hydroxyl group. The hydrogen atoms bridging two or three neighboring, low-coordinated Mg atoms of the surface are suggested to be responsible for the complex bands in the 1130-880 cm^{-1} region. No IR band corresponds to the calculated frequencies of the H atoms bound to the single 4- and 5-coordinated Mg and O atoms, suggesting that the stabilization of H species at these sites is unfavorable compared to the 3-coordinated sites. The calculated adsorption energies confirm that dissociative adsorption can occur only when 3-coordinated cations and/or anions are involved. The ionic bond $(Mg)H^{-}\cdot\cdot H^{+}(O)$ between the formed H species strongly contributes to the exothermic effect of these reactions.

1. INTRODUCTION

The chemical reactivity of MgO is usually related to surface defects (corners, edges, steps, etc.) with low-coordinated Mg and O sites [1-3]. These sites have raised significant interest since they can play an important role in many catalytic processes on MgO. The hydrogen molecule is considered to be a very efficient probe for IR spectroscopic characterization of these sites. H_2 molecules dissociate at the defect sites of MgO producing several surface species whose structure cannot be unuambiguously determined only on the basis of their spectroscopic features.

In this work we consider some preliminary results obtained by coupling the IR spectroscopic analysis and ab initio HF calculations. The defect sites are modeled by bare stoichiometric clusters of MgO. This approach has been successfully used for quantitative description of frequency shifts of CO and CH_3CN probe molecules coordinated at the defect sites of MgO, silica, and zeolites [2,4,5].

2. EXPERIMENT

MgO ex-hydroxide (MgO-h) was prepared by thermal decomposition of the parent $Mg(OH)_2$ under vacuum conditions directly inside the IR chamber. The hydroxide was slowly decomposed *in vacuo* at *ca.* 523 K and finally outgassed at 1123 K. This procedure gives MgO with high specific surface area (SSA_{BET}= 200 $m^2 \cdot g^{-1}$) which is assumed to be completely dehydroxylated, as no OH stretching vibration bands were observed in the background IR spectrum. IR spectra of the adsorption of H_2 at room temperature were obtained by a Bruker IFS 48 instrument; the resolution was 4 cm^{-1}. The IR chamber, linked to a vacuum pump, allowed both the thermal pretreatment and the adsorption-desorption experiments to be performed "in situ". The spectra are reported in absorbance, the background spectrum of the MgO sample before H_2 absorption being subtracted.

3. CALCULATIONS

Calculations were performed with the GAUSSIAN-94 package [6]. In order to choose a suitable basis set, we checked the effect of the basis functions on the computed O-H and Mg-H stretching frequencies of HMgOH (**1**) and $HO(MgH)_2OH$ (**2**) molecules (Figure 1). When using an adequate basis set, the HF method should correctly describe the difference in frequency of a chemical group between closely related compounds [7]. Therefore the convergence of the differences $\Delta\nu_{Mg\text{-}H} = \nu_{Mg\text{-}H}[HMgOH]-\nu_{Mg\text{-}H}[HO(MgH)_2OH]$ and $\Delta\nu_{O\text{-}H} = \nu_{O\text{-}H}[HMgOH]-\nu_{O\text{-}H}[HO(MgH)_2OH]$ was considered as an index of adequacy for the basis set. The results (Table 1) show that the convergence is reached at the 6-31G(d) level, the inclusion of d-functions being indispensable for the convergence. Therefore in all the calculations we used a combined basis set: the 6-31G(d) basis set for the H atoms, the Mg and O atoms involved in the adsorption, and the O atoms of the first coordination sphere of these Mg atoms and the 3-21G basis set for the rest of the models.

Table 1
Frequency differences (see text)

Basis set	$\Delta\nu_{Mg\text{-}H}$ (cm^{-1})	$\Delta\nu_{O\text{-}H}$ (cm^{-1})
3-21G	116	63
6-31G	131	63
6-31G(d)	100	59
6-31G(dp)	110	57
6-311G(d)	96	52

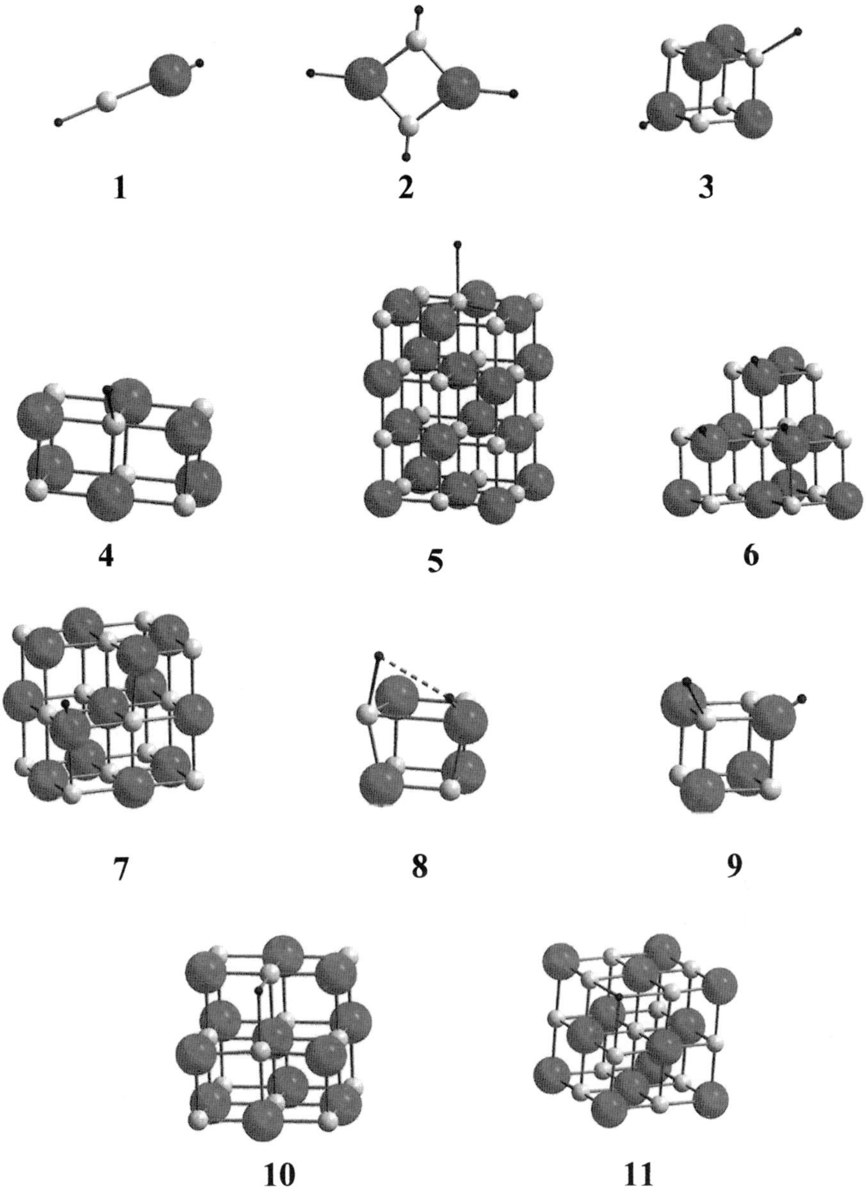

Fig.1. Cluster models

Table 2 shows the stretching frequencies of the O_{3c}-H and Mg_{3c}-H groups calculated with the Mg_4O_4 (**3**), $Mg_{10}O_{10}$ (**6**), and $Mg_{13}O_{13}$ (**7**) cluster models.

Table 2

Calculated frequencies against cluster size.

Cluster type	ν_{O-H}	ν_{Mg-H}
Mg_4O_4 /Combined basis set	3713	1326
Mg_4O_4 /6-31G (d)	3716	1336
$Mg_{10}O_{10}$ /Combined basis set	3699	1335
$Mg_{13}O_{13}$ /Combined basis set	3718	

The frequencies insignificantly change upon increase of the model size. On the basis of these results in the following considerations the effect of the cluster size on the predicted frequencies can be neglected to an accuracy of about 10 cm^{-1}.

For all models only the positions of the adsorbed H atoms and of the O and Mg surface atoms involved in the adsorption were optimized; the position of all the other atoms corresponded to the MgO bulk structure (R_{MgO} = 2.1 Å [8]).

The HF stretching frequencies were calculated using force constants determined by a two-degree polynomial fit of the energies at three points around the optimum geometry. For O_{nc}-H and Mg_{nc}-H (n=3, 4, and 5) species these three points correspond to the R_{min} and $R_{min}\pm0.01$ Å bond lengths, where R_{min} is the optimum length of the O-H and Mg-H bonds in these structures. For the bridging H atoms at the $2Mg_{nc}$ and $3Mg_{nc}$ sites, these three points were chosen on the symmetry axis of these species crossing the H minimum energy position. The harmonic HF frequencies ν^{HF} were corrected for the basis set incompleteness and the electron correlation neglect by the scaling procedure $\nu = k\nu^{HF}$, where k is equal to 0.9049 for OH and 0.9517 for MgH. These scaling factors were defined by fitting the computed HF harmonic frequencies to the experimental values for the $Si(OH)_4$ [9] and $Mg\text{-}H^+$ [10] molecules. This approximation was shown to reproduce the OH stretching frequencies of Si-OH, P-OH, B-OH, Al-OH, and Si-OH-Al within 10 cm^{-1} [11].

4. RESULTS AND DISCUSSION

4.1. IR spectra

Figure 2 shows the FTIR spectra of the hydrogen adsorption on MgO ex-hydroxide at room temperature. Two sets of bands are present: the first includes two bands at 3712 and 3460 cm^{-1}, the second both a sharp peak at 1325 cm^{-1} and broader absorptions at 1131, 1067 (shoulder), 959, 886, and 861 (shoulder) cm^{-1} (Fig. 2,a). The components at 3460, 1325, and 886 cm^{-1} disappear after evacuation of H_2 at room temperature, whereas the other signals remain essentially unchanged (Fig. 2,b). A subsequent H_2 readmission restores the initial spectral pattern (Fig. 2,c).

Based on literature data, the bands observed in the 3750-3400 cm^{-1} and 1350-850 cm^{-1} ranges can be attributed to the stretching vibrations of hydroxyls groups (a bending mode of which produces the small component at 886 cm^{-1}) and hydride species, respectively [12-14]. Moreover, as Coluccia et al. [12,13] inferred from an analogy between the IR spectra of H species on MgO and MgH_2, the broad bands in the 1200-800 cm^{-1} region should be attributed to Mg_2H or/and Mg_3H surface structures, with the H atom bridging the neighboring, low-coordinated Mg surface sites. The sharp band at 1325 cm^{-1} was assigned by Knözinger et al. [14] to the Mg_{3c}-H surface species. The above data show that two different types of H_2 heterolytic dissociations occur.

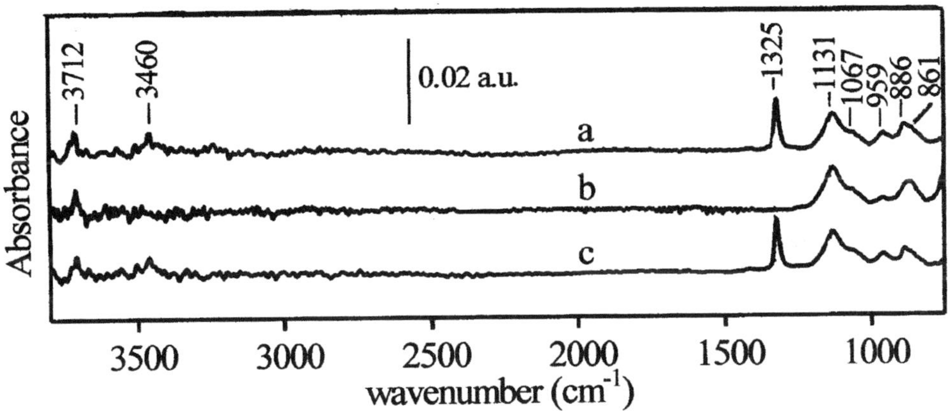

Fig. 2: IR spectra following hydrogen adsorption at room temperature on MgO pre-outgassed at 1123 K: a) in presence of 200 Torr H_2; b) after outgassing 10 min at room temperature; c) after readmission of 200 Torr H_2.

The two processes and the related most significant bands can be summarised as follows on the basis of the reversibility at room temperature.

reversible splitting	$H_2 \rightleftharpoons H^+ + H^-$	$\nu_{Mg\text{-}H}$	1325 cm^{-1}
		$\nu_{Mg\text{-}OH}$	3460 cm^{-1}
irreversible splitting	$H_2 \rightarrow H^+ + H^-$	$\nu_{Mg\text{-}H}$	1131, 1067, 959, 861 cm^{-1}
		$\nu_{Mg\text{-}OH}$	3712 cm^{-1}

It is impossible on the basis of the spectroscopic data alone to relate each hydride and hydroxyl species to specific Mg^{2+} and O^{2-} sites in low coordination onto the surface. Therefore, we have attempted to calculate the frequencies of the species which can be generated on the sites modelled in Fig. 1, considering both single and multiple Mg^{2+} and O^{2-} sites as done previously on studying the adsorption of CO [2,4] and CD_3CN [5].

4.2. Adsorption at the single Mg and O sites

The coordination of hydrogen at the Mg_{3c} and O_{3c} corner sites, Mg_{4c} and O_{4c} edge sites, and Mg_{5c} and O_{5c} sites of the (001) planes was modeled by clusters **3**, **4**, and **5** respectively. The results are shown in Table 3.

Table 3

Stretching frequencies calculated for adsorption on single sites and closest experimental peaks (if any)

Bond type	ν_{cal} (cm^{-1})	ν_{exp} (cm^{-1})
Mg_{3c}-H	1326	1325
O_{3c}-H	3713	3712
Mg_{4c}-H	985	
O_{4c}-H	3512	
Mg_{5c}-H	914	
O_{5c}-H	3644	

The calculated 1326 and 3713 cm^{-1} frequencies of the Mg_{3c}-H and O_{3c}-H species are in excellent agreement with the 1325 and 3712 cm^{-1} bands in the IR spectra (Figure 2). This result reinforces the earlier assignment of the 1325 cm^{-1} band to the Mg_{3c}-H species [14] and suggests O_{3c}-H species to be responsible for the 3712 cm^{-1} band.

However, it has to be noticed that the 1326 cm^{-1} band is associated with a reversible species, whereas that at 3712 cm^{-1} belongs to a irreversible one and, consequently, cannot be considered associated partners produced in the same heterolytic process.

The predicted stretching frequencies at 3512 (O_{4c}-H), 3644 (O_{5c}-H), 985 (Mg_{4c}-H), and 914 (Mg_{5c}-H) cm^{-1} do not correspond to any IR band in the spectra. On the basis of these results one can exclude the stabilization of H species onto 4- and 5-coordinated single Mg and O sites. This is in perfect agreement with the experiments that showed that hydrogen dissociates only on MgO samples pre-outgassed at the highest temperature and, therefore, exposing 3-coordinated sites [15]. The possible role of multiple sites has to be considered.

4.3. Adsorption at the multiple Mg and O sites

The cluster models **7**, **10**, and **11** mimic the stabilization of an H atom at different double and triple Mg and O sites of the surface. The O-H bond is too short to enable the formation of the bridging O-H-O species at the step, the distance between two neighboring O_{4c} atoms being about 3 Å. However, the O_{4c}-H species produced by H bonding at one of the O atoms of these double $2O_{4c}$ sites is influenced by the other O atom; this causes the O_{4c}-H group to bend towards the surface. By the same geometrical reason, neither $3O_{3c}H$ nor $2O_{3c}H$ bridging species appear to be stable. For the adsorption of hydrogen at the triple $3O_{3c}$ site (**7**) the only energy minimum corresponds to the coordination of the H atom at a single O_{3c} oxygen of this triple site, resulting in the formation of the O_{3c}-H species. The computed frequency 3718 cm^{-1} of this species is very close to the frequency of the analogous species produced by the adsorption at the single O_{3c} corner site (3713 cm^{-1}, see above). This means that the stabilizations of hydrogen at single O_{3c} and triple $3O_{3c}$ sites can not be distinguished by means of IR spectroscopic methods. It has to be noticed that no ν_{OH} band can be predicted, at this stage of our modelling, close the experimental one at 3640 cm^{-1}.

As distinct from the multiple O sites, on the multiple Mg sites (see **10** and **11**) the adsorbed H atom bridges all the neighboring Mg atoms, with equal Mg-H bond lengths. Since the characteristic Mg-H bond length is close to the length of the Mg-O bond, the adsorbed H atom nearly fits the crystallographic position of the "extracted" O atom at these Mg defect sites.

One can not expect the computed frequencies for the H atoms at the multiple $2O_{4c}$, $2Mg_{3c}$, $2Mg_{4c}$, and $3Mg_{3c}$ sites to have quantitative accuracy, owing to the use of scaling factors defined for the single O-H and Mg-H bonds. An important qualitative result of these calculations is that the stretching frequencies of the H atoms at the multiple Mg sites are lower than the stretching frequency of the Mg_{3c}-H species (1325 cm^{-1}). This allows for the conclusion that the 1131, 1067, 959, and 886 cm^{-1} bands should be associated with the symmetric and asymmetric stretching modes of the hydrogen stabilized at the multiple Mg sites.

4.4. Absorption energies

The HF adsorption energies at different sites are reported in Table 4. The ionic pair O_{3c}-Mg_{3c} is generally considered to be the most active site in the chemisorption of H_2 on MgO. Our results confirm this hypothesis, showing that the adsorption on two nearest three-coordinated ions is highly exothermic, with an energy gain of about 25 kcal/mol. (According to a previous theoretical study [16] the exothermicity of the reaction is estimated to be in the range 22-33 kcal/mol). The involved structure is strongly stabilized by the interaction between the two hydrogen atoms: when forcing the O-H and the Mg-H bonds to lay on opposite sides of the diagonal of the $(MgO)_4$ cluster, the exothermic energy decreases to 6.3 kcal/mol.

Table 4
Adsorption energies at different sites.

Adsorption site	E_{ads}^{a} (kcal/mol)
O_{3c}-Mg_{3c} (8)	24.7
O_{3c}-Mg_{3c} (9)	6.3
O_{3c}-Mg_{4c}	13.6
O_{4c}-Mg_{3c}	12.3
O_{4c}-Mg_{4c}	-5.9
O_{5c}-Mg_{5c}	-39.1

[a] $E_{ads} = E_{sub} + E_{ads} - E_{comp}$

The dissociative adsorptions of H_2 at the O_{4c}-Mg_{3c} and O_{3c}-Mg_{4c} sites are energetically favourable: the exothermic effects of the reactions are 12 and 14 kcal/mol, respectively. Our energies calculations conform to the trend of the site activity suggested by an experimental TPD (thermal programmed desorption) study [16]:

O_{4c}-Mg_{3c} < O_{3c}-Mg_{4c} < O_{3c}-Mg_{3c}

When both the surface atoms of a Mg-O site have coordination number equal or higher than 4, the adsorption is endothermic (Table 4).

5. CONCLUSIONS

The IR spectroscopic and ab-initio quantum chemical results show that both the reversible and irreversible H_2 splitting occur heterolitically on defect sites where at least one of the partners of the acid-base couple is a single cation or anion in corner position. In particular, the H^- moiety resulting from the reversible splitting is stabilized on a single Mg_{3c} site, while the H^+ moiety produced through the irreversible dissociation is stabilized on a O_{3c} site.

At present, the obtained results do not allow an unambiguous recognition of the sites responsible for the stabilisation of the counterpart moieties resulting from these two processes (giving origin to reversible hydroxyls and irreversible hydride species, respectively). Nevertheless, some general findings about the nature of such sites have been obtained. The irreversible IR bands in the 1150-850 cm^{-1} region are compatible with the symmetric and asymmetric modes of H^- stabilised at multiple Mg sites ($2Mg_{3c}$, $2Mg_{4c}$, or $3Mg_{3c}$), which should be adjacent to the O_{3c} ions where the irreversible OH species absorbing at 3712 cm^{-1} are formed. On the other hand, the H^+ moieties resulting from the reversible dissociation of H_2 are not stabilised on single O_{3c} or O_{4c} sites, but, probably, on multiple O_{lc} (l=3,4) sites, where the OH species formed on one of the oxygen atoms can be influenced by the other O atom.

Our energy calculations support the O_{4c}-Mg_{3c} < O_{3c}-Mg_{4c} < O_{3c}-Mg_{3c} trend of the acid-base pairs reactivity proposed by experimental TPD studies [16]. The adsorption on Mg-O sites, where the coordinations of both Mg and O are higher than 3, appears to be endothermic.

REFERENCES

1. S.Coluccia, M.Baricco, L.Marchese, G.Martra, and A.Zecchina, Spectrochim. Acta, 49A (1993) 1289.

2. A.Pelmenschikov, G.Morosi, A.Gamba, and S.Coluccia, J.Phys.Chem.B,. 102 (1998) 2226.

3. M.A.Nygren and L.G.M.Pettersson, J.Chem.Phys., 105 (1996) 9339.

4. A.Pelmenschikov, G.Morosi, A.Gamba, and S.Coluccia, J.Phys.Chem., 99 (1995) 15018.

5. A.Pelmenschikov, G.Morosi, A.Gamba, S.Coluccia, G.Martra, and E.Paukshtis, J.Phys.Chem., 100 (1996) 5011.

6. Frisch, M. J.; Trucks, G. W.; Schlegel, H. B.; Gill, P. M. W.; Johnson, B. G.; Robb, M. A.; Cheeseman, J. R.; Keith, G. A.; Petersson, D. A.; Montgomery, J. A.; Al-Laham, M. A.; Zakrzewski, V. G.; Ortiz, J. V.; Foresman, J. B.; Cioslowski, J.; Stefanov, B. B.; Nanayakkara, A.; Challacombe, M.; Peng, C. Y.; Ayala, P. Y.; Chen, W.; Wong, M. W.; Andres, J. L.; Replogle, E. S.; Gomperts, R.; Martin, R. L.; Fox, D. J.; Binkley, J. S.; Defrees, G. J.; Baker,

J.; Stewart, J. P.; Head-Gordon, M.; Gonzalez, C.; Pople, J. A. GAUSSIAN 94 (Revision A.1); Gaussian, Inc.: Pittsburgh, PA, 1995

7. W.Hehre, L.Radom, P.Schleyer, and J.Pople, Ab-initio Molecular Orbital Theory, Wiley-Interscience, New York, 1986.

8. R.Wyckoff, The Structure of Crystals, American Chem. Soc., New York, 1931.

9. H.Boem and H.Knözinger, in: Catalysis-Science and Technology, Vol. 4, J.Anderson and M.Boudart (eds.) Springer, Berlin, 1983, p. 39.

10. G.Herzberg, Molecular Spectra and Molecular Structure, Vols. I-II, Krieger Pub. Company, Malabar, FL, 1950.

11. A.G.Pelmenschikov, G.Morosi, and A.Gamba, J.Phys.Chem., 95, (1991) 10037.

12.S. Coluccia, F. Boccuzzi, G. Ghiotti, and C. Mirra, Z. Physik. Chem., N.F. 121 (1980) 141

13. S. Coluccia, F. Boccuzzi, G. Ghiotti, and C. Morterra, J. Chem. Soc., Faraday Trans. I 78 (1982) 2111

14. E. Knözinger, K.H. Jacob, and P. Hofmann, J. Chem. Soc., Faraday Trans. 89 (1993) 1101.

15. a) E. Garrone, A. Zecchina, and F. S. Stone, Phil. Mag., B42 (1980) 683; b) S. Coluccia, L. Marchese, S. Lavagnino, and M. Anpo, Spectrochim. Acta, 43A (1987) 1573.

16. T.Ito, M.Kuramoto, M.Yoshioka, and T.Tokuda, J.Phys.Chem, 87 (1983) 4411.

Studies in Surface Science and Catalysis 140
A. Gamba, C. Colella and S. Coluccia (Editors)

Modification of redox and catalytic properties of Keggin-type, Sb-doped P/Mo polyoxometalates in the selective oxidation of isobutane to methacrylic acid: control of preparation conditions

F. Cavani, R. Mezzogori, A. Pigamo, F. Trifirò

Dipartimento di Chimica Industriale e dei Materiali, Viale Risorgimento 4, 40136 Bologna, Italy

Keggin-type phosphomolybdates (POM's) were tested as catalysts for the selective oxidation of isobutane to methacrylic acid. Doping the POM with antimony improves the catalytic performance especially at isobutane-lean conditions, since a redox reaction between Sb^{3+} and Mo^{6+} leads to the development of a reduced POM which is stable even in an oxidizing environment, and which is more selective to methacrylic acid. Another important factor to control the reactivity is the pH of precipitation of the POM. When the preparation of the catalyst is carried out via the formation of a lacunary precursor, the time necessary to reach steady catalytic performance ("equilibration time") is considerably less than that for POM's prepared conventionally at strongly acid pH. An hypothesis about the nature of the active sites is formulated.

1. INTRODUCTION

The current industrial production of methylmethacrylate by the "acetone-cyanohydrin" process suffers from a number of drawbacks, which make it environmentally unfriendly. In particular, it makes use of a very toxic reactant (HCN) and intermediate (acetone cyanohydrin), and coproduces large amounts of impure ammonium sulphate, contaminated with organic compounds. Among the several alternative synthetic routes which have been proposed, particularly interesting from both the practical and scientific points of view is the single-step oxidation of isobutane to methacrylic acid, intermediate in the synthesis of methylmethacrylate. Several industrial companies have studied this reaction (and the selective oxidation of propane to acrylic acid, as well), and it has been established that the most active and selective catalysts are those which are based on Keggin-type polyoxometalates (POM's), containing phosphorus and molybdenum as the main components [1-18].

An important aspect of this reaction is that the processes claimed in all patents make use of fuel-rich conditions, thus with sub-stoichiometric oxygen [1-5]. Under these conditions the conversion of isobutane is necessarily low (in the best cases, not higher than 25%), and therefore recycle of the unconverted reactant becomes necessary. It has been proposed that the reason for this is that the catalyst is selective only provided molybdenum in the POM can be kept at an average reduced state lower than that typical of the calcined POM [17]. This can be obtained only provided a reducing, hydrocarbon-rich gas-phase is employed as the feedstock to the reactor. Some authors have reported catalyst preparations which lead to the development of reduced compounds that perform better than POMs prepared conventionally [10,12,13]. However, in all cases very few indications are given about the possibility of maintaining these performances for prolonged lifetimes. Indeed, the main question is whether this reduced state of the POM can be maintained under reaction conditions, or is fated to evolve inside the reactor.

In a previous work we found that the addition of small amounts of Sb^{3+} to the POM-based catalyst makes possible the occurrance of a redox reaction between Mo^{6+} in the primary POM structure and antimony, with development of a reduced compound which is stable even under oxidizing conditions [19]. The objective of the work reported here was twofold: 1) to evaluate the catalytic performance of this Sb-doped POM catalyst in the title reaction, and 2) to analyze the effect of the pH of precipitation of the POM on its chemical-physical features and catalytic performance, as a possible parameter to control the reactivity of these compounds.

2. EXPERIMENTAL

Ammonium salts of phosphorus/molybdenum Keggin type POM's and related lacunary compounds were prepared using the following procedure: the compound ("precursor") was precipitated by the addition of HCl (until the desired pH value) to a solution containing dissolved $(NH_4)_6Mo_7O_{24}{\cdot}4H_2O$ and H_3PO_4 (initial pH 4.5), in the relative amounts as required by the stoichiometry. In the case of the Sb-doped POM, $SbCl_3$/HCl was also added to the solution, in the desired amount. The precipitate was dried at 120°C overnight (with solvent evaporation), and then calcined up to 350°C in static air, for 6 hours. The effective Sb content was determined by dissolving the sample in a basic medium, and analysis of Sb using the atomic absorption technique (Phillips PU 9100). The amount of Sb was found to be equal to 0.23 atoms per Keggin unit: $(NH_4)_3PMo_{12}O_{40}/Sb_{0.23}O_x$. The carbon and nitrogen contents were determined on the solids, using a Carlo Erba EA 1110 CHNS-O Instrument. Powder XRD data were obtained with a Phillips PW 1050/81 diffractometer, controlled by a PW1710 unit using Ni-filtered CuK_α radiation. The surface areas (BET single point) were determined by N_2 adsorption at 77K, using a Carlo Erba Sorpty 1826 apparatus. Diffuse Reflectance UV-Vis spectra were recorded at room temperature using a Perkin-Elmer Lambda 19 spectrometer, equipped with a 60-mm integrating sphere coated with barium sulphate reflective paint.

The catalytic tests were carried out in a stainless-steel continuous flow reactor, at atmospheric pressure. The feed composition was the following: isobutane between 1 (isobutane-lean) and 26 (isobutane-rich) mol.%, oxygen 13%, steam 12%, remainder helium. Each series of catalytic tests was carried out using 1.5 g of catalyst, granulated into particles ranging from 0.4 to 0.5 mm in size. Unless otherwise specified, the residence time was equal to 3.6 s. The reactor outlet was kept at 200°C, to prevent product condensation and methacrylic acid polymerization. A volume of the gas phase was sampled on-line by means of a sampling valve, and analyzed by gas chromatography. A Carbosieve S column was utilized for CO and CO_2 analysis, with a programmed increase in oven temperature from 40 to 240°C (TCD). A GP 10% SP-1200/1% H_3PO_4 on Chromosorb WAW (FID) was utilized for the analysis of the other products. After the catalytic tests, the catalysts were unloaded by cooling the reactor under a helium atmosphere, and then characterized.

3. RESULTS AND DISCUSSION

3.1 Effect of Sb on the redox state and catalytic performance of a working POM

The catalysts exhibit an initial period during which the catalytic performance is not stable. Several hours of "equilibration" are necessary before a steady performance is reached;

in the case of $(NH_4)_3PMo_{12}O_{40}$ prepared at pH < 1, the steady state is reached in 80 hours of reaction under isobutane-rich conditions (26% isobutane, 13% oxygen in the feed), while a shorter period is necessary when the reaction is carried out under isobutane-lean conditions (1% isobutane, 13% oxygen).

Figure 1 compares the conversion of isobutane and the selectivity to the different reaction products (methacrolein, methacrylic acid, acetic acid, carbon monoxide and carbon dioxide) for an "equilibrated" catalyst (one that has reached a steady catalytic performance), under different reaction conditions: isobutane-lean and isobutane-rich. The catalyst was prepared by precipitation of the "precursor" of composition $(NH_4)_3PMo_{12}O_{40}$ at strongly acid pH (< 1), followed by thermal treatment at 350°C.

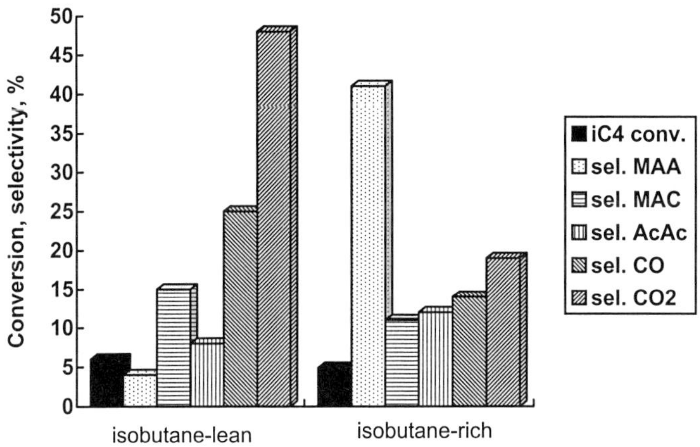

Figure 1. Comparison of the catalytic performance of an "equilibrated" $(NH_4)_3PMo_{12}O_{40}$ catalyst under isobutane-lean (T 350°C) and isobutane-rich (T 352°C) conditions. MAA: methacrylic acid; MAC: methacrolein; AcAc: acetic acid.

Figure 2 compares the ex-situ UV-Vis DR spectra of catalysts, unloaded after reaction at the two different conditions, and of the fresh, calcined catalyst, before reaction. A large absorption band is observed in the 250 to 500 nm range. This band is indeed constituted of several components, due to the different energies associated with $O^{2-} \rightarrow Mo^{6+}$ charge-transfers in the Keggin anion. A first component is present in the 270 to 320 nm range, a second one lies between 330 and 400 nm, and a third one is centred above 400 nm. The bands are strongly convoluted. It has been reported that the energy associated with the absorption band above 400 nm (centred at ≈ 430-460 nm) is a function of the cationic composition of the Keggin secondary framework, and possibly of other structural and morphological features as well (i.e., the crystallinity of the compound) [20]. Moreover, it has been reported that the Low-energy Charge-Transfer band energy can be affected by the oxidation potential of the oxometal [21].

The main difference in the spectra concerns the presence of an intense absorption band at around 700 nm, which can be attributed to Intervalence Charge-Transfer between Mo^{5+} and Mo^{6+} in the Keggin unit [22,23], thus indicating the presence of a partially reduced POM for the catalyst which has been unloaded after reaction under isobutane-rich conditions (spectrum b in Figure 2). This means that in the latter case the gas phase is reducing towards the fully oxidized POM, and that the equilibrium attained at steady state involves the development of a partially reduced POM. On the contrary, under isobutane-lean conditions the gas phase behaves towards the POM in the same way as does the atmosphere employed for the calcination treatment (air), and the POM therefore is fully oxidized (spectrum c in Figure 2). Indeed, the electronic spectrum is very similar to that of the calcined catalyst (spectrum a in Figure 2). Figure 1 shows that the equilibrated catalyst is much more selective to methacrylic acid under isobutane-rich conditions than under isobutane-lean conditions. This can be attributed to the fact that a more reduced POM under steady state is more selective for the transformation of isobutane to the products of partial oxidation rather than to carbon oxides [17,24]. On the contrary, when the catalyst is fully oxidized, the formation of carbon oxides is the prevailing reaction.

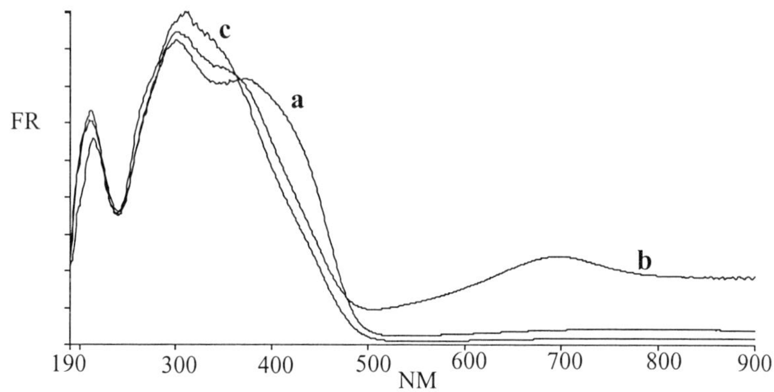

Figure 2. UV-Vis-DR spectra of calcined $(NH_4)_3PMo_{12}O_{40}$ (a) precipitated at pH < 1, of the same catalyst after reaction under isobutane-rich conditions (b), and after reaction under isobutane-lean conditions (c).

The effect of residence time on isobutane conversion and on selectivity to the various products at the temperature of 320°C, under isobutane-rich conditions, is illustrated in Figure 3. The data indicate that methacrolein, methacrylic acid, and carbon dioxide are all formed through direct, parallel reactions; acetic acid and possibly carbon monoxide are instead formed through consecutive reactions. Methacrolein undergoes consecutive reactions of transformation to acetic acid, to carbon oxides and possibly in part also to methacrylic acid. Indeed the selectivity to the latter product seems to increase slightly with increasing isobutane conversion.

In previous works, it was found that doping with Sb^{3+} affects the redox properties of P/Mo POMs' [19]. This is shown in Figure 4, which reports the electronic spectra of a POM of composition $(NH_4)_3PMo_{12}O_{40}/Sb_{0.23}O_x$. It is assumed that the antimony is fully extraframework from the Keggin anion: in fact it has been established that Sb ions replace part of the ammonium cations in the secondary framework of the POM [19]. In agreement, the amount of ammonium cations per Keggin unit was found to be equal to 2.7, lower than the theoretical one, 3.0. The calcined POM is reduced, as suggested by the intense IVCT band at 700 nm. This can be attributed to the redox reaction occurring between Mo^{6+} and Sb^{3+}:

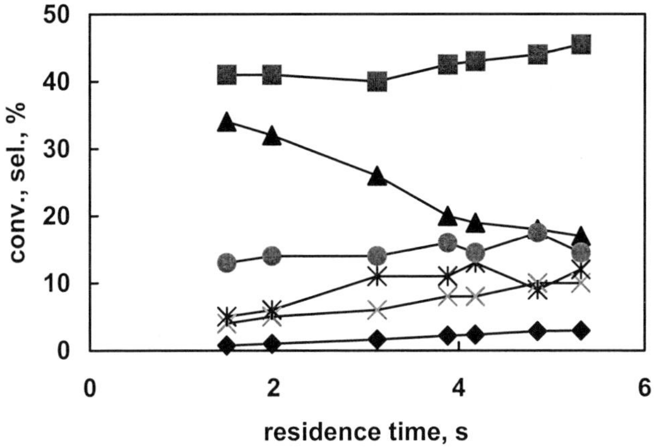

Figure 3. Effect of the residence time on the catalytic performance of the $(NH_4)_3PMo_{12}O_{40}$ catalyst, prepared at pH < 1. T 320°C; isobutane-rich feed composition. Symbols: isobutane conversion (◆), sel. to methacrylic acid (■), to methacrolein (▲), to acetic acid (✖), to carbon monoxide (✱) and to carbon dioxide (●).

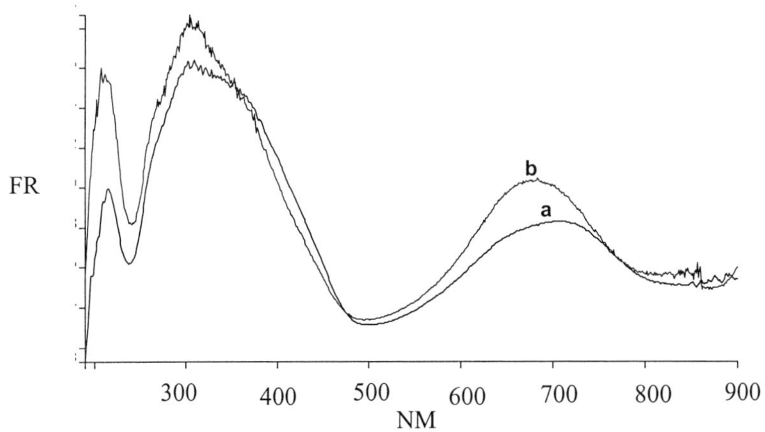

Figure 4. UV-Vis DR spectra of the $(NH_4)_3PMo_{12}O_{40}/Sb_{0.23}O_x$ catalyst prepared at pH < 1 after calcination (a), and unloaded after 25 h time-on-stream under isobutane-rich conditions (b).

$2\ Mo^{6+} + Sb^{3+} \Leftrightarrow 2\ Mo^{5+} + Sb^{5+}$

Indeed the electrons furnished to the POM are not fully localized over defined Mo sites, but rather are delocalized over the entire Keggin anion. This reduced state is stable even under oxidizing conditions, i.e., under the conditions of the calcination treatment, and under the isobutane-lean conditions employed for reaction.

The effect of isobutane partial pressure on catalytic performance for the Sb-doped POM is illustrated in Figure 5. The catalyst is now selective even under isobutane-lean conditions, while the undoped POM under these conditions is fully unselective (see Figure 1). A small improvement in selectivity to methacrylic acid with respect to the undoped POM is also observed under isobutane-rich conditions. These data indicate that the development of a reduced POM, stable even under reaction conditions which are oxidizing and typically non-selective (isobutane-lean conditions), makes it possible to obtain a selective catalyst. The improvement in selectivity under isobutane-rich conditions is much smaller due to the fact that under these conditions the catalyst is already selective, even in the absence of the Sb dopant.

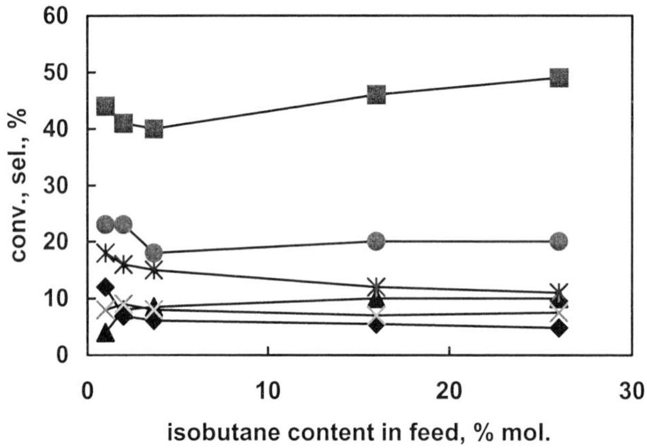

Figure 5. Catalytic performance of $(NH_4)_3PMo_{12}O_{40}/Sb_{0.23}O_x$ prepared at pH < 1 as a function of isobutane content in the feed. T 350°C, τ 3.6 s, feed composition: 13% O_2, 12% H_2O, remainder He. Symbols as in Figure 3.

The effect of the residence time on the catalytic performance under isobutane-rich conditions for the Sb-doped compound is shown in Figure 6. The initial selectivity to methacrylic acid is higher than for the undoped POM, but it decreases for increasing values of residence time. The decrease in selectivity corresponds to an increase in carbon dioxide and carbon monoxide. The selectivity to methacrolein is lower than that obtained for the undoped POM. It is significant that the sum of methacrolein plus methacrylic acid is approximately the same with the two catalysts. This means that under these conditions the effect induced by the presence of Sb mainly concerns the ratio between the two parallel reactions of transformation of isobutane to methacrolein and to methacrylic acid. With this catalyst, too, acetic acid and carbon monoxide are secondary products, obtained by consecutive reactions.

In conclusion, the addition of Sb as a dopant has the following main effects on the POM features and reactivity:

a) It furnishes electron(s) to the POM, yielding a reduced compound which is stable under oxidizing conditions.

b) The reduced POM is selective to methacrylic acid even under isobutane-lean conditions (thus conditions which are not reducing towards the POM), while the oxidized, undoped POM is not selective under these conditions.

c) The reduced Sb-doped POM is also slightly more selective than the undoped one to methacrylic acid under isobutane-rich conditions; however, the selectivity to methacrolein plus methacrylic acid is approximately the same for the Sb-doped and for the undoped POM.

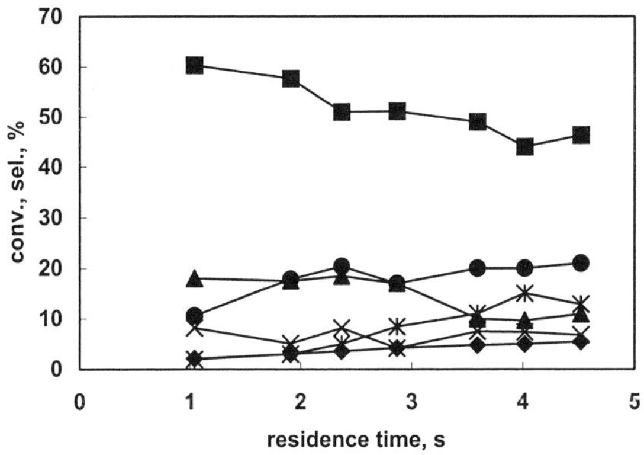

Figure 6. Catalytic performance of $(NH_4)_3PMo_{12}O_{40}/Sb_{0.23}O_x$ prepared at pH < 1 as a function of residence time. T 350°C; feed composition: 26% isobutane, 13% O_2, 12% H_2O, remainder He. Symbols as in Figure 3.

3.2 Effect of the pH of precipitation on the chemical-physical features and reactivity of $(NH_4)_3PMo_{12}O_{40}$

POM's are usually prepared at strongly acid pH values, thus at conditions at which the Keggin anion $PMo_{12}O_{40}^{3-}$ is formed; in the presence of ammonium cation, the corresponding unsoluble salt precipitates. However, the preparation of POM's can be carried out at higher pH, i.e., between 1 and 7. In this range, lacunary compounds develop, where the atomic ratio between the heteroatom and the oxometal is higher than in intact Keggin anions. We prepared unsoluble ammonium salts at pH between strongly acidic (< 1) and mildly acidic (4.0). The FT-IR spectra of the compounds obtained, after drying at 120°C, are reported in Figure 7 (left).

The results obtained show that with increasing pH, spectra develop where the absorption bands are shifted and doubled with respect to those of the intact Keggin unit, as a consequence in decrease of the anion symmetry. Limit spectra are those obtained at strongly acid pH (< 1 and 1.8) and at pH equal to 4.0; for the intermediate value of pH (3.2), the spectrum corresponds to a convolution of the two limit spectra. According to literature indications, the spectrum of the compound obtained at pH 4 might correspond to that of the POM having the composition $(NH_4)_7PMo_{11}O_{39}$ [25]. A confirmation for this comes from the analysis of the N content of the compound, which is 5.1 wt.%, and thus corresponds to 6.7 $(NH_4)^+$ cations per $PMo_{11}O_{39}^{7-}$ formula, close to the theoretical value 7. X-ray diffraction patterns for the same samples are reported in Figure 7 (right). In this case, the presence of a non-negligible amount of lacunary compound is already detected in the sample prepared at pH 1.8.

FT-IR and X-ray diffraction patterns of the same compounds after calcination at 350°C are reported in Figure 8. Surprisingly, in all cases the thermal treatment of these compounds at 350°C leads to the formation of a Keggin POM. If we consider the stoichiometry of a possible transformation of the lacunary to the intact POM we have the following reaction:

$$12\ (NH_4)_7PMo_{11}O_{39} \rightarrow 11\ (NH_4)_3PMo_{12}O_{40} + (NH_4)_3PO_4 + 48\ NH_3 + 24\ H_2O$$

Indeed, the XRD pattern of the calcined compound precipitated at pH 4.0 indicates that the transformation of the lacunary precursor also involves the formation of small amounts of MoO_3.

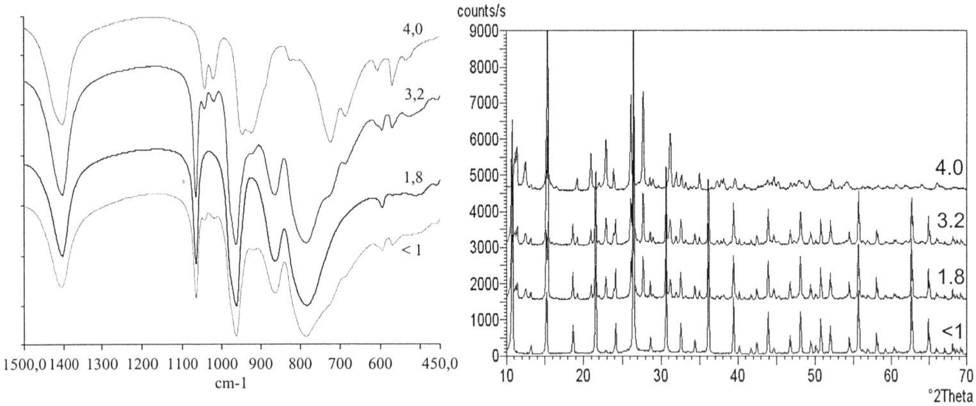

Figure 7. FT-IR spectra (left) and X-ray diffraction patterns (right) of catalysts precursors prepared at increasing values of pH.

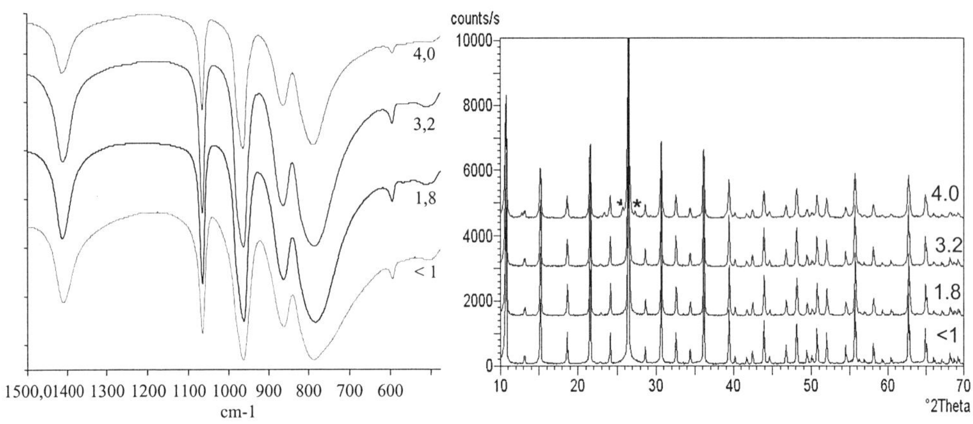

Figure 8. FT-IR spectra (left) and X-ray diffraction pattern (right) of catalysts prepared at increasing values of pH and calcined at 350°C. * MoO_3.

When loaded into the reactor, all the POM's exhibit an initial unsteady catalytic behavior before reaching the so-called "equilibrated" state. This is shown in Figure 9, for the calcined compound precipitated at strongly acid pH (<1). During this equilibration period, the main effects are i) a progressive increase in catalytic activity, and ii) a progressive increase in the selectivity to methacrylic acid, with a corresponding decrease in the selectivity to carbon dioxide. Along with these modifications, changes also are found in the chemical-physical features, and specifically:

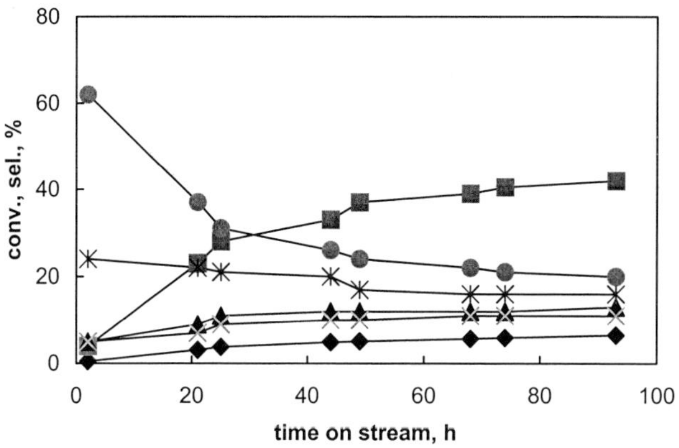

Figure 9. Catalytic performance of $(NH_4)_3PMo_{12}O_{40}$ prepared by precipitation at pH < 1 as a function of time-on-stream. T 380°C, τ 3.6 s; feed composition: 26% isobutane, 13% O_2, 12% H_2O, remainder He. Symbols as in Figure 3.

a) The extent of POM reduction increases progressively, as evidenced by the progressive increase in the intensity of the band at 700 nm in the electronic spectrum for increasing values of elapsed time-on-stream under isobutane-rich conditions (Figure 10). On the basis of that discussed above, this might explain, at least in part, the progressive increase in the selectivity to methacrylic acid during the equilibration time [24].
b) FT-IR spectra confirm that the POM structure has been mostly maintained (Figure 11, left); however, small amounts of MoO_3 are detected in the X-ray diffraction pattern (Figure 11, right) of the catalysts unloaded after reaction at isobutane-rich conditions.
c) The ammonium cation content decreases. In fact, the amount of N in the catalyst, as determined analytically, decreases from the initial amount corresponding to 3.0 molecules per Keggin unit to the final 2.3. This difference can not be justified by the formation of the small amounts of MoO_3, and this means that the negative charges of the Keggin anion are in part compensated for by other cation types, either by protons, or by Mo cations, possibly as oxo or hydroxo species. The migration of molybdenum ions from the anionic position in the Keggin structure into the cationic framework position of the still intact framework has been proposed to occur also during the incipient thermal structural decomposition of the Keggin unit, before the extensive formation of crystalline MoO_3 [26]. This explains the low analytical content of ammonium in the compound, and the development of Mo-O-Mo species (with one Mo cation in the Keggin anion, and one in the secondary framework), as has been evidenced by EPR spectroscopy [26].

It has been proposed that the development of a compound where a fraction of Mo ions are located in the secondary framework of the POM leads to a more active catalyst in the oxidehydrogenation of isobutyric acid to methacrylic acid [26]. In agreement, in the present case, possible explanations for the progressive increase in activity during equilibration time under isobutane-rich conditions are not only the increase in the POM reduction, but also the development of more active sites in the POM, as a consequence of the redistribution of Mo species in the POM framework. This may also positively affect the selectivity to methacrylic acid, contributing to its progressive increase during the equilibration period.

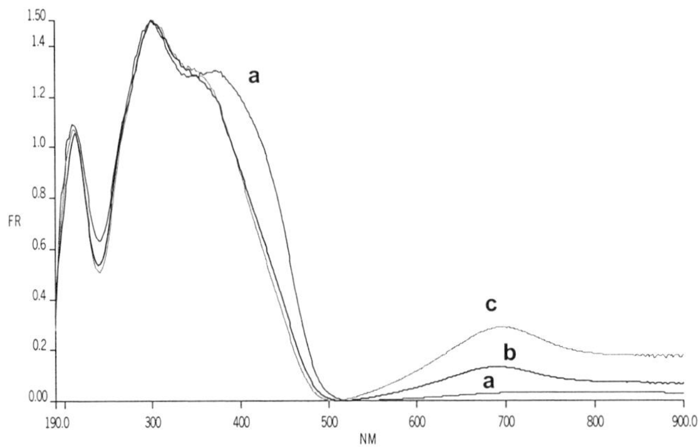

Figure 10. UV-Vis-DR spectra of $(NH_4)_3PMo_{12}O_{40}$ after calcination (a), after reaction at isobutane-rich conditions for 25 h (b), and after reaction for 100 h (equilibrated catalyst) (c).

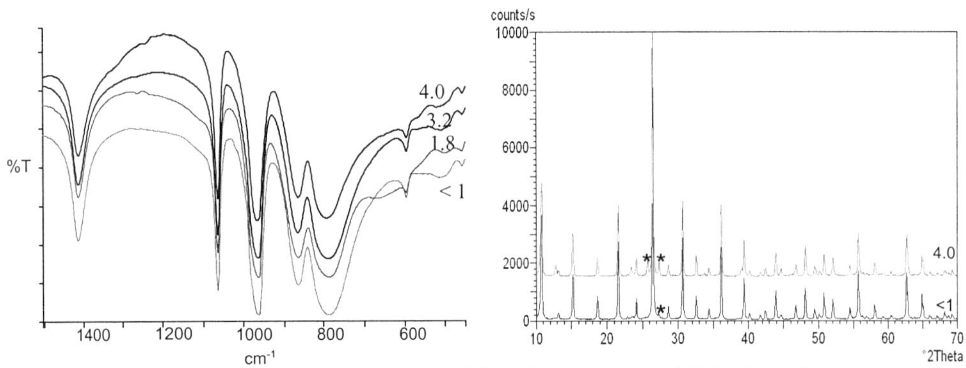

Figure 11. FT-IR spectra (left) and X-ray diffraction patterns (right) of catalysts prepared at increasing values of pH, unloaded after reaction under isobutane-rich conditions. * MoO_3.

Figure 12 shows the variation of catalytic performance during time-on-stream (equilibration time) for the catalyst prepared by precipitation at pH 4.0. In this case the conversion is constant during service time (around 7%), and the selectivity to methacrylic acid exhibits a lower increase (from the initial 21% up to 40% for the equilibrated catalyst after approximately 50 hours time-on-stream) than for the compound prepared at strongly acid pH (see Figure 9). In this case, the only chemical-physical features which are modified during the equilibration time are i) the extent of POM reduction, which progressively increases, as demonstrated by the electronic spectrum of the unloaded catalyst as compared to that of the calcined catalyst, and ii) the amount of crystalline MoO_3, which slightly increases with respect to the amount present in the calcined catalyst (Figure 11, right).

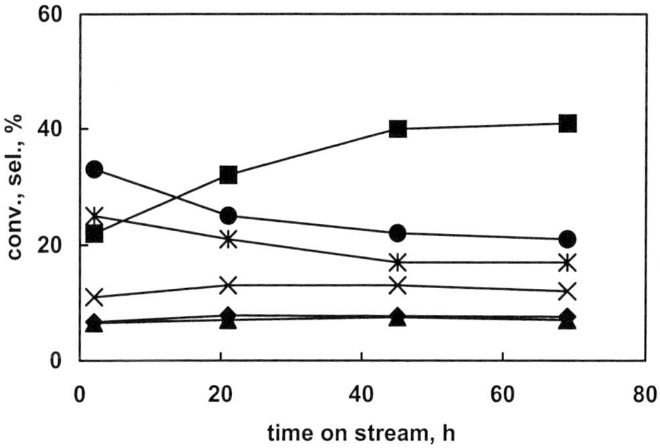

Figure 12. Catalytic performance of $(NH_4)_3PMo_{12}O_{40}$ prepared by precipitation at pH 4.0 as a function of time-on-stream. T 350°C, τ 3.6 s; feed composition: 26% isobutane, 13% O_2, 12% H_2O, remainder He. Symbols as in Figure 3.

All these data indicate that the progressive increase in activity during service time observed for the catalyst prepared at strongly acid pH (Figure 9) is due to the development of an active species during equilibration, which occurs as a consequence of partial structural decomposition of the POM and redistribution of molybdenum in the framework. This active species can be hypothesized to include Mo ions located in the cationic position of the framework. An additional phenomenon is the increase in the extent of reduction of the POM, which is in part responsible for the progressive increase in selectivity.
The same change in activity is not observed in the case of the catalyst prepared at pH 4.0 (Figure 12) because the catalyst, already after calcination, possesses the mentioned active species, which are formed during the thermal treatment of the lacunary precursor and its transformation into the Keggin compound. In this case the only change in catalytic performance observed during equilibration is an increase in selectivity to methacrylic acid (less important than in the case of the catalyst prepared at strongly acid pH), likely related to a progressive increase in the extent of reduction of the POM.

In conclusion, the pH of precipitation of the POM is one parameter which may affect the performance of the final catalyst during the "equilibration" time. In particular, it is possible to considerably shorten the time-on-stream necessary to reach a steady performance, by preparation of the precursor at mildly acidic conditions (pH 4.0). This procedure yields a final compound that already contains the specific sites active for the title reaction. When instead the preparation is prepared conventionally at strongly acid pH, generation of the active species occurs during permanence in the reaction environment under isobutane-rich conditions, by partial decomposition of Keggin units and redistribution of the cationic composition in the POM framework.

Acknowledgements
Elf Atochem is gratefully acknowledged for financial support

References

1. H. Krieger and L.S. Kirch, US Patent 4,260,822 (1981), assigned to Rohm and Haas Co.
2. S. Yamamatsu and T. Yamaguchi, Eur. Patent 425,666 (1989), assigned to Asahi Chem. Co.; Jap Patent 02-042,032 (1990)
3. K. Nagai, Y. Nagaoka, H. Sato and M. Ohsu, EP 418,657 (1990), assigned to Sumitomo Chem. Co.
4. K. Nagai, Y. Nagaoka and N. Ishii, EP 495,504 A2 (1992), assigned to Sumitomo Chem. Co.
5. T. Kuroda and M. Okita, Jap. Patent 04-128,247 (1991), assigned to Mitsubishi Rayon Co.
6. S. Paul, V. Le Courtois and D. Vanhove, Ind. Eng. Chem., Research, 36 (1997) 3391
7. N. Mizuno, M. Tateishi and M. Iwamoto, Appl. Catal., A: General, 118 (1994) L1.
8. N. Mizuno, W. Han, T. Kudo and M. Iwamoto, Stud. Surf. Sci. Catal., 101 (1996) 1001
9. N. Mizuno, M, Tateishi and M. Iwamoto, J. Catal., 163 (1996) 87
10. N. Mizuno, D.-J. Suh, W. Han and T. Kudo, J. Molec. Catal., A: Chemical, 114 (1996) 309
11. L. Jalowiecki-Duhamel, A. Monnier, Y. Barbaux and G. Hecquet, Catal. Today, 32 (1996) 237
12. W. Ueda, Y. Suzuki, W. Lee and S. Imaoka, Stud. Surf. Sci. Catal., 101 (1996) 1065
13. W. Li and W. Ueda, Stud. Surf. Sci. Catal., 110 (1997) 433
14. F. Cavani, E. Etienne, M. Favaro, A. Galli, F. Trifirò and G. Hecquet, Catal. Lett., 32 (1995) 215
15. G. Busca, F. Cavani, E. Etienne, E. Finocchio, A. Galli, G. Selleri and F. Trifirò, J. Molec. Catal., 114 (1996) 343
16. F. Cavani, E. Etienne, G. Hecquet, G. Selleri and F. Trifirò, in "Catalysis of Organic Reactions", R.E. Malz (Ed.), Marcel Dekker, (1996), p. 107
17. F. Cavani, 4ème Conférence sur la Catalyse "Paul Sabbatier", Twelfth European Conference on Catalysis, 5-9 July 1999, Strasbourg, France, Abstracts p. 22
18. F. Cavani, A. Lucchi, A. Tanguy and F. Trifirò, DGMK-Conference "C4 Chemistry-Manufacture and Use of C4 Hydrocarbons", W. Keim, B. Lücke, J. Weitkamp (Eds.), Tagungsbericht 9705 Hamburg 1997, Proceedings of the Conference, Aachen, October 1997, p. 173
19. F. Cavani, A. Tanguy, F. Trifirò and M. Koutyrev, J. Catal., 174 (1998) 231
20. M. Fournier, C. Louis, M. Che, P. Chaquin and D. Masure, J. Catal., 119 (1989) 400.
21. M.T. Pope, E. Papaconstantinou, G.M. Varga and D.R. Wexell, in "Progress in Coordination Chemistry", M. Cais (Ed.), Elsevier, New York, 1968, p. 322
22. G.M. Varga, E. Papaconstantinou and M.T. Pope, Inorg. Chem., 9 (1970) 662
23. H. So and M.T. Pope, Inorg. Chem., 11 (1972) 1441
24. F. Cavani, R. Mezzogori, A. Pigamo and F. Trifirò, C R Acad. Sci., Série II, Chimie, 3 (2000) 523
25. J.B. Moffat and G.B. McGarvey, J. Molec. Catal., 69 (1991) 137
26. S. Albonetti, F. Cavani, F. Trifirò, M. Gazzano, M. Koutyrev, F.C. Aissi, A. Aboukais and M. Guelton, J. Catal., 146 (1994) 491

Studies in Surface Science and Catalysis 140
A. Gamba, C. Colella and S. Coluccia (Editors)

Evaluation of Italian phillipsite and chabazite as cation exchangers for Ba^{2+} and Co^{2+}

A. Colella[a] and B. de Gennaro[b]

[a] Dipartimento di Scienze della Terra, Università Federico II, Via Mezzocannone 8, 80134 Napoli, Italy

[b] Dipartimento di Ingegneria dei Materiali e della Produzione, Università Federico II, Piazzale V. Tecchio 80, 80125 Napoli, Italy

The cation exchange properties of Na-exchanged phillipsite and chabazite for Ba^{2+} and Co^{2+}, two cations present in many nuclear power plants wastewaters, were investigated at 25°C and 0.1 total normality, from either a kinetic or a thermodynamic points of view.

Kinetic curves showed a rather fast exchange reaction for both cations, as equilibrium was mostly reached in about one day. Results of $2Na^+ \rightleftharpoons Ba^{2+}$ and $2Na^+ \rightleftharpoons Co^{2+}$ equilibria, interpreted in terms of distribution coefficient (K_d) and separation factor (α_B^A) and discussed on the basis of the known structural features of the two zeolites, evidenced a moderate to good selectivity of chabazite and phillipsite for Ba, which preludes to possible utilizations of Italian tuffs in the removal of this cation from wastewaters of nuclear origin. On the contrary, selectivity for Co was demonstrated to be very poor.

1. INTRODUCTION

Environmental pollution is nowadays matter of deep concern. Air, soil and water are the main natural resources subject to pollution. Water in particular, due to its high affinity to ionic or ionizable compounds, is prone to dissolve and store several noxious or toxic species, especially cations, spreading them around in soil and subsequently in plants and in animals. A strict control is therefore necessary on the polluting charge of wastewaters of either municipal or industrial origin, before recycling or discharging them in water bodies.

Wastewaters of nuclear power plants contain various dangerous cationic species, the most common being ^{137}Cs, ^{90}Sr, ^{133}Ba and ^{60}Co [1,2]. Cs and Sr are present in remarkable concentration and are the most dangerous cations, since they present long half lives. ^{133}Ba originates from ^{137}Cs decay through β and α radiations, but it has a half life shorter than Cs. During Ba-Cs transformation γ radiations may be produced too [3]. ^{90}Co derives from ^{58}Ni, used as building material of the reactor, by acquisition of two neutron. It may also result from neutronic activation of corrosion products and other impurities in the cooling plant [4].

During the past few decades natural zeolites have been shown to exhibit a remarkable cation exchange selectivity for certain polluting cations and to be therefore suitable cation exchangers for removing the same cations from wastewaters [5]. Phillipsite and chabazite, in particular, which are very common minerals in Italy being the main constituents of the huge formations of volcanic tuffs [6,7], couple large availability and good cation exchange selectivity for several polluting cations [8], e.g., Cs^+ and Sr^{2+}, that are often part of the nuclear wastes [9].

In the frame of an investigation aimed at evaluating the ion-exchange features of these two zeolites towards some common radionuclides, the aim of the present paper is to study the thermodynamics and kinetics of the ion-exchange reactions of phillipsite and chabazite for Ba^{2+} and Co^{2+}.

2. EXPERIMENTAL

2.1. Materials

Both materials used for studying equilibrium and kinetics of the ion-exchange reactions came from the huge formation of Neapolitan yellow tuff [10]. One of them, collected from a tuff quarry in Marano (Naples), was rich in phillipsite, whereas the other, coming from a drilling core in Parco Margherita (Naples), was rich in chabazite. Rock mineralogy was investigated by X-ray powder diffraction using a Philips PW1730, equipped with a Philips 3710 count unit. Zeolite grade was obtained through the Reference Intensity Ratio (RIR) procedure [11].

Enriched phillipsite and chabazite samples, practically free from other cation exchanger phases, were obtained from the parent rocks by enrichment processes (grinding and sieving, separation by heavy liquids or magnetic table, etc.), based on the greater friability and lower density of zeolite relative to other rock constituents [12]. Chemical analysis of the two zeolites in the enriched products was performed by electron microprobe analysis (CAMECA SX50). Water content was measured by thermogravimetry (Netzsch STA 409 thermoanalyzer).

The two zeolite-rich samples, in their original form or after pre-exchange in Na^+ form, were stored for long time (at least one week) at room temperature over saturated $Ca(NO_3)_2$ solution (relative humidity near 50%) before using them for ion-exchange runs.

2.2. Measurement of cation exchange capacity

Cation exchange capacity (CEC) was measured by the cross-exchange procedure [13,14]. Accordingly, weighed amounts of zeolite-containing materials were passed through by 1 M KCl or 1 M NaCl solutions until the concentrations in the eluate of the parent rock's cations, measured by atomic absorption spectrophotometry (AAS, Perkin Elmer 100), became lower than 0.1 mg/l. Two experimental values of cation exchange capacity, CEC_1 and CEC_2, were therefore obtained, the former calculated by summing up the amounts of K^+, Ca^{2+} and Mg^{2+} eluted by the Na^+ solution and Na^+ amount eluted by the K^+ solution (referred to as K_1, Ca_1, Mg_1 and Na_2, respectively), and the second by summing up Na^+, Ca^{2+} and Mg^{2+} amounts eluted by the K^+ solution and the K^+ amount eluted by Na^+ solution (referred to as Na_2, Ca_2, Mg_2 and K_1, respectively). In other words $CEC_1 = K_1 + Ca_1 + Mg_1 + Na_2$ and $CEC_2 = K_1 + Ca_2 + Mg_2 + Na_2$. The average value of these two values, $(CEC_1 + CEC_2)/2$, was assumed to be the experimental CEC of the sample.

2.3. Ion exchange runs

The two zeolite-rich samples were previously Na-exchanged by percolation with a 1 M Na^+ solution, prepared from reagent grade Carlo Erba NaCl, up to complete elution of the parent cations, other than Na^+, i.e., K^+, Mg^{2+}, Ca^{2+}. In order to minimize the amount of residual K^+ impurities in the zeolite, elution was improved using a solution made with extra pure NaCl (Aldrich Chemicals, purity: 99.999%).

Kinetic runs were performed by contacting various Na-exchanged phillipsite-rich and chabazite-rich samples, under continuous stirring, with different Ba^{2+} or Co^{2+} 0.1 N solutions, prepared from reagent grade Carlo Erba $BaCl_2{\cdot}2H_2O$ and $Co(NO_3)_2{\cdot}6H_2O$, respectively, at a solid to liquid ratio of 1/200.

Kinetic curves were then analyzed using a Langmuir-type equation [15,16]:

$$\frac{C}{C_{max}} = \frac{kt}{1 + kt}, \tag{1}$$

where C represents the concentration of the ingoing cation, t the time, C_{max} the equilibrium concentration of the ingoing cation and k is a constant. Eq. 1 can be linearized as follows:

$$\frac{t}{C} = \frac{1}{k\,C_{max}} + \frac{t}{C_{max}}, \tag{2}$$

from which k and C_{max} can easily be obtained.

In order to obtain ion-exchange equilibrium data, the Na-exchanged zeolite samples were allowed to react at 25°C for three days, which was shown to be sufficient for attaining equilibrium, in sealed Teflon bottles with solutions containing different amounts of Na^+ and the competing cation (Ba^{2+} or Co^{2+}) at 0.1 total normality, using the same reagents of the kinetics runs. Solid-to-liquid ratio ranged between 1/100 and 1/500. Reversibility tests of ion exchange were performed following the recommendations of Fletcher and Townsend [17].

Ingoing and outgoing cations were measured either in the liquid or in the solid phase. Ba and Co concentrations were estimated with EDTA using erio-T, NH_3-NH_4^+ buffer at pH 10 and Mg-EDTA for the former; murexide and 1M ammonia solution for the latter [18]. The possible presence in solution of other cations, coming from the solid phase, was checked by AAS. Solids were analyzed by dissolving them with a hydrofluoric-perchloric attack. Concentrations of the various cations in the obtained solution were measured by AAS.

To measure the maximal exchange level, i.e., which amount of the total CEC was available for cation exchange, Na-exchanged zeolites were allowed to react with 0.1 N solutions of the two cations (Ba^{2+} or Co^{2+}) for one week, substituting the exhausted solution for a fresh one every 12 hours.

Through the analysis of the equilibrium of the ion-exchange reactions, the selectivity of the zeolites for the examined cation was calculated. Accordingly, given a general cation exchange equation:

$$z_B A_{(s)}^{z_A+} + z_A B_{(z)}^{z_B+} \rightleftharpoons z_B A_{(z)}^{z_A+} + z_A B_{(s)}^{z_B+}, \tag{3}$$

where A is a cation of valence z_A, B is a cation of valence z_B, and subscripts s and z denote solution and zeolite, respectively, the selectivity can be defined through the distribution coefficient K_d and the separation factor α_B^A:

$$K_d = \frac{X_{A(z)}}{X_{A(s)}}, \tag{4}$$

$$\alpha_B^A = \frac{X_{A(z)}\ X_{B(s)}}{X_{B(z)}\ X_{A(s)}}, \qquad (5)$$

where $X_{A(z)}$, $X_{B(z)}$ and $X_{A(s)}$, $X_{B(s)}$ are the equivalent fractions of the cations A and B in zeolite and in solution, respectively.

If the value of K_d or α_B^A >1, the exchanger is selective for A, if it is <1 the exchanger is unselective for A (and selective for B), whereas if it is = 1 there is no particular preference exhibited by the zeolite for each of the two cations. It is worth reminding that selectivity depends on (a) the ion charge density, (b) the total and peculiar concentration of the ions in solution, (c) the structural features of the exchanger, (d) the interaction energy between ion and exchanger and (e) hydration energy of the competing ions.

3. RESULTS AND DISCUSSION

3.1. Materials characterization

The RIR analysis, performed on the zeolite tuffs and on the enriched samples in order to estimate the effectiveness of the enrichment processes, showed that in the mineralogical composition of the phillipsite- and chabazite-enriched materials minor contents of other cation exchanging phases were present. These, however, given their exiguity, were considered negligible (Table 1).

Table 1
Mineralogical composition of phillipsite- and chabazite-enriched materials

Mineral	Phillipsite-rich tuff	Chabazite-rich tuff
Phillipsite	75.20	73.80
Chabazite	2.50	1.29
Analcime	-	1.74
Feldspar	5.50	6.74
Mica	-	0.58
Glass	16.80	15.85

Table 2 reports the chemical composition and the formula of pure phillipsite and chabazite, obtained by microprobe analysis. Phillipsite appears to be rich in K, followed by Na and Ca, with minor amounts of Mg, chabazite is rich K and Ca with minor contents of Na and Mg, which is typical for the Italian sedimentary zeolites [6,7].

Measured CEC, obtained through the cross exchange method (see Experimental), turned out to be 2.72 meq/g and 2.70 meq/g for the phillipsite-rich and for the chabazite-rich materials, respectively.

3.2. Cation exchange equilibria

Figures 1 and 2 show, as an example, the kinetic curves of Ba^{2+} and Co^{2+} exchange for Na^+-exchanged phillipsite and Na^+-exchanged chabazite, respectively, obtained applying the Langmuir model to experimental data.

The two curves, confirming the good affinity of the model (lines) with experimental data (points), show that both exchange reactions are very quick and suggest that a 3-day time is enough to reach equilibrium in the thermodynamic runs.

Processing the experimental data with the Langmuir model enabled to obtain the Langmuir parameters (see Equations 1-2). In particular C_{max} turned out to be 2.65 and 1.90 for the systems reported in Figs. 1 and 2, respectively.

Table 2
Chemical analysis of phillipsite and chabazite in the zeolite-enriched tuffs

	Phillipsite %	Chabazite %
SiO_2	52.15	51.40
Al_2O_3	18.56	17.20
Fe_2O_3	0.20	0.14
MgO	0.20	0.15
CaO	2.35	4.74
Na_2O	3.30	0.85
K_2O	7.54	7.07
H_2O	15.73	18.45
Si/Al	2.38	2.53
E%	1.04	- 4.60

Formulae
Phillipsite: $(Na_{1.38}K_{2.08}Ca_{0.54}Mg_{0.06})[Al_{4.72}Fe_{0.03}Si_{11.26}O_{32}]\cdot 11.34H_2O$
Chabazite: $(Na_{0.28}K_{1.50}Ca_{0.85}Mg_{0.04})[Al_{3.38}Fe_{0.02}Si_{8.57}O_{24}]\cdot 10.27H_2O$

E% is a measure of the unbalance between the content of the trivalent framework cations (essentially Al) and that of the extraframework cations.

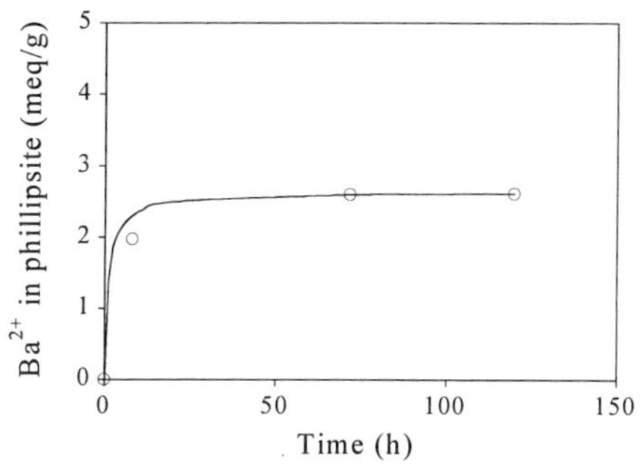

Fig. 1 - Kinetics of Ba^{2+} uptake by Na-phillipsite from a 0.1 N Ba^{2+} solution at 25°C (solid-to-liquid ratio = 1/200). Curve obtained from the Langmuir model interpolating the experimental points.

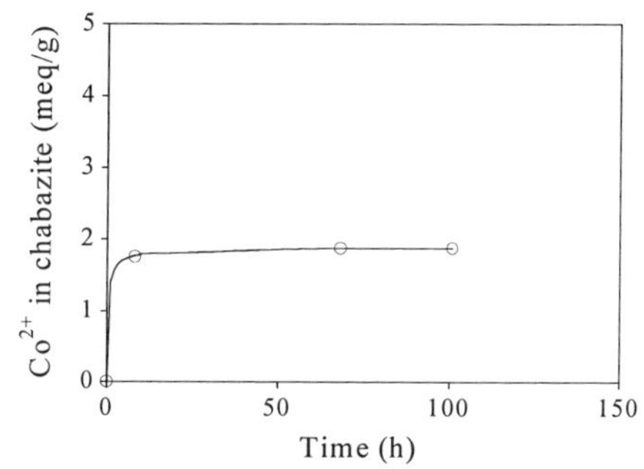

Fig. 2 - Kinetics of Co^{2+} uptake by Na-chabazite from a 0.1 N Co^{2+} solution at 25°C (solid-to-liquid ratio = 1/200). Curve obtained from the Langmuir model interpolating the experimental points.

Figures 3-10 are a representation of the selectivity of phillipsite and chabazite towards Ba^{2+} and Co^{2+}, in terms of distribution factor K_d (Figs. 3-4 and 7-8) and of separation coefficient α_B^A (Figs. 5-6 and 9-10).

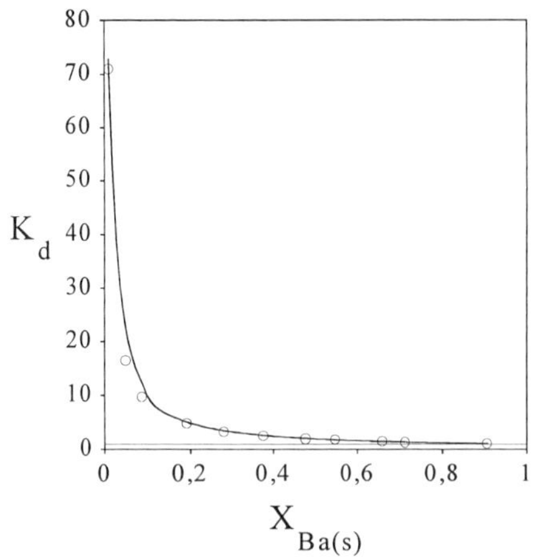

Fig. 3 - Distribution coefficient of Ba^{2+} in Na-exchanged phillipsite, as a function of the cation concentration in solution.

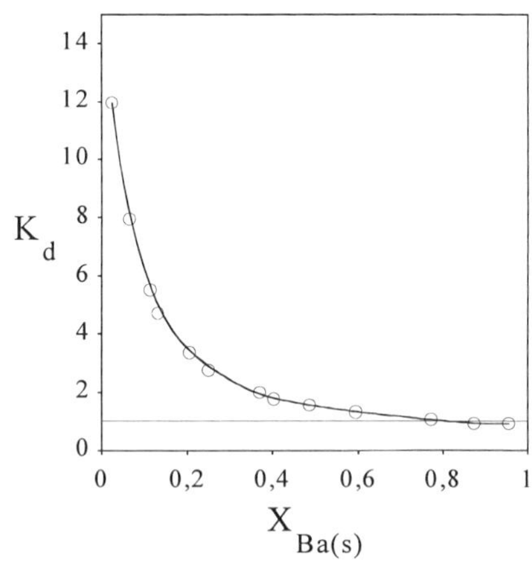

Fig. 4 - Distribution coefficient of Ba^{2+} in Na-exchanged chabazite, as a function of the cation concentration in solution.

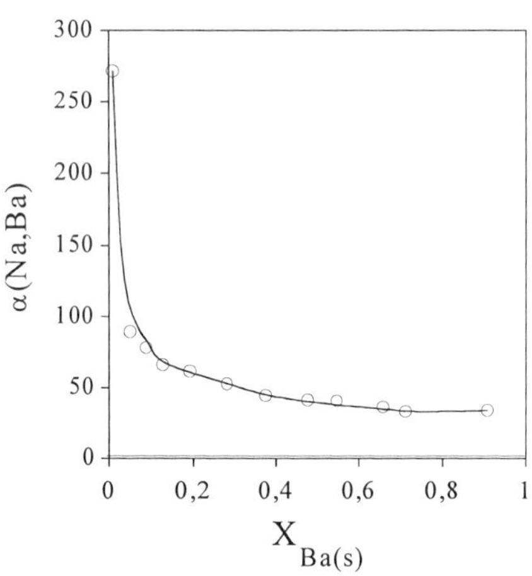

Fig. 5 - Separation factor of Ba^{2+} in Na-exchanged phillipsite, as a function of the cation concentration in solution.

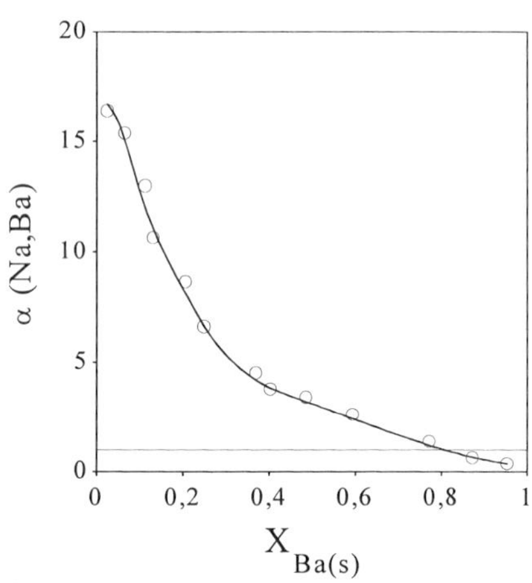

Fig. 6 - Separation factor of Ba^{2+} in Na-exchanged chabazite, as a function of the cation concentration in solution.

The trend of K_d in Figs. 3 and 4 suggests a good selectivity of phillipsite and chabazite for Ba^{2+} in most part of the composition range of liquid phase. In fact, K_d is constantly >1 for phillipsite or mostly >1 for chabazite, which corresponds to a marked preference of the exchangers for the examined cation. For both zeolites the selectivity rapidly decreases, increasing the Ba^{2+} concentration in solution, which is a common occurrence for any exchange reaction. Figs. 5-6, in which the separation factor α_{Na}^{Ba} is reported as a function of Ba^{2+} concentration in solution, fully confirm the above considerations.

The better selectivity shown by phillipsite for Ba^{2+}, compared to chabazite, is not surprising, considering that a barium zeolite, isostructural with phillipsite, occurs somewhat frequently in nature (i.e., harmotome) [19,20]. On the other hand, the results collected for phillipsite perfectly conform to those reported for the same equilibrium by Barrer and Munday [21], although the sedimentary phillipsite used by these authors was more siliceous. It addition Barrer and Munday reported a substantial irreversibility of the 2Na $\rightarrow$ Ba exchange reaction, which has not been observed in the present investigation. Quite similar data on the 2Na $\rightleftarrows$ Ba exchange in a sedimentary phillipsite from the Canary Islands have been reported by Garcia Hernandez et al. [22].

Also the data collected for chabazite are in agreement with previous investigations by Barrer and coworkers [23,24], even though these authors have experimented in different conditions of temperature or cation concentration in solution.

An interpretation of the excellent behaviour of Na-phillipsite and Na-chabazite towards Ba is possible on the basis of the known structural features of these two zeolites.

According to the most recent structure refinement [25,26], there are two types of cation sites in phillipsite: type I site is on the mirror plane (010) and is usually populated totally or partially by large-size cations; type II site, whose occupancy does not exceed 50%, is located near the intersection of the two sets of channels and is usually populated by smaller cations, but can host also large cations. There are two other positions, close to site II, labelled sites II' and II" [25,26], which are usually empty, unless the number of available cations is larger than that sites I and II may accommodate. The great selectivity of phillipsite for Ba suggests that this cation, in substantial accordance with data reported previously [25], can occupy every extra-framework cationic sites.

Chabazite framework presents two different kinds of exchange sites [27-28], the so-called C2-C4 sites, situated in the large cavity (chabazite cage) and accessible through eight-membered rings and the C1 site, situated in the center of the double exagonal ring, that is accessible through the six-membered ring. It is therefore likely that the former sites are easily accessible to barium, what is in agreement with the good selectivity exhibited by chabazite at the lower concentrations in the liquid phase (Figs. 4 and 6), whereas the latter unselective site, is accessible only when the ionic strength of Ba in solution increases.

The results collected for the exchange of Na-phillipsite- and Na-chabazite-rich tuff with Co^{2+} (Figs. 7-10) show a clear unselectivity of both exchangers for this cation, the values of Kd and α_B^A being constantly <1 in the whole range of Co^{2+} concentration in solution.

The unselectivity showed by phillipsite and chabazite for Co^{2+} is hardly explainable as a difficulty of access to sites, in fact the size of this cation should enable it to easily enter most sites of both zeolites. This behaviour can be therefore ascribed either to the high hydration energy of Co^{2+} (E_{hyd} = -1996 kj/mole), which would favour its stay in solution [29-30] and/or to the difficulty of this cation to assume the right coordination of the extra framework cationic sites of the two zeolites. Accordingly, as an example, Co^{2+} might occupy more or less forcedly site II of the phillipsite structure, whereas accessibility would be practically hindered to site I (exchange limit close to 50%).

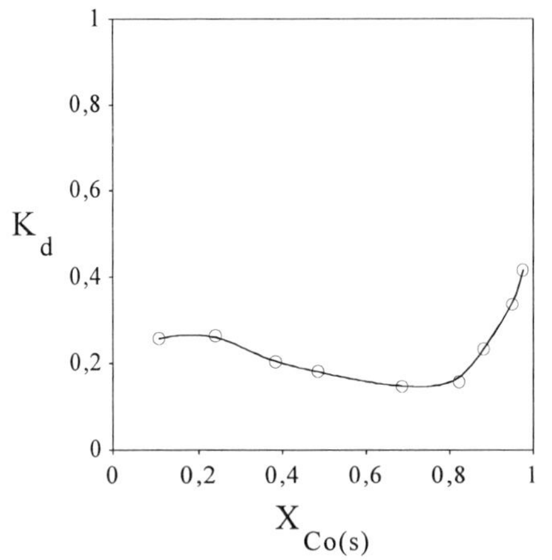

Fig. 7 - Distribution coefficient of Co^{2+} in Na-exchanged phillipsite, as a function of the cation concentration in solution.

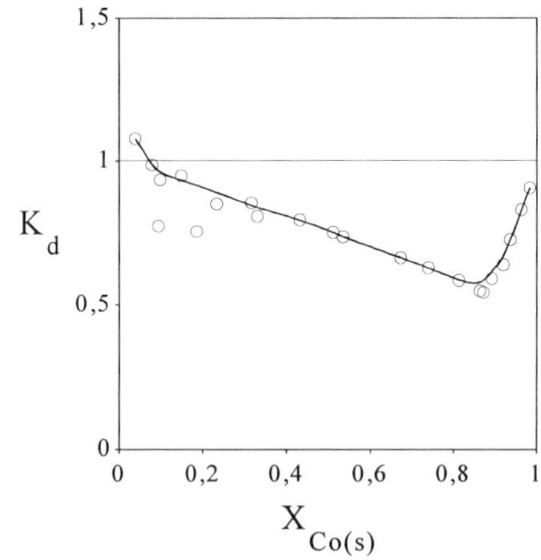

Fig. 8- Distribution coefficient of Co^{2+} in Na-exchanged chabazite, as a function of the cation concentration in solution.

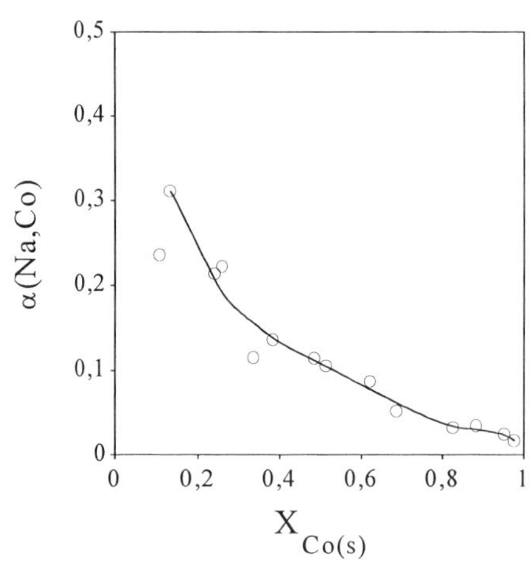

Fig. 9 - Separation factor of Co^{2+} in Na-exchanged phillipsite, as a function of the cation concentration in solution.

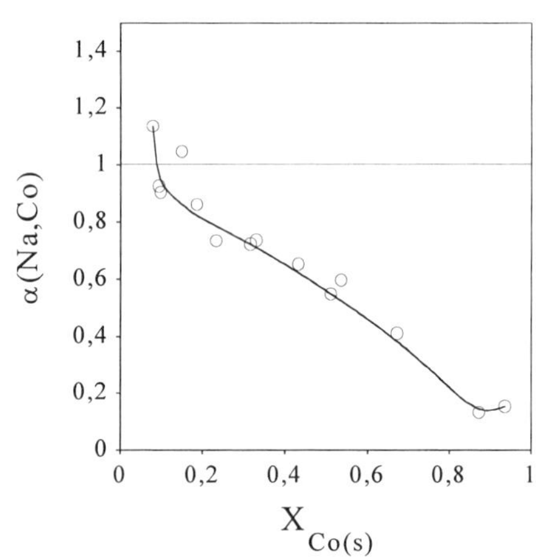

Fig. 10 - Separation factor of Co^{2+} in Na-exchanged chabazite, as a function of the cation concentration in solution.

4. CONCLUSION

The present investigation contributes to a better knowledge of the cation exchange selectivity series of phillipsite and chabazite [5]. The collected results are consistent with those of previous investigations, in which good selectivity for large cations and unselectivity for smaller cations were demonstrated by both zeolites [8].

As regards in particular the two pollutant cations examined, it is of interest that both zeolites show a moderate to good selectivity for Ba in the whole range of cation composition in solution.

Seeing that both zeolites are ordinary constituents of most Italian zeolite-rich tuffs [10], it may be concluded that Italian tuffs have excellent possibilities to be used to remove Ba from wastewaters of nuclear origin. On the other hand applications appear practicable also for Co, if discontinuous removal processes (addition of zeolites to the waste solution) are selected, followed by procedures for stabilizing the pollutant-loaded zeolitic sludge [8].

Research is now directed at confirming the above results also in dynamic condition, collecting data of fixed-bed column tests.

5. ACKNOWLEDGEMENTS

This research was carried out with the financial support of the Ministry of University and Scientific and Technological Research (MURST), Progetti di Ricerca di Interesse Nazionale (Cofin '98).

REFERENCES

1. M. Sitting, Pollutant Removal Handbook, Noyes Data Corporation, Park Ridge, New Jersey, 1973, p. 478.
2. G. Bianucci and E. Ribaldone Bianucci, Hoepli, Milan, Italy, 1977, p. 381.
3. E. Chmielewska-Horvathova and J. Lesny, Polish Journal of Environmental Studies, 5(2) (1996) 13.
4. E. Chmielewska-Horvathova and J. Lesny, J. Radioanal. Nucl. Chem., Letters, 166 (1992) 41.
5. C. Colella, Mineralium Deposita, 31 (1996) 554.
6. M. de' Gennaro, C. Colella, E. Franco and R. Aiello, Industrial Minerals, No. 186 (1983) 47.
7. M. de' Gennaro, C. Colella, R. Aiello and E. Franco, Industrial Minerals, No. 204 (1984) 97.
8. C. Colella, C., in Natural Microporous Materials in Environmental Technology, P. Misaelides, F. Macasek, T. J. Pinnavaia and C. Colella, eds., NATO Sciences Series, Series E: Appl. Sc., Vol. 362, Kluver A.P., Dordrecht, The Nertherlands, 1999, p. 207.
9. M. Adabbo, D. Caputo, B. de Gennaro, M. Pansini and C. Colella, Microporous and Mesoporous Materials, 28 (1999) 315.
10. M. de' Gennaro and A. Langella, Mineral. Deposita, 31 (1996) 452.
11. S.J. Chipera and D.L. Bish, Powder Diffraction, 10 (1995) 47.
12. M. de'Gennaro and E. Franco, Ind. Min. (Rome), 30 (1979) 329.
13. C. Colella, M. de' Gennaro, E. Franco and R. Aiello, Rend. Soc. Ital. Min. Petr., 38(3) (1982) 1423.

14. M. Pansini, C. Colella, D. Caputo, M. de' Gennaro, and A. Langella, Microporous and Mesoporous Materials, 5 (1996) 357.
15. G.E. Boyd, J. Schubert and A.W. Adamson, Jour. Amer. Chem. Soc., 69 (1947) 2818.
16. H. Mimura, F. Tachibana and K. Akiba, K., J. Nuclear Science Techn., 29(2) (1992) 184.
17. P. Fletcher and R.P. Townsend, J. Chem. Soc., Faraday Trans. I, 77 (1981) 497.
18. G. Schwarzenbach and H. Flaschka, Complexometric Titration, Methuen, London, (1969).
19. R. Rinaldi, J.J. Pluth and J.V. Smith, Acta Crystal., B30 (1974) 2426.
20. G. Gottardi and E. Galli, Natural Zeolites, Springer-Verlag, Berlin, 1985, p. 134.
21. R.M. Barrer and B.M. Munday, J. Chem. Soc. A, (1971) 2904.
22. J.E. García Hernández, R. Diaz, J.S. Notario del Pino and M.M. Gonzáles Martín, Appl. Clay Sci. 9 (1994) 129.
23. R.M. Barrer and D.C.R. Sammon, J. Chem. Soc., (1955) 2838.
24. R.M. Barrer, J.A. Davies and L.V.C. Rees, J. Inorg. Nucl. Chem., 31 (1969) 219.
25. A.F. Gualtieri, E. Passaglia and E. Galli, in Natural Zeolites for the Third Millenium, C. Colella and F.A. Mumpton, eds., De Frede Editore, Naples, Italy, 2000, p. 93.
26. A.F. Gualtieri, D. Caputo and C. Colella, Microporous and Mesoporous Materials, 32 (1999) 319.
27. M. Calligaris, G. Nardin, L. Randaccio and P. Comin Chiaromonti, Acta Cryst., B38 (1982) 602.
28. A. Alberti, E. Galli, G. Vezzalini, E. Passaglia and P.F. Zanazzi, Zeolites, 2 (1982) 303.
29. G. Eisenman, Biophys. J., 2 (1962) 259.
30. H.S. Sherry, in Ion exchange: A series of advances, J.A. Marinsky, ed., Marcel dekker, New York, 2 (1969) 89.

Studies in Surface Science and Catalysis 140
A. Gamba, C. Colella and S. Coluccia (Editors)

The role of Al- and Fe-oxy-hydroxides in determining surface properties of soil ped models, with emphasis on phosphorus sorption/desorption phenomena

E. Coppola[a], M. L. Ambrosino[b], S. Deiana[c] and A. Buondonno[a]

[a] Dipartimento di Scienze Ambientali, Polo Scientifico di Caserta, Seconda Università di Napoli, via Vivaldi 43, 81100 Caserta, Italy.

[b] Dipartimento di Scienza degli Alimenti, Università di Napoli "Federico II", via Università 100, 80055 Portici, Italy.

[c] Dipartimento di Scienze Ambientali, Agrarie e Biotecnologie Agroalimentari, viale Italia 39, Sassari, Italy.

An investigation was carried out to elucidate the role of *Al* or *Fe* poorly-ordered oxy-hydroxides in determining the surface properties of soil ped models formed by interaction with bentonite clay (*C*) and polyphenol (*pp*) in acidic *milieu*, with emphasis on phosphorus (*P*) sorption/desorption phenomena. *P* was chosen because of its important involvement on pedogenic processes as well as on soil fertility. *P* was added to models (P_a) at the rate of 10 mg·g^{-1}. P_a was then extracted from the models at different aggregation states (suspended, freeze dried, air-dried, re-suspended), by using the electro-ultrafiltration technique (*EUF-P*). The cumulative amount *EUF-P* extracted after 40 min, representing the loosely-bound immediately-exchangeable *P* (P_i) and the maximum desorbable *EUF-P*, representing the slowly released *P* (P_m) were considered. The models were analyzed for their Cation Exchange Capacity (*CEC*), Specific Surface Area (*SSA*), Specific Cationic Surface Charge (*SCSC*), Titratable Acidity (*TA*) to pH = 7.0 or pH = 8.0 by KOH or by $Ca(OH)_2$, and Phosphorus Sorption Coefficient (*PSC*). The *C-Fe-pp* model was characterized by an elevated surface reactivity, showing values of *CEC* (1777 μeq·g^{-1}), *SSA* (483.2 m^2·g^{-1}), *SCSC* (3.678 μeq·m^{-2}), *TA* (from 572 to 804 μeq·g^{-1}) and *PSC* (92.4%), which were considerably larger than the respective determined for the *C-Al-pp* model (*CEC* = 1173 μeq·g $^{-1}$, *SSA* = 420.0 m^2·g^{-1}, *SCSC* = 2.793 μeq·m^{-2}, *TA* from 36 to 176 μeq·g^{-1}, and *PSC* = 76.8%). Contrastingly, both the amounts *EUF* P_i and P_m desorbable from the same *C-Fe-pp* model were much larger than the respective determined for the *C-Al-pp* model. With reference to the aggregation state of the prepared materials, P_i decreased on the average as suspended > freeze-dried > air-dried ≥ re-suspended. The desiccation *via* air-drying definitively stabilized the *P* pool, especially in the *C-Al-pp* model. Such stabilization was irreversible, since both P_i and P_m for *C-Al-pp*, and P_i for *C-Fe-pp*, did not substantially change in the re-suspended materials. With regard to kinetics aspects, the *EUF-P* desorption from *C-Al-pp* always followed a 2nd reaction order, while that from *C-Fe-pp* followed zero or 1st reaction order from suspended and freeze-dried, or from air-dried and re-suspended materials, respectively. The achieved data clearly demonstrated that the different interactions of bentonite clay and polyphenol with *Al-* or *Fe-*hydrolytic products led to the formation of aggregates with peculiar surface properties and physical-chemical behavior, with distinct mechanisms controlling the interactions of *P* with *C-Al-pp* or *C-Fe-pp* in both sorption and desorption phenomena.

1. INTRODUCTION

Peds are the fundamental structural units of soil [1,2,3,4,5,6]. As organo-mineral aggregates, they are almost stable bridge-systems "*C-polyM-SOM*" formed by interactions among clay (*C*), polyvalent metal cations (*polyM*), and Soil Organic Matter (*SOM*) [3,4,5]. The soil forming factors and processes prevailing in a given pedoclimatic environment determine the peculiar constitution and development of soil peds and their surface properties, such as surface area, charge, morphology, functional groups. The surface reactivity, *viz.* the resultant of the various ped surface properties, is probably the most relevant expression of the soil character. In fact, the surface reactivity governs the exchanges of matter and energy between the soil and its environment, including the soil thermodynamic behavior, nutrient or toxic adsorption, water absorption, soil strength, soil transport processes and ion exchange phenomena [7,8], thus influencing the evolution of the soil itself.

Previous researches based on a modeling approach showed that the nature of the *polyM* bridge cation binding *C* with *SOM* dramatically affected the aggregate properties [9,10]. In particular, poorly ordered *Al* or *Fe* oxy-hydroxides exerted a contrasting effect on mineralogical, physical and chemical properties of ped models prepared in acidic *milieu* [10]. The present work aims to analyze the consequences of such different features on the surface activities of the ped models, devoting emphasis on their effects on phosphorus (*P*) sorption/desorption phenomena. Investigations on phosphorus (*P*) dynamics in soil are of a great interest for basic and applied studies in Environmental Pedology, *i.e.* to refine the comprehension of some pedogenic processes, or to manage soil *P* fertility. *P* is an essential nutrient for the living organisms. Its availability in soil to biota and to plants is controlled by several abiotic-chemical and biological factors and processes, often interrelated each other. As a rule, the behavior of phosphorus in soil is strictly related to the pedoclimatic conditions, and it trustworthily reflects the pedogenic history of soil. In neutral-acidic non-calcareous soils, the chemical-abiotic processes of *P* dynamics are mostly controlled by the dominant *Al-* and *Fe-* amorphous compounds which are essential components of organo-mineral aggregates in soils as Alfisols, Andisols, Oxisols, Spodosols [11]. In particular, the *P* sorption capacity represents a diagnostic character for Andisols [11]. Several mechanisms, including precipitation/dissolution and sorption/desorption, control the *P* mobility in soil. Albeit the overall process can be regarded as a continuum between precipitation and surface reactions [12], *P*-desorption isotherms studies are the favorite tools to describe the P release from soil particles and the relationships between *P* in solution and *P* retained by the soil matrix [12,13,14,15]. A useful technique to obtain both quantitative and kinetics information on *P* desorption is the electro-ultrafiltration (*EUF*), first introduced by Németh [16,17] as a multiple-element soil test to determine nutrient availability to plants. This technique involves a combination of electro-dialysis and ultrafiltration to remove electrolytes such as *P* from aqueous soil suspensions [18,19,20,21] or from rock materials suspensions [22]. By performing stepwise extractions under controlled electrical field strength and temperature, *P-*

release isotherms can be achieved plotting the cumulative amounts extracted against the respective extraction time. Simple calculations can provide the main kinetics parameters and the extent of reaction allows estimating the maximum amount of *P* extractable from a given matrix. The *EUF* then offers some advantages compared with classical ion-exchange isotherms: (1) the use of water as reaction medium avoids undesired side-effects by foreign electrolytes; (2) whereas the ratio soil/suspension is kept constant, the solution is continuously diluted, so that the precipitation of sparingly-soluble salts and the aggregation of soil particles are prevented; and (3), most relevant, the *EUF*-desorption process is induced by continuous disequilibrium conditions, so that the elements release is not mediated by a specific ion-exchange reaction, but it is essentially dependent on the strength of the element/matrix interactions; from this standpoint, the *EUF*-kinetics parameters let to appropriately determine the matrix *buffering capacity* (*BC*) [22], which represents the ability of the matrix to replenish into the solution the aliquots of a given element displaced away by the extraction process. All this renders the performance of the EUF-extraction isotherms free from the typical constraints of the electrolytic nature of the medium and of ion selectivity and competition in the release process by exchange reactions.

In this paper, the *EUF* technique is applied to study the quantitative and kinetics of *P* release as affected by surface properties of soil ped models formed in acidic *milieu* by interaction among montmorillonitic clay, *Al* or *Fe* oxy-hydroxides and polyphenol [10]. To enhance differences between ped models, various aggregation states (suspensions or solid phases) were considered.

2. MATERIAL AND METHODS

2.1. Ped model preparation

Ped models were prepared following the procedures described by Buondonno and Coppola [10], starting from suspensions at pH 5.5 containing montmorillonitic clay (*C*), *Al*- or *Fe*-chloride, and a polyphenol (*pp*) representing soil humic substances ($C_{72}H_{42}O_{56}$, mw 1701.2 D). The final *C* : *Al*- : *pp* and *C* : *Fe*- : *pp* ratios were both 1 g : 6 mmol : 0.6 mmol. The ped models prepared in the presence of *Al*- or *Fe*- will be indicated as *C-Al-pp*, or *C-Fe-pp*, respectively.

2.2. Physical-chemical properties

The following main properties were determined:

- Cation Exchange Capacity (*CEC*), by $BaCl_2/MgSO_4$ exchange [23], as $\mu eq \cdot g^{-1}$ sample ;
- Specific Surface Area (*SSA*), by ethylene glycol monoethylether (*EGME*) adsorption [24], as $m^2 \cdot g^{-1}$ sample; the Specific Cationic Surface Charge (*SCSC*) was also calculated as *CEC/SA*, and expressed as $\mu eq \cdot m^{-2}$ sample;
- Titratable Acidity (*TA*), by titrating up to pH = 7.0 and pH = 8.0 by 2 m*N* KOH and by 2 m*N* $Ca(OH)_2$ suspensions containing 50 mg sample in 25 ml of 1 *N* KCl or in 1 *N* $CaCl_2$, respectively [25]; *TA* was expressed per both mass unit, as $\mu eq \cdot g^{-1}$ sample, and surface unit, as $\mu eq \cdot m^{-2}$.

2.3. Phosphorus Sorption Coefficient (*PSC*)

The *PSC* was determined following the procedures described by Nunozawa and Tanaka [26], here modified in accordance with the specific aim of the investigation: 250 mg of sample was suspended in 25 ml of 426 $mg \cdot l^{-1}$ $(NH_4)_2HPO_4$ solution, with starting *P* concentration (*Ps*) = 100 $mg \cdot l^{-1}$ and initial pre-equilibration pH ($pre\text{-}pH_{PSC}$) = 7.79. After 24 h shaking, the final post-equilibration pH ($post\text{-}pH_{PSC}$) was recorded, the residual *P* concentration in solution (*Pr*) was determined, and the *PSC* was calculated as:

$$PSC\% = [100\ (Ps - Pr)]\ /\ Ps \tag{1}$$

The *P* sorbed after 24 h was also expressed per mass unit, as $\mu g\ P \cdot g^{-1}$ sample, and per surface unit, as $\mu g\ P \cdot m^{-2}$.

2.4. Preparation of materials containing *P*

Ten milligram of *P* was added (P_a) per 1 g of ped models by dispersing suitable aliquots of each model in a *P*-solution (426 $mg \cdot l^{-1}$ $(NH_4)_2HPO_4$; *P* concentration = 100 $mg \cdot l^{-1}$ and initial pH = 7.79); for each model, materials at three different aggregation states were obtained:
a) – suspended: 250 mg of sample in 25 ml *P*-solution, kept under shaking for 24 h at 25 °C before use;
b) – solid phase/air dried: 1 g of sample in 100 ml *P*-solution, shaken 24 h, dried at 40 °C for 15 days, homogenized and screened to 0.250 mm;
c) – re-suspended: 250 mg of solid material obtained by *b)* procedure was re-suspended in 25 ml water and kept under shaking for 24 h at 25 °C before use.
d) – solid phase/freeze-dried: 1 g of sample in 100 ml *P*-solution, shaken 24 h, freeze-dried, homogenized and screened to 0.250 mm.

2.5. *P* desorption isotherms by electro-ultrafiltration (*EUF-P*)

A three-cell apparatus (Vogel 724S) was used in electro-ultrafiltration [22]. The middle cell containing the material suspension had a stirrer and a water inflow. Each side of the middle cell was provided with a micropore filter attached to the platinum electrodes that separated the middle cell from the two outside chambers. These chambers had vacuum connections. Therefore, the ions accumulating at the respective electrodes were washed away by the continuous stream of the solution from the suspension in the middle cell to the eluate in the collecting tanks combined with the side chambers. A steady disequilibrium between the *P* on the materials and the respective soluble form in solution occurred and the materials were forced to replenish the *P* displaced away with the eluate. The concentration of *P* in the eluates was measured by UV-VIS spectrophotometry as phosphomolybdate [23].

Twenty-five milliliters of the suspended *a)* or re-suspended *c)* materials, each containing 250 mg of ped model, or 250 mg of solid phase/air-dried *b)* or freeze-dried *d)* was used to perform a batch of eight medium-energy *EUF-P* fractionated extractions. The following parameters were set to obtain the suitable release conditions: running model/solution ratio in the middle cell: 1 : 10 w/v; stepwise extraction time: 5 minutes; voltage: 400 V max; current: 40 mA max; temperature: 80 °C.

The *EUF-P* release rates were evaluated assuming that the experimental data fit the general kinetic equation [19,20,22]:

$$\partial y/\partial t = k\ (Y-y)^n, \tag{2}$$

where $\partial y/\partial t$ = desorption rate for P (mg·100g^{-1}·min^{-1}), y = cumulative *c-EUF-P* (mg·100g^{-1}) extracted at t time, t = extraction time (min), Y = maximum desorbable *m-EUF-P* (mg·100g^{-1}), k = constant rate, and n = order of reaction.

Zero order, first order and second order kinetics models were fitted to the experimental data of *EUF-P* isotherms. Equation (2) was then developed for n = 0, 1, or 2; after integrating and solving for y, the Eq. (2) reads:

$$y = kt \quad (3)$$
$$y = Y\,(1-e^{kt}) \quad (4)$$
$$y = Yt/[t+(1/(kY))] \quad (5)$$

for zero, first, and second order kinetics models of *P* release, respectively.

The cumulative amount *c-EUF-P* extracted after 40 min and considered to represent the loosely bound immediately-exchangeable *P* was designated as P_i; the maximum desorbable *m-EUF-P*, assumed to represent the slowly released *P* was designated as P_m [22]. The buffering capacity (*BC*), *i.e.*, the *P* release rate, expressed as mg·100g^{-1}·min^{-1}, was the first derivative of Eq. (2), calculated at t = 0.

Each determination was replicated not less than three times. All data were expressed on the dry-weight (105°C) basis.

3. RESULTS

3.1. Surface properties

The presence of *Al-* or *Fe-* dramatically affected the surface properties of the prepared models. Both models showed very high *CEC* and *SSA* (Table 1). This was clearly due to the presence of polyphenol as organic component of the aggregates. In fact, the *CEC* and *SSA* were 785 μeq·g^{-1} and 160.4 m^2·g^{-1} for bentonite, respectively, but they increased up to 2420 μeq·g^{-1} and 620.2 m^2·g^{-1} for polyphenol.

Table 1
Cation Exchange Capacity (*CSC*, μeq·g^{-1}), Specific Surface Area (*SSA*, m^2·g^{-1}), and Specific Cationic Surface Charge (*SCSC*, μeq·m^{-2}) of *C-Al-pp* and *C-Al-pp* models

Model	*CSC*	*SSA*	*SCS*
C-Al-pp	1173	420.0	2.793
C-Fe-pp	1777	483.2	3.678

Indeed, the model *C-Fe-pp* exhibited not only the largest *CEC* and *SSA*, but also the largest *SCSC* with respect to *C-Al-pp* (Table 1). This is consistent with previous findings [10], which revealed that *C-Al-pp* model formed small aggregates more compact and stable than *C-Fe-pp*, and denotes that *C-Fe-pp* developed a wider reactive surface, whereas *Al-* hydrolytic products were more able in binding and neutralizing the negative charges of clay and organic matter. A further clue for such a conjecture is provided by the measurement of the titratable acidity

(Table 2), which adequately reflects the anion sorptivity of the models. To take into account the different role of clay mineral and organic matter in cation exchange phenomena, acidity titration was performed in 1 *N* $CaCl_2$ and KCl solutions. In fact, it is well known that organic matter binds Ca more strongly than K ions, whereas 2:1 clay minerals such as bentonite are more selective for K than Ca ions [25]. As a consequence, the presence of K^+ favored the exchange of the clay-sorbed and interlayer *Al-* or *Fe-*, while the presence of Ca^{2+} destabilized the organically-bound *Al-* and *Fe-*.

Table 2
Titratable acidity up to pH = 7 and pH = 8 by 2 m*N* KOH in 1 *N* KCl and by 2 m*N* $Ca(OH)_2$ in 1 *N* $CaCl_2$ (data are expressed as $\mu eq \cdot g^{-1}$ sample, and as $\mu eq \cdot m^{-2}$ sample)

Model	titrating solution	$\mu eq \cdot g^{-1}$ pH=7.0	$\mu eq \cdot g^{-1}$ pH=8.0	$\mu eq \cdot m^{-2}$ pH=7.0	$\mu eq \cdot m^{-2}$ pH=8.0
C-Al-pp	KOH	40	144	0.095	0.342
C-Al-pp	$Ca(OH)_2$	36	176	0.086	0.419
C-Fe-pp	KOH	572	804	1.184	1.664
C-Fe-pp	$Ca(OH)_2$	532	798	1.109	1.651

The data in Table 2 clearly show that the model *C-Fe-pp* needed much more base than *C-Al-pp* to be neutralized (about 14 times) or to be brought to pH = 8.0 (about 5 times). Accordingly, also the anion surface receptivity of *C-Fe-pp* was noticeably larger than that of *C-Al-pp*: about 13 times at pH = 7.0 and about 4 times at pH = 8.0. For both models, however, the titration end points were not significantly affected by the nature of the environment dominant cation, K^+ or Ca^{2+}. This could suggest the absence of a prevailing selective binding of *Al-* or *Fe-* by the inorganic or the organic components of the model.

3.2. *P* sorption

In agreement with the observations above, larger amount of *P* was sorbed by *C-Fe-pp* than *C-Al-pp* (Table 3). Interestingly, the post-reaction $post\text{-}pH_{PSC}$ was neutral for *C-Al-pp*, but strongly acid for *C-Fe-pp*. This clearly denotes the higher surface reactivity of *C-Fe-pp*, and confirms that this model behaves as an acid stronger than *C-Al-pp*. Such consideration also implies that *P* interacted with *C-Al-pp* or with *C-Fe-pp via* distinct physical-chemical mechanisms, which deserve clarification in further development of the present research.

Table 3
Phosphorus Sorption Coefficient (*PSC%*), amounts of *P* sorbed (as $\mu g\ P \cdot g^{-1}$ sample, and as $\mu g\ P \cdot m^{-2}$ sample), pre-equilibration ($pre\text{-}pH_{PSC}$) and post-equilibration ($post\text{-}pH_{PSC}$) pH

Model	*PSC%*	*P* sorbed $\mu g\ P \cdot g^{-1}$	*P* sorbed $\mu g\ P \cdot m^{-2}$	$pre\text{-}pH_{PSC}$	$post\text{-}pH_{PSC}$
C-Al-pp	76.8	7676	18.3	7.79	7.05
C-Fe-pp	92.4	9236	19.1	7.79	4.43

3.3. Quantitative and kinetics of *EUF-P* release

Both quantitative and kinetics parameters of *P* release by electro-ultrafiltration (*EUF-P*) were substantially affected by the presence of *Al-* or *Fe-* in the models, as well as by the various aggregation states of the investigated materials (Tables 4, 5; Figs. 1, 2).

Table 4
Quantitative and kinetics parameters of *P* release isotherms by electro-ultrafiltration (*EUF-P*)*

Model	*aggregation state*	P_i mg·100g^{-1}	P_m mg·100g^{-1}	*n* *order*	*BC* mg·100g^{-1}·min^{-1}
C-Al-pp	suspended	475.9	537.2	2	148.2
C-Fe-pp	suspended	556.3	601.7	1	37.5
C-Al-pp	freeze-dried	106.9	110.0	2	44.7
C-Fe-pp	freeze-dried	267.3	440.5	1	10.5
C-Al-pp	air-dried	24.4	31.0	2	2.3
C-Fe-pp	air-dried	205.7	*n.e.*	0	5.6
C-Al-pp	re-suspended	21.6	25.3	2	3.0
C-Fe-pp	re-suspended	197.7	*n.e.*	0	5.3

* P_i = loosely-bound immediately-exchangeable *P* (mg·100g^{-1}); P_m = maximum desorbable *EUF-P* (mg·100g^{-1}); *n* = kinetics order; *BC* = buffering capacity (mg·100g^{-1}·min^{-1}); (*n.e.* = not estimable, does not apply to zero order kinetics)

Table 5
Kinetics equations and parameters for *P* release isotherms by electro-ultrafiltration (*EUF-P*)*

Model	*a.s.*	equation	parameters
C-Al-pp	*s*	$y = 1000-\{Yt/[t+(1/(kY))]\}$	$y = 1000-\{537.2t/[t+(1/(5.1 \cdot 537.2))]\}$
C-Fe-pp	*s*	$y = 1000-[Y(1-e^{kt})]$	$y = 1000-[601.7(1-e^{600.0t})]$
C-Al-pp	*f.d.*	$y = 1000-\{Yt/[t+(1/(kY))]\}$	$y = 1000-\{110.0t/[t+(1/(36.9 \cdot 110.0))]\}$
C-Fe-pp	*f.d.*	$y = 1000-[Y(1-e^{kt})]$	$y = 1000-[440.5(1-e^{238.0t})]$
C-Al-pp	*a.d.*	$y = 1000-\{Yt/[t+(1/(kY))]\}$	$y = 1000-\{31.0t/[t+(1/(24.0 \cdot 31.0))]\}$
C-Fe-pp	*a.d.*	$y = 1000-(kt)$	$y = 1000-(5.6t)$
C-Al-pp	*r.s.*	$y = 1000-\{Yt/[t+(1/(kY))]\}$	$y = 1000-\{25.3t/[t+(1/(46.7 \cdot 25.3))]\}$
C-Fe-pp	*r.s.*	$y = 1000-(kt)$	$y = 1000-(5.3t)$

* *a.s.* = aggregation state; *s* = suspended; *f.d.* = freeze-dried; *a.d.* = air dried; *r.s.* = re-suspended; *y* = cumulative *c-EUF-P* (mg·100g^{-1}) extracted at *t* time; *t* = extraction time (min); *Y* = maximum desorbable *m-EUF-P* (mg·100g^{-1}); *k* = constant rate (x 10^4); 1000 = amount of *P* initially added (mg·100g^{-1})

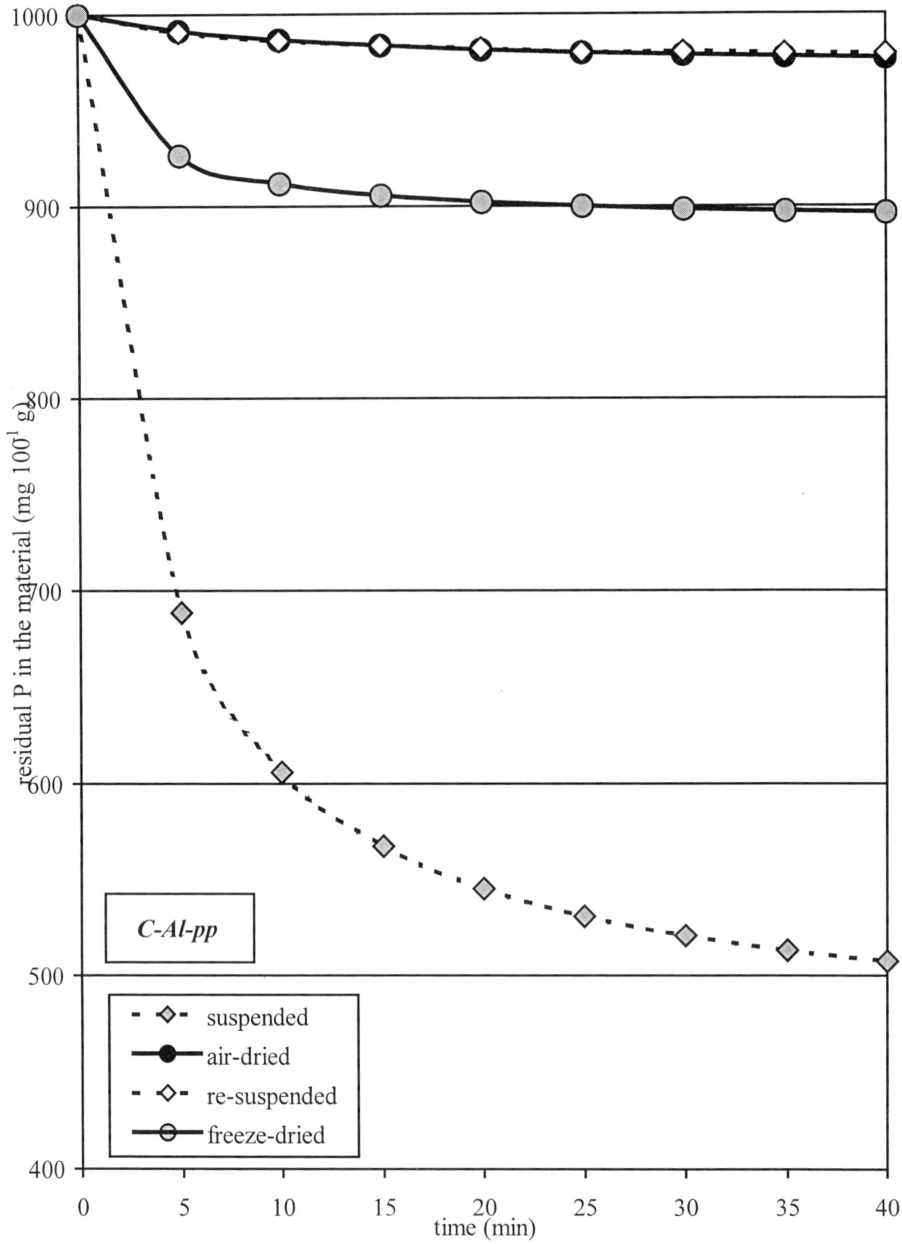

Fig. 1. *EUF-P* release from *C-Al-pp* model at different aggregation states.

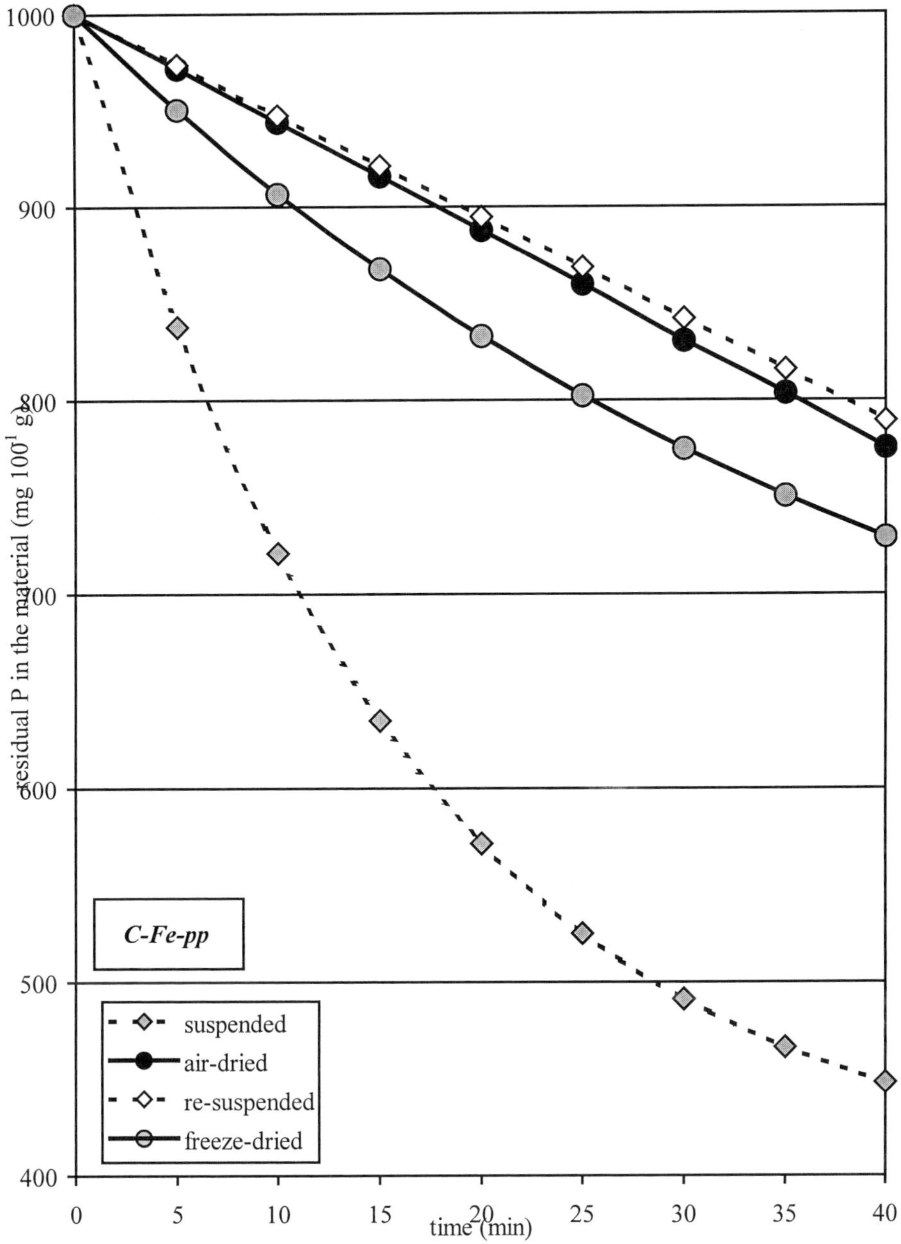

Fig. 2. *EUF-P* release from *C-Fe-pp* model at different aggregation states.

As a rule, *C-Fe-pp* samples released amounts of P_i always larger than those determined for *C-Al-pp*, from about 1.2 times (suspended material) to 9.2 times (re-suspended material). With reference to the aggregation state of the prepared materials, P_i decreased on the average as suspended > freeze-dried > air dried ≥ re-suspended, with a ratio of 4.7 : 1.7 : 1.05 : 1.00.

In any case, all the materials appeared to substantially retain the initially added P_a. In fact, the maximum P_m extractable by EUF did not exceed 53.7% and 60.2% of P_a for suspended *C-Al-pp* and *C-Fe-pp* materials, respectively, and it decreased to 11.0% and 44.1% for the respective *C-Al-pp* and *C-Fe-pp* freeze-dried materials, respectively.

The desiccation *via* air-drying definitively stabilized the *P* pool, especially in the *C-Al-pp* model; for this sample, the P_i and P_m where about 2 or 3% of P_a, respectively, while the P_i of *C-Fe-pp* represented about the 20% of P_a (please, note that P_m cannot be estimated for *C-Fe-pp*).

Evidently, the air-drying induced a much more hard-wearing aggregation than freeze-drying did, so that the individual surface properties of the materials were overwhelmingly influenced. Indeed, such stabilization was irreversible, since both P_i and P_m for *C-Al-pp*, and P_i for *C-Fe-pp*, did not substantially change in the re-suspended materials.

With regard to kinetics aspects, *EUF-P* release isotherms with distinct equation parameters and following various reaction orders were obtained (Tables 4, 5).

The *EUF-P* desorption from *C-Al-pp* always followed a 2nd reaction order, while that from *C-Fe-pp* followed zero or 1st reaction order from suspended and freeze-dried or from air-dried and re-suspended materials, respectively (Tables 4, 5; Figs. 1, 2).

For suspended and freeze dried materials, the initial *EUF-P* release rate (*BC*) from *C-Al-pp* was higher than that from *C-Fe-pp* (Table 4), but it rapidly decreased within the 40-min extraction period (Fig. 1). On the contrary, more *EUF-P* was slowly but steadily released from the respective suspended and freeze dried *C-Fe-pp* materials (Table 4, Fig. 2). For both air-dried and re-suspended materials, the *BC* was particularly poor, in the order of about 2–3 $mg \cdot 100g^{-1} \cdot min^{-1}$ for *C-Al-pp*, and of about 5–6 $mg \cdot 100g^{-1} \cdot min^{-1}$ for *C-Fe-pp*.

It is also interesting to note that, in accordance with the observations above, the equation parameters for *P* desorption from air-dried *C-Al-pp* model were almost similar to those obtained for the respective re-suspended model (Table 5). An analogous behavior was observed for the *C-Fe-pp* model (Table 5). Consequently, for whichever *C-Al-pp* and *C-Fe-pp* models, the *P* release isotherms curves from the respective air-dried or re-suspended materials were practically coincident (Figs. 1, 2).

4. CONCLUSION

The achieved data clearly demonstrate that the different interactions of bentonite clay and polyphenol with *Al* or *Fe* hydrolytic products led to the formation of aggregates with peculiar surface properties and physical-chemical behavior, with particular reference to the *P* sorption/desorption phenomena.

The *C-Fe-pp* model exhibited Cation Exchange Capacity, Specific Surface Area, Specific Cationic Surface Charge, Titratable Acidity, and *P* sorption capacity higher than *C-Al-pp*. For both model, the amounts of P desorbable by *EUF* were much lower than the amounts of *P* initially added. However, the *C-Fe-pp* model was less able than *C-Al-pp* in retaining the initially added *P*. Evidently, distinct mechanisms controlled the interactions of *P* with *C-Al-pp* or *C-Fe-pp* in both sorption and desorption phenomena. In particular, the presence of *Al-* or

Fe- exerted different influence on quantitative as well as on kinetics aspects of *P* release. Also the state of aggregation of the models substantially affected the *P*-release reactions. The desiccation by air-drying dramatically lowered the extractable *EUF-P*, likely retaining *P via* an inter-particulate trapping/occlusion in the consolidated aggregates.

On the whole, our findings validate the modeling approach as a reference to better understand the factor and processes of *P* dynamics in soil and stimulate to go deep into research aiming to further elucidate the mechanism of *P* interaction with soil organo-minerals aggregates.

REFERENCES

1 E.A. FitzPatrick, Soils. Their formation, classification and distribution, Longman, New York, 1983.
2 O.A. Chadwick and R. C. Graham, Pedogenic processes. In: M. E. Sumner (ed.), Handbook of Soil Science, CRC press, Boca Raton, FL, USA, 2000, p. A 41.
3 M.H.B. Hayes and F.L. Himes, Nature and properties of humus-mineral complexes. In: P.M. Huang and M. Schnitzer (eds.), Interactions of soil minerals with natural organics and microbes, SSSA Sp. Publ. No. 17, Soil Sci. Soc. Am. Ed., Madison, WI, USA, 1986, p. 103.
4 J.A. McKeague, M.V. Cheshire, F. Andreux and J. Berthelin, Organo-mineral complexes in relation to pedogenesis. In: P.M. Huang and M. Schnitzer (eds.), Interactions of soil minerals with natural organics and microbes, SSSA Sp. Publ. No. 17, Soil Sci. Soc. Am. Ed., Madison, WI, USA, 1986, p. 549.
5 F.J. Stevenson, Humus chemistry. Genesis, composition, reactions, J. Wiley and Sons, New York, 1982.
6 B.D. Kay and D.A. Angers, Soil Structure. In: M. E. Sumner (ed.), Handbook of Soil Science, CRC press, Boca Raton, FL, USA, 2000, p. A229.
7 J.M. Skopp, Physical properties of primary particles. In: M. E. Sumner (ed.), Handbook of Soil Science, CRC Press, Boca Raton, FL, USA, 2000, p. A3.
8 D.L. Sparks, Kinetics and mechanisms of chemical reactions at the soil mineral/water surface. In: D.L. Sparks (ed.), Soil physical chemistry, 2nd ed., CRC Press, Boca Raton, FL, USA, 1998, p. 135.
9 A. Buondonno, M.L. Ambrosino, E. Coppola, D. Felleca, F. Palmieri, P. Piazzolla and A. De Stradis, Models of organo-mineral aggregates with polyphenols. 2. Stability of bentonite-Ca-gallotannin interaction products. In: N. Senesi and T.M. Miano (eds.), Humic substances in the global environment and implications on human health, Elsevier, Amsterdam, 1994, p. 1101.
10 A. Buondonno and E. Coppola, Modeling soil ped formation: properties of aggregates formed by montmorillonitic clay, Al- or Fe- poorly-ordered oxides and polyphenol in acidic *milieu*, this volume.
11 N. van Breemen and P. Buurman, Soil Formation, Kluwer Academic Pub., Dordrecht, The Netherlands, 1998.
12 A.N. Sharpley, Phosphorus availability. In: M. E. Sumner (ed.), Handbook of Soil Science, CRC Press, Boca Raton, FL, USA, 2000, p. D18.
13 N.J. Barrow, Adv. Agronom. 38 (1985) 183.

14 P.E. Fixen and J.H. Grove, Testing Soils for Phosphorus. In: R.L. Westerman (ed.), Soil Testing and Plant Analysis, 3rd ed., SSSA Book Series No. 3, SSSA Inc. Madison, WI, U.S.A., (1990) 141.
15 E.M. Frossard, M. Brossard, M.J. Hedley and A. Metherell, Reactions controlling the cycling of phosphorus in soils. In: H. Tiessen (ed.), Phosphorus in the global environment: transfers, cycles and management. J. Wiley and Sons, New York, 1995, p. 107.
16 K. Németh, Plant and Soil, 64 (1982) 7.
17 K. Németh, Plant and Soil, 83 (1985) 1.
18 K. Németh, A. Buondonno, and G. Mentasti, Relationships between EUF-P, chemical P extractions and some properties of typical soils of Southern Italy. In: K. Németh (ed.), Cost Reduction and Environmental Protection, Proc. 3rd Int. EUF-Symposium, Sudzucker, Mannheim, Germany, 1988, p. 525.
19 H. Grimme and K. Németh, Plant and Soil 64, (1982) 43.
20 A. Buondonno, S.A. Bufo, M.D.R. Pizzigallo, and D. Felleca, Kinetics of phosphorus release from Italian Alfisol on volcanic Inceptisol as influenced by phosphate addition to soils and by temperature. In: K. Németh (ed.), Cost Reduction and Environmental Protection, Proc. 3rd Int. EUF-Symposium, Sudzucker, Mannheim, Germany, 1988, p. 512.
21 A. Buondonno, E. Coppola, D. Felleca, P. Violante, and K. Németh, Agrochimica 35 (1991) 221.
22 A. Buondonno, C. Colella, E. Coppola, M. de' Gennaro and A. Langella, Quantitative and kinetics of K and P release from Italian zeolitized tuffs. In: C. Colella and F.A. Mumpton (eds.), Natural Zeolites for the Third Millennium, E. De Frede Editore, Napoli, Italy, 2000, p. 449.
23 P. Buurman, B. van Lagen and E.J. Velthorst (eds.), Manual for soil and water analysis. Backhuys Publishers, Leiden, The Netherlands, 1996.
24 D.L. Carter, M.M. Mortland and W.D. Kemper, Specific surface. In: A. Klute (ed.), Methods of soil analysis, Part 1, A.S.A. – S.S.S.A., Madison, WI, USA, 1986, p. 413.
25 A. Buondonno and A. Violante, Can. J. Soil Sci., 71 (1991) 285.
26 F. Nunozawa and A. Tanaka, Soil Sci. Plant Nutr., 30 (1984) 51.

Studies in Surface Science and Catalysis 140
A. Gamba, C. Colella and S. Coluccia (Editors)

Doped zirconia catalysts for the dehydration of 4-methylpentan-2-ol

M.G. Cutrufello, I. Ferino, R. Monaci, E. Rombi, V. Solinas

Dipartimento di Scienze Chimiche, Università di Cagliari, Complesso Universitario di Monserrato, s.s. 554 bivio Sestu, 09042 Monserrato (CA), Italy

High surface zirconia (140 m^2/g) was prepared from $ZrOCl_2 \cdot 8H_2O$. Li-, K-, Ca-, and Ba-doped samples were obtained by an impregnation procedure. The samples were tested as catalysts for the dehydration of 4-methylpentan-2-ol. Catalytic runs, lasting up to 80 hours, were carried out at atmospheric pressure in a fixed-bed flow microreactor. Pure ZrO_2 mainly leads to the undesired 2-alkene. Doping it with low amounts of any oxide results in a 1-alkene selective catalyst. A high dopant content is not useful or even detrimental. The acid-base properties of the catalysts were assessed by microcalorimetry, using ammonia and carbon dioxide as probe molecules. Present results indicate that the addition of the doping oxide modifies the original acidic character of zirconia thus shifting selectivity towards the desired 1-alkene.

1. INTRODUCTION

Zirconia-based catalysts are known to possess a high selectivity towards the formation of 1-alkenes in the dehydration of secondary alcohols [1-3]. A useful application seems to be the dehydration of 4-methylpentan-2-ol, which could represent an alternative route to the preparation of 4-methylpent-1-ene, starting material for manufacturing thermoplastic polymers of high technological properties. Besides the desired 1-alkene, the alcohol dehydration always leads to the formation of 4-methylpent-2-ene, often accompanied by skeletal isomers of C_6-alkenes. In addition, dehydrogenation to 4-methylpentan-2-one can occur simultaneously with dehydration. Previous results on several oxide [4-8] and zeolite [9] catalysts showed that the competition between the two reactions and the 1-/2-alkene ratio strongly depend on the fine tuning of the acid-base properties of the catalyst. Pure zirconia did not show a good catalytic behaviour, leading preferentially to 2-alkenes, while a high selectivity to 1-alkene could be reached by using Na-doped zirconia [4].

In the present work the behaviour of zirconia samples doped with oxides of alkali metals and alkaline-earth metals was investigated, in order to better understand the role of both the nature and the amount of the doping cation. Li-, K-, Ca-, and Ba-doped zirconia samples were prepared. Their surface acid-base properties were assessed by means of adsorption microcalorimetry, using ammonia and carbon dioxide as probe molecules. Their catalytic activity for the 4-methylpentan-2-ol dehydration was tested in a flow microreactor.

2. EXPERIMENTAL

2.1. Sample preparation

Pure zirconium oxide was prepared by a method, based on the hydrous zirconia precipitation from an aqueous solution of inorganic zirconium salts, which takes and adapts some indications reported in the literature [10,11]. $ZrOCl_2 \cdot 8H_2O$ (Aldrich), which in the same experimental conditions had proved to give an oxide with a higher surface area than $ZrO(NO_3)_2 \cdot 2H_2O$ [10], was used as precursor. An aqueous solution (ca. 1 M) of $ZrOCl_2 \cdot 8H_2O$ was heated under reflux up to 373 K; a diluted (1:1) ammonia (Aldrich) solution was then added drop by drop, in slight excess, up to pH 10. The gelatinous precipitate of zirconium hydroxide was digested for 48 h at the precipitation temperature, washed with warm distilled water until the complete elimination of chloride ions, vacuum filtered and dried overnight at 383 K. Pure zirconium oxide of high surface area (S_{BET} = 140 m^2/g) was then obtained by calcining for 2 h at 773 K the hydroxide.

Li-, K-, Ca-, and Ba-doped zirconia samples were obtained by impregnating different portions of the dried zirconium hydroxide with aqueous solutions of the corresponding acetates, drying overnight at 383 K and calcining at 773 K for 2 h. The doped samples are listed in Table 1.

Table 1
Doped zirconia samples. Figures in square brackets refer to the concentration of dopant oxide ($mmol_{oxide}/g_{ZrO_2}$). Figures in parenthesis represent the specific surface area S_{BET} (m^2/g).

Li_2O[0.2]/ZrO_2 (81)	K_2O[0.2]/ZrO_2 (109)	CaO[0.2]/ZrO_2 (96)	BaO[0.2]/ZrO_2 (106)
Li_2O[0.7]/ZrO_2 (99)	K_2O[0.6]/ZrO_2 (126)	CaO[0.7]/ZrO_2 (168)	BaO[0.7]/ZrO_2 (175)
Li_2O[1.8]/ZrO_2 (42)		CaO[1.9]/ZrO_2 (193)	

2.2. Microcalorimetric characterisation

Tian-Calvet heat flow equipment (Setaram) was used for microcalorimetric measurements. Every sample was evacuated (10^{-3} Pa) overnight at 673 K before the successive introduction of small doses of the probe gas (ammonia or carbon dioxide). The equilibrium pressure relative to each adsorbed amount was measured by means of a differential pressure gauge (Datametrics). The run was stopped at a final equilibrium pressure of 133.3 Pa. The adsorption temperature was maintained at 353 K, in order to limit physisorption. After overnight outgassing at the same temperature, a second run was carried out up to 133.3 Pa. The adsorption and calorimetric isotherms were obtained from each adsorption/readsorption experiment. The amount of probe gas irreversibly adsorbed was calculated by subtracting from the first adsorption isotherm the second one, carried out after outgassing the sample. For each catalyst, a plot of the differential heat of adsorption as a function of the adsorbed amount was drawn, which gives information on the influence of surface coverage on the energetics of the adsorption.

2.3. Catalytic tests

Catalytic runs were carried out at atmospheric pressure in a quartz-made fixed-bed flow microreactor (10 mm i.d.). All the stainless-steel equipment devices had been passivated by hot HNO_3 treatment before the assembly. The catalyst was activated in situ (6 h at 773 K under CO_2-free air flow). 4-Methylpentan-2-ol was fed in with an N_2 stream (partial pressure, $P_{0,alcohol}$ = 19.3 kPa; time factor, W/F = 0.54 $g_{cat}{\cdot}h/g_{alcohol}$). On-line capillary GC analysis conditions were: Petrocol DH 50.2 column, oven temperature between 313 and 473 K, heating rate 5 K/min. Products identification was confirmed by GC-MS. For each catalyst a run in which several reactor temperatures were checked was carried out, in order to study the influence of the thermal history of the sample on its catalytic behaviour and to reach an appropriate conversion level (ca. 50%) at which the selectivities values of the different catalysts can be compared. Then, a new run with a fresh portion of the same sample was started at the desired temperature and carried out isothermally for 80 h. Further runs, where both the flow rate and the catalyst amount were considerably changed, while keeping the same W/F value, were also carried out; no significant differences in conversion were observed, which rules out the occurrence of external diffusion limitations.

3. RESULTS

3.1. Acid-base properties

Figures 1-5 show the differential heat of adsorption, Q_{diff}, as a function of the irreversible uptake of the probe molecule (NH_3 or CO_2) for all the catalysts. Pure ZrO_2 (Figure 1) possesses a little portion of very strong acid sites, as indicated by the very high initial value of Q_{diff} (337 kJ/mol). It anyway rapidly decreases to much lower (but still high) values at low uptake, decreasing then more gradually and reaching a quite high uptake value, which indicates a large number of acid sites. The curve relative to the CO_2 adsorption decreases gradually, always laying under the NH_3 adsorption curve and ending at rather low uptake. The basic sites are therefore neither as strong nor as many as the acid ones.

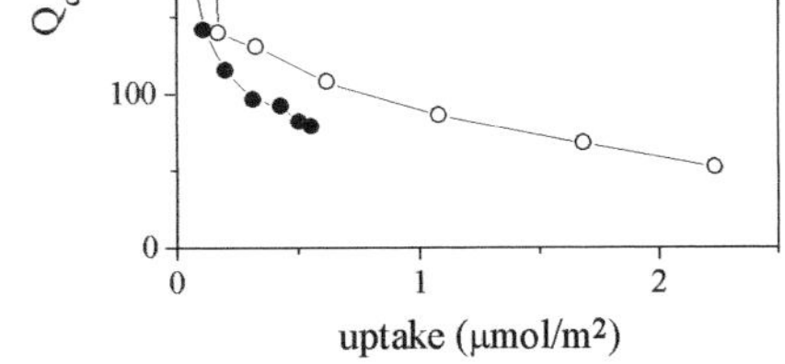

Figure 1. Differential heat of adsorption for ZrO_2 catalyst.

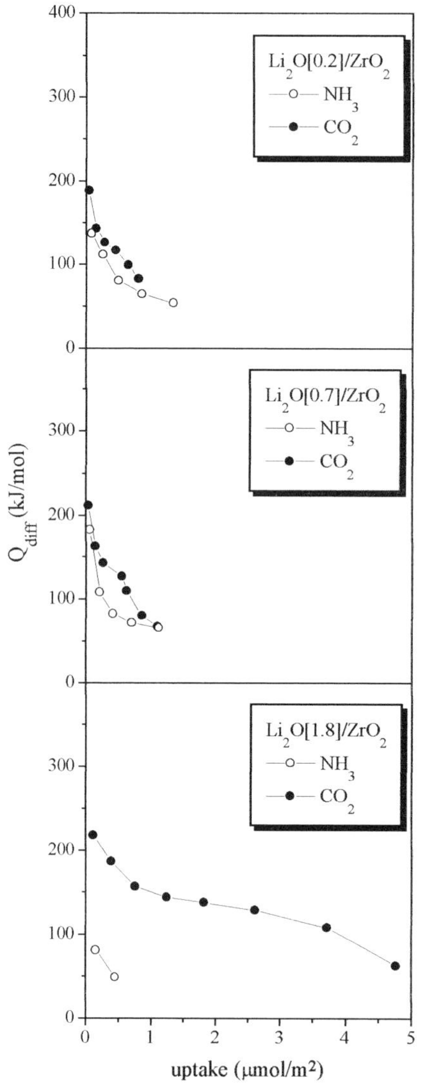

Figure 2. Differential heat of adsorption for Li-doped ZrO_2 catalysts.

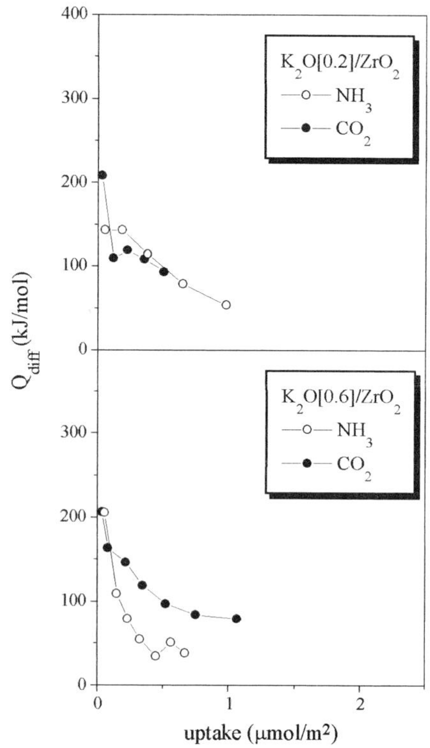

Figure 3. Differential heat of adsorption for K-doped ZrO_2 catalysts.

The addition of Li_2O (Figure 2) produces a decrease in both the number and the strength of the acid sites. The sites with a very high differential heat of adsorption disappear and the number of acid sites decreases as the dopant concentration increases. The opposite trend is observed for the basic sites. Even with the lowest amount of Li_2O, the basic sites become stronger than the acid ones; at low uptake, a change in the slope reveals the presence of a small number of sites belonging to the same family. This becomes more evident in the sample containing a higher percentage of lithium oxide which exhibits a short plateau at about 140 kJ/mol. The sample with the highest Li amount is strongly basic: the plateau is quite wide and the whole curve corresponding to the basic sites is very high in comparison with that for the acid ones.

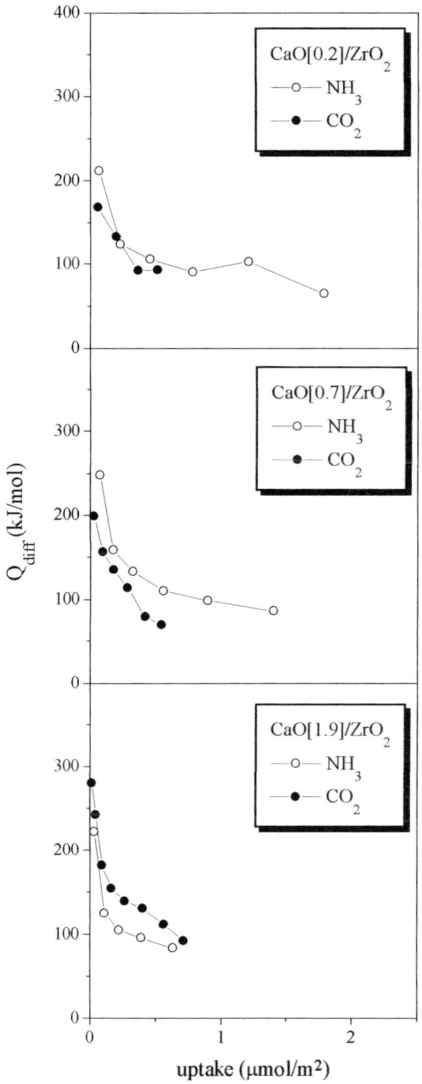

Figure 4. Differential heat of adsorption for Ca-doped ZrO_2 catalysts.

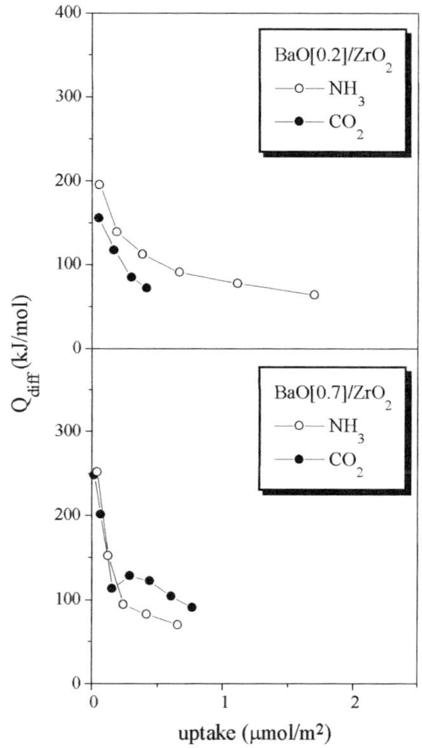

Figure 5. Differential heat of adsorption for Ba-doped ZrO_2 catalysts.

The effect of the addition of the other oxides (Figures 3-5) is similar to that of Li_2O. The very strong acidic sites found in the ZrO_2 sample are no more present. Increasing the dopant content, the number of acid sites decreases; at the same time, the basic sites increase in number and strength, the relevant curves becoming higher than those relative to the NH_3 adsorption. Comparison between $Li_2O[1.8]/ZrO_2$ and $CaO[1.9]/ZrO_2$ shows that the presence of calcium, even at the highest amount, does not induce such a high basicity as lithium does.

3.2. Catalytic behaviour

Catalytic runs carried out changing the temperature showed, for all the catalysts, that while conversion increases with the increasing temperature, selectivities do not depend on it. The temperature at which a 50% conversion is attained was 543 K for ZrO_2. Upon dopant addition a decrease in conversion was observed for all the catalysts, the temperature for 50% conversion being in the range 593-673 K.

Both initial (extrapolated at 0 h on-stream) and long-run (averaged after 24 h on-stream) selectivity values for isothermal runs are reported in Table 2. A remarkable difference between the initial and long-run data can be observed for the samples with the lowest amount of doping oxides. Such a difference becomes less marked for $Li_2O[0.7]/ZrO_2$ and

Table 2
Selectivities of the samples: a, initial values (extrapolated at 0 h on-stream); b, values after stabilisation (averaged after 24 h on-stream).

Catalyst	Selectivity (mol%)									
	1-A		2-A		C_6		K		HK	
	a	b	a	b	a	b	a	b	a	b
ZrO_2	34	34	60	60	4	4	2	2	-	-
$Li_2O[0.2]/ZrO_2$	51	84	47	15	2	-	-	1	-	-
$Li_2O[0.7]/ZrO_2$	34	42	19	22	1	1	45	34	1	1
$Li_2O[1.8]/ZrO_2$	3	3	2	2	2	1	72	69	21	25
$K_2O[0.2]/ZrO_2$	64	80	29	15	1	4	5	1	1	-
$K_2O[0.6]/ZrO_2$	65	66	13	11	-	1	22	22	-	-
$CaO[0.2]/ZrO_2$	39	82	59	18	1	-	1	-	-	-
$CaO[0.7]/ZrO_2$	53	62	42	32	3	2	2	4	-	-
$CaO[1.9]/ZrO_2$	81	81	16	14	1	1	2	4	-	-
$BaO[0.2]/ZrO_2$	50	86	50	13	-	1	-	-	-	-
$BaO[0.7]/ZrO_2$	70	73	23	19	1	7	6	1	-	-

1-A: 1-alkene (4-methylpent-1-ene); 2-A: 2-alkene (4-methylpent-2-ene); C_6: skeletal isomers of C_6-alkenes; K: ketone (4-methylpentan-2-one); HK: heavy ketones (deriving from the condensation of the ketone)

$CaO[0.7]/ZrO_2$, being negligible for all the other catalysts. Anyway, after 24 h on-stream all the catalysts appeared to be quite stable. In this light, their behaviour can be compared referring to the selectivity values after stabilisation (b column in Table 2).

On pure zirconium oxide, 4-methylpent-2-ene is favoured compared to the alkene with the double bond in terminal position, which is the desired product. The skeletal isomerisation reaction occurs to a rather low extent and the dehydrogenation is almost negligible.

The presence of lithium oxide as dopant deeply modifies the behaviour of zirconia. Only dehydration takes place on the sample containing the lowest amount of Li_2O and it is strongly directed towards the 1-alkene. Higher dopant contents induce on the catalysts a notable dehydrogenation character; the dehydration reaction is still favoured, and mainly – although less markedly – directed to the 1-alkene formation, for the $Li_2O[0.7]/ZrO_2$ sample, but it is practically negligible in comparison with the dehydrogenation for $Li_2O[1.8]/ZrO_2$. On this catalyst dehydrogenation is accompanied by a considerable formation of heavy products, deriving from the ketone condensation.

On $K_2O[0.2]/ZrO_2$ dehydration is by far predominant, the dehydrogenation activity being negligible. The increase in the K_2O content ($K_2O[0.7]/ZrO_2$ sample) is accompanied by an increase in the selectivity to dehydrogenation at the expense of the dehydration, which however is still the main reaction.

The $CaO[0.2]/ZrO_2$ sample is exclusively a dehydration catalyst with a clear predominance of the alkene with the double bond in terminal position. On both $CaO[0.7]/ZrO_2$ and $CaO[1.9]/ZrO_2$, the dehydrogenation reaction is rather modest, but not negligible; on the latter, 4-methylpent-1-ene is particularly favoured among the dehydration products.

On the Ba-doped catalysts, dehydrogenation products can be observed only in trace amounts; dehydration is mainly directed towards 4-methylpent-1-ene.

4. DISCUSSION

The catalytic results are interesting from an applicative point of view. They show the possibility of obtaining stable and selective catalysts for the dehydration of 4-methylpentan-2-ol into 4-methylpent-1-ene by doping zirconium oxide with very low amounts of alkali or alkaline-earth metals oxides. It is noteworthy that the nature of the dopant oxide does not seem to be determinant in the development of such interesting catalytic properties; the addition of 0.2 mmol/g of any of the oxides has the same effect: a very high selectivity, higher than 80%, towards the alkene with the double bond in terminal position, associated with a complete absence or a very modest presence of the parasite reaction of dehydrogenation. Dopant oxide contents higher than 0.2 mmol/g proved to be useless or even detrimental, depending on the nature of the metal cation.

Previous results [4-8] for the same reaction on zirconia-, ceria-, and lanthana-based catalysts showed a correlation between the acid-base properties (in terms of both the concentration and the strength of the sites) of the samples and their catalytic behaviour. When the number of the acid sites is sensibly higher than that of the basic ones, the main product is the undesired 2-alkene, which forms through an E1 mechanism (Saytzeff orientation). The alcohol adsorbs on the oxide surface through the interaction between the OH group and an acid site; the rupture of the C–OH bond leads to the carbocation which mainly gives the alkene with the double bond in internal position. It can also undergo skeletal isomerisation on the most acidic sites. When the number of acid and basic sites is comparable, a two-point adsorption of the alcohol molecule on a acid-base pair can set in. It can then evolve in different ways, depending on the relative strength of the sites. If the acid and base sites have similar strength, a concerted E2 mechanism takes place, leading mainly – through an alkene-like activated complex – to the 2-alkene (Saytzeff orientation). On the other hand, if the basic sites are stronger than the acid ones, either an E2 mechanism with a carbanion-like transition state, or even an E1cB mechanism giving a carbanion species (through the rupture of the C–H bond of the methyl group) can occur, both leading mainly to the 1-alkene (Hofmann orientation). When the number of basic sites is significantly higher than that of the acid ones, a dehydrogenation pathway prevails. This involves the adsorption of the alcohol molecule on the catalyst surface – through the formation of a hydrogen bond between the OH group and the basic site – followed by the $C–H_\alpha$ bond rupture and H_2 elimination.

To check the validity of such a reaction scheme, the ratio between the concentration of the basic (n_B) and the acid sites (n_A) was obtained from Figures 1-5. To evaluate the strength of the sites, the fraction of sites with a differential heat of adsorption higher than 80 kJ/mol ($n_{A>80}/n_A$ and $n_{B>80}/n_B$) was also calculated. Data are reported in Table 3.

Table 3
Acid-base properties of the catalysts.

Catalyst	n_B/n_A	$n_{B>80}/n_B$ (%)	$n_{A>80}/n_A$ (%)
ZrO_2	0.25	97	57
$Li_2O[0.2]/ZrO_2$	0.60	100	37
$Li_2O[0.7]/ZrO_2$	0.98	78	43
$Li_2O[1.8]/ZrO_2$	10.47	32	–
$K_2O[0.2]/ZrO_2$	0.51	100	65
$K_2O[0.6]/ZrO_2$	1.59	95	34
$CaO[0.2]/ZrO_2$	0.29	100	82
$CaO[0.7]/ZrO_2$	0.48	76	67
$CaO[1.9]/ZrO_2$	1.13	100	56
$BaO[0.2]/ZrO_2$	0.24	83	60
$BaO[0.7]/ZrO_2$	1.17	100	71

In Figure 6 the initial selectivity towards 4-methylpent-1-ene (S_{1-A}) is reported as a function of the n_B/n_A ratio for the present catalysts; results for other oxides previously tested [4-8] are also reported. The two sets of data are fitted by the same volcano-shaped curve. The maximum of S_{1-A} is reached when n_B/n_A is around 1 (either Hofmann-oriented E2 or E1cB mechanism). At both lower and higher n_B/n_A values the selectivity to the 1-alkene is low; in the former case 4-methylpent-2-ene is the preferred product (E1 mechanism), while in the latter dehydrogenation is favoured at the expense of dehydration. It is worthy of note that only dehydrogenation takes place on $Li_2O[1.8]/ZrO_2$ (Table 2), in agreement with its very high n_B/n_A ratio (Table 4); this catalyst does not appear in Figure 6, its abscissa being out of scale.

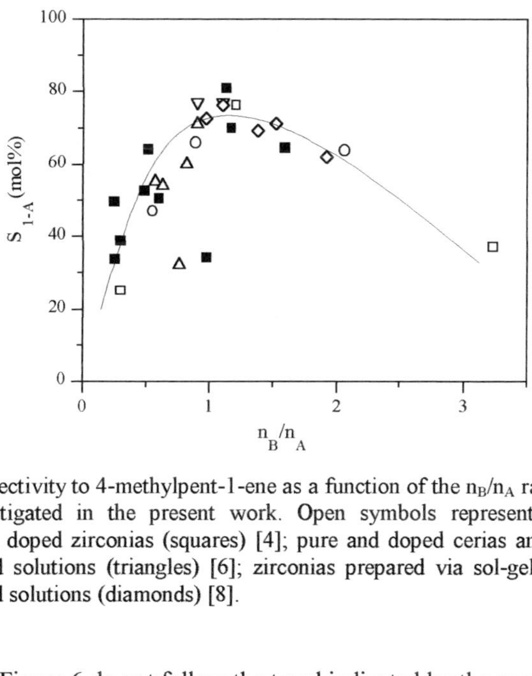

Figure 6. Initial selectivity to 4-methylpent-1-ene as a function of the n_B/n_A ratio. Full symbols refer to the samples investigated in the present work. Open symbols represent previously investigated catalysts: pure and doped zirconias (squares) [4]; pure and doped cerias and lanthanas (circles) [5]; ceria-zirconia solid solutions (triangles) [6]; zirconias prepared via sol-gel (reversed triangles) [7]; ceria-lanthana solid solutions (diamonds) [8].

Two points in Figure 6 do not follow the trend indicated by the curve. The open triangle at $n_B/n_A = 0.76$ is relevant to a mixed CeO_2-ZrO_2 catalyst which gave mainly (68%) the 2-alkene and other C_6-alkenes with the double bond in internal position. The behaviour of this catalyst has already been explained [6] considering that, unlike the other systems with similar n_B/n_A values, the strength of its acid and basic sites is comparable ($n_{A>80}/n_A = 61\%$ and $n_{B>80}/n_B = 55\%$). The E1cB mechanism cannot thus be operating on this catalyst. More difficult to understand is the behaviour of the $Li_2O[0.7]/ZrO_2$ catalyst (the full square at $n_B/n_A = 0.98$ in Figure 6). In this system the basic sites are sensibly stronger than the acid ones (Table 4) and it is therefore expected to give mainly the 1-alkene through the E1cB mechanism. Instead, mainly ketone formation occurs. It might be speculated that, due to its small size, the Li^+ cation might be somewhat shielded by the large oxygens, and the alcohol molecule cannot efficiently interact with the acid-base pair. Further work is needed to elucidate this point. The change in selectivity occurring with time-on-stream, which is particularly noticeable in the case of the catalysts with the lowest amount of doping oxides (Table 2), also deserves further investigation. It cannot be ascribed to the deposition of reaction products on the surface, as a regenerative treatment of the catalysts in flowing air at 773 K did not restore the initial behaviour. Some kind of irreversible modification of the surface seems hence to occur during the run.

REFERENCES

1. T. Yamaguchi, H.Sasaki, K. Tanabe, Chem. Lett., (1973) 1017.
2. M. Araki, K. Takahashi, T. Hibi, Sumitomo Chemical Co., Eur. Pat. Appl., 0 150 832 (1985).
3. M. Araki, T. Hibi, Sumitomo Chemical Co., Eur. Pat. Appl., 0 222 356 (1986).
4. A. Auroux, P. Artizzu, I. Ferino, V. Solinas, G. Leofanti, M. Padovan, G. Messina, R. Mansani, J. Chem. Soc., Faraday Trans., 91 (1995) 3263.
5. A. Auroux, P. Artizzu, I. Ferino, R. Monaci, E. Rombi, V. Solinas, G. Petrini, J. Chem. Soc., Faraday Trans., 92 (1996) 2619.
6. M.G. Cutrufello, I. Ferino, V. Solinas, A. Primavera, A. Trovarelli, A. Auroux, C. Picciau, Phys. Chem. Chem. Phys., 1 (1999) 3369.
7. I. Ferino, M.F. Casula, A. Corrias, M.G. Cutrufello, R. Monaci, G. Paschina, Phys. Chem. Chem. Phys., 2 (2000) 1847.
8. M.G. Cutrufello, I. Ferino, R. Monaci, E. Rombi, G. Colón, J.A. Navío, Phys. Chem. Chem. Phys., submitted.
9. A. Auroux, P. Artizzu, I. Ferino, R. Monaci, E. Rombi, V. Solinas, Microporous Mater., 11 (1997) 117.
10. D. Tichit, D. El Alami, F. Figueras, Appl. Catal. A: General, 145 (1996) 195.
11. G.K. Chuah, S. Jaenicke, S.A. Cheong, K.S.Chan, Appl. Catal. A: General, 145 (1996) 267.

Studies in Surface Science and Catalysis 140
A. Gamba, C. Colella and S. Coluccia (Editors)

Carbon tetrachloride hydrodechlorination with organometallics based platinum and palladium catalysts on MgO : EXAFS characterization and catalytic studies

V. Dal Santo*, C. Dossi[#,§], R. Psaro*, S. Recchia[#] and L. Sordelli*

*CNR Research Centre "CSSCMTBSO" and Dipartimento di Chimica Inorganica, Metallorganica e Analitica, Università degli Studi di Milano, Via Venezian 21, 20133 Milano, Italy.

[#] Dipartimento di Scienze Chimiche, Fisiche e Matematiche, Università dell'Insubria, Via Valleggio 11, 22100 Como, Italy

Hydrodechlorination of carbon tetrachloride was performed over Pd and Pt on MgO catalysts obtained from different precursors.
The Pd samples show low activity and high selectivity to deep hydrogenolysis of C-Cl bonds to CH_3(ads) fragments leading to C_1 – C_{7+} hydrocarbons, following a typical Schulz-Flory distribution for alkyl polymerization.
On the other hand Pt catalysts show high and stable conversion levels, but totally different selectivities. $CHCl_3$ and in lesser extent methane are the major products.
The structural modifications of Pt catalysts under reaction conditions were characterised by EXAFS spectroscopy.
The amount and the nature of chlorine species preadsorbed on catalysts play an important role contributing to stabilise the properties of metallic particles.

1. INTRODUCTION

Carbon tetrachloride finds different industrial uses as solvent, as dopant in semiconductor industries, as raw material [1-3].
However, together with other chlorinated compounds like PCBs and CFCs, it possesses noxious effects on environment, particularly a high ozone potential depletion. It was, in fact, classified in the IV group among CFCs and so its use was banned, in developed countries, from 1996.
Among the more diffuse CFCs disposal methods, like thermal combustion [4,5], catalytic combustion [6], catalytic hydrogenation is one of the more promising, owning to the low reaction temperature and the production of useful compounds, without pollutants by-products like dioxins, CO, Cl_2 and $COCl_2$.

[§]Corresponding author

It has been reported that catalysts obtained from H_2PtCl_6 on MgO [7, 8] showed good conversion and selectivities for the hydrodechlorination of CCl_4 to $CHCl_3$, being less prone to deactivation than Pt/Al_2O_3 at lower H_2/CCl_4 ratio. This behavior was ascribed to the basicity of MgO that retards the coking.
On the other hand, Pd catalysts were used, with good results, for the partial dehalogenation of chloro-fluorocarbons to hydrofluorocarbons [9-11], and of chlorobenzenes [12], but at present there are few studies on chlorinated aliphatic substrates like CCl_4.
The aim of this work is to test different Pt and Pd catalysts supported on basic MgO in the hydrodechlorination of CCl_4. Particular attention was devoted to the choice of the organometallic precursor, in order to maintain the basic properties of activated MgO and consequently to increase metal support interactions. A specific metal-support interaction on $Pd(acac)_2/MgO^{500}$ catalysts was observed by previous studies [13].

2. EXPERIMENTAL

2.1. Catalyst preparation

$Pd(acac)_2$ was synthesized in our laboratory according to the literature procedure [14]. $[Pd(C_3H_5)Cl]_2$ was obtained from Strem Chemicals; $Pt(acac)_2$, and $Pt(CH_3CN)_2Cl_2$ from Aldrich and used as received without further purification.
The MgO support material (Merck 97% batch n° 308 TA 390565) was refluxed in deionized water for 3 hours, and then dried in air overnight at 110°C. Highly dehydroxylated and decarbonated magnesium oxide (MgO^{500}) was obtained by heating to 500°C in air, and further evacuation ($P=10^{-5}$ mbar) at this temperature overnight. After this treatment, BET surface area is around 300 m^2g^{-1}. Measurements of BET surface areas were carried out using the single point technique with a Micromeritics Pulse Chemisorb 2700 instrument.
Toluene and CH_2Cl_2 (Aldrich) were dried over activated molecular sieves in inert atmosphere. The oxide supports were impregnated at room temperature under Argon atmosphere with a diluted solution of metal precursor in anhydrous solvent, obtaining a 2 wt.% metal loading for each sample. The solvent was removed in vacuo, and the sample dried overnight in vacuo.
Metal loadings were determined by ICP/AES (Pt) and AAS (Pd) after dissolution of the catalyst in a HCl/HNO_3 mixture.

2.2 Catalytic tests

Prior the catalytic runs, the impregnated catalysts, were reduced in flowing hydrogen from 25 to 500°C at 10°C/min and kept at 500°C for 1 h. Oxidative regeneration of used catalysts was done by heating in O_2 flow (50 mL/min) from 25 to 500°C at 10°C/min; the sample was then left at 500°C for 1 h in O_2, cooled down in flowing Ar, and reduced by standard treatment.
The hydrodechlorination of CCl_4 (Merck GR batch n° k22643522605) was performed in a continuos flow, fixed bed glass microreactor, working at atmospheric pressure, with

H_2/ CCl_4 ratios varying from 5 to 20. Gases are all reagent grade (99.999% obtained from SIAD).

The reaction mixture was obtained by bubbling H_2 or H_2/Ar in CCl_4 thermostatted at various temperatures in order to obtain the desired H_2/ CCl_4 ratio. Flow rates of the gas mixture vary from 22.0 mL/min to 33.0 mL/min depending on the desired contact time; namely, for 100 mg of 2% Pd catalyst, for $\tau = 9$ s total flow rate is 33.0 mL/min.

Quantitative analyses of reactants and products were carried out on an on-line gas-chromatograph (Carlo Erba Instruments HRGC 5160) fitted with a thermostatted sampling valve and FID, using a 50 m, 0.2 mm cross-linked methylsilicone (Hewlett-Packard PONA) capillary column.

The HCl evolution vs. time on stream (t.o.s.) was monitored by an on-line potentiometric cell, as described previously [15].

2.3 EXAFS characterization

X-ray absorption spectra were collected at the D42 beamline of LURE (Orsay-France) operating at 2 GeV of positron energy and at a storage ring current of 200 mA. The experimental station was equipped with a Si(331) channel-cut crystal monochromator. Second harmonics are absent due to the extinction rule of the Si(331); third harmonic are neglectable due to the low photon flux at that energy. Pt metal foil has been used for the angle/energy calibration. All spectra were recorded at 25°C in transmission mode at the Pt L_{III}-edge (11.56 keV) over the range 11.35 – 12.50 keV.

The powder samples were loaded under inert atmosphere in the catalysis-EXAFS cell (Lytle type). Sample homogeneity was checked by taking low-exposition photographs of the samples. The spectra of $Pt(acac)_2/MgO^{500}$ impregnated from toluene and CH_2Cl_2 solutions were recorded after in-situ reduction at 500°C and after the catalytic tests at 150°C in H_2/CCl_4= 10.

The single scattering data analysis was performed with the «EXAFS pour le Mac» [16] software. Experimental $\chi(k)$ data were extracted from the absorption data by a conventional procedure (linear pre-edge background subtraction; 5th polynomial fit for the atomic-like contribution, subtracted following the procedure proposed by Lengeler and Eisenberger [17] normalization to edge height). The k^3-weighted $\chi(k)$ data were Fourier transformed (Kaiser window, τ=2.5) in a typical k range of 3-14 $Å^{-1}$, and the main contributions to the Fourier transform modulus were filtered in order to obtain metal nearest-neighbor shells. The so obtained filtered contributions were analyzed using the non-linear least-squares fitting programs developed by Michalowicz [18]. Experimental phase shift and amplitude functions of the scattering oxygen and platinum atoms used in the fit of the supported samples were extracted from pure $Pt(acac)_2$ and Pt foil, respectively.

3. RESULTS AND DISCUSSION

Different Pd and Pt catalysts supported on MgO^{500} were synthesized by impregnation of the pretreated support from anhydrous solutions (in toluene or CH_2Cl_2) of the organometallic precursors ($Pd(acac)_2$, $Pt(acac)_2$, $[Pd(C_3H_5)Cl]_2$, $Pt(CH_3CN)_2Cl_2$).

This method of synthesis allows us to obtain a high degree of interaction between the metal precursor and the support, as already reported for Pd/MgO catalysts [13, 19]. Good dispersion of the metal phase is thus obtained after reduction, without altering the basic properties of MgO^{500} as it would otherwise occur using aqueous solutions of the inorganic salts conventionally used for impregnation.
Already at a first insight of Figure 1, they show completely different behavior: Pt catalysts reach high conversion values that are maintained for long reaction periods, whereas catalytic conversion of Pd catalysts is initially high, but quickly decreases during the catalytic run.
The selectivity is also different. Pt based catalysts primarily produce $CHCl_3$, with some methane and, in lower amount, ethane.
Pd based catalysts promote instead the formation of hydrocarbons (from C_1 to C_7). Here, methane, ethane and, in lesser extent, propane and butane are the major reaction products whereas hydrochlorocarbons of general formula CH_xCl_{3-x} and chlorocarbons dimers as C_2Cl_6, C_2Cl_4 are produced in a very low amount.

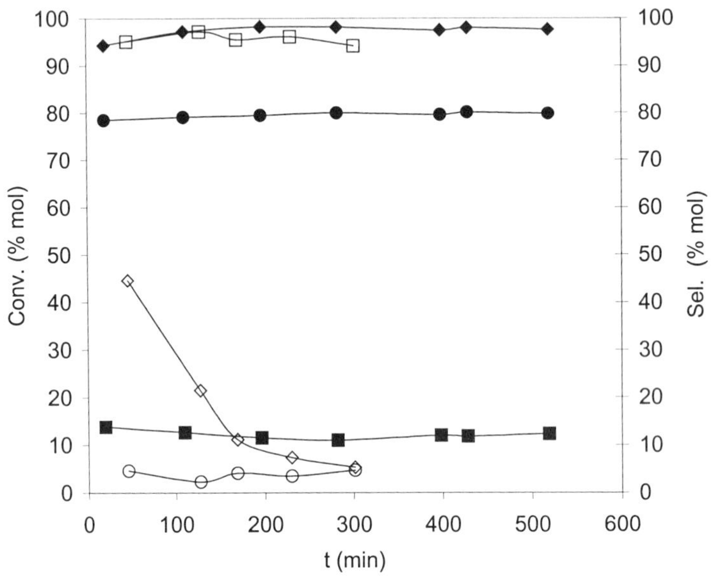

Fig 1. Typical catalytic performances at 150°C of Pt/MgO^{500} obtained from $Pt(acac)_2$ in CH_2Cl_2 (solid symbols), and Pd/MgO^{500} obtained from $Pd(acac)_2$ in CH_2Cl_2 (open symbols) at (Conversion ◊,♦; Selectivity to hydrocarbons □,■; selectivity to $CHCl_3$ + CH_2Cl_2 ○,●).

In order to maximize the production of hydrocarbons with Pd-based materials, and to minimize the formation of chlorinated byproducts, like hexachloroethane or tetrachloroethene, it was decided to operate at relatively low temperature (115°C) and high values of the H_2/CCl_4 ratios (around 20). In the Table 1, it is shown the distribution of products for $Pd(acac)_2/MgO^{500}$ (Pd03) and $[Pd(C_3H_5)Cl]_2/MgO^{500}$ (Pd23) impregnated in CH_2Cl_2. The main difference between the two materials is the molar Cl/Pd ratio, of 0.3 for Pd03 catalyst, and 2.3 for the Pd23 material.

Table 1
Catalytic performances of Pd based catalysts after 150 min on stream

Cat.	T [°C]	H_2/CCl_4	Conv. (% mol)	Sel. (% mol) C_1	C_2	C_3	C_4	C_5	CH_2Cl_2	$CHCl_3$	C_2Cl_6
Pd03	150	5	4.1	17.6	30.6	14.0	4.0	1.1	3.1	6.8	3.5
	115	20	6.9	29.6	38.0	15.7	3.5	1.3	2.3	1.5	0.7
Pd23	115	20	3.9	37.2	22.2	1.4	5.4	0	0	0	16.0

The major effect of the organometallic precursor, and consequently of Cl/Pd ratio, is the high selectivity to dimerization by-reactions to hexachloroethane, which is an undesirable product in the reaction. Instead, a high H_2/CCl_4 ratio combined with low temperature resulted in a slight increase of conversion and of selectivity to hydrocarbons. Distribution of hydrocarbons for $Pd(acac)_2/MgO^{500}$ catalysts fits a Schulz-Flory model [20, 21] for alkyl polymerization mechanism. Typical values for α_s α_i are comprised between 0.25 and 0.15 depending on reaction conditions (H_2/CCl_4 ratio, τ, t.o.s.), and the major difference between them, during the same catalytic run, is around 0.04.
Selectivity to methane is strongly depressed at low H_2/CCl_4 ratios (R), below 5; this is indirectly confirming the formation of methane by hydrogenolysis of heavier hydrocarbons (Table 2). In fact, the hydrodechlorination of CCl_4 with Pd-based catalysts requires an excess of hydrogen to convert all CCl_4 into hydrocarbons. The similarity of R=10 and R=20 is to be attributed to the fact that no effects of partial pressure of H_2 are occurring once the stoichiometric ratio is exceeded.

Table 2
Hydrocarbon selectivities vs. H_2/CCl_4 (R) ratio.

$Pd(acac)_2/MgO^{500}$	Sel (mol %) R = 5	R = 10	R = 20
CH_4	13.6	38.4	36.5
C_2H_6	29.2	35.5	35.7
C_3H_8	14.0	10.0	10.2

T = 150 °C, τ = 9 s, selectivities values are taken at similar conversion (around 8%).

This peculiar hydrogenolytic activity of our Pd/MgO catalysts leading to the complete breaking of C-Cl bonds is to be ascribed to the electron-rich Pd phase in strong contact

with O^{2-} sites of MgO [13, 19], giving a strong hydridic character to chemisorbed hydrogen atoms.
On Pt catalysts, instead, the hydrogenolytic mechanism, which is predominantly leading to $CHCl_3$ formation, is well known in the literature [8, 22], and it is no more discussed in the paper.
During the catalytic test, even working at low conversion, there is a progressive deactivation that could be due to chlorine poisoning and/or formation of high molecular weight hydro(chloro)carbons that cover active sites.
The major drawback of these Pd/MgO catalysts is, however, related to the practical impossibility of regenerating spent catalysts by conventional O_2 treatments at high temperatures. Under these oxidative conditions, sublimation of $PdCl_2$ takes place from the reaction between metallic Pd and Cl_2 evolved from combustion of hydrochlorocarbons. As the consequence, extensive leaching of palladium results, and the catalytic activity is lost.
On the contrary, all Pt catalysts on MgO^{500} at different Cl contents showed very high activity and stability in CCl_4 hydrodechlorination (Table 3).

Table 3
Effect of Cl content on Pt/MgO catalysts at T = 150 °C, H_2/CCl_4 = 10 τ = 6 s

Sample	Cl/Pt (mol/mol)	Conv. (mol %)	Sel. (mol%) CH_4	C_2H_6	$CHCl_3$	C_2Cl_6
Pt00	0.0	88.0	15.8	3.8	79.5	0
Pt09	0.9	95.0	12.6	0.4	80.7	0
Pt26	2.6	87.3	21.0	1.4	75.9	0

Pt00: $Pt(acac)_2$/MgO impregnated from toluene; Pt09: $Pt(acac)_2$/MgO impregnated from CH_2Cl_2; Pt26: $Pt(CH_3CN)_2Cl_2$ impregnated from CH_2Cl_2.

Unexpectedly, the higher activity is observed with the Pt09 catalyst, having an intermediate Cl/Pt ratio. The selectivity to $CHCl_3$ is also higher at a Cl/Pt ratio of 0.9.
Increasing chlorine content, catalytic activity tends to decrease again, with a parallel increase in the hydrocarbon selectivity.
In order to investigate the modification of catalysts during the hydrodechlorination reaction, we performed an EXAFS analysis of the more significant Pt samples.
Classical chemisorption measurements to estimate particle sizes cannot be profitably applied to spent catalysts, because the presence of chlorine and/or organic residues on metal particles reduces the number of sites for the hydrogen adsorption, underestimating the dispersion.
EXAFS spectroscopy overcomes these problems, moreover it gives useful information about the morphology and the state of metal particles.
The Pt00 and Pt09 catalysts were characterized by EXAFS spectroscopy after the reductive activation and after the catalytic runs (fit results reported in Table 4).

Table 4
Curve fitting results of Pt L_{III} edge EXAFS data of Pt00 and Pt09 catalysts before and after a catalytic run

Sample	shell	CN	R (Å)	$\Delta\sigma$(Å)	ΔE_o(eV)
Pt00 reduced	O	2.7 ± 0.1	2.030 ± 0.004	0.076 ± 0.004	0.7 ± 0.4
	Pt	1.6 ± 0.5	2.60 ± 0.01	0.09 ± 0.01	10 ± 4
Pt00 after catalysis	Cl	0.9 ± 0.2	2.31 ± 0.01	0.09 ± 0.02	6 ± 1
	Pt	4.6 ± 0.4	2.703 ± 0.008	0.096 ± 0.05	6 ± 1
Pt09 reduced	O	2.3 ± 0.4	2.07± 0.02	0.12 ± 0.01	4 ± 1
	Pt	4.7 ± 0.5	2.757 ± 0.006	0.083 ± 0.003	0 ± 1
Pt09 after catalysis	Pt	9.9 ± 0.9	2.753 ± 0.006	0.087 ± 0.005	-3.7 ± 0.6

Pt09 after reduction exhibits small Pt clusters (CN(Pt) = 4.6), in close contact with the oxygen atoms of support (CN(O) = 2.3), and no presence of chlorine in contact with the metal phase.
The mean particle diameter of metal particles could be extimated, in the hypothesis of fcc packing and cubocthaedral geometry, around 11 Å, consisting of 30 atoms clusters [23]. Chlorine is adsorbed on MgO support as Cl^- ions, probably close to the metallic particles; a similar behavior is already reported in literature for analogous systems based on Pd [19]. After the catalytic test, a considerable increase in the particle size is observed (CN(Pt) = 9.9). In fcc packing and in the hypothesis of cuboctahedral particles, such a CN value corresponds to a mean particle diameter of about 30 Å, which means a maximum of 30% surface atoms and an even smaller fraction of metal atoms in contact with the support surface.
The small oxygen and/or chlorine contribution to the signal is thus covered by the heavy metal scatterers. The Pt00 catalyst after reductive activation exhibits a dispersion almost at monoatomic level (CN(Pt)=1.6) and, after catalytic run, a moderate increase in particle size (CN(Pt) = 4.6) and the presence of chlorine (CN(Cl) = 0.9) on the metallic phase.
These results cannot be rationalized on the basis of the structure insensitivity of the hydrodechlorination reaction, since the most active catalyst, Pt09, shows a much higher size of Pt particles. Catalytic activity seems, therefore, to be related to the stability of metal phases under reaction conditions.
The good stability of Pt phase, and the absence of appreciable coking allows to perform regeneration by a simple oxidation-reduction treatments of the used catalysts without altering the catalytic properties, as it is shown in table 5. The strong metal/support interactions prevent any leaching of volatile Pt species, as it was instead observed for Pd.

Table 5
Catalytic properties of regenerated samples by O_2/H_2 treatments at 500°C

Sample	conv	sel. (mol %)				
	(mol %)	CH_4	C_2H_6	$CHCl_3$	C_2Cl_4	C_2Cl_6
Pt00	82.1	15.0	1.6	80.6	0.7	0
Pt09	93.5	11.2	0.9	79.5	0	2.3
Pt26	76.7	18.9	1.7	77.1	0	0

Upon such regenerative treatments, all the samples underwent minor decreases in conversion levels, and, for the Pt09 sample, the appearance of slight dimerization activity.

Finally, we undertook some investigations on the nature of the support phase under steady-state catalytic conditions. It was in fact observed that magnesium oxide undergoes some extensive chemical modifications at the beginning of the catalytic test. A sharp increase in the temperature of the catalytic bed (up to 40°C in most cases for chlorine-free materials), and the absence of any evolution of HCl were observed, although the conversion of CCl_4 to $CHCl_3$ + hydrocarbons was still higher than 80% (Figure 2).

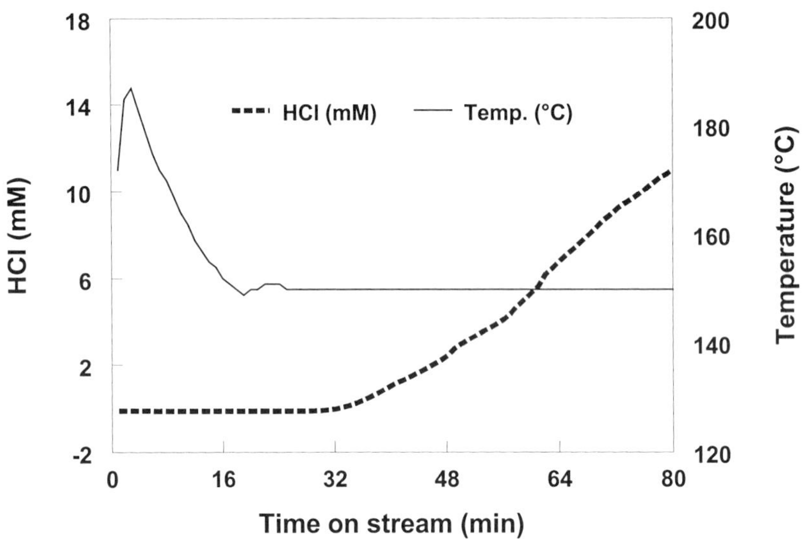

Fig. 2. Evolution of reactor temperature and HCl concentration as a function of time on stream on CCl_4 hydrodechlorination at 150°C with Pt00 catalyst.

After this induction period the temperature of the catalytic bed reaches steady state conditions, and evolution of HCl takes place with a rate predictable on the basis of the distribution of reaction products.

After catalytic runs, all catalysts show a chlorine/magnesium ratio equal to 0.33. This change of phase leads to a mixed magnesium oxo-chloride, whose formula, according to literature data [24, 25], could be ascribed to a Sorel-type $MgO\text{-}Mg(OH)_2\text{-}5MgO\ MgCl_2\ xH_2O$, which is the actual support phase during steady-state catalytic reaction.

It may be concluded that the initial chlorine doping of the Pt/MgO material has a profound influence on the catalytic activity, preventing detrimental overheating phenomena at the beginning of the catalytic run. However, a too high chlorine content results in a decreased metal-support interaction, and lower steady-state conversions.

4. CONCLUSIONS

Pd and Pt supported catalysts show completely different properties, due to different reaction mechanisms. The hydridic character of adsorbed H atoms and the easy cleavage of C-Cl bonds species lead to almost complete hydrogenation of CCl_4 to hydrocarbons on Pd/MgO materials. Instead, Pt catalysts show very high activity and selectivity to partial hydrodechlorination to trichloromethane, with methane as major byproduct.

The initial chlorine doping of MgO support upon catalyst preparation plays an important role in preventing excessive overheating phenomena at the beginning of the hydrodechlorination reaction, when MgO reacts with forming HCl to form a Sorel-type magnesium oxo-chloride phase, which is stable under reaction conditions.

REFERENCES

1. M. J. Molina and F. S. Rowland, Nature, 249 (1974) 810.
2. N. I. Buchar, T. F. Kuech and G. J. Scilla, Cryst. Growth, 110 (1991) 405.
3. J. Tokuda, M. Takai, H. Nakai, K. Gamo and S. Nawaba, J. Opt. Soc. Am. B, 4 (1987) 267.
4. J. V. Michael, K. P. Lim, S. S. Kumaran and J. H. Kiefer, J. Phys. Chem., 97 (1993) 1914.
5. P. H. Taylor, B. Dellinger and D. S. Tirey, Int. J. Chem. Kinet., 26 (1991) 1051.
6. H. Windaw and M. Wyatt, Platinum Met. Rev., 37 (1993) 186.
7. S. Y. Kim, H. C. Choi, O. B. Yang, K. H. Lee, J. S. Lee and Y. G. Kim, Chem. Commun., (1995) 2165.
8. H. C. Choi, S. H. Choi, O. B. Yang, J. S. Lee, K. H. Lee and S. Y. Kim, J. Catal., 161 (1996) 790.
9. Z. Karpinski, K. Early and J. L. d'Itri, J. Catal., 164 (1996) 378.
10. B. S. Ahn, S. C. Lee, J. D. Moon and B. G. Lee, J. Mol. Catal., 106 (1996) 83.
11. B. Coq, J. M. Coignon, F. Figuéras and D. Tournigant, J. Catal., 141 (1993) 21.
12. B. Coq, G. Ferrat and F. Figuéras, J. Catal. 101 (1986) 434.
13. V. Dal Santo, L. Sordelli, C. Dossi, S. Recchia, E. Fonda, G. Vlaic and R. Psaro, J. Catal., 198 (2001) 000.

14. "Gmelin Handbuch der Anorganischer Chemie", Palladium, Verlag Chimie, Weinheim, 1947, p. 302.
15. C. Dossi, S. Recchia, A. Fusi, Fresenius J. Anal. Chem., 367 (2000) 416.
16. A., Michalowicz, J. Phys. IV France, 7C2-235, (1997).
17. B. Lengeler and E. P. Eisenberger, Phys. Rev. B, 21 (1980) 4507.
18. F. James and M. Roos, Comput. Phys. Commun., 10 (1975) 343.
19. M. Kappers, C. Dossi, R. Psaro, S. Recchia and A. Fusi, Catal. Lett., 39 (1996) 183.
20. A. G. Somorjai, Introduction to Surface Chemistry and Catalysis, John Wiley and Sons inc., 1994, p. 495.
21. P. J. Flory, Principles of Polymer Chemistry, Cornell University Press, Ithaca, N. Y. 1953.
22. A. H. Weiss, B. S. Gambhir and R. B. Leon, J. Catal., 22 (1971) 245.
23. H. G. Fritsche and R. E. Benfield, Z. Phys. D. S., (1993) 15.
24. L. Urwongse and C. A. Sorrel, J. Am. Ceram. Soc., 63 (1980) 501.
25. C. A. Sorrel and C. R. Armstrong, J. Am. Ceram. Soc., 59 (1976) 51.

Studies in Surface Science and Catalysis 140
A. Gamba, C. Colella and S. Coluccia (Editors)

Effect of Ti insertion in the silicalite framework on the vibrational modes of the structure: an *ab initio*, and vibrational study

A. Damin[a], G. Ricchiardi[a], S. Bordiga[a], A. Zecchina[a], F. Ricci[b], G. Spanò[b] and C. Lamberti[a,c]

[a]Dipartimento di Chimica IFM, Via P. Giuria 7, I-10125 Torino, Italy
[b]EniChem S.p.A., Centro Ricerche Novara, "Istituto Guido Donegani", Via G. Fauser 4, 28100 Novara, Italy
[c]INFM Unità di Ricerca di Torino Università

We report a thorough theoretical study based on both "conventional" cluster and embedded cluster models on the effect induced on the vibrational modes of the MFI framework by the isomorphous insertion of a Ti atom. On an industrial ground, this insertion has generated one of the most important catalyst of the last two decades: titanium silicalite (TS-1). To allow a direct comparison of TS-1 with the parent Ti-free structure (silicalite), quantum chemical calculations on both cluster models $Ti[OSi(OH)_3]_4$ and $Si[OSi(OH)_3]_4$ have been performed. In both cases we have employed the B3LYP/6-31G(d) level of theory in order to provide the basis for the assignment of the main vibrational contributions. On an experimental ground, dehydrated TS-1 exhibits a IR spectrum characterized by a well defined band located at 960 cm^{-1} and a Raman spectrum showing two components at 960 and 1125 cm^{-1}, being the latter enhanced in case of resonant Raman effect achieved using an UV laser source. In this work, the enhancement of the intensity of the 1125 cm^{-1} feature and the invariance of the 960 cm^{-1} feature in UV-Raman experiments, are discussed in terms of resonant Raman selection rules. The resonance-enhanced 1125 cm^{-1} mode is unambiguously associated with a totally symmetric vibration of the TiO_4 tetrahedron, achieved through in-phase antisymmetric stretching of the four connected Ti-O-Si bridges. This vibration can also be described as an in-phase stretching of the four Si-O bonds pointing towards Ti. The resonance enhancement of this feature is explained in terms of the electronic structure of the Ti-containing moiety. Asymmetric stretching modes of TO_4 units show distinct behavior when T is occupied by Si or Ti, or when the oxygen atom belongs to OH groups (such as in terminal tetrahedra of cluster models and in real defective zeolites). Asymmetric SiO_4 and TiO_4 stretching modes appear above and below 1000 cm^{-1} respectively, when they are achieved through antisymmetric stretching of the T-O-Si bridges, and around 800 cm^{-1} (in both SiO_4 and TiO_4) when they involve symmetric stretching of the T-O-Si units. In purely siliceous models, the transparency gap between the main peaks at 800 and 1100 cm^{-1} contains only vibrational features associated with terminal Si-OH groups, while in Ti- containing models it contain also the above mentioned asymmetric TiO_4 modes, which in turn are strongly coupled with Si-OH stretching modes. Calculations on periodic models of silicalite and TS-1 free of OH groups using the QMPOT embedding method, correctly reproduce the transparency gap of silicalite and the appearance of asymmetric TiO_4 vibrations at 960 cm^{-1} in TS-1.

1. INTRODUCTION

Titanium silicalite-1 (TS-1) is a synthetic zeolite [1] in which a small number of Ti atoms substitute tetrahedral Si atoms in a purely siliceous framework with the MFI structure. It is an active and selective catalyst in a number of low-temperature oxidation reactions with aqueous H_2O_2 as the oxidant [2,3]. For this reason, it has been one of the most studied materials in heterogeneous catalysis in the last years. Although the long range structure of the material is well known [4,5] the structure of the active site is not clear. Several experimental [4-17] and computational [6'18-25] results demonstrate that the substitution of Si by Ti is isomorphous, and it is generally believed that the distribution of Ti over the available framework sites is at least partially disordered [5,15]. A variety of techniques has been developed, able to detect and discriminate tetra-coordinated framework titanium from extraframework titanium atoms with higher coordination. Among these we cite vibrational spectroscopies (both IR and Raman) [6-11], UV-Vis [11-13], EXAFS and XANES [11-15], spectroscopies and powder diffraction experiments (using both x-rays and neutrons) [4,5]. The first two methods in particular, due to their ease of use, have become standard analysis techniques for catalysts containing titanium. The infrared feature which is more clearly associated with the presence of tetrahedrally coordinated framework Ti is a relatively strong absorption appearing, in dehydrated samples, at 960 cm^{-1}. This work is focused on the vibrational peculiarities of the TS-1 material, as a consequence, a particular attention will be devoted to the history of the 960 cm^{-1} band.

The earlier interpretation of the 960 cm^{-1} band [7] was based on the discussion of the modes of isolated SiO_4 tetrahedra as compared with tetrahedra neighboring with a TiO_4 unit. Assuming a higher ionicity of the Ti-O bond as compared with the Si-O one (later confirmed by quantum mechanical calculations), the Si-O stretching mode was expected to shift downwards due to the interaction with the Ti cation ($Si\text{-}O^{\delta-}...Ti^{\delta+}$). Based on quantum mechanical calculations on cluster models, it was later proposed, by de Man and Sauer, [20], that the mode is a simple antisymmetric stretching of the Si-O-Ti bridge. Indeed the two assignments may be seen as coincident, because they describe the same physical mode by focusing on different "building units" [7,8,26]. Smirnov and van de Graaf [21] calculated the vibrational spectra of a periodic model of TS-1 with molecular dynamics techniques. Their results support the localized Ti-O-Si nature of the 960 cm^{-1} vibration, but they also put into evidence the inequivalence of the Si-O and Ti-O bonds and show that the Si-O stretching gives the greater contribution to the vibration. Taking this into consideration, the assignment of the band to a $Si\text{-}O^-$ vibration in defects, proposed by Camblor *et al.* [27] is not exclusive of the previous assignments. In this work we will also show that the vibrational features of framework Ti and of defect sites are strongly coupled.

Raman spectroscopy revealed that the 960 cm^{-1} band is not the only vibrational feature appearing upon insertion of Ti in the MFI framework. Scarano *et al.* and the group of Jacobs [8] have observed a Raman active mode at 1125 cm^{-1} not present in the Ti-free silicalite and attributed to the insertion of Ti in the framework. This band was not observed in the IR spectra because totally overshadowed by the intense antisymmetric stretching of the Si-O-Si unit. More recently, Li *et al.* [10] have reported that the intensity of the 1125 cm^{-1} Raman band is remarkably enhanced using an UV laser as scattering source. The reason of this effect is due to the strong Ligand to Metal Charge Transfer (LMCT) absorption that TS-1 exhibits in the UV region of the electromagnetic spectrum, due to the $O^{2-}Ti^{4+} \rightarrow O^{-}Ti^{3+}$ transition. Li *et al.* have so operated in resonant Raman conditions. In such condition, no enhancement of the 960 cm^{-1} band was observed.

Resonant Raman spectroscopy is defined as a Raman experiment in which the exciting wavelengths coincides or is near to the wavelength of an electronic adsorption of the sample. This condition guarantees a high energy transfer to the sample, with minor disturbance by fluorescence. If the electronic absorption is due to a localized center, like a transition metal atom, excitation is also partially localized and the vibrational features of the immediate vicinity of the absorbing atom can be enhanced, if they meet the appropriate selection rules. This enhancement can be of several orders of magnitude. Besides the normal Raman selection rules, peculiar selection rules apply to the intensity enhancement. In particular, two kinds of vibrations are enhanced: a) totally symmetric vibrations with respect to the absorbing center, and b) vibrations along modes which cause the same molecular deformation caused by the electronic excitation [28].

In order to understand why the 1125 cm^{-1} band is enhanced in resonant Raman conditions, while the 960 cm^{-1} band is not, a careful comparison of the IR, Raman and resonant-Raman spectra of TS-1 samples at increasing Ti loading x (including a Ti-free silicalite sample, $x = 0$), coupled with quantum mechanical calculations on model compounds is needed. This study will give information on the symmetry of the modes involving the Ti atom.

2. METHODS: CLUSTERS AND EMBEDDED CLUSTER MODELS

Geometry optimization and vibrational frequency calculation of cluster models were performed both with the Hartree-Fock and density functional methods using the B3LYP functional. The basis set adopted is the standard 6-31G(d) in all cases. The program used was Gaussian98 [29] with Molden as the graphical interface [30] Because our HF frequencies are in good agreement with the B3LYP ones when scaled in the usual way (scale factor 0.9), in the following we will discuss only the unscaled B3LYP results. In models containing several dangling OH groups, the SiOH bending vibrations were strongly coupled with the Si-O and Ti-O stretching modes. In order to simplify the analysis of the latter, in some cases we substituted ^{1}H with ^{3}H in OH groups in order to decouple their bending vibrations. This led to small shifts of the Si-O and Ti-O vibrations (2-5 cm^{-1} was typical). For sake of completeness we report both the results obtained with terminal ^{1}H and ^{3}H atoms.

Embedded cluster calculations were performed with the QMPOT method [31], coupling a quantum mechanical description of a core set of atoms with a molecular mechanics description of the periodic solid. The QM cluster was treated at the Hartree-Fock level with a split valence SVP basis set [32]. The periodic solid was described with a shell model potential parametrized on QM calculations using the same level of theory. The choice of different methods and basis sets for the cluster and embedded cluster calculations was dictated by the availability of the forcefield for Ti-O interactions. The programs used for the calculation of the subsystems were Gaussian98 [29] and GULP [33], InsightII [34] was the graphical interface program. Periodic vibrational frequencies were calculated for k=0 (858 modes) and converted into an approximate density of states by assigning a gaussian shape with constant width (10 cm^{-1}) and area, and summing over all modes.

3 EXPERIMENTAL: SYNTHESIS AND SPECTROSCOPY

Samples with increasing Ti content have been synthesized in EniChem laboratories following a procedure described in the original patent [1]. The total insertion of Ti atoms in the MFI framework has been witnessed by comparison of the amount of Ti found in the samples by chemical analysis with the cell volumes obtained by Rietveld refinement of powder XRD data, see Ref. [4] for more details.

For IR spectra a Brucker IFS 66 FTIR spectrometer equipped with an HgCdTe cryodetector has been used. For all spectra a resolution of 2 cm^{-1} has been adopted. Samples have been studied in transmission mode on thin self-supported wafers. Adsorbates have been dosed in gas phase through a vacuum manifold directly connected with the measure cell.

The "conventional" Raman spectra were obtained on a Perkin Elmer 2000 NIR-FT Raman spectrometer equipped with an InGaAs detector. The lasing medium was an Nd-YAG crystal pumped by a high-pressure krypton lamp, resulting in an excitation wavelength of 1064 nm (9385 cm^{-1}). The power output was ca. 1000 mW. Samples were examined as such or contacted with solutions.

UV resonance Raman spectra were obtained using a Renishaw Raman System 1000 by exciting with a frequency doubled Ar^{+}-laser, operating at 244 nm (40984 cm^{-1}). The photons scattered by the sample where dispersed by the monochromator and simultaneously collected on an UV-enhanced CCD camera. The collection optic was a x40 objective. A laser output of 12 mW was used, which resulted in a maximum incident power at the sample of approximately 2 mW. An exposure time of 240 s per spectrum was used.

4. RESULT AND DISCUSSION

4.1. *Ab initio* study

Cluster models have been previously used to interpret the vibrational features of titanosilicates. In particular, Sauer and de Man [20] have calculated the vibrational spectra of a variety of small clusters, demonstrating that antisymmetric Ti-O-Si vibration may appear at frequencies around 960 cm^{-1}. However, the occurrence and exact frequency of these vibrations vary widely with model size and connectivity, in a manner which has not yet been explained. The interpretation of the vibrational structure of cluster models containing open chains is much complicated by the strong coupling of the T-O stretching modes with the bending modes of terminal hydroxyls. This problem can be overcome by adopting clusters formed by closed rings: however, if small or condensed rings are adopted (*e.g.* four-membered), the strain imposed by ring closure completely alter the vibrational features in the region of interest (700-1200 cm^{-1}). In this work, we focus on one type of cluster - a central tetrahedron surrounded by four complete tetrahedra, also known as "shell-3" model because it contains three complete shells of neighbors of the central atom - and refine the discussion of ref. [20] in three directions. First, we have improved the accuracy of the calculations for the isolated cluster, by using Density Functional theory and the 6-31G(d) basis set. Secondly, we have simplified the interpretation of the results by using isotopic substitution (^{1}H substituted with ^{3}H) to eliminate the coupling of Si-O-H bending modes with Si-O and Ti-O stretching modes. Finally, we have attempted the calculation of the vibrational spectrum of a true periodic model of TS-1, by embedding the above-mentioned cluster into a periodic lattice treated with empirical potentials.

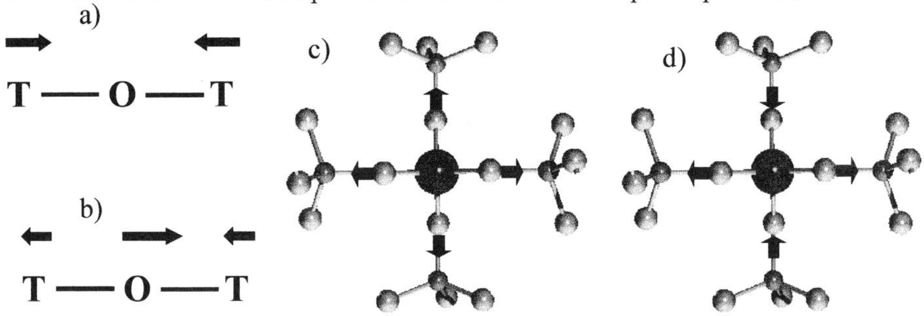

Scheme I. Representation of the symmetric and antisymmetric stretching modes of the T-O-T unit, parts (a) and (b) respectively. Symmetric and antisymmetric stretching modes of the TO_4 unit, parts (c) and (d) respectively.

$Si(OSi(OH)_3)_4$ and $Ti(OSi(OH)_3)_4$ clusters (named *SiSi4* and *TiSi4* in the following) have been used. The calculated frequencies in the 800-1200 cm^{-1} are reported in Table 1 and 2 respectively. The calculated modes can be analyzed according to the symmetry of the vibration around the central atom of each tetrahedron or according to the symmetry of the T-O-T bridge deformation, *i.e.* whether the two bonds stretch in phase or not (Scheme Ia,b). In the case of this second classification, we can further separate the modes according to the symmetry of the deformation of the four T-O-T bridges around the central cluster atom. An example is given in Scheme I. In Scheme Ic) the totally symmetric stretching mode of the internal tetrahedron is represented, which can be also seen as an in-phase combination of four T-O-Si antisymmetric modes, or if we neglect the central atom or consider it like a simple perturbation like an in-phase stretching of the four neighbouring Si-O oscillators. In Scheme Id) one of the asymmetric modes of the central tetrahedron is represented, which can be also described as an out-of-phase combination of four T-O-Si bridges or neighboring Si-O bonds. We will see later that these modes have different responses to the Si/Ti substitution. The main spectroscopic features of the *SiSi4* and *TiSi4* models are described in Tables 1 and 2 together with their assignment in terms of vibrations of TO_4 or T-O-Si units.

In the computed spectra of both of *SiSi4* and *TiSi4*, clusters, the highest modes are represented by stretching modes of the central atom of the cluster. In the case of *TiSi4*, only one stretching of the central tetrahedron falls in this region, which is also an in-phase antisymmetric stretching of the four TiOSi bridges (1097 cm^{-1}). Due to his symmetry around the Ti atom, this mode is the only candidate for the assignment of the 1125 cm^{-1} Raman band observed experimentally, see Introduction and *vide infra.* The mode also appear at similar frequencies in closed ring clusters (not reported) and can be regarded as the main Raman fingerprint of tetrahedral titanium. It is completely IR-inactive due to its symmetry. The analogous mode for *SiSi4* is observed at 1159 cm^{-1}. When the symmetric deformations of TO_4 involve symmetric T-O-Si stretching, modes at 827 and 814 cm^{-1} are computed for *SiSi4* and *TiSi4* respectively.

Table 1. Computed stretching frequencies for the $Si[OSi(OH)_3]_4$ model (ν), and symmetry of the corresponding mode with respect to the central SiO_4 and Si-O-Si unit. A = antisymmetric and S = symmetric.

ν (cm^{-1})	SiO_4	Si-O-Si	Comments
1159	S	A	See text
1105,1096	A	A	See text
996, 984, 971	-	-	Si-O stretching modes of terminal silanols involved in hydrogen bonds with neighboring OH as donors.
993	-	-	Si-O stretching modes of terminal silanol (no H-bond)
917	-	-	Si-O stretching modes of terminal silanol in which O is hydrogen bond acceptor.
876, 855	A	S	See text
827	S	S	See text

Table 2. Computed stretching frequencies for the $Ti[OSi(OH)_3]_4$ model (ν), and symmetry of the corresponding mode with respect to the central TiO_4 and T-O-T unit. A = antisymmetric and S = symmetric.

ν (cm^{-1})	TiO_4	Ti-O-Si	Comments
1097	S	A	See text
987, 964	-	-	Si-O stretching modes of terminal silanols involved in hydrogen bonds with neighboring OH as donors.
986, 968, 965,909, 897	A	A	Ti-O stretching modes strongly coupled with Si-O stretching modes of terminal silanols involved in H-bonds with neighboring OH as donors.
913	-	-	Si-O stretching modes of terminal silanols involved in hydrogen bonds with neighboring OH as acceptors
829, 824	A	S	See text
814	S	S	See text

If we consider now the asymmetric deformations of the central tetrahedron (occupied by Ti or Si), the two models are much different. The asymmetric deformations of SiO_4 in *SiSi4* produce features at 1105 and 1096 cm^{-1} when they occur through antisymmetric Si-O-Si stretching. On the contrary, TiO_4 asymmetric vibrations in *TiSi4*, appear below 1000 cm^{-1} (multiplet at 986, 968, 965, 909 and 897 cm^{-1}). When the asymmetric deformations of TO_4 involve symmetric T-O-Si stretching, modes at 876, 855 cm^{-1} and at 829, 824 cm^{-1} occur for *SiSi4* and *TiSi4* respectively. This fact lead us to the discussion of the bands appearing in the central region of the spectra (850-1000 cm^{-1}), *i.e.* in the experimentally observed "gap" between the high-frequency (1250-1050 cm^{-1}) antisymmetric Si-O-Si vibrations and the low frequency (around 800 cm^{-1}) symmetric Si-O-Si vibrations, see Figs. 1,2. This region is of much interest, because it contains: i) the highly debated 960 cm^{-1} band observed in both IR and Raman spectra; ii) the features associated with framework defects.

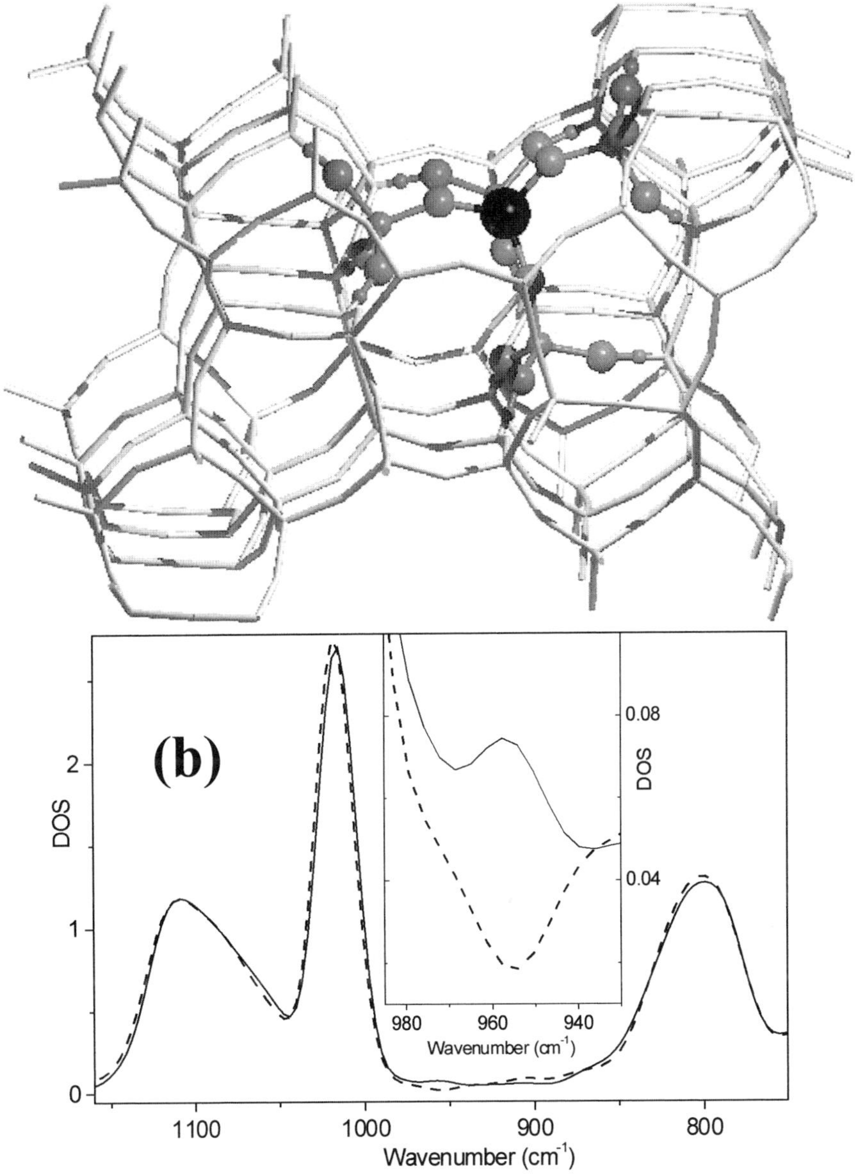

Fig. 1. Part (a): graphical representation of the periodic model used in this study. The shell-3 clusters has been embedded into the crystallographic T5 position. Part (b): the so obtained DOS for the silicalite (dashed line) and for TS-1 (full line). The inset reports a magnification of the of the two curves in the gap between the symmetric and antisymmetric stretching modes of the T-O-T units, highlighting the appearance of the 960 cm^{-1} band for TS-1, otherwise hardly visible.

The calculated spectrum of *SiSi4* in this region contains only Si-O stretching modes associated with terminal Si-OH groups of the models. These modes are spread over a wide range (917-996 cm^{-1}) because of the different hydrogen bonds involving the terminal OH groups of the cluster. These modes are good candidates to interpret the IR modes appearing in the gap of defective silicates (*e.g.* defective silicalite) [27,35,36]. No stretching modes of the central tetrahedron appear in this region. This is confirmed by our calculation on the periodic model (*SiSi4* embedded in the periodic structure), which correctly reproduce the experimentally observed gap between 850 and 1000 cm^{-1} (Fig. 1b *vs.* dashed line in Fig.2).

In the case of *TiSi4* cluster, the Si-OH stretching modes are strongly coupled with asymmetric TiO_4 stretching modes. These modes are out-of-phase combinations of the four Ti-O-Si antisymmetric stretching. They also span a wide frequency range (897-986 cm^{-1}), and no single mode can be associated individually to the 960 cm^{-1} band. This clearly show on one hand, that our cluster is inadequate to model a non-defective titanosilicate, and on the other hand, that in defective titanosilicates the Si-OH and asymmetric TiO_4 vibrations are strongly coupled. This is why spectroscopic features in this range are still under debate [10].

The modeling a non-defective titanosilicate thus requires either larger clusters or a periodic model. We have embedded the shell-3 clusters into the crystallographic T5 position of a unit cell of the MFI framework (Fig. 1a). In the case of the embedding of the *SiSi4* cluster, the resulting structure is purely siliceous, while for the *TiSi4* cluster, we obtain a structure with one Ti atom per unit cell. No hydroxy groups are present in either model. The choice of the T5 position is partly arbitrary, and dictated by the availability of the calculated structure from a previous work [25]. Fig. 1b reports the calculated density of states of Silicalite and TS-1. The DOS of silicalite correctly reproduce the two main groups of framework vibrations centered at 800 and 1100 cm^{-1}, and the transparency region between them. The DOS of Ti-silicalite is very similar but differs for a small, expected [1,6-9,11], feature at 960 cm^{-1}. Graphical analysis of the individual normal modes contributing to the DOS at this frequency shows four contributions: two modes involving antisymmetric stretching of Si-O-Si bridges far from the Ti atom and two modes which can be classified as out-of-phase asymmetric stretching of the Ti-O-Si bridges. The former have no counterpart in cluster calculations and may be artifacts due to a faulty modeling of the interface between the QM and MM regions, while the latter are in full agreement with the cluster calculations and with the assignment of the 960 cm^{-1} band presented above. This result, although not conclusive due to the limitations of the model, indicates that the embedding of the *TiSi4* cluster into a non-defective periodic framework causes the disappearance of the scattered (897-987 cm^{-1}) bands due to coupled Ti-O-Si-OH vibrations and the appearance of a single narrow band at 960 cm^{-1}.

Quantum mechanical calculations not only provides vibrational frequencies to be compared with the experimental ones, but also describe the electronic structure of the clusters, which has deep implications in the Resonant Raman experiments. In particular, the structure and symmetry of the LUMO of the Ti-containing clusters have to be considered (an accurate calculation of the true electronic structure of the excited state is beyond the scope of the present work). The LUMO of the *TiSi4* cluster is almost totally symmetric around the Ti atom and two-fold degenerate. The LUMO is formed by an antibonding combination of O *p* orbitals with a totally symmetric combination of Ti *d* orbitals ($d(z^2)$, $d(x^2-y^2)$).

It is reasonable to deduce that the observed UV absorption at 205 nm (about 49000 cm^{-1}) involves a totally symmetric excited state in which all four Ti-O bonds are symmetrically stretched with respect to the ground state (due to the Ti-O antibonding character of the LUMO). Thus the symmetric TiO_4 stretching meets both selection rules a) and b) (see the introduction) for the resonance enhancement of its Raman intensity. We therefore conclude that the observed band at 1125 cm^{-1} (calculated at 1100 cm^{-1} in the *TiSi4* cluster) can be attributed to this kind of vibration. Based on the discussion above, this vibration can be also described as the in-phase combination of the four Ti-O-Si stretching modes, or as the in-phase combination of the four $Si\text{-}O^{\delta-}$ bonds perturbed by Ti.

On the contrary, none of the vibrations appearing in the "gap" meets either of the selection rules for Raman enhancement. This explain why the 960 cm^{-1} band is not enhanced. Indeed, it apparently disappears in Resonant experiments due to the enormous intensity of the strongly enhanced 1125 cm^{-1} band. Nevertheless, the Ti-sensitive nature of the 960 cm^{-1} band is confirmed by non-resonant experiments, as discussed in the previous sections.

4.2. Evidence of the correlation of the 1125 and 960 cm^{-1} bands with Ti content

In this experimental section we will prove that both the IR active 960 cm^{-1} band and the Raman active 1125 cm^{-1} bands have an intensity which is linearly correlated with the Ti content of TS-1 samples. The use of resonant Raman will testify that the latter band is enhanced according to the assignments discussed in our quantum chemical study.

Fig. 2 reports, in a wide spectroscopic range (3800-700 cm^{-1}), the IR feature of both silicalite (dashed spectrum) and TS-1 (solid spectrum). The two spectra look almost identical, being the only significant difference the narrow 960 cm^{-1} band visible in the upper right corner of the figure and magnified in the inset (which also allow a direct comparison with the computed DOS reported in Fig. 1b). It becomes now evident that the 960 cm^{-1} band is a small, although significant, feature of the overall spectrum of TS-1. This band lies in the gap between the symmetric and asymmetric stretching modes of the T-O-T units, the latter totally saturated. At higher frequencies, the stretching of hydroxyls are visible: sharp peak at 3737 cm^{-1} and broad component around 3475 cm^{-1}, which are due to isolated and H-bonded Si-O-H (and probably to Ti-O-H) species respectively. In particular, the latter component implies that TS-1 is a defective material (as silicalite is), hosting an important fraction of T vacancies, generating internal hydroxylated cavities [35-37]. At 2005, 1180 and 1640 cm^{-1} the overtone and combination bands are also visible.

To validate the attribution of the vibrational modes done in section 4.1, Raman, resonant Raman and IR spectra on a set of TS-1 samples characterized by an increasing Ti loading (x) are reported in Figure 3. Part a) of the figure reports spectra collected on three samples (x = 3.0, 1.4 and 0.0 *i.e.* silicalite) with a FTIR Raman spectrometer [λ_{laser} = 1064 nm (9385 cm^{-1})], while part b) of the figure shows spectra obtained with a dispersive instrument working with a λ_{laser} = 244 nm (40984 cm^{-1}). With the latter experimental setup we are in resonant Raman conditions for the $O^{2-}Ti^{4+} \rightarrow O^{-}Ti^{3+}$ LMCT. In fact, for tetrahedral coordinated Ti^{4+} this LMCT exhibits an absorption band centered around 205 nm (about 49000 cm^{-1}) with a low energy tail extending down to 250 nm (40000 cm^{-1}) [11,12], see inset in Fig. 3b. The Raman spectrum of silicalite sample is very similar independently on the used laser source, showing bands at 1230, 1085, 975 cm^{-1} and a complex absorption centered at 820 cm^{-1}. In fact no resonant enhanced effect is expected

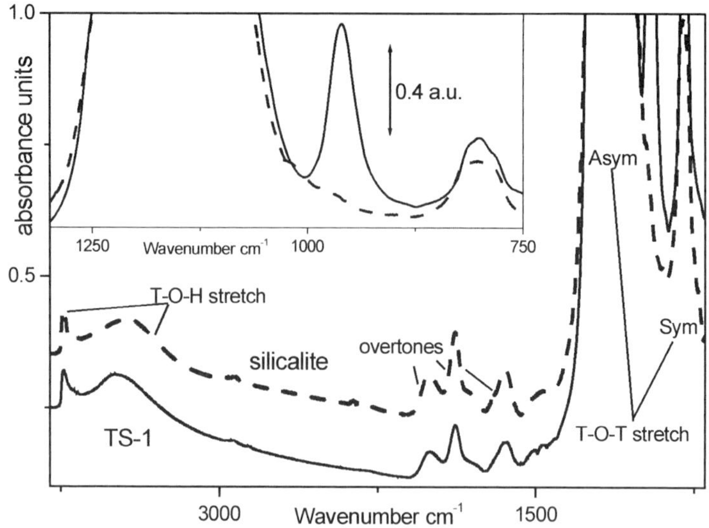

Fig. 2. Comparison between the IR spectra in the 3800-700 cm^{-1} range of silicalite (dashed line) and TS-1 (full line). The inset reports the magnification of the framework stretching region, to be compared to DOS reported in Fig. 1b.

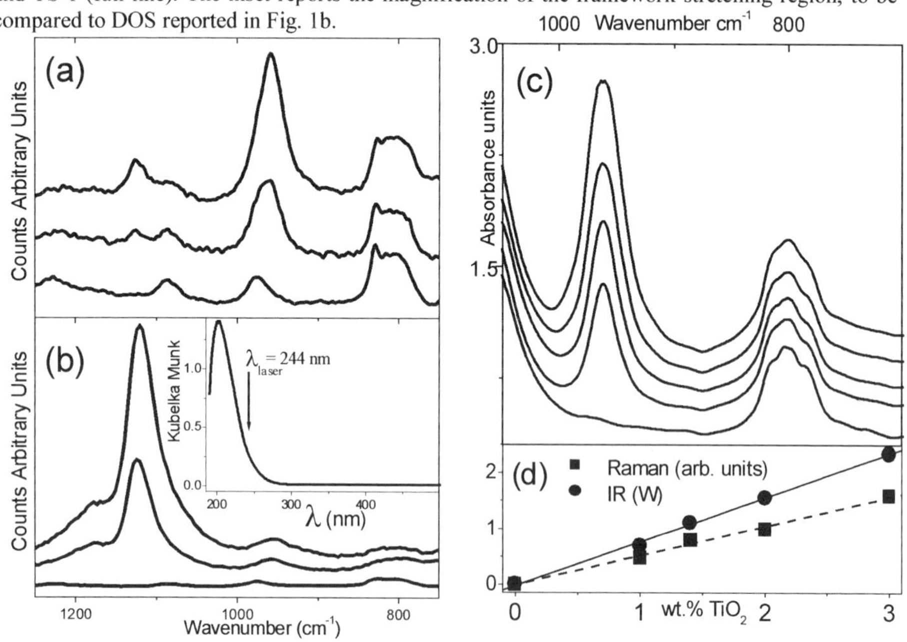

Fig. 3. Part (a): Raman spectra of TS-1, from top to bottom $x = 3.0$, 1.4 and 0.0 (silicalite). Part (b): as part for resonant Raman. The inset reports the DRS spectra of TS-1; the vertical arrow shows the wavelength of the UV laser source. Part (c): IR spectra of TS-1, from top to bottom $x = 3.0$, 2.0, 1.4, 1.0 and 0.0. Part (d) intensity of the 960 cm^{-1} band vs. x, in (non resonant) Raman (■) and IR (●) spectra. For the latter, the height W at FWHM of band is reported, see text.

for a sample transparent at the excitation wavelength, occurring the $O^{2-}Si^{4+} \rightarrow O^{-}Si^{3+}$ LMCT at much higher energies. The bands at 1230 and 1085 cm^{-1} have been associated to $\nu_{asym.}$(Si-O-Si) (Raman inactive) [8a] while the stronger absorption at 975 cm^{-1} has been assigned to $\nu_{asym.}$(Si-O-Si*) where Si* is a silicon atom connected with an hydroxy-group $Si[OSi]_3OH$ [8a]. The fact that this component, characteristic of defective silicalite, occurs close to the 960 cm^{-1} band of TS-1 is probably the origin of the animated debate on the nature of this band (see Introduction). Finally the complex absorption centered at 820 cm^{-1} has been associated with the $\nu_{sym.}$(T-O-T) (Raman active) [8a]. Moving now to TS-1 samples we observe, two main differences: i) the appearance of a new band at 1125 cm^{-1} which is enhanced by resonant Raman effect (compare Fig. 3b *vs.* 3a); ii) the growth of a band at 960 cm^{-1} which intensity is proportional to x, see Fig. 3d. For both resonant (Fig. 3b) and non-resonant Raman (Fig. 3a), the constant value of I(1125)/I(960) ratio (0.25 and 11 respectively, independently to x) indicates that the two bands should be related to two different spectroscopic manifestations of the same phenomenon: insertion of Ti in the MFI framework.

Even if the 960 cm^{-1} band is generally recognized as a proof that Ti(IV) heteroatoms are isomorphically inserted in the zeolite structure [6-8,11], a quantitative correlation between the Ti content (x) and the intensity of the 960 cm^{-1} IR band is very critical. On an experimental ground this is due to its great intensity at high Ti content and to the fact that it is modified, in both intensity and position, by adsorbed molecules. For this reasons, great care has been made in order to prepare TS-1 pellets as thin as possible and to measure TS-1 samples that have undergone the same dehydration treatment. Fig. 3c reports the IR spectra of five TS-1 samples containing increasing Ti content x, from x = 3.0 down to x = 0.0, pure silicalite (bottom curve). The spectra have been normalized on the overtone crystal modes, *i.e.* considering the bands in the 2000-1500 cm^{-1} range (see Fig. 2). The quality of the normalization procedure is confirmed by the good superposition of the spectra in the 750-850 cm^{-1} range, which is also not influenced by Ti content. The common method of dilution with KBr can not be used for reducing the absorbance intensities of the framework modes, since it does not allow thermal treatments, which are needed in order to remove the adsorbed water. In fact, the 960 cm^{-1} band is affected in its shape, frequency and intensity according to the amount of weakly adsorbed basic molecules.

We assume that: (i) the 960 cm^{-1} band does not saturate for the TS-1 sample with lowest Ti content; and (ii) the "true" Full Width at Half Maximum (FWHM) of this band is constant over all set of TS-1 samples. The former assumption is supported by the fact that the intensity of the 960 cm^{-1} band, for the x = 1.0 sample, is less than 1.5 in absorbance units, while the latter is exactly what expected for a band associated to different concentrations of a unique species. Assumption (i) allows us to measure the "true" FWHM of the 960 cm^{-1} band in TS-1 samples (27 cm^{-1}). Now, following assumption (ii), we were able to estimate for all samples reported in Fig. 3, the absorbance (W) at the height where the width of the band corresponds to 27 cm^{-1}. This method minimize the errors due to the instrument sensibility and allows a quantitative estimation of the band intensity. Such obtained values plotted against x in the Fig. 3d (• data), give a high linear correlation (r = 0.9998) and validate assumption (i) and (ii). A further, and definitive, validation comes from the similar linearity found for the intensities of the Raman band, which can not be affected by saturation problems, see Fig. 3d (■ data). These results implies that the 960 cm^{-1} band is well a fingerprint of the insertion of Ti in the zeolitic framework, even if not enhanced by Raman resonant spectroscopy for the reasons outlined in our theoretical study (section 4.1). Thus the original interpreatation of our group [6-8,11] is confirmed and the assignment of the 960 cm^{-1} band to the presence of defects [10] is rejected.

5. CONCLUSIONS

This work discusses the vibrational features associated with the insertion of tetrahedral Ti in the MFI zeolite lattice. The understanding of these features forms the base for the technical characterization of Ti containing silicate catalysts using spectroscopic methods, which is of capital importance in industrial catalysis. A combination of spectroscopic and computational techniques is used in order to assign the main vibrational features of Ti-silicalite, also taking into account the presence of hydroxylated defects.

A set of experiments on TS-1 samples with variable Ti content, synthesized and treated in a reproducible way, has allowed to proof the quantitative correlation between Ti content and the intensity of the 960cm^{-1} IR feature. The linear correlation between the intensity of the 960cm^{-1} band and x has been also confirmed by Raman spectroscopy.

Raman experiments on silicalite and TS-1 with excitation wavelengths of 1064 nm (non resonant) and 244 nm (resonant) show that: (i) the main features associated with Ti insertion in the lattice are vibrations at 1125 and 960 cm^{-1}; the former being drastically enhanced by UV-resonance, while the latter is not; (ii) a mode is observed at 978 cm^{-1} on defective silicalites and TS-1, which we attribute to the Si-O stretching in silanols. The proximity of the 960 and 978 cm^{-1} modes has prompted us to re-examine IR spectroscopy in the same region in order to distinguish the 960 cm^{-1} band from defect modes.

Quantum mechanical calculations of the vibrational frequencies and electronic structure of shell-3 cluster models allow to assign the main vibrational features. The 1125 cm^{-1} peak is undoubtedly assigned to the symmetric stretching vibration of the TiO_4 tetrahedron, achieved through in-phase antisymmetric stretching of the four connected Ti-O-Si oscillators. According to its symmetry and to the electronic structure of the Ti moiety, this is the only vibration fulfilling the resonant Raman selection rules. This assignment is equivalent to the assignment to the in-phase stretching of the four Si-O bonds surrounding Ti. The asymmetric vibrations of TO_4 of our cluster models appear above 1000 cm^{-1} for SiO_4 and between 897 and 986 cm^{-1} for TiO_4. These modes are strongly coupled with the Si-O stretching of the silanols in the outer part of the cluster. Embedding of the Ti-containing cluster in a periodic silicalite framework makes the mode coalesce into a band centered at 960 cm^{-1}. In the purely siliceous model, the only modes contributing to this region are the Si-O vibrations of the silanols in the outer sphere. Based on these results, we confirm the assignment of the 960 cm^{-1} band to the asymmetric stretching of the TiO_4 unit, which can equivalently be described as the out-of-phase antisymmetric stretching of the four connected Ti-O-Si oscillators, or as the out-of-phase stretching of the four Si-O bonds pointing towards Ti.

REFERENCES

[1] M. Taramasso, G. Perego and B. Notari, US Patent No. 4410501 (1983).
[2] B. Notari, Adv. Catal. 41 (1996) 253, and references therein.

[3] G. Bellussi, A. Carati, G.M. Clerici, G. Maddinelli and R. Millini, J. Catal., 133 (1992) 220. M.A. Mantegazza, G. Leofanti, G. Petrini, M. Padovan, A. Zecchina and S. Bordiga, Stud. Surf. Sci. Catal. 82 (1994) 541; M.A. Mantegazza, G. Petrini, G. Spanò, R. Bagatin and F. Rivetti, J. Mol. Catal. A 146 (1999) 223.
[4] R. Millini, E. Previdi Massara, G. Perego and G. Bellussi, J. Catal., 137 (1992) 497; C. Lamberti, S. Bordiga, A. Zecchina, A. Carati, A.N. Fitch, G. Artioli, G. Petrini, M. Salvalaggio, G.L. Marra, J. Catal., 183 (1999) 222
[5] G. L. Marra, G. Artioli, A. N. Fitch, M. Milanesio and C. Lamberti, Microporous Mesoporous Mater., 40 (2000) 85; C.Lamberti, S.Bordiga, A.Zecchina, G. Artioli, G.L. Marra and G. Spanò, J. Am. Chem. Soc., 123 (2001) in press, and references therein.
[6] G. Tozzola, M.A. Mantegazza, G. Ranghino, G. Petrini, S. Bordiga, G. Ricchiardi, C. Lamberti, R. Zulian and A. Zecchina, J. Catal., 179 (1998) 64.
[7] M.R. Boccuti, K.M. Rao, A.Zecchina, G. Leofanti and G. Petrini, Stud. Surf. Sci. Catal., 48 (1989) 133.
[8] a) D. Scarano, A. Zecchina, S. Bordiga, F. Geobaldo, G. Spoto, G. Petrini, G. Leofanti, M. Padovan and G. Tozzola, J. Chem. Soc. Faraday Trans., 89 (1993) 4123; b) G. Deo, A. Turek, I.E. Wachs, D.R.C. Huybrechts and P. A. Jacobs Zeolites 13 (1993) 365.
[9] D.R. Huybrechts, P.L. Buskens and P.A. Jacobs, J. Mol. Catal. 71 (1992) 129; A.J. van der Pol, and H.P. van Hooff, Appl. Catal. A, 92 (1992) 93.
[10] C. Li, G. Xiong, Q. Xin, J. Liu, P. Ying, Z. Feng, J. Li, W. Yang, Y. Wang, G. Wang, X. Liu, M. Lin, X. Wang, and E. Min, Angew. Chem. Int. Ed., 38, (1999) 2220.
[11] S. Bordiga, S. Coluccia, C. Lamberti, L. Marchese, A. Zecchina, F. Boscherini, F. Buffa, F. Genoni, G. Leofanti, G. Petrini, and G. Vlaic J. Phys. Chem., 98 (1994) 4125; S. Bordiga, F. Boscherini, S.Coluccia, F.Genoni, C. Lamberti, G. Leofanti, L. Marchese, G. Petrini, G. Vlaic and A. Zecchina, Catal. Lett. 26 (1994)195.
[12] C. Lamberti, S. Bordiga, D. Arduino, A. Zecchina, F. Geobaldo, G. Spanò, F. Genoni, G. Petrini, A. Carati, F. Villain and G. Vlaic, J. Phys. Chem. B, 102 (1998) 6382.
[13] D. Trong On, L. Le Noc and L. Bonneviot, Chem. Comm., (1996) 299.
[14] D. Gleeson, G. Sankar, C.R.A. Catlow, J.M. Thomas, G. Spanó, S. Bordiga, A. Zecchina, C. Lamberti, Phys. Chem. Chem. Phys., 2 (2000) 4812.
[15] V. Bolis, S. Bordiga, C. Lamberti, A. Zecchina, A. Carati, F. Rivetti G. Spanò and G. Petrini, Langmuir, 15 (1999) 5753; V. Bolis, S. Bordiga, C. Lamberti, A. Zecchina, A. Carati, G. Petrini, F. Rivetti and G. Spanò, Microporous Mesoporous Mater., 30 (1999) 67.
[16] R. Millini and G. Perego, Gazzetta Chimica Ital., 126 (1996) 133, and references therein.
[17] G.N. Vayssilov, Catal. Rev. Sci. Engl., 39 (1997) 209, and references therein.
[18] A. Jentys and C.R.A. Catlow, Catal. Lett., 22 (1993) 251.
[19] R. Millini, G. Perego and K. Seiti, Stud. Surf. Sci. Catal., 84 (1994) 2123.
[20] A.J.M. de Mann, and J. Sauer, J. Phys. Chem., 100 (1996) 5025.
[21] K.S. Smirnov and B. van de Graaf, Microporous Mater., 7 (1996) 133.
[22] P.E. Sinclair, G. Sankar, C.R.A. Catlow, J.M. Thomas and T. Maschmeyer, J. Phys. Chem. B, 101 (1997) 4237.
[23] C.M. Zicovich-Wilson, R. Dovesi and A.Corma, J. Phys. Chem. B, 103 (1999) 988.
[24] P.E. Sinclair and C.R.A. Catlow, J. Phys. Chem. B, 103 (1999) 1084.
[25] G. Ricchiardi, A.J.M. de Man and J. Sauer, Phys. Chem. Chem. Phys., 2 (2000) 2195.
[26] A.Miecznikowski and J. Hamuza, Zeolites, 7 (1987) 249; A.J.M. de Man, B.W.H. van

Beest, M. Leslie and R.A. van Santen, J. Phys. Chem., 94 (1990) 2524.
[27] M.A. Camblor, A. Corma and J. Perez-Pariente, J. Chem. Soc. Chem. Commun., (1993) 557.
[28] Y. Nishimura, A.Y. Hirakawa and M. Tsuboi, in R.J.H. Clark and R.E. Hester, (Eds.) Advances in Infrared and Raman Spectroscopy, Heyden & Son, London, 1978; P.R. Carey, Biochemical Applications of Raman and Resonance Raman Spectroscopies, Academic Press, New York, 1982 (and references therein).
[29] M.J. Frisch, G.W. Trucks, H.B. Schlegel, G.E. Scuseria, M.A. Robb, J.R. Cheeseman, V.G. Zakrzewski, J.A. Montgomery Jr., R.E. Stratmann, J.C. Burant, S. Dapprich, J.M. Millam, A.D. Daniels, K.N. Kudin, M.C. Strain, O. Farkas, J. Tomasi, V. Barone, M. Cossi, R. Cammi, B. Mennucci, C. Pomelli, C. Adamo, S. Clifford, J. Ochterski, G.A. Petersson, P.Y. Ayala, Q. Cui, K. Morokuma, D.K. Malick, A.D. Rabuck, K. Raghavachari, J.B. Foresman, J. Cioslowski, J.V. Ortiz, A.G. Baboul, B.B. Stefanov, G. Liu, A. Liashenko, P. Piskorz, I. Komaromi, R. Gomperts, R.L. Martin, D.J. Fox, T. Keith, M.A. Al-Laham, C.Y. Peng, A. Nanayakkara, C. Gonzalez, M. Challacombe, P.M.W. Gill, B. Johnson, W. Chen, M.W. Wong, J.L. Andres, C. Gonzalez, M. Head-Gordon, E.S. Replogle and J.A. Pople, Gaussian 98, Revision A.7, Gaussian, Inc., Pittsburgh PA, 1998.
[30] G.A. Petersson, A. Bennett, T.G. Tensfeldt, M.A. Al-Laham, W.A. Shirley and J. Mantzaris, J. Chem.Phys, 89 (1996) 2193.
[31] U. Eichler, C. Kölmel and J. Sauer, J. Comput. Chem., 18 (1997) 463; U. Eichler, M. Brändle and J. Sauer, J. Phys. Chem. B, 101 (1997) 10035.
[32] A. Schäfer, H. Horn and R. Ahlrichs, J. Chem. Phys., 97 (1992) 2571.
[33] J. Gale, J. Chem. Soc. Faraday Trans., 93 (1997) 629.
[34] InsightII 4.0.0 molecular modelling system. Molecular Simulation Inc., San Diego.
[35] A. Zecchina, S. Bordiga, G. Spoto, L. Marchese, G. Petrini, G. Leofanti and M. Padovan, J. Phys. Chem., 96 (1992) 4985; A. Zecchina, S. Bordiga, G. Spoto, L. Marchese, G. Petrini, G. Leofanti and M. Padovan, M. J. Phys. Chem., 96 (1992) 4991; A. Zecchina, S. Bordiga, G. Spoto, L. Marchese, G. Petrini, G. Leofanti, M. Padovan, C. Otero Areàn, J. Chem. Soc. Faraday Trans., 88 (1992) 2959; G.L. Marra, G. Tozzola, G. Leofanti, M. Padovan, G. Petrini, F. Genoni, B. Venturelli, A. Zecchina, S. Bordiga, G. Ricchiardi, Stud. Surf. Sci. Catal., 84 (1994) 559.
[36] S. Bordiga, P. Ugliengo, A. Damin, C. Lamberti, G. Spoto, A. Zecchina, G. Spanò, R. Buzzoni, L. Dalloro, F. Rivetti, Topics in Catal., (2001), in press; S. Bordiga, I. Roggero, P. Ugliengo, A. Zecchina, V. Bolis, G. Artioli, R. Buzzoni, G. L. Marra, F. Rivetti, G. Spanò and C. Lamberti, J. Chem. Soc. Dalton Trans., (2000) 3921.
[37] G. Artioli, C. Lamberti and G. L. Marra, Acta Cryst. B, 56 (2000) 2.

Studies in Surface Science and Catalysis 140
A. Gamba, C. Colella and S. Coluccia (Editors)

Surface Properties of Mesoporous Ti-MCM-48 and their Modifications Produced by Silylation.

V. Dellarocca[a], L. Marchese[b*], M.L. Peña[c], F. Rey[c], A. Corma[c] and S. Coluccia[a]

[a] Dipartimento di Chimica IFM, Università di Torino, Via P. Giuria 7, 10125 Torino, Italy.

[b] Dipartimento di Scienze e Tecnologie Avanzate, Università del Piemonte Orientale 'A. Avogadro', c.so Borsalino 54, 15100 Alessandria, Italy.

[c] Instituto de Tecnología Química, CSIC-UPV, Universidad Politécnica de Valencia, Avenida de los Naranjos s/n, 46022 – Valencia, Spain.

The catalytic activity of Ti-MCM-48 catalysts may be improved by increasing the hydrophobic character of the surface by a silylation post-treatment with hexamethyldisilazane (HDMS) as silylating agent. In this way the poisoning of the catalysts is noticeably decreased. The silylation procedure led to the transformation of silanols, $Si(OSi)_3(OH)$, into $Si(OSi)_3[OSi(CH_3)_3]$, which was confirmed by ^{29}Si-MAS-NMR and diffuse reflectance (DR) UV-Vis spectroscopies.

1. INTRODUCTION

Porous materials are very attractive for applications in the field of heterogeneous catalysis. The success of these materials consists in their very high surface area and in the possibility to exhibit "shape selectivity", due to the regular size of pores and channels. For example, in zeolites, which are crystalline microporous materials, the pores are in the range of 5-12 Å. By introducing proper elements in these materials, it is possible to obtain catalysts with peculiar chemical properties. In particular, the introduction of transition metals (Fe, V, Ti, etc.) into siliceous structures generates redox properties. For example, the isomorphic substitution of Si atoms with Ti(IV) ions in the zeolitic framework originates highly active catalysts for a number of reactions [1], i.e. oxidation using hydroperoxide as oxidising agent [2]. The first Ti-zeolite, with the structure of the MFI zeolite, was prepared by Enichem in 1983 *via* hydrothermal synthesis and was named TS-1 [3]; it appeared to be active and selective in oxidative reactions with H_2O_2 at low temperature. However, even in zeolites with large channels, like Ti-β, synthesised by Camblor et al. [4], the diffusion of reactants is limited to molecules smaller than 7 Å.

In order to avoid this limitation, the interest in synthesising mesoporous materials is nowadays increasing in catalysis as the high pore dimension (10-150 Å) allows the access to bulky organic molecules. In 1991, the Mobil Oil Company synthesised a mesoporous alumino-silicate through a basic synthesis using alumino-silica gel and alkyl-

* to whom correspondence should be addressed marchese@ch.unito.it

trimethylammonium bromide/hydroxide with long alkyl chain as structuring agent. This material, characterised by a regular structure with hexagonal array of pores, was named MCM-41 [5]. Although the walls of these siliceous systems are amorphous, the long-range structure is highly ordered. Later on, cubic and lamellar structures were synthesised and respectively named MCM-48 and MCM-50 [6]. The presence of pores with larger dimension (15-100 Å) in mesoporous materials, upon introduction of Ti ions, expands the opportunity of heterogeneously catalysed oxidative reactions. The first Ti-based mesoporous material was synthesised by Corma *et al.* with a MCM-41-type structure [7]. The channel size allowed both the oxidation of bulky organic molecules and the use of organic hydroperoxide as oxidising agent instead of H_2O_2. The catalytic activity was suggested to be strictly connected to the presence of tetrahedral Ti(IV) sites, which was confirmed by EXAFS and UV-Vis spectroscopy measurements [8,9], although it was not excluded the presence of small amounts of TiO_2-type clusters, successively confirmed by UV-Vis spectroscopy [10].

Ti ions have been also introduced in MCM-48 materials. The cubic structure of this support, having two perpendicular arrays of channels, is more interesting for catalytic purposes as the diffusion of reactants inside channels appeared less limited [11-13]. Ti-MCM-48 resulted to be highly active in the epoxidation of cyclohexene and 1-hexene [14].

Noteworthy, a remarkable increase in the activity of Ti-containing mesoporous catalysts was obtained by anchoring trimethylsilyl groups on the silanols, which led to an increase of the hydrophobicity of the surface [15,16]. In this paper we report on peculiar aspects of the preparation and characterisation of mesoporous Ti-functionalised MCM-48 silicas. The spectroscopic and catalytic properties of Ti-MCM-48 catalysts upon a post treatment of silylation with hexamethyldisilazane (HDMS) as silylating agent, are also reported.

2. EXPERIMENTAL

2.1. Synthesis

Ti-MCM-48 samples were synthesised using amorphous silica, (Aerosil 200, Degussa) as silicon source, titanium tetraethoxide (Alpha Products), cetyltrimethylammonium bromide /hydroxide (CTABr/OH) and tetramethylammonium hydroxide (TMAOH). The molar composition was: SiO_2 : x $Ti(OEt)_4$: 0.306 CTAOH : 39.5 H_2O, (where x was adjusted to yield Ti loading 1, 2 and 3 wt% as TiO_2) and the preparation procedure was as in ref. 15. Figure 1 shows a ternary diagram reporting the synthesis gel composition of the bare MCM-48 and Ti-MCM-48: in this diagram, the proportions of reactants to be used for obtaining the cubic phase instead of the lamellar or the hexagonal ones, can be inferred. It appears that the range of SiO_2:H_2O:CTA ratios is more

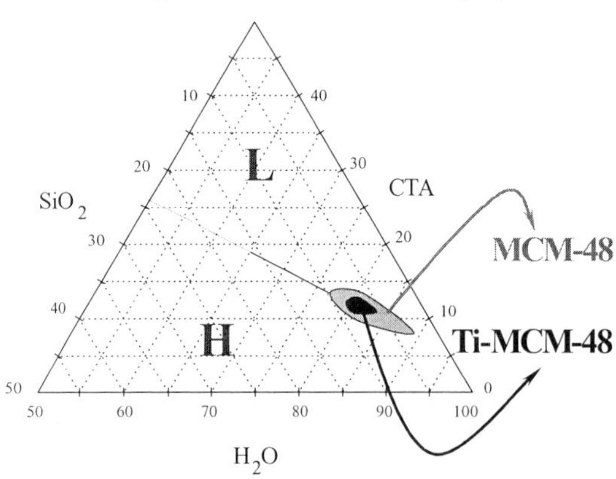

Figure 1. Ternary diagram representing the composition of the synthesis gel (CTA = cetyltrimethylammonium bromide/hydroxide)

restricted in the case of the Ti-containing material with respect to MCM-48; in fact the addition of the Ti precursor causes a change in the pH with an increase in the difficulty to obtain the cubic phase. Crystallisation was carried out in stainless steel autoclaves, with autogeneous pressure, at 150°C for 6h. The final product was then filtered and washed until neutral pH value and dried at 60ºC overnight. The samples were calcined at 540°C first in nitrogen and then in air flow in order to remove the organic fraction. Finally, Ti-MCM-48 was silylated with hexamethyldisilazane (HDMS) as silylating agent, using a HMDS/SiO_2 ratio of 0.25. The silylation was performed at 120°C in toluene solution under argon atmosphere and the silylated samples were outgassed at 250°C before running the spectra. The silylated material was labelled Ti-MCM-48S.

2.2. Characterisation and catalytic activity

A Philips X'Pert MPD diffractometer equipped with a PW3050 goniometer (Cu K_α radiation) was used to perform X-Ray powder diffraction. Textural characterisation (N_2 and Ar adsorption) was performed with an automatic instrument ASAP 2010 from Micromeritics.

^{29}Si-MAS-NMR spectra were collected at ambient temperature on a Varian VXR 400S WB spectrometer at 79.459 MHz, using a CP/MAS Varian probe with zircona rotors. The spectra were recorded with pulses of 4.5 μs of 1/3 π rad and recycle delay of 50 s. ^{29}Si chemical shifts were referred to tetramethylsilane.

After removing the organic fraction in nitrogen and air flow at 540°C, the samples were put in quartz cells in order to collect diffuse reflectance (DR) UV-Vis-NIR spectra. The calcined samples were pre-treated and activated in these cells, permanently connected to a vacuum line (residual pressure $\leq 10^{-5}$ Torr, 1 Torr=133.33 Pa), by oxidation at 550°C for 10 hours (100 Torr O_2) and subsequent outgassing. The silylated samples were simply outgassed at 250°C. DR UV-Vis-NIR spectra were collected with a Perkin Elmer (Lambda 19) spectrometer equipped with an integrating sphere with $BaSO_4$ as reference. Photoluminescence spectra were recorded with a Spex Fluorolog-2 FL212 spectrometer.

The Ti content was determined by atomic absorption on a Varian Spectra A-10 Plus spectrometer. C, H, and N analysis were performed on a Fissons EA-1108 elemental organic analyser.

The catalytic tests were done on calcined and silylated Ti-MCM-48 materials. The reaction of epoxidation of cyclohexene was tested using *tert*-butylhydroperoxide (TBHP) as oxidant with olefin/TBHP ratio = 4 at a reaction temperature of 60ºC. The reaction was monitored by Gas Chromatography using a 5 % phenylsilicone column (HP-5) of 25 meters length.

3. RESULTS AND DISCUSSION

3.1. XRD powder diffraction and adsorption measurements

X-Ray diffraction of calcined Ti-MCM-48 with 2wt% Ti content as TiO_2 (Fig. 2,a), shows a main (211) reflection at 2θ= 2.42°, and other weaker peaks at higher 2θ values, as expected for a material with cubic Ia3d symmetry [6]. Using Bragg's equation and the geometrical properties of the cubic cell, it is possible to calculate "a", the unit cell parameter; the results are reported in Table 1. Upon silylation, the main XRD peaks are similar to those of the calcined sample, as shown in Figure 2, curve b, indicating that the structure kept unit cell parameters close to the calcined sample (Table 1). However, the slightly less defined pattern at high 2θ indicates some loss of regularity in the array of channels. As for the

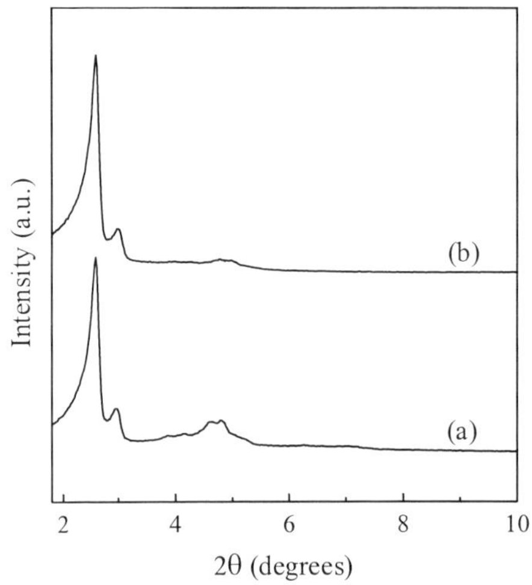

Figure 2. XRD spectra of Ti-MCM-48 calcined (a) and silylated (b). The spectra are recorded in air.

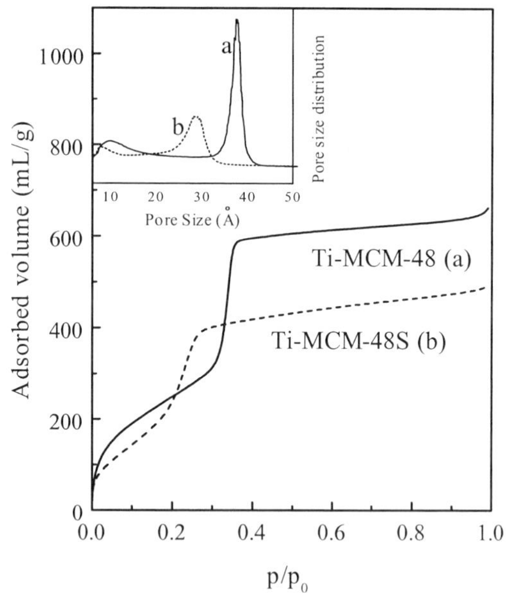

Figure 3. Argon adsorption isotherms of calcined (a) and silylated (b) Ti-MCM-48 materials. The pore size distributions of the two samples are reported in the inset. The samples were activated by degassing at 400°C.

porosity, Argon adsorption measurements (Fig. 3) showed a shift of the position of the flex at lower p/p_0 value, corresponding to a decrease of the pore size. The pore size distribution of the silylated Ti-MCM-48 material, as calculated with the Horwath-Kawazoe formalism, decreased with respect to the calcined sample from 37.5 to 28.5 Å (see inset of Figure 3). This was suggested to be due to the steric hindrance of the trimethylsilyl groups anchored on the inner surface of the pore of the Ti-MCM-48 structure [17]. All these results are summarised in Table 1.

Table 1. Structural and textural properties of calcined and silylated Ti-MCM-48.

Samples	Ti (wt%)	Unit cell parameter "a" (Å)	Pore diam. (Å)	Pore Vol. (mL/g)	BET - SSA (m^2/g)
Ti-MCM-48 (calcined)	2.1	83.2	37.5	0.89	1047
Ti-MCM-48S (silylated)	1.9	83.1	28.5	0.66	956

3.2. ^{29}Si-MAS NMR spectroscopy

Figure 4 shows solid-state ^{29}Si-MAS NMR of calcined (curve a) and silylated (curve b) 2wt% Ti-MCM-48 materials. Three distinct resonances are present in the spectrum of the calcined sample, at -110, -101 and -91 ppm (curve a). They result to be sharper than those found in the as-synthesised sample (spectrum not shown). As these bands are similar to those found in purely siliceous MCM-41 and MCM-48, accordingly to the literature [6,12,18], they are assigned to Si(OSi)$_4$ [Q4], Si(OSi)$_3$OH [Q3] and Si(OSi)$_2$(OH)$_2$ [Q2] groups, respectively. The band at -91 ppm is very weak, suggesting the presence of a small amount of geminal-hydroxyl groups on the silica surface of the mesopore walls. Upon silylation treatment with hexamethyldisilazane (HMDS), the NMR spectrum showed two resonances at +14 and –108 ppm, whilst the Q2 and Q3 peaks have nearly completely disappeared. The new band at +14 ppm is straightforwardly associated with the Si atom in Si(CH$_3$)$_3$(OSi) groups. The disappearance of Q2 and Q3 indicates that the shell of Si atoms, originally Si(OSi)$_3$OH and Si(OSi)$_2$(OH)$_2$, have changed upon silylation. As reported in Scheme 1, silylation results in the substitution of OH with Si(CH$_3$)$_3$ groups, and then the magnetic environment of Si in the so generated Si(OSi)$_3$[OSi(CH$_3$)$_3$] and Si(OSi)$_2$[OSi(CH$_3$)$_3$]$_2$ become similar to that of Q4 species Si(OSi)$_4$ As a result, the signal of all these Si species merges into the peak at –108 ppm [19-21].

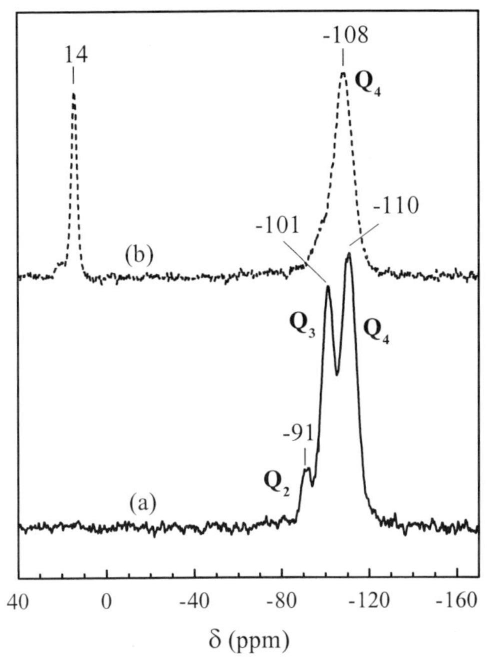

Figure 4. Solid-state ^{29}Si-MAS NMR of Ti-MCM-48 calcined (a) and Ti-MCM-48 silylated (b). The spectra are recorded at ambient temperature.

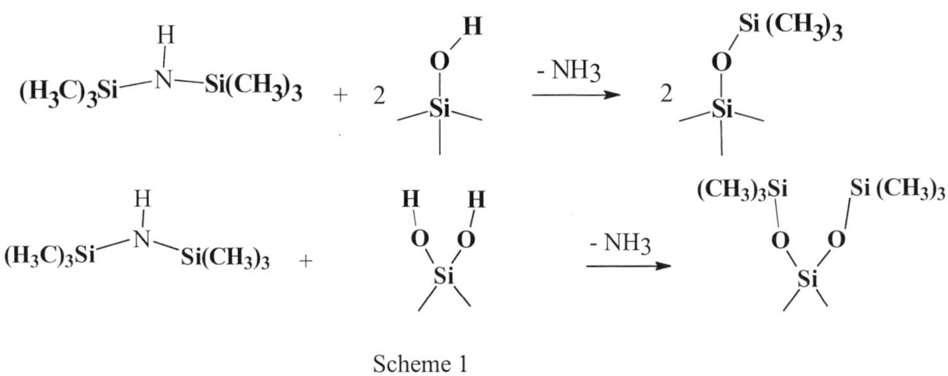

Scheme 1

3.3. Diffuse Reflectance UV-Vis spectroscopy

UV-Vis spectroscopy, in both diffuse reflectance and luminescence modes, is useful to investigate the local environment of the titanium sites. Figure 5 shows DR UV-Vis spectra of calcined (curve a) and silylated (curve b) 2wt% Ti-MCM-48 samples. In the case of the calcined sample (curve a) a main band centred at 215 nm, asymmetric towards higher wavelength, is observed, similar to the signal present in Ti-MCM-41 spectra [10,22,23]. Bands in the 200-240 nm range were assigned to oxygen to isolated tetrahedral titanium(IV) ions charge transfer (CT) transitions [23,24,25]. The position of the maximum of this band is strictly dependent on the electronic features of the oxygen atoms bonded to the Ti centres, as higher the electronegativity of these oxygen atoms, higher the energy of the CT transition where they are involved, and then lower the wavelength of the corresponding absorption band. As a consequence, $Ti(OSi)_4$ groups are responsible of CT transition at lower wavelength than $Ti(OSi)_3OH$ or $Ti(OSi)_2(OH)_2$ sites, due to the higher electron density of the oxygen atoms of siloxy- respect to hydroxy-ligands [23,26]. Noticeably, the band observed for the silylated sample (curve b) is sharper and centred at lower wavelength (210 nm). Considering that, upon silylation, besides silanols (see scheme 1), also titanols $Ti(OSi)_3OH$ are almost completely transformed into trimethylsilylated $Ti(OSi)_3[OSi(CH_3)_3]$ species, the spectral changes observed indicate that the component at 230 nm in the spectrum of calcined Ti-MCM-48 is due to titanol species. It may be of interest that geminal $Ti(OSi)_2(OH)_2$ titanols were found to be less likely to occur [9]. The electronic CT band of the new $Ti(OSi)_3[OSi(CH_3)_3]$ species merges with that of $Ti(OSi)_4$ centres, producing the 210 nm band [27].

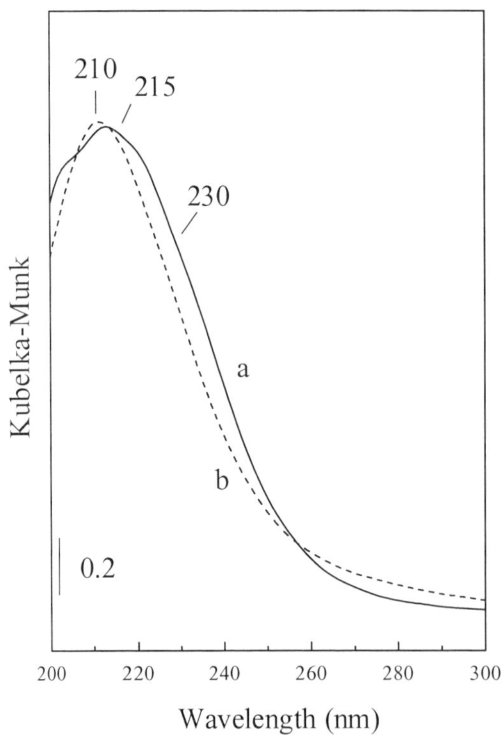

Figure 5. DR UV-Vis spectra of 2% Ti-MCM-48 calcined (a) and silylated (b).

Figure 6A illustrates the effect of water adsorption on 2 wt% calcined and silylated Ti-MCM-48 samples, monitored by DR UV-Vis spectroscopy. The characteristic UV-Vis CT band of the calcined sample at 215 nm (curve a) becomes broader and shifts to higher wavelength (curve a'). As reported in the literature, this behaviour monitors the fact that tetrahedral Ti centres expand their coordination sphere by H_2O adsorption (Scheme 2, on passing from structure (a) to (b) and (c)) [23 and references therein]. The spectrum in the NIR range of the calcined sample (Fig. 6B, curve a) shows a main band at 1366 nm, corresponding to the

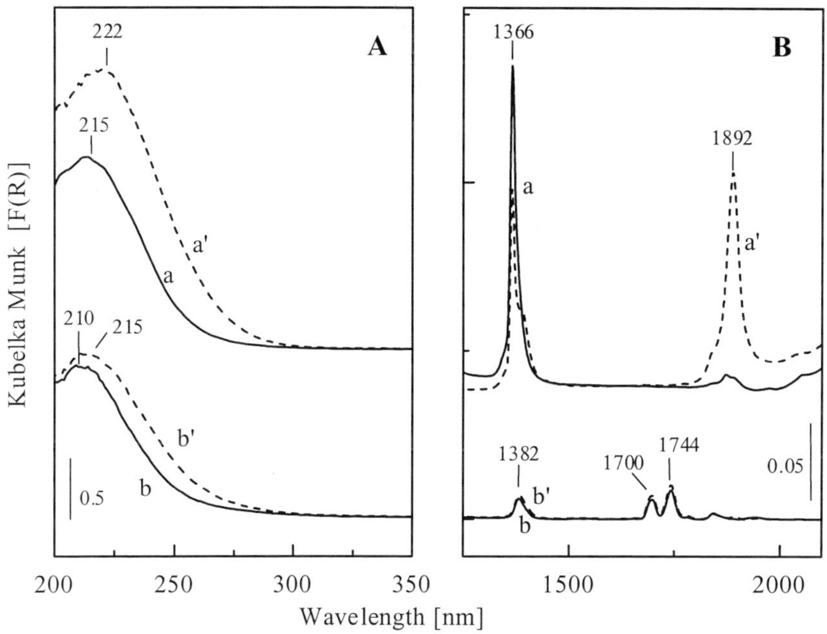

Figure 6. DR UV-Vis (section A) and NIR (section B) spectra of Ti-MCM-48 calcined (a) and silylated (b). H_2O adsorption (vapor pressure) is also shown (curves a' and b').

overtone stretching of isolated silanols ($2\nu_{OH}$). After water adsorption, the intensity of this band decreased and the combination band ($\nu_{OH}+\delta_{OH}$) of adsorbed water appeared at 1892 nm (curve a'). This behaviour was not observed in the case of the silylated sample: after water adsorption, the CT UV-Vis band (curve b, section A) underwent just a small modification (curve b') and in the NIR range, where weak bands at 1382 nm, due to residual silanols, and at 1700 and 1744 nm, due to overtones/combination bands of trimethylsilyl groups, are present, the signal due to adsorbed water molecules did not appear (curves b and b', section B). These results confirm the higher hydrophobicity of silylated Ti-MCM-48 samples with respect to the analogous calcined catalyst; the hydrophobicity was revealed on both Ti(IV) centres and the inner surface of the silica.

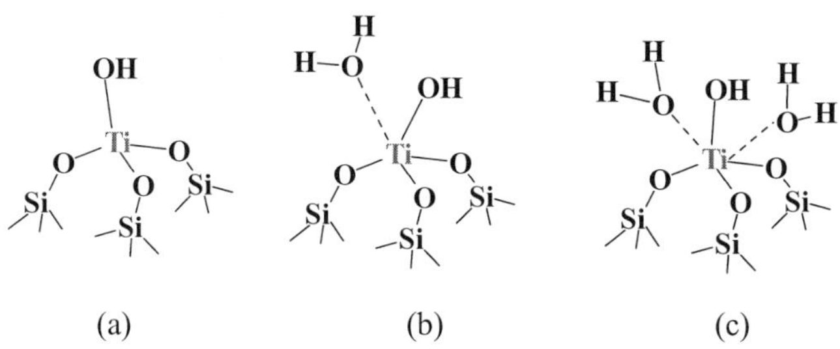

Scheme 2

3.4. Photoluminescence spectroscopy

Photoluminescence spectra of calcined Ti-MCM-48 samples with increasing Ti(IV) loading (1, 2 and 3 wt% as TiO_2) were recorded in order to investigate the dependence of the dispersion of Ti(IV) centres on the amount of Ti introduced.

The spectra, reported in Figure 7, are similar to those found in Ti-MCM-41 [10,22,23], and showed two overlapped peaks at 440 and 500 nm. These components were both attributed to tetrahedral Ti(IV) sites with different environment, i.e., bonded to oxygen atoms with different electron charge density [10]. Photoluminescence was found to be a unique feature of Ti(IV) in tetrahedral coordination, as Ti(IV) in octahedral coordination produces only a very weak emission. The shape and the intensity of the bands are strictly dependent on the Ti(IV) loading. The spectrum of 1% (curve a) Ti-MCM-48 presents a low intensity, due to the presence of a low concentration, although well dispersed, of Ti(IV) tetrahedral sites. The 2% sample showed a spectrum with the highest intensity of the bands (curve b), suggesting that this sample contains the highest amount of tetrahedral and isolated Ti(IV) sites. The overall intensity of the photoemission spectra was much lower in the case of the 3 wt % sample (curve c), and this can be attributed either to self-quenching effects for the higher Ti concentration and/or to the presence of small TiO_2-type clusters, which are less luminescent [10].

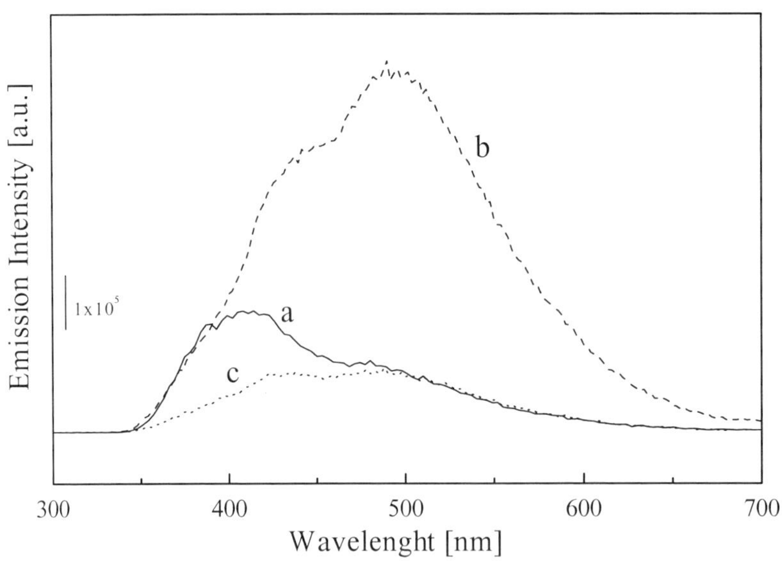

Fig. 7. Photoemission spectrum of Ti-MCM-48 1% (a), 2% (b) and 3% (c), λ_{ex}= 250 nm. The samples were activated by oxidation at 540°C.

3.5. Catalytic activity

Ti-MCM-48 is a catalyst for the epoxidation of olefins, the activity of which being strictly connected to the presence of tetrahedral Ti(IV) sites. The accepted mechanism, in the case of olefin epoxidation with H_2O_2 is reported in Scheme 3 [1].

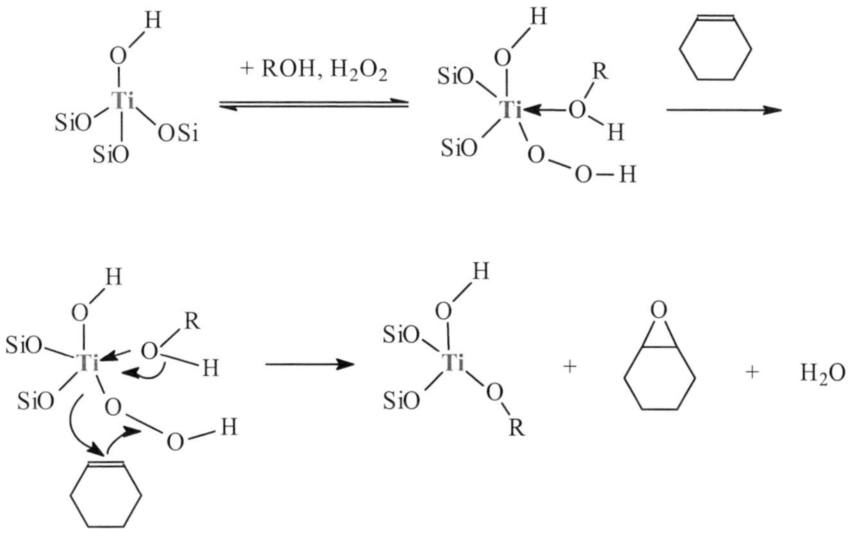

Scheme 3

Silylation plays an important role in the activity of the catalyst. The catalytic performances of two samples (1 and 2% Ti content as TiO_2) calcined and silylated, in epoxidation reaction with *tert*-butylhydroperoxide, are reported in figure 8. The silylated samples exhibit higher conversion and higher selectivity to epoxide with respect to the non-silylated ones for both Ti concentrations. This was suggested to be due to the higher degree of hydrophobicity of the silica surface of silylated samples which, thanks to the absence of the OH groups transformed into trimethylsilyl groups, prevents water adsorption. In fact, water, produced by the reaction and possibly present in organic solvents used for the epoxidation, poisons the catalyst by formation of glycols which may irreversibly coordinate to the titanium catalytic centres, causing the deactivation of the catalyst, as shown in Scheme 4. Increasing the hydrophobic character of the surface by silylation of both silanols and titanols, reduces this problem, and indeed, a remarkable increment of the catalytic performance for epoxidation of olefins was obtained upon silylation of Ti-MCM-48 [15,16,18,21,28].

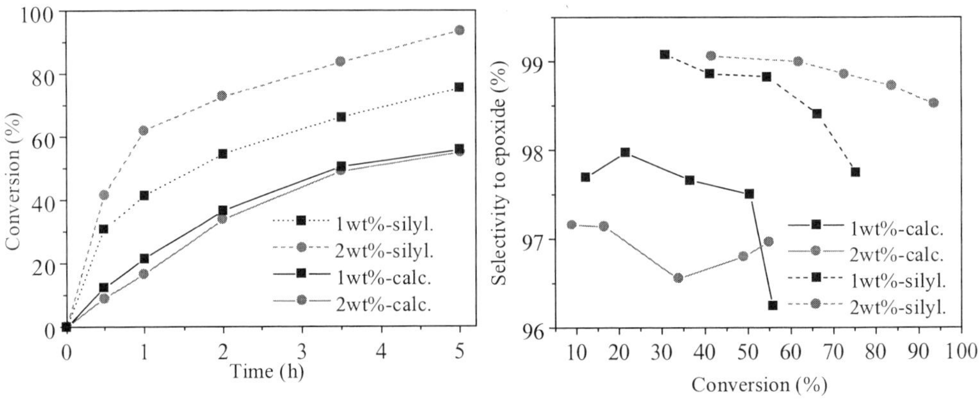

Fig.8. Cyclohexene epoxidation with tertbutyl hydroperoxide of Ti-MCM-48 1% (squares) and 2% (circles) calcined (solid lines) and silylated (dashed lines).
[Olefin/TBHP= 4, T = 60°C, catalyst = 0.5 wt%].

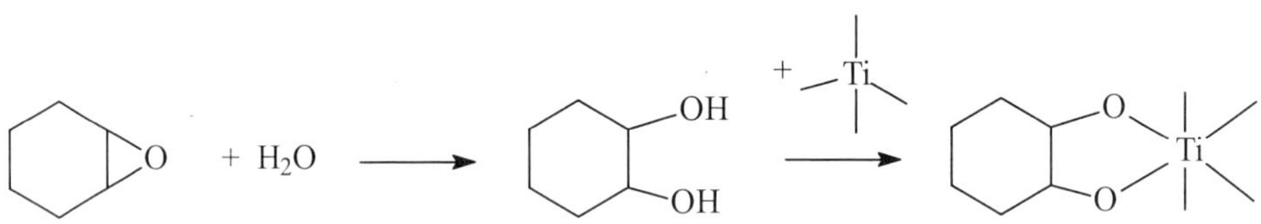

Scheme 4

4. CONCLUSIONS

Ti-MCM-48 mesoporous silica are active catalysts in reactions of epoxidation of olefins with *tert*-butylhydroperoxide and hydrogen peroxide. Systems with different Ti(IV) loading were synthesised using cetyltrimethylammonium hydroxide surfactant as template. All samples were then silylated using hexametyldisilazane (HDMS) as silylating agent. This resulted in the transformation of most of the surface silanols and titanols species into trimethylsilyl groups by reaction with HDMS, increasing the hydrophobicity of the silica surface. As a consequence, the catalytic activity resulted noticeably improved due to the fact that the high hydrophobicity reduces the interaction of the catalytic sites with water molecules, which would deactivate the Ti(IV) centres.

ACKNOWLEDGEMENTS

Financial supports from Italian MURST (Ministero dell'Università e della Ricerca Scientifica e Tecnologica), MAT97-1207-CO3-01 and MAT97-1016-CO2-01 Spanish projects and the Italian-Spain bilateral action HI1998-0101 are acknowledged. M.L.P thanks the Ministerio de Educación y Cultura for the Doctoral fellowship.

REFERENCES

1. B. Notari, Advanced in Catalysis, 41 (1996) 253.
2. G. Perego, G. Bellussi, C. Cordo, M. Taramasso, F. Buonuomo and A. Esposito, Stud. Surf. Sci. Catal., 28 (1986) 129.
3. M. Taramasso, G. Perego and B. Notari, US Pat. 4, 410 (1983) 501.
4. M.A. Camblor, A. Corma, A. Martinez and J. Pérez-Pariente, J.Chem. Soc. Chem. Commun., (1992) 589.
5. C.T. Kresge, M.E. Leonowicz, W.J.Roth, J.C. Vartuli and J.S. Beck, Nature, 359 (1992) 710.
6. J.C. Vartuli, K.D. Schmitt, C.T. Kresge, W.J. Roth, M.E. Leonowicz, S.B. McCullen, S.D. Hellring, J.S. Beck, J.L. Schlenker, D.H. Olson and E.W. Sheppard, Chem. Mater., 6 (1994) 2317.
7. A. Corma, M.T. Navarro and J. Pérez Pariente, J. Chem. Soc., Chem. Commun., (1994) 147.
8. T. Maschmeyer, F. Rey, G. Sankar and J.M. Thomas, Nature, 378 (1995) 159.
9. P.E. Sinclair, G. Sankar, C.R.A. Catlow, J.M. Thomas and T. Maschmeyer, J. Phys. Chem. B, 101 (1997) 4232.
10. L. Marchese, E. Gianotti, V. Dellarocca, T.Maschmeyer, F. Rey, S. Coluccia and J.M. Thomas, Phys. Chem. Chem. Phys, 1 (1999) 585.
11. K.A. Koyano and T. Tatsumi, Chem. Commun., (1996) 145.
12. A. Corma, Chem. Rev. 97, (1997) 2373.
13. A. Corma, Q. Kan and F. Rey, Chem. Commun., (1998) 579.
14. A. Corma, J.L. Jordá, M.T. Navarro and F. Rey, Chem. Commun., (1998) 1899.
15. T. Tatsumi, K.A. Koyano and N. Igarashi, Chem. Commun., (1998) 325.
16. A. Corma, M. Domine, J.A. Gaona, J.L. Jordá, M.T. Navarro, F. Rey, J. Pérez-Pariente, J. Tsuji, B. McCulloch and L.T. Nemeth, Chem. Commun., (1998) 2211.
17. The calculated surface coverage per trimethylsilyl group is 47.6Å/molecule. Insight II Molecular User guide, San Diego, Bisym/MSI (1995).
18. K. Koyano, T. Tatsumi, Y. Tanaka and S. Nakata, J. Phys. Chem. B, 101 (1997) 9436.
19. B.A. Morrow and A.J. McFarlan, Langmuir, 7 (1991) 1695.
20. D.W. Sindorf and G.E. Maciel, J. Am. Chem. Soc., 105 (1983) 3767.

21. M.L. Peña, V. Dellarocca, F. Rey, A. Corma, S. Coluccia and L. Marchese, Microp. Mesop. Materials, 2001, in press.
22. L.Marchese, T. Maschmeyer, E. Gianotti, S. Coluccia and J.M. Thomas, J. Phys.Chem. B, 101 (1997) 8836.
23. L. Marchese, E. Gianotti, T. Maschmeyer, G. Martra, S. Coluccia and J.M. Thomas, Il Nuovo Cimento, 19D (1997) 1707.
24. S. Bordiga, S. Coluccia, C. Lamberti, L. Marchese, A. Zecchina, F. Boscherini, F. Buffa, F. Genoni, G. Leofanti, G. Petrini and G. Vlaic, J. Phys. Chem., 98 (1994) 4125.
25. M.R. Boccuti, K.M. Rao, A. Zecchina, G. Leofanti and G. Petrini, Stud. Surf. Sci. Catal., 48 (1989) 133.
26. C.K. Jørgensen, Prog. Inorg. Chem., 12 (1970) 101.
27. V. Dellarocca, M.L. Peña, F. Rey, A. Corma, S. Coluccia and L. Marchese, J. Phys. Chem, 2001, in press.
28. A. Corma, J.L. Jordá, M.T. Navarro, J. Pérez-Pariente, F. Rey and J. Tsuji, Stud. Surf. Scien. Catal., 129 (2000) 169

Studies in Surface Science and Catalysis 140
A. Gamba, C. Colella and S. Coluccia (Editors)

Computer simulations of ethane sorbed in an aluminophosphate molecular sieve

P. Demontis, J. Gulìn Gonzàlez*, G. B. Suffritti and A. Tilocca†

Dipartimento di Chimica, Università degli Studi di Sassari, Via Vienna 2, I-07100 Sassari, Italy

Classical Molecular Dynamics (MD) simulations have been carried out to study the dynamic properties of ethane sorbed in an $AlPO_4$-5 aluminophosphate. The main purpose is to gain new insight into the diffusive regime controlling the motion of ethane in such one-dimensional microporous structure. In recent experimental studies a standard MSD vs. *t* linear dependence (normal diffusion) was found, while earlier experiments seemed to reveal a single-file regime for this system. The present calculations show that ethane diffusion follows the standard regime on the time scale of MD simulations: particle passing (leading to normal unidirectional diffusion) is rather frequent over observation times of several nanoseconds. Consequently a single file regime *for this system* could only be achieved through the formation of stable clusters of ethane molecules that, once formed, move as single larger species in a single file regime. This possibility has been tested by calculating the lifetimes of ethane clusters of different size: the larger lifetimes observed are less than 1 ns, and are further reduced when the flexibility of the host lattice is incorporated in the model. Therefore these simulation support the experimental observation that normal diffusion dominates the motion of ethane in the $AlPO_4$-5.

1. INTRODUCTION

The sorption and catalytic properties of zeolites are largely affected by the way in which the confinement in the pores affect the diffusive properties of guest molecules [1]. It is well known from the current literature [2] that correlation effects can lead to unusual features, and special diffusive regimes can emerge in some cases. The microporous structure of zeolite $AlPO_4$-5 is made of straight nonconnected pores with 7.3 Å diameter running along the *c* crystallographic axis (Figure 1). The confinement of molecules in such one-dimensional structures sometimes leads to deviations from standard (Fickian) behavior. Diffusion takes place in one single direction, and if the diffusing molecules are of such dimensions that they cannot pass each other in the channels, subsequent movements of different particles will be highly correlated and the motion becomes of single-file type. The following general expression represents the time dependence of the mean square displacement:

$$\langle z^2(t) \rangle = 2Ft^{\alpha} \qquad (1)$$

* On leave from: Higher Polytechnic Institute JAE, Dept. of Physics, Marianao, La Havana, Cuba.
† Present address: Dipartimento di Scienze Matematiche, Chimiche e Fisiche, Università degli Studi dell'Insubria, Via Lucini 3, I-22100 Como, Italy

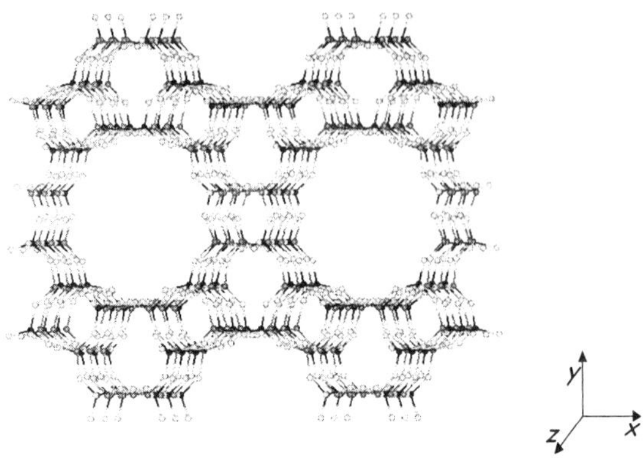

Fig.1. Structure of $AlPO_4$-5; two parallel channels are shown.

In equation (1) the exponent α is equal to 1 for normal diffusion (in which case F coincides with the diffusion coefficient D), while $\alpha = 0.5$ in the case of single file diffusion [3-6]. In recent quasielastic neutron scattering (QENS) experimental investigations [7] a standard MSD vs. t linear dependence was found for the diffusion of ethane in $AlPO_4$-5, while earlier pulsed-field gradient NMR experiments [4] seemed to reveal a single-file behavior for this system. The discrepancy between the experimental observations could arise from the different time scales sampled (~ 1 ns for QENS and ~ 1 ms for NMR), in particular when different crystals are used in the experiments. Indeed, the presence of defects may cause blockages in the channels which slow down the diffusive motion at large time scales [7]. Some subsequent stochastic simulations [8] were based on the assumption of single-file behavior for ethane in $AlPO_4$-5. However, at the present there are not clear evidences as to whether the diffusion of ethane in such structure shows normal or single file behavior. In this work we present the results of Molecular Dynamics simulations aimed at illuminating the diffusive regime that controls the motion of ethane in a non-defective $AlPO_4$-5 structure.

2. MOLECULAR DYNAMICS SIMULATIONS

We recently proposed an effective harmonic potential for representing the structural and dynamical properties of the $AlPO_4$-5 framework [9]. The calculation of interatomic forces within a lattice represented through this potential requires only a small computational effort, thus it is particularly suitable for the very long simulations performed in this study, together with a flexible united-atom model for ethane molecules sorbed in zeolites [10,11]. In the latter potential a Morse function represents the C-C bond vibration, and a Lennard-Jones (12-6) potential models the interaction between two intermolecular sites as well as the interaction between a site and a framework oxygen. In Table 1 all the parameters involved in the potentials are collected. Compared to the original parameters [10] (developed for the ethane-silicalite system) only the $\varepsilon_{CH3\text{-}O}$ parameter has been changed, so as to reproduce closely the experimental heat of adsorption of ethane in $AlPO_4$-5 [12]. Ethane loading was varied between 1 and 4 (the largest possible concentration) molecules/unit cell by fixing the total number of sorbed molecules to 12 and varying the number of cells included in the simulation box accordingly. MD trajectories of 35 ns were carried out in all cases. In order to study the infinite dilution case a single molecule was included in the simulation box, and a 155 ns run was carried out. The run temperature was around 300 K in all cases.

Table 1
Potential parameters

	$AlPO_4$-5 harmonic potential interactions: $V(r) = 0.5\ k\ (r\text{-}r_0)^2$				
	Al – O	P – O	O - (Al) - O	O - (P) – O	Al – P
k (kJ mol^{-1} nm^{-2})	292 880	376 560	92 048	92 048	2 092
r_0 (nm)	0.172	0.1516	0.280873	0.247560	0.32
Ethane-zeolite and ethane-ethane intermolecular (LJ) potential: $V(r) = 4\varepsilon\ [(\sigma/r)^{12}\text{-}(\sigma/r)^6]$					
		(CH_3)-O_{zeo}		(CH_3)-(CH_3)	
σ (nm)		0.3461		0.3775	
ε (kJ mol^{-1})		1.0		0.867	
Ethane intramolecular (Morse) potential: $V(r) = D_e\{[1\text{-}\exp(\text{-}\beta\ (r\text{-}r_{eq}))]^2\ \text{-}1\}$					
D_e (kJ mol^{-1})	351.381				
β (nm^{-1})	18.408				
r_{eq} (nm)	0.1536				

3. RESULTS AND DISCUSSION

3.1 Mean Square Displacement

Figure 2 shows the log-log plots of the MSDs for the different ethane concentrations.

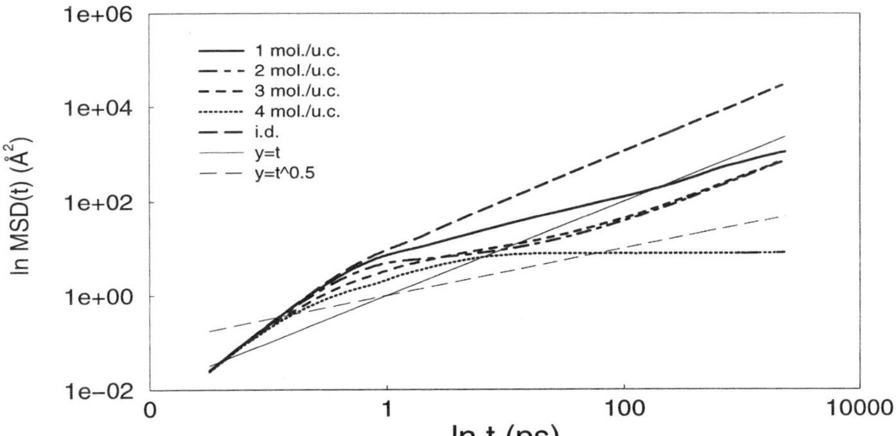

Fig. 2. Log-log plots of MSD curves. The curves corresponding to normal diffusion ($z^2(t) \propto t$) and single file ($z^2(t) \propto t^{0.5}$) are also drawn for reference.

Every linear region in this type of plots denotes a distinct diffusive regime, represented by a power law like $y= Ax^{\alpha}$; e.g., the Fickian diffusive regime is described by a straight line of slope 1. Then it is possible to obtain the characteristic time and space length of every regime, and to draw some conclusion about the diffusion mechanism. The short-time portion of the curves in all cases correspond to a "quasi-free" motion of the molecule which moves in the almost constant potential at the bottom of the adsorption well (newtonian dynamics $z(t)$ = vt, corresponding to a slope of two). The actual diffusive regime starts at longer times, as it is identified by the long-time limit of the mean square displacement. Therefore the nature of the diffusive regime can be established by looking at the slope of the logarithmic curves in Figure 2 at times above ~ 100 ps. Indeed, the transition from the newtonian dynamics to the true diffusive behavior, determining the decrease in slope, may require up to several hundredth picoseconds for the studied systems. By comparing the long time slope of the MSD curves to the two reference curves corresponding to normal and single file regime, a clear picture emerges. The long time dependence of the MSD curve is obviously linear at infinite dilution, where a single file behavior is clearly impossible. Also at 2 and 3 molecules/u.c. the logarithmic curves become parallel to the normal diffusion reference after a relatively long transition period. At 1 molecule/u.c. both single-file and normal-diffusion regimes are seemingly present: the limiting slope of the MSD curve is intermediate between the normal-diffusion and single file diffusion references, and actually appears to be closer to the single file one. Indeed the α exponent in Equation 1 can be estimated as 0.66 for this system. A turnover from single file to normal diffusion can be observed for guest systems (like ethane in $AlPO_4$-5) where the mutual passages are relatively hindered and occur only occasionally [13]. By direct inspection of the MD trajectories we verified that two ethane molecules *can* pass each other, albeit not very often, during the simulation. At times long enough that several such passages have been observed, the usual linear MSD vs. t behavior is achieved. At 1 mol./u.c. the frequency of the passages is low, and the turnover to normal diffusion is not reached on the time scale reported in Figure 2. However the deviation of the α exponent from the ideal 0.5 value is indicative of a diffusive regime where the particle crossings begin to drive the system towards the normal diffusion behavior. On the other hand, at higher loading, due to the larger crossing frequency, the turnover time is very short and hardly noticeable, so that only the linear regime of normal diffusion is apparent. These results hence support the observation that when sorbed in an $AlPO_4$-5 non-defective crystal ethane should diffuse "normally", if the molecular migration is followed for a long enough time, so that during the measurement many mutual passages of two molecules are observed. The diffusion coefficients estimated by the long-time slope of the MSD curves are reported in Table 2, together with the corresponding relaxation time needed to reach the diffusive regime in each case. The latter times appear to increase with increasing loading; however it should be remarked that the value reported for 1 mol./u.c. is the time required to reach the intermediate diffusive regime (MSD $\propto t^{0.66}$) described above. According to the preceding discussion, for this system a further turnover to normal diffusion is actually expected to occur over longer times (> 2 ns) than those reported in Figure 2. The estimated diffusion coefficients can be compared with the experimental measurements of Jobic *et al.* [7] which obtain $D = 1.4 \cdot 10^{-9}$ $m^2\ s^{-1}$ at 0.7 mol./u.c., $1.1 \cdot 10^{-9}$ $m^2\ s^{-1}$ at 1.1 mol./u.c. and $0.9 \cdot 10^{-9}$ $m^2\ s^{-1}$ at 1.6 mol./u.c.; the agreement with our data is reasonable, as well as the general tendency to decrease with loading. The flat center of mass MSD curve for 4 molecules/u.c. shows the absence of any significant translational motion of the guest species.

Table 2
Diffusion coefficients (10^{-9} m^2 s^{-1}) and relaxation times (ps)

loading (mol./u.c.)	*D*	*t*
0 (i.d.)	66	8
1	1.8	20
2	1.6	180
3	1.6	400
4	~ 10^{-3}	-

At this loading the ethane molecules can only undertake small oscillations around the adsorption sites, but as all the neighboring sites are occupied a true diffusive motion cannot occur. This situation is shown in Figure 3, where we report a configuration taken from the MD trajectory with 4 molecules/u.c.; note that the relative arrangement of the molecular centers of mass remains almost unchanged during the whole trajectory.

3.2 The cluster hypothesis

According to the results discussed in the previous section the motion of ethane molecules falls in the normal diffusion regime over the time scale of MD simulations, i.e., several nanoseconds. The only way in which a single file behavior can be achieved *over longer time scales* is through the formation of stable clusters which, once formed, moves as single large molecules [14].As two clusters cannot pass each other, a single file regime should naturally emerge. The formation of such stable structures would probably require much longer simulations to be observed. Nevertheless a useful way to verify this hypothesis through MD simulations is to build several stable clusters inside the channels by energy minimization and to look at the following dynamics.

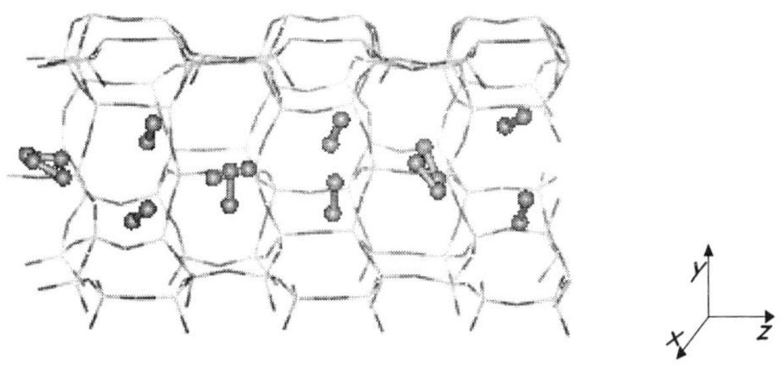

Fig. 3. A configuration taken from the run at 4 molecules/u.c. showing the "frozen" arrangement discussed in the text.

In particular we are interested in measuring the lifetime of such structures, in order to assess if they can persist long enough to give rise to a single file behavior.

We considered clusters formed by 2, 4 and 6 molecules; after finding the energy-minimized structure for each cluster size, 20 such structures were randomly distributed in 12 cells aligned along *z*. The interactions between molecules belonging to two different clusters were turned off, i.e., each cluster is *isolated* in the sense that its dynamics is determined only by the interaction with the channel walls (apart from the intra-cluster interactions). Each cluster does not "see" the others: in this way we can follow the *independent* trajectories of 20 clusters and a good statistical accuracy is achieved with a moderate computational effort (the MD run lasted 3 ns for each cluster size). The $AlPO_4$-5 lattice was considered rigid in order to avoid its excessive distortion induced by the simultaneous presence of 20 clusters, which in turn could introduce fictitious indirect interactions between them. After thermalizing each cluster at 300 K, its dynamics was followed until two adjacent molecules became separated by more than 12.7 Å: at this point the cluster dissociation was considered to have occurred [8]. The cluster was then removed and another one was "created" in a random axial position along the channel. The distribution of cluster lifetimes (Figure 4) shows that the stability of clusters increases with the number of molecules per cluster. It is interesting to note that this trend is different from that observed in the MD simulations of Sholl [8], where the stability appears to decrease by increasing the number of molecules per cluster from 2 to 5. The internal structure and flexibility of ethane molecules in our model, as opposed to the spherical structureless LJ centers in the Sholl's model, are probably the main source of this difference. Notwithstanding the increased stability of larger clusters the basic point is that the cluster lifetimes observed in our simulations are always lower than 1 ns. Moreover, when some particularly stable clusters were selected from the fixed-framework simulations and inserted into the flexible lattice model, the lifetime of the cluster shows a further, considerable decrease.

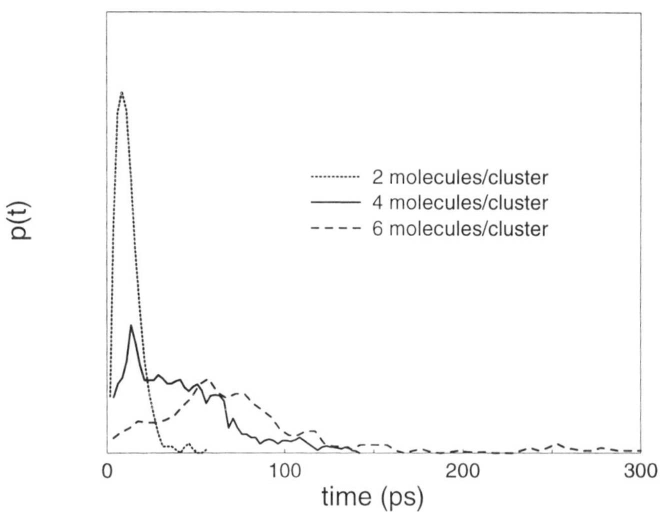

Fig. 4. Distributions of cluster lifetimes

These lifetimes are far too short to support the hypothesis made at the beginning of this section. In our opinion the possibility that clusters formed by more than six molecules can attain lifetimes of several nanoseconds is rather unlikely; anyway, work is in progress to probe the behavior of such larger cluster, whose simulation is more CPU-demanding. The present preliminary data seem to exclude the concerted diffusion of ethane cluster as a possible source of single file behavior over long time scales: the clusters simply do not live enough to cover such times.

4. CONCLUSIONS

The simulations presented in this paper provide clear evidences on the diffusion mechanism which controls the motion of ethane in $AlPO_4$-5. The frequency of particle crossings leads to normal diffusion when the observation time falls in the nanosecond time scale; the apparent deviations at low loading are basically due to a low crossing frequency, which entails a long turnover time before the normal diffusion can occur. A thorough analysis of the stability of cluster structures formed by several ethane molecules allowed to exclude the possibility of a transition to the single file regime at longer (~ 1 ms) times: at room temperature the clusters tend to dissociate rather rapidly, so that a concerted cluster diffusion can be ruled out for this system.

REFERENCES

1. J. Kärger and D. M. Ruthven, Diffusion in Zeolites and Other Microporous Solids, Wiley, New York, 1992.
2. J. Kärger, G. Fleischer and U. Roland, in Diffusion in Condensed Matter, J. Kärger, P. Heitjans, R. Haberlandt (eds.), Friedr. Vieweg & Sohn, 1998.
3. J. Kärger, Phys. Rev. A, 45 (1992) 4173.
4. V. Gupta, S. S. Nivarthi, A. V. McCormick and H. T. Davis, Chem. Phys. Lett., 247 (1995) 596.
5. K. Hahn, J. Kärger and V. Kukla, Phys. Rev. Lett., 76 (1996) 2762.
6. V. Kukla, J. Kornatowski, D. Demuth, I. Girnus, H. Pfeiffer, L. V. C. Rees, S. Schunk, K. K. Unger and J. Kärger, Science 272 (1996) 702.
7. H. Jobic, K. Hahn, J. Kärger, M. Bée, A. Tuel, M. Noack, I. Girnus and J. K. Gordon, J. Phys. Chem. B., 101 (1997) 5834.
8. D. S. Sholl, Chem. Phys. Lett., 305 (1999) 269.
9. P. Demontis, J. Gulín González, G. B. Suffritti and A. Tilocca, Microporous Mesoporous Mater., in the press.
10. P. Demontis, G. B. Suffritti and A. Tilocca, J. Chem. Phys. 105 (1996) 5586.
11. J. Kärger, P. Demontis, G. B. Suffritti and A. Tilocca, J. Chem. Phys. 110 (1999) 1163.
12. F. Eder and J. A. Lercher, J. Phys. Chem., 100 (1996) 16460.
13. H. L. Tepper, J. P. Hoogenboom, N. F. A. van der Vegt and W. J. Briels, J. Chem. Phys., 110 (1999) 11511.
14. D. S. Sholl and K. A. Fichthorn, Phys. Rev. Lett., 79 (1997) 3569.

Studies in Surface Science and Catalysis 140
A. Gamba, C. Colella and S. Coluccia (Editors)

Stabilisation of nanostructured CeO_2-ZrO_2 solid solutions by addition of Al_2O_3: a suitable way for production of thermally stable oxygen storage/release promoters for three-way catalysts

R. Di Monte, P. Fornasiero, J. Kašpar, M. Graziani.

Dipartimento di Scienze Chimiche, Università di Trieste, 34127 Trieste, Italy

Abstract

By impregnating γ-Al_2O_3 with cerium/zirconium citrate solutions and subsequent calcination nanostructured $Ce_mZr_{1-m}O_2$ mixed oxides supported on Al_2O_3 are obtained, which feature remarkably high oxygen storage even after a calcination at 1373 K for 24 h. Mutual thermal stabilisation between alumina and solid solutions has been observed, which prevents formation of α-alumina and sintering effects after a severe ageing.

1. Introduction

Ceria-based materials find very important applications in industrial catalysis (*1*). In particular, CeO_2 and more recently CeO_2-ZrO_2 mixed oxides are a key component of the catalytic devices for automotive pollution control (*2*). This has lead to an intense interest in the properties of these materials in the scientific literature (*3-7*). Recent environmental regulations which were proposed/issued by both US and European authorities (see for example US TIER II regulations issued in December 1999), require for an urgent improvement of the durability and efficiency of the automotive catalytic converters. Thus, durability as high as 130.000 miles will be phased-in beginning in 2004 by US TIER II regulation.

The conversion efficiency of a three-way catalyst (TWC) is strictly related to the so-called oxygen storage/release capacity (OSC), besides other factors. The OSC is the ability to attenuate the negative effects of rich/lean oscillations of exhaust gas composition through the Ce^{4+}/Ce^{3+} redox process. By maintaining a stoichiometric composition at the catalyst, the highest conversion efficiency of the exhaust is attained. Accordingly, the efficiency of the OSC is monitored by the vehicle emission on-board diagnostics (EOBD); decline of OSC being an indication of TWC failure. Therefore development of new highly stable OSC systems is of paramount importance.

Previous observations (*8,9*) and patent claims (see for example US 5,945,369 issued on August 31st, 1999) clearly indicated the unsuitability of impregnation of CeO_2 on Al_2O_3 for production of effective OSC systems, because the high dispersion and intimate contact of the CeO_2 component with Al_2O_3 leads upon ageing to easy formation of $CeAlO_3$ that deactivates the OSC component. Accordingly, it is usual practice to employ pre-formed CeO_2 or CeO_2-ZrO_2 particles to make the TWC. These particles are then supported (wascoated), together with the other components (noble metals and Al_2O_3) on the honeycomb. The present paper indicates that by impregnation of CeO_2-ZrO_2 solid solutions (CZ) over Al_2O_3 highly efficient and thermally stable OSC system are obtained, provided that high ZrO_2 content is employed.

2. Experimental

$Ce_mZr_{1-m}O_2$ (13 wt%)/γ-Al_2O_3 (m = 1, 0.6, 0.2, 0) were prepared by using a modified citrate complexation method (*10*), by wet impregnating the resulting citrate-containing solution on γ-Al_2O_3 (BET surface area 186 m^2 g^{-1}, pore volume 1.03 ml g^{-1}). Hereafter the samples will be indicated as CZXX/Al_2O_3 where XX indicates the CeO_2 content relative to ZrO_2 (100, 60, 20 or 0 mol%). The preparation of the materials is carried out as follows. $Ce(NO_3)_3{\bullet}6H_2O$ (99.99%, Aldrich) was dissolved in water and mixed with a water solution of $ZrO(NO_3)_2$ (20 % ZrO_2, MEL Chemicals), then a water solution of citric acid (99.7%, Prolabo) was added. The ratio metal cation to ligand was 1 to 2.1. The resulting solution was stirred at 348 K for 5 hrs, then at room temperature (rt.) for 12 h, and finally concentrated to carry out an "incipient wetness" impregnation of the support. The material was dried at 393 K for 12 hrs, heated up to 773 K at a heating rate of 3 K min^{-1} and then calcined at this temperature for 5 hours to obtain a yellow powder. Hereafter these samples are indicated as fresh ones. Catalysts were aged by calcination at 1273 or 1373 K for 5, 24 or 48 h. Temperature programmed reduction (TPR) was carried out in a conventional instrument (*11*). Dynamic-OSC was measured by alternately pulsing every 70 s CO (100 μl) and O_2 (100 μl) over the sample (30-100 mg, maintained in a flow of Ar of 25 ml min^{-1}) (*11*). Powder XRD spectra were collected on a Siemens Kristalloflex Mod.F Instrument (Ni-filtered CuKα). The profile fitting of the XRD patterns was performed by a Rietveld analysis program (RIETAN94). The peak shape was assumed to be a modified pseudo-Voigt function with asymmetry.

3. Results and discussion

Table 1 summarises the structural characterisation of Al_2O_3 and $Ce_mZr_{1-m}O_2/Al_2O_3$ samples after various treatments in air in the range of temperature of 773-1373 K.

Calcination progressively transforms γ-Al_2O_3 into θ-Al_2O_3 and α-Al_2O_3, the latter being the only phase detected after 24 h at 1373 K. XRD analysis by Rietveld refinement of CZ20/Al_2O_3 calcined at 1373 K for 5 h revealed the presence of a single phase CZ solid solution. A c/a = 1.018 ratio was calculated for the pseudo-cubic cell, which is typical of a *t*-phase (*12*). Noticeably, the particle size, as calculated from the Schrerrer formula, is significantly decreased after calcination at 1373 K by supporting $Ce_{0.2}Zr_{0.8}O_2$ on Al_2O_3. For CZ20/Al_2O_3 calcined at 1373 K a crystallite size of 11 nm is evaluated from the XRD analysis which increases to 35 nm on the unsupported sample (Table 1). The nano-structured nature of the CZ20/Al_2O_3 was confirmed by HRTEM (*13*). Some CeO_2 non incorporated into the mixed oxide was detected in the case of CZ60/Al_2O_3, in agreement with previous observation that compositional non-homogeneity is favoured at intermediate compositions. The nanosized nature of the CZ component seems attributable to the presence of zirconia rather then ceria, since significantly higher particle size was detected in CZ100/Al_2O_3.

Table 1 Structural characterisation of the catalysts.

Sample	Calcination		Phase composition / %				CZ Crystallite size / nm
	Temperature / K	Time / h	Al_2O_3			CZ^a	
			γ	θ	α		
Al_2O_3	973	5	100			-	-
	1373	5		56	44	-	-
	1373	24			100	-	-
$Ce_{0.2}Zr_{0.8}O_2/Al_2O_3$	773	5	100			-	-
	1273	5		100		*t* (100)	7
	1273	48		100		*t* (100)	9
	1373	5		100		*t* (100)	11
	1373	24		98	2	*t* (100)	11
$Ce_{0.6}Zr_{0.4}O_2/Al_2O_3$	773	5	100			-	
	1273	5		100		c (90) *t* (10)	12 (*c*) 7 (*t*)
	1273	48		100		c (90) *t* (10)	12 (*c*) 7 (*t*)
	1373	5		100		c (74) *t* (26)	14 (*c*) 7 (*t*)
	1373	24		80	20	*c* (75) *t* (25)	18 (*c*) 10 (*t*)
CeO_2/Al_2O_3	773	5	100			-	-
	1373	5		100		*c* (100)	20
	1373	24		34	66	*c* (100)	28
$Ce_{0.2}Zr_{0.8}O_2$	773	5	-	-	-	*t* (100)	6
	1373	5	-	-	-	*t* (100)	35
$Ce_{0.6}Zr_{0.4}O_2$	773	5	-	-	-	*t*'' (100)	5
	1373	5	-	-	-	*c* (85) *t* (15)	20 (*c*) 15 (*t*)

[a] *t*'', *t*' and *t* are tetragonal phases (space group $P4_2/nmc$) with pseudo-cubic cell parameter ratio (c/a) respectively of 1.000, 1.002-1.01 and 1.02; *c* phase: cubic fluorite type of lattice (space group Fm3m). When a mixture of phases was detected approximate composition were evaluated: *c* phase: $Ce_{0.8}Zr_{0.2}O_2$; *t* phase $Ce_{0.2}Zr_{0.8}O_2$.

There is a synergic stabilisation between the CZ and Al_2O_3 component, since the transformation process of Al_2O_3 is also retarded by the presence of the CZ component, particularly at high ZrO_2 content, as shown in Figure 1. As a result of this interaction the surface area results significantly stabilised compared to pure Al_2O_3 with respect to thermal ageing. This effect is noticeable after calcination at 1373 K, 24 h. This is consistent with previous studies which shows that CeO_2 (*14,15*) or La_2O_3 (*16*) prevents the formation of α-Al_2O_3, which is associated with a drastic collapse of surface area. It has been suggested that formation of micro-domains of $CeAlO_3$ or $LaAlO_3$ on the alumina surface may inhibit surface diffusion of species responsible for sintering (*16*). However, it should be noted that also ZrO_2 stabilises alumina (*17*), suggesting that a more general scenario should be considered.

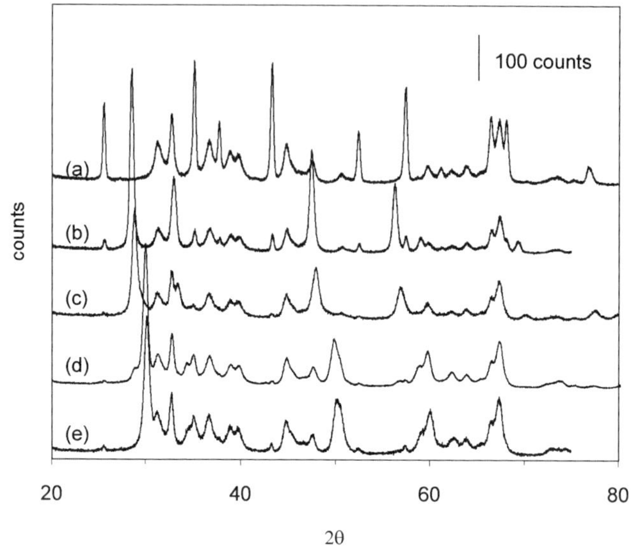

Figure 1. XRD spectra of the samples calcined at 1273 K for 5 h. (a) Al_2O_3, (b) CeO_2/Al_2O_3, (c) $Ce_{0.6}Zr_{0.4}O_2/Al_2O_3$, (d) $Ce_{0.2}Zr_{0.8}O_2/Al_2O_3$ and (e) ZrO_2/Al_2O_3.

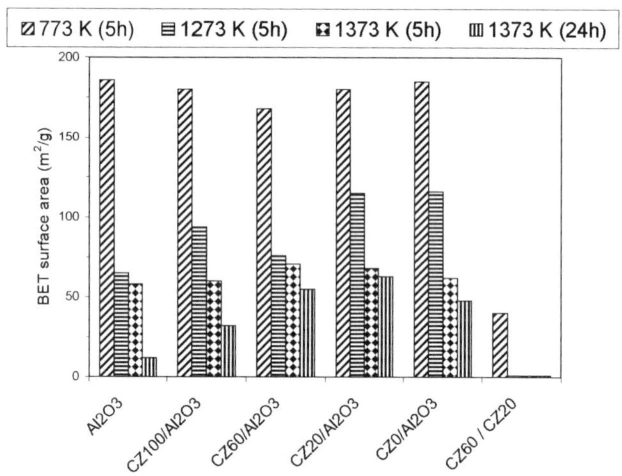

Figure 2. Effect of calcination on BET surface areas of the investigated CZ/Al_2O_3 samples.

Figure 3 compares the TPR profiles of $CZ20/Al_2O_3$ and $CZ100/Al_2O_3$ calcined at 1273 K. Consistently with the relatively high CeO_2 particle size, the reduction of $CZ100/Al_2O_3$ occurs at high temperature. The TPR profile consists of three peaks, the first two (I and II in fig. 3) have been attributed to reduction of well dispersed ceria crystallite and bulk ceria, respectively (*2,18*). The third peak has been mainly associated with formation of $CeAlO_3$. Accordingly, the X-ray diffraction pattern obtained after TPR and a mild oxidation at 700 K (fig. 4, trace b) shows the presence of cerium alluminate. In contrast, reduction at relatively low temperature persists in CZ20/Al_2O_3. Consistently with the lack of reduction at 1250 K, no formation of $CeAlO_3$ was detected in the CZ20/Al_2O_3 sample (Fig. 4, traces c-d). This indicates that incorporation of ZrO_2 into CeO_2 may prevent this undesirable deactivation pathway. The more ZrO_2 is added to CeO_2, the more effective stabilisation is achieved, highlighting the crucial role of ZrO_2 in improving the stability of these systems. Phase separation into CeO_2-rich and ZrO_2-rich phases was detected after calcination at 1373 K in CZ60/Al_2O_3, leading to a partial formation of $CeAlO_3$ after the TPR/oxidation at 700 K treatment.

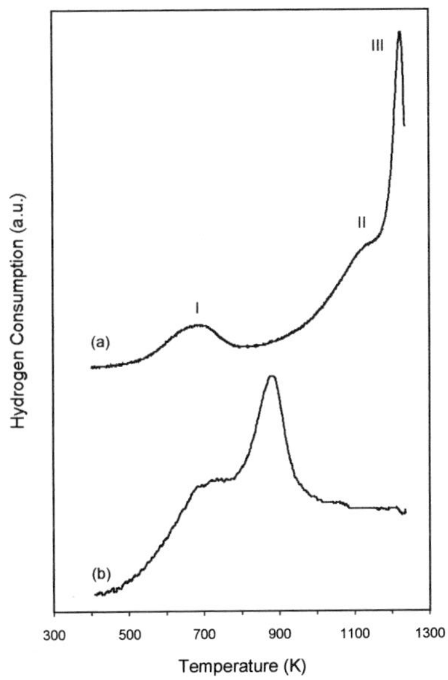

Figure 3. TPR profiles of (a) CeO_2/Al_2O_3 and (b) $Ce_{0.2}Zr_{0.8}O_2/Al_2O_3$ calcined at 1273 K for 5h.

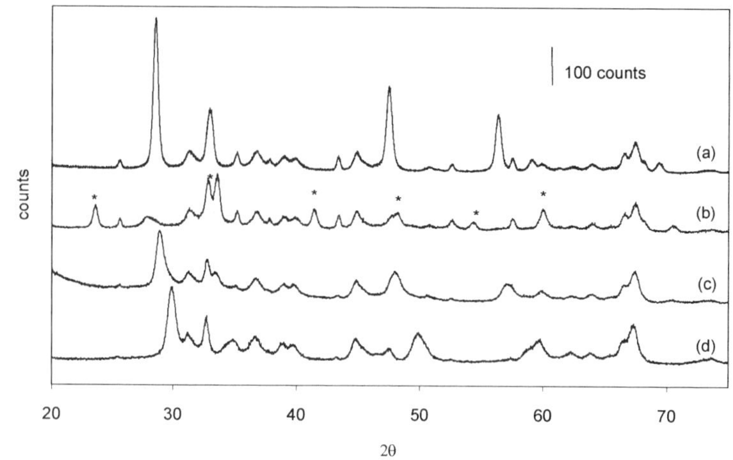

Figure 4. Powder X-ray diffraction profiles of (a) CeO_2/Al_2O_3 calcined at 1273 K 5 h, (b) CeO_2/Al_2O_3 calcined at 1273 K 5 h, subjected to a TPR up to 1273 K followed by an oxidation at 700 K, (c) $Ce_{0.6}Zr_{0.4}O_2/Al_2O_3$ calcined 1273 K 5 h, subjected to a TPR up to 1273 K followed by an oxidation at 700 K, and (d) $Ce_{0.2}Zr_{0.8}O_2/Al_2O_3$ calcined 1273 K 5 h, subjected to a TPR up to 1273 K followed by an oxidation at 700 K. (*) peaks belonging to $CeAlO_3$.

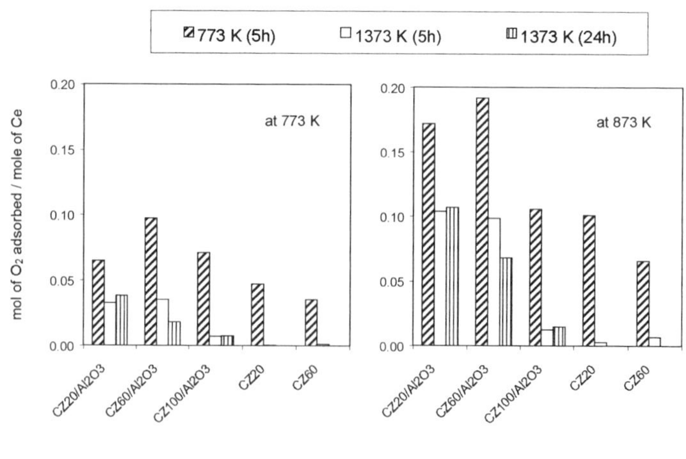

Figure 5. Dynamic oxygen storage measured at (a) 773 and (b) 873 K over CZ/Al_2O_3 samples calcined at 773 and 1373 K.

Dynamic OSC was measured over all the synthesised samples using CO as reducing agent. For sake of consistency, all the measured values are reported mol (O_2) mol (CeO_2)$^{-1}$. As illustrated in Figure 5, the response of the system to the effects of calcination is remarkably different according whether Al_2O_3-supported or unsupported samples are considered. For the latter samples negligible OSC was measured after the calcination at 1373 K, even though significant OSC was measured in the fresh CZ20 and CZ60. This highlights the crucial role of Al_2O_3 preserving the OSC with respect the effects of high temperature calcination. It is not clear whether the improvement of the OSC is simply related to the lower particle size of the CZ component in the Al_2O_3 supported samples, e.g. their nano-structured nature, or other effects are operating. In fact, at 773-873 K both surface and bulk of the CeO_2-ZrO_2 mixed oxides may contribute to the CO-OSC (*19*) suggesting a complex nature of the promoting effect. A perusal of Figure 5 reveals also the crucial role of high contents of ZrO_2 in improving the thermal stability of the OSC property. The comparison of the values measured for CZ20/Al_2O_3 and CZ60/Al_2O_3 after calcination at 1373 K for 5 and 24 h, respectively, shows that no decline of OSC was observed for the former sample when the calcination time is increased from 5 to 24 h. In contrast the OSC of CZ60/Al_2O_3 progressively declined which we attribute to the increasing degree of phase separation as detected by XRD. This is an interesting point since it contradicts the previous suggestion that micro-domain type of mixed oxides are more efficient as OSC systems compared to single phase solid solution (*20*).

4. Conclusions

The addition of Al_2O_3 to CeO_2-ZrO_2 solid solutions strongly improves the dynamic-OSC in the $Ce_mZr_{1-m}O_2/Al_2O_3$ samples compared to the unsupported ones. This effect is noticeable in freshly synthesised samples, but, more remarkably, the presence of Al_2O_3 prevents a strong deactivation of the OSC property even after calcination at 1373 K. In contrast such ageing leads to an almost complete deactivation of the dynamic-OSC in unsupported $Ce_mZr_{1-m}O_2$. The stability of the $Ce_mZr_{1-m}O_2$ phase and of the OSC depends on the sample composition, zirconia rich compositions being more thermally stable. The intimate contact between the $Ce_mZr_{1-m}O_2$ and Al_2O_3 stabilises high dispersion of the $Ce_mZr_{1-m}O_2$ component and, in addition provides an effective stabilisation of the transition aluminas with respect to transformation to the α-Al_2O_3.

Acknowledgements. Financial support from University of Trieste, MURST (Roma) "Progetti di Ricerca di Rilevante Interesse Nazionale - 1998", Regione Friuli Venezia-Giulia, Fondo regionale per la ricerca L.R. 3/1998 and “Progetto Finalizzato MSTA II” of the National Research Council are gratefully acknowledged.

5. References

1. S. Bernal, J. Kaspar, and A. Trovarelli, "Recent progress in catalysis by ceria and related compounds", S. Bernal, J. Kaspar, and A. Trovarelli (eds.), vol. 50, Catal. Today, Elsevier Science, Amsterdam, 1999, pp.173-443.
2. J. Kaspar, M. Graziani, and P. Fornasiero, "Ceria-Containing Three Way Catalysts" in: "Handbook on the Physics and Chemistry of Rare Earths: The Role of Rare Earths in Catalysis.", K.A. Gschneidner Jr. and L. Eyring (Eds.), Chap.184, Elsevier Science B.V., Amsterdam, 2000, pp.159-267.
3. T. Murota, T. Hasegawa, S. Aozasa, H. Matsui, and M. Motoyama, J. Alloys Comp., 193 (1993) 298.
4. M. Ozawa, M. Kimura, and A. Isogai, J. Alloys Comp., 193 (1993) 73.

5. G. Ranga Rao, J. Kaspar, R. Di Monte, S. Meriani, and M. Graziani, Catal. Lett., 24 (1994) 107.
6. A. Trovarelli, F. Zamar, J. Llorca, C. de Leitenburg, G. Dolcetti, and J.T. Kiss, J. Catal., 169 (1997) 490.
7. N. Izu, T. Omata, and S. Otsuka-Yao-Matsuo, J. Alloys Comp., 270 (1998) 107.
8. J.Z. Shyu, W.H. Weber, and H.S. Gandhi, J. Phys. Chem., 92 (1988) 4964.
9. T. Miki, T. Ogawa, A. Ueno, S. Matsuura, and M. Sato, Chem. Lett., (1988) 565.
10. C. Marcilly, P. Courty, and B. Delmon, J. Amer. Ceram. Soc., 53 (1970) 56.
11. N. Hickey, P. Fornasiero, J. Kaspar, M. Graziani, G. Blanco, and S. Bernal, Chem. Commun., (2000) 357.
12. M. Yashima, H. Arashi, M. Kakihana, and M. Yoshimura, J. Amer. Ceram. Soc., 77 (1994) 1067.
13. R. Di Monte, P. Fornasiero, J. Kaspar, M. Graziani, J.M. Gatica, S. Bernal, and A. Gomez Herrero, Chem. Commun., (2000) 2167.
14. S. Humbert, A. Colin, L. Monceaux, F. Oudet, and P. Courtine, "Simultaneous Atmosphere and temperature cycling of three-way automotive exhaust catalysts" in: "Catalysis and Automotive Pollution Control III", A. Frennet and J.M. Bastin (Eds.), Elsevier, Amsterdam, 1995, pp.829-839.
15. M.H. Yao, R.J. Baird, F.W. Kunz, and T.E. Hoost, J. Catal., 166 (1997) 67.
16. F. Oudet, P. Courtine, and A. Vejun, J. Catal., 114 (1988) 112.
17. T. Horiuchi, Y. Teshima, T. Osaki, T. Sugiyama, K. Suzuki, and T. Mori, Catal. Lett., 62 (1999) 107.
18. H.C. Yao and Y.F. Yu Yao, J. Catal., 86 (1984) 254.
19. M. Boaro, C. de Leitenburg, G. Dolcetti, and A. Trovarelli, J. Catal., 193 (2000) 338.
20. T. Egami, W. Dmowski, and R. Brezny, SAE paper 970461 (1997).

Studies in Surface Science and Catalysis 140
A. Gamba, C. Colella and S. Coluccia (Editors)

Chemistry and Photochemistry of H_2 on MgO surfaces

O. Diwald, T. Berger, M. Sterrer and E. Knözinger[a]*,

[a] Institut für Physikalische und Theoretische Chemie der Technischen Universität Wien, c/o Vet. Med. Universität, Veterinärplatz 1/GA,
A- 1210, Wien, Austria

On nanometer-sized MgO particles emerging from chemical vapour deposition an enormously high surface concentration of low-coordinated ions and vacancies is observed. They are intimately related to a variety of chemical and photochemical surface reactions initiated by H_2: (1) heterolytic splitting on ion pairs in the dark gives rise to the appearance of vicinal OH and MgH groups, whereas UV-induced homolytic splitting on coordinatively unsaturated O^{2-} anions provides OH and surface mobile H• radicals; (2) UV-induced charge transfer from MgH to surface anion vacancies F_S^{2+} which are thus transformed into colour centres F_S^+, is another source for surface mobile H• radicals; (3) the H• radicals previously formed then contribute as reducing agent to additional surface colour centre formation (F_S^+); (4) molecular oxygen finally bleaches the surface colour centres and becomes complexed as superoxide anion (O_2^-) by Mg^{2+}.

All participants of the reactive happening on the MgO surface except H• may directly be traced by an array of spectroscopic tools: (1) UV diffuse reflectance (coordinatively unsaturated O^{2-}); (2) infrared (MgH, OH), and (3) electron paramagnetic resonance spectroscopy (F_S^+ and O_2^- complexed by Mg^{2+}). Owing to the non-equilibrium character of chemical vapour deposition as MgO production technique, more or less broad distributions of low coordinated sites and F_S^{2+} species were expected with inherent complications for the respective UV, IR and EPR spectra. In fact, very limited numbers of species in well-defined geometries were found: (1) two ion pairs for heterolytic H_2 splitting in the dark; (2) one type of oxygen anion for UV-induced homolytic H_2 splitting; (3) two predominant F_S^{2+} centres for colour centre formation; (4) two types of Mg^{2+} cations where O_2^- is complexed after colour centre bleaching.

The different reaction channels starting with H_2 splitting on low coordinated surface sites and finalized by the bleaching process of F_S^+ centres by O_2, are not independent from each other. The connectivity between them is mediated by the surface mobile H•radicals that end up as OH protons in close vicinity to colour centres or O_2^- species. — The OH groups attached to paramagnetic species such as F_S^+ or O_2^- are seen by EPR via the magnetic interaction (superhyperfine splitting) between the proton and the electron spin centre. On the other hand, these OH groups also probe the paramagnetic species via electronic interaction. Thus the colour centre and superoxide anion species on the MgO surface become indirectly detectable also by IR spectroscopy

1. Introduction

Highly dispersed MgO as a result of non-equilibrium techniques of production hosts comparatively high concentrations of low coordinated (LC) ions and vacancies on its surface [1-3]. The sites involved are coordinatively unsaturated and, therefore, exhibit — in comparison to equilibrium surfaces — unusual reactivity in the dark as well as under the influence of UV light [4,5]. This is particularly true for the simplest reactant, namely, molecular hydrogen (H_2) which gives rise to a puzzling manifold of surface reactions on high surface area MgO [6-8].

In fact, two different surface sites were found where heterolytic H_2 splitting takes place at room temperature and in the dark [9,10]: low coordinated ion pairs ($O^{2-}{}_{LC}$ /$Mg^{2+}{}_{LC}$) embedded in a specific local geometry. In the course of the reaction they are transformed into a vicinal pair of a hydroxyl (OH) and a hydride (MgH) group which are both IR active:

$$O^{2-}_{LC} \text{ - } Mg^{2+}_{LC} + H_2 \longrightarrow O^{2-}(H^+) \text{ - } Mg^{2+}(H^-) \qquad (1)$$

The anionic constituents of these low coordinated sites must be involved in electronic transitions studied by UV diffuse reflectance [11-13] and photoluminescence spectroscopy [14-16]. Even UV-induced charge transfer or ionization on low coordinated O^{2-} ions is feasible [17-19]. The resulting electron hole centres are paramagnetic and, thus, observable by electron paramagnetic resonance spectroscopy.

$$O^{2-}_{LC} + h\nu \longrightarrow O^{-}_{LC} + e^{-} \qquad (2)$$

They react with molecular hydrogen according to homolytic H_2 splitting:

$$O^{-}_{LC} + H_2 \longrightarrow (O_{LC}H)^{-} + \dot{H} \qquad (3)$$

The fate of the evolving electron in eq.2 has so far not been traced unambiguously by experiments. Most likely, however, it is captured by closely spaced electron deficient sites such as anion vacancies [7,20].

In a previous study the local environment of one of the the two types of ion pairs where heterolytic H_2 splitting occurs was characterized in detail [21-23]: it may be visualized by the removal of a low coordinated O^{2-} anion from a corner or edge site [21,24]. As a result pairs of a 3- and 4-coordinated neighbouring Mg^{2+} cations and anions become available (Fig.1a). In addition to the H_2 splitting capability this type of site exhibits another interesting property: it may undergo the following UV induced redox reaction (Fig.1a):

$$O^{2-}_{LC} - Mg^{2+}_{LC} \vee \overset{+H_2}{\rightarrow} O^{2-}_{LC}(H^+) - Mg^{2+}_{LC}(H^-) \vee \overset{+h\nu}{\rightarrow} O^{2-}_{LC}(H^+) - Mg^{2+}_{LC} \dot{\vee} + \dot{H} \qquad (4)$$

The transformation of the anion vacancy $F_S{}^{2+}$ ($\vee$ in eq.4) into a colour centre $F_S{}^{+}$ ($\dot{\vee}$ in eq.4) is reflected by a colour change of the MgO sample from white to blue. The reaction is

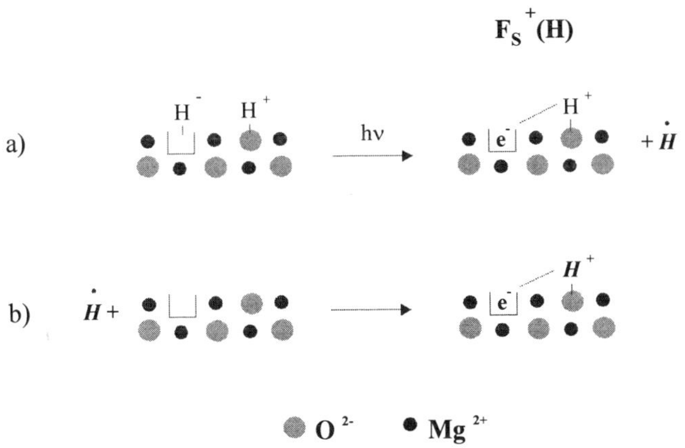

Fig 1: Schematic representation of two reactions schemes providing F_S^+(H) centres. These paramagnetic surface defects interact directly with protons of nearby hydroxyl groups which may either originate from a) heterolytic H_2 splitting or from b) oxidation of neutral H atoms. The latter comparatively mobile species is formed in the course of homolytic H_2 splitting, but also emerges as a product of hydride oxidation as indicated in (a)

monitored by IR (observation of OH and MgH groups), EPR spectroscopy (observation of colour centres) and DR UV-Vis spectroscopy (electronic excitation). Most interestingly there is significant electronic and magnetic interaction between an OH proton and the unpaired electron of the colour centre (F_S^+). As a result there is a significant shift of the OH stretching band to lower wavenumbers [25] and superhyperfine (shf) splitting of the EPR signal components [26,27], respectively. This type of surface complex is generally designated F_S^+(H).

If the MgO sample is exposed to H_2 gas and UV light at the same time, the reaction according to eq.4 is superimposed by that of eq.3. In both processes (eqs.3 and 4) a H· radical appears as product. It is likely to travel to remote anion vacancies and to transform them into further colour centres (Fig.1b) [28], which also interact with a closely spaced OH group. Therefore, they have also to be considered as F_S^+(H) centres. This is intimately related to an interesting problem. There are OH groups of different history — OH out of heterolytic H_2 chemisorption, out of UV-induced homolytic H_2 splitting or out of hydride oxidation — and there is quite a manifold of different anion vacancy geometries. Presently, e.g., single vacancies in corners or edges [21,24,27] and double vacancies in tub or pit form [29] are discussed. The question which type of OH group interacts with which colour centre type is, however, still subject to a controversial discussion [21,22].

Therefore, particular emphasis was laid in the present study on site selectivity of H_2 splitting (heterolytic and homolytic), of colour centre formation, and of colour centre bleaching by addition of O_2 gas. Monitoring of educt consumption and product formation was observed by IR, UV and/or EPR spectroscopy. The variation of the H_2 pressure applied in the introductory reaction step (heterolytic or homolytic H_2 splitting) introduces relevant trends in the abundance of the subsequently generated surface species. The evaluation of these trends clearly helps to unravel the puzzling connectivity between the processes of H_2 splitting and surface colour formation.

2. Experimental

All experiments were carried out with the same type of MgO obtained by chemical vapour deposition in a flow system [2]. High purity Mg pieces as educt were supplied by Johnson Matthey GmbH. The specific surface area of the resulting MgO material determined by BET (LN_2) measurements is around 400 m^2/g. In order to guarantee a totally dehydroxylated surface, the sample was gradually annealed at 1173 K under dynamic vacuum ($<10^{-5}$ mbar) before each experiment. This leads to a reduction of the specific surface area to 300 m^2/g. The rate of temperature increase for the annealing steps was 10K/ min. All samples were treated at 870 K with oxygen in order to burn organic contaminants originating from the oil of the vacuum pumps used in the flow system. The gases H_2 (99.999%) and $^{16}O_2$ (99.998%) for adsorption studies were provided by Messer Griesheim. - A 300W Xe lamp (Oriel) was applied for UV irradiation. The light beam passes through a water filter in order to avoid sample heating by IR irradiation. All UV excitation experiments on MgO were carried out at room temperature. The exposure time amounted to 10-30 minutes in hydrogen atmosphere.

The IR and EPR sample cells are connected to an appropriate high vacuum pumping rack. It allows thermal activation of the sample at less than 10^{-5} mbar and adsorption/ desorption experiments with diverse gases. For the IR experiments small quantities of MgO powder (20-30 mg) were pressed into selfsupporting pellets. The pressure applied was less than 10 bar and did not initiate any change of the specific surface area. For the EPR experiments similar amounts of the MgO sample batch, also used for IR spectroscopy, were filled in EPR tubes.

The IR spectra were recorded using a Fourier-transform IR spectrometer model IFS 113v (Bruker Optik GmbH). The resolution was 3 cm^{-1}. 300 interferogram scans were averaged in order to guarantee a reasonable signal-to-noise ratio. The reference for the absorbance spectra is a MgO sample previously subjected to thermal activation and then cooled down to room temperature. - The EPR spectra were recorded using a Bruker EMX 10/12 spectrometer system in the X band. The spectra presenting paramagnetic oxygen species required a reduction of the temperature to 77K in order to avoid line broadening. 10 coadded spectra sufficed to obtain a satisfactory signal-to-noise ratio. The DPPH signal as well as the lines originating from traces of Mn^{2+} in the sample were applied for g value calibration.

For UV diffuse reflectance measurements CVD powder was placed in a high vacuum tight quartz cell with optical windows made of Suprasil. Thermal activation, as well as adsorption and desorption steps were performed by connecting the cell to a conventional vacuum line that also guarantees pressures less than 10^{-5} mbar. The UV spectra were acquired using a Perkin Elmer Lambda 15 spectrophotometer equipped with an integrating sphere. $BaSO_4$ powder was used as a reference.

3. Results and discussion

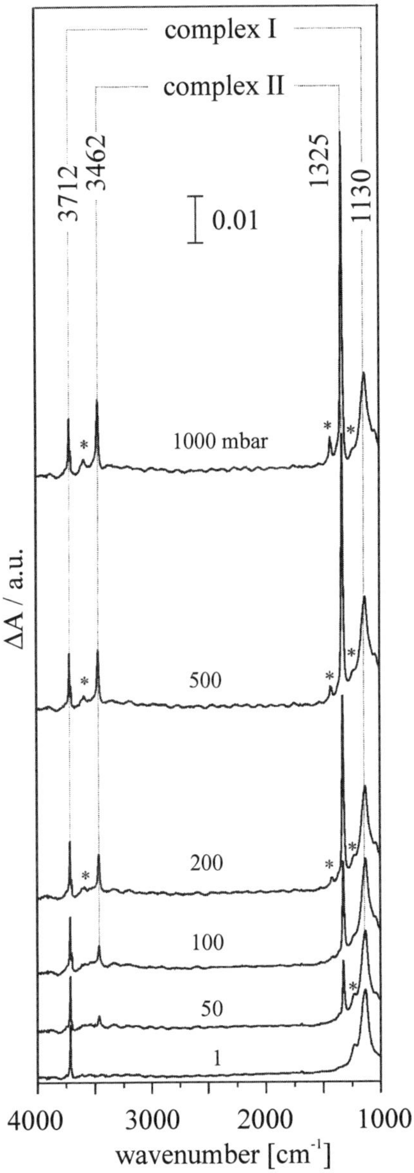

Fig. 2: IR spectra of H_2 chemisorption complexes obtained after H_2 addition (1mbar≤p (H_2)≤1000 mbar) to a previously dehydroxylated sample of MgO nanoparticles.

H_2 gas does not exhibit any UV excitation in the wavelength range above 200 nm which might explain UV induced surface chemistry on MgO. Thus, the only paths along which H_2 may enter the game of UV induced processes on MgO surfaces is the interaction of H_2 with UV excited surface sites or the UV excitation of the products of H_2 chemisorption. As mentioned previously (see introduction) there are two such chemisorption complexes (Fig.2) [6,9,22]. One of them, the chemisorption complex I (ν(OH)= 3712 cm^{-1}, ν(MgH)= 1130 cm^{-1}) exhibits a limited abundance which is reached already at H_2 pressures below 1 mbar. Furthermore it cannot be removed completely from the MgO surface by high vacuum conditions at room temperature and is, therefore, called the "irreversible" chemisorption complex [22]. The respective site is immediately attached to an anion vacancy in edge or corner position (see introduction, eq.4). On raising the H_2 pressure to more than 1 mbar another H_2 chemisorption complex (II, ν(OH)=3462 cm^{-1}, ν(MgH)=1325 cm^{-1}) becomes visible (Fig.2). At room temperature it is perfectly reversible. Different from chemisorption complex I, the constituents of the respective ion pair are most likely 4-coordinated within a (1,1,0) microplane [10,30]. (Absorption bands of minor intensities at 3596, 1424 and 1230 cm^{-1} – see asterisk's in Fig. 2 – have to be attributed to unspecified reversible H_2 chemisorption complexes of comparatively low abundance).

UV irradiation of a MgO sample hosting exclusively the chemisorption complex I (high vacuum conditions guarantee the absence of the reversible complex II) induces the following spectral changes [22]: (1) an intensity decrease of the hydride band at 1130 cm^{-1}; (2) blue colouring of the originally white sample which is attributed to surface colour centre formation ($F_S^+(H)$ according to eq. 4, see introduction); (3) the appearance of a broad and unspecific IR absorption background between 4000 and 2500 cm^{-1} which — as the blue colour in (2) — originates from the electronic excitation of the colour centre $F_S^+(H)$; (4) three OH stretching bands appearing at 3632, 3602 and 3528 cm^{-1} and exhibiting essentially the same intensities. Their positions are considerably below those of isolated surface OH groups (~3690 cm^{-1}). Obviously they are subjected to an H-bond like interaction with a closely spaced $F_S^+(H)$ electron as proton acceptor [25]. — The interpretation of these observations was corroborated by bleaching the colour centres via O_2 admission: The blue colour as well as the colour centre related IR features vanish instantaneously.

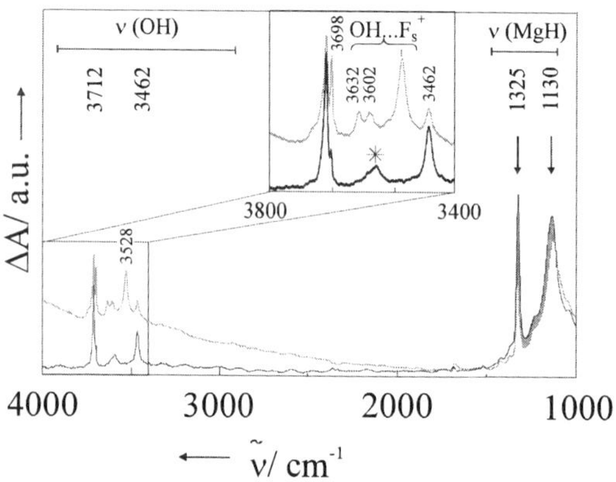

Fig. 3: IR spectra of MgO in the presence of 100 mbar H_2 before (black curve) and after 10 minutes of UV irradiation (grey curve). A blow up of the OH stretching region is given in the insert. The asterisk therein indicates an absorption attributed to an additional chemisorption complex of minor abundance [25].

After UV irradiation of a MgO sample exposed to 100 mbar H_2 the IR spectrum should provide an insight in how far the chemisorption complex II is involved as educt in the subsequent UV-induced processes (Fig.3). In fact, not only the hydride band of chemisorption complex I (1130 cm^{-1}), but also that of chemisorption complex II (1325 cm^{-1}) loses intensity. This indicates that both hydrides are oxidized in the course of electron transfer processes. It has, however, to be admitted that an acceptor for the electron from the hydride group of complex II has not yet been specified. Also in this case a surface mobile H· radical is emerging during the process (eq.3). Different from the experiment based on the hydride of chemisorption complex I the band at 3528 cm^{-1} now strongly dominates those at 3632 and 3602 cm^{-1} (inset in Fig.3). A closer inspection of the spectral interval where isolated surface OH groups absorb (>3690cm^{-1}) provides a relevant clue for a so far unattended reaction channel. Without any doubt the UV irradiation of MgO in the presence of H_2 creates a so far unknown isolated surface OH group which absorbs at 3698 cm^{-1} [19]. An intensity correlated counterpart in the MgH stretching region — as an evidence of a heterolytic H_2 splitting mechanism — has not been observed. At room temperature it is absolutely not affected by high vacuum conditions. Nor is it sensitive to O_2 — as the other colour centre related spectral features that vanish when the colour centres are destroyed by bleaching with O_2.

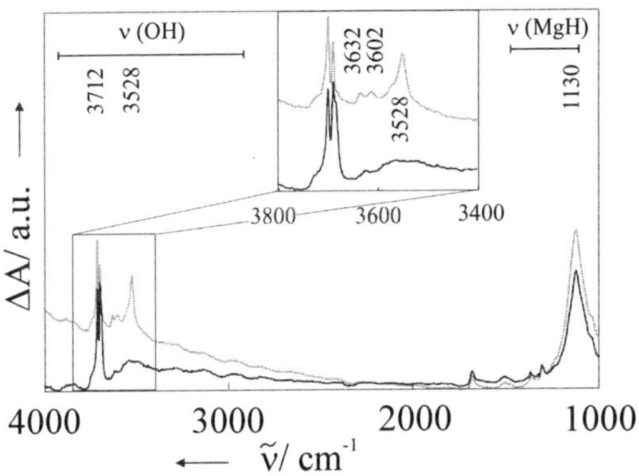

Fig. 4: IR spectra of MgO before (grey curve) and after colour centre bleaching via the addition of O_2 gas (black curve). The colour centres were produced by simultaneous exposure of MgO to 100 mbar H_2 and UV light. Before the addition of O_2 high vacuum was applied. A blow up of the OH stretching region is given in the insert.

It was mentioned previously (see introduction) that UV light ($200 \leq \lambda \leq 300$ nm) initiates the transformation of low coordinated oxygen anions into O^- radicals (eq.2) which were identified by EPR spectroscopy [17-19]. Subsequently they react with H_2 according to eq.3. The result of this homolytic H_2 splitting process is the OH group absorbing at 3698 cm^{-1}. The simultaneously emerging H· radical (eq.3) is mobile on the surface and reacts — as the H· radical out of the UV induced charge transfers from hydride I and II to anion vacancies (Fig.1b) — with remote anion vacancies which are most likely not involved in one of the heterolytic H_2 splitting processes. Obviously there are two reaction channels on MgO surfaces exposed to H_2 and UV light both ending up with the formation of colour centres $F_S^+(H)$:

- Those originating from the reduction of low coordinated anion vacancies are designated here $F_S^+{}_A$ (H). Their spin centre interacts with the hydroxyl group of the chemisorption complex I (Fig.1a).

- Those originating from reduction of so far unspecified anion vacancies, which are remote from the H_2 splitting site, by the surface mobile H· radicals (Fig.1b, eqs. 3). They are largely dominating the $F_S^+{}_A$ (H) centres in abundance at 100 mbar H_2 and are designated here $F_S^+{}_B$ (H).

As a matter of fact EPR spectroscopy does not allow for a reliable proof of the existence of two different colour centre species $F_S^+{}_A$(H) and $F_S^+{}_B$(H) [22]. On the other hand, only recently ENDOR was successful in discriminating them [27]. Most peculiarly, IR spectroscopy yields indirect, but relevant evidence for their presence: Chemisorption complex I is saturation limited. $F_S^+{}_A$(H) is derived from it in a subsequent reaction (eq.4). The related IR and/or EPR signals must, therefore, also be subject to the saturation criterion. In fact, this is true for the IR bands at 3632 and 3602 cm^{-1} which are attributed to isolated OH groups interacting with the colour centre $F_S^+{}_A$(H) [22]. On the other hand, the intensity of the IR band at 3528 cm^{-1} grows with H_2 pressure — in agreement with the behaviour of the $F_S^+{}_B$(H) signal obtained by EPR spectroscopy [26]. Obviously the abundance of $F_S^+{}_B$(H) is considerably larger than that of $F_S^+{}_A$(H) which is — for whatever reasons — not visible by EPR.

The strongest argument supporting the presence of the two different colour centre types stems from bleaching experiments. On adding O_2 gas to the MgO sample which was previously subjected to the above described process of colour centre formation and then to high vacuum conditions, both $F_S^+(H)$ centres are effectively destroyed by the following redox process (Fig.4):

$$O^{2-}_{LC}(H^+) - Mg^{2+}_{LC}\overset{\bullet}{\vee} + O_2 \rightarrow O^{2-}_{LC}(H^+) - Mg^{2+}_{LC}(O_2^-)\vee \tag{6}$$

The emerging EPR active superoxide anion species O_2^- is complexed by a cation (which must not necessarily be a constituent of the anion vacancy) and interacts now with the IR active OH group previously involved in the H bond with the colour centre electron [22,25]. The g_{zz} signal component of O_2^- is sensitively influenced by the local crystal field of the complexing cation [31,32]. On the other hand the OH stretching frequency depends on the strength of the interaction of OH with O_2^-. Thus the redox reaction described by eq.6 may be monitored by EPR and IR spectroscopy. In Fig.5 the respective curves are compared for fundamentally different experimental conditions:

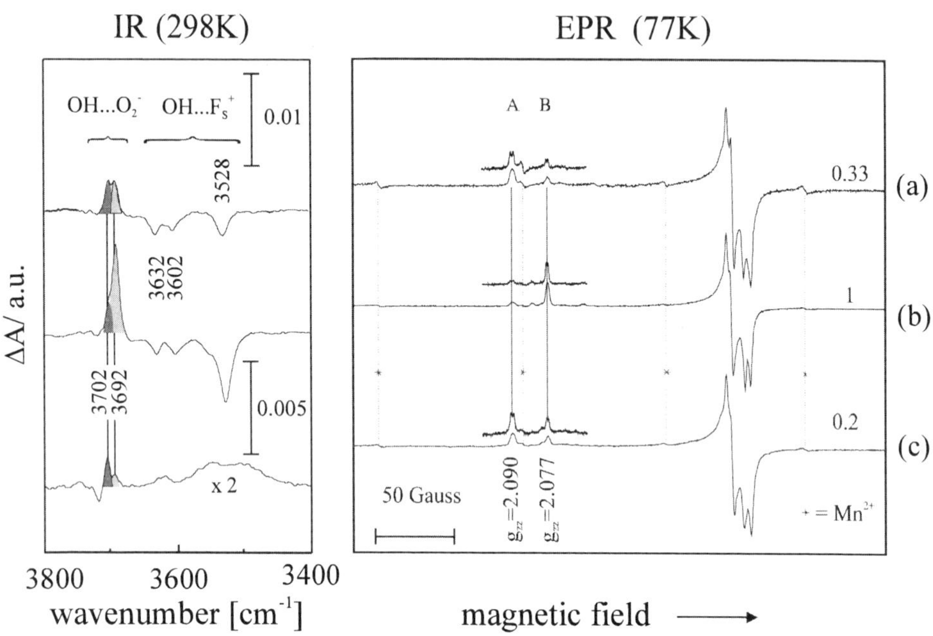

Fig. 5: IR difference (left) and EPR spectra (right) related to the colour centre bleaching process initiated by O_2 gas on MgO surfaces. The colour centres were produced by simultaneous exposure of the MgO sample to (a) 1 mbar H_2 and UV light and (b) 100 mbar H_2 and UV light. Before the addition of O_2 high vacuum was applied. The IR difference spectra were obtained by subtracting the grey curve of Fig.4 from the black one. For comparison (c) the respective IR and EPR spectra related to the oxidation of irreversibly formed hydride groups (H_2 chemisorption complex I) by O_2 gas are also shown. (The ordinate values of the EPR spectra have to be multiplied by the numbers at the right limit – 0.33, 1, 0.2 – in order to provide the true ordinate scaling.)

(a) UV irradiation of the MgO sample under 1 mbar H_2 followed by evacuation (HV) and O_2 addition (10 mbar): only the irreversible heterolytic H_2 chemisorption and the homolytic process become effective.
(b) UV irradiation of the MgO sample under 100 mbar H_2 followed by evacuation (HV) and O_2 addition (10 mbar): all H_2 splitting processes — heterolytic I and II as well as UV-induced homolytic H_2 chemisorption — become effective.
(c) in addition, curve c demonstrates the effect occurring when H_2 chemisorption complex I reacts with 10 mbar O_2 in the absence of UV light.

The IR spectra are presented in the difference mode: the absorbance data obtained before colour centre bleaching were subtracted from those obtained thereafter. As the bleaching process is complete, the "colour centre" IR bands (OH stretching bands at 3632, 3602 and 3528 cm^{-1}) appear as a negative pattern (Fig.5 a and b, left). It is intensity correlated with two positive OH stretching bands at 3702 and 3692 cm^{-1} [22] which are closely spaced to the stretching bands of isolated free OH groups. The slight shift to smaller wavenumbers (from 3712 to 3702/3692 cm^{-1}) originates from specific interactions with the respective neighbouring O_2^- species. Obviously the saturation phenomena with respect to H_2 pressure, observed for chemisorption complex I (see Fig.2) and for the negative "colour centre" IR bands at 3632 and 3602 cm^{-1} (Fig. 5a and b, left), is again revealed by the colour centre bleaching process which ends up with the formation of the OH...O_2^- surface complex related to the absorption at 3702 cm^{-1}. On the other hand, the second positive OH band related to the bleaching process (3692 cm^{-1}) grows with H_2 pressure (during the preceding colour centre formation) as the IR band at 3528 cm^{-1}. The growth of this latter band with H_2 pressure is intimately related to the increasing abundance of H• radicals originating from both homolytic H_2 splitting (eq.3) and photolysis of chemisorption complex II.

The surface site specific EPR features of the O_2^- species are the g_{zz} signal components [31,32]. According to the EPR spectra in the right half of Fig. 5a and b there are —irrespective of the H_2 pressure applied (100 or 1 mbar, respectively) — only two dominating O_2^- species, namely, A and B. Since for 1 mbar H_2 the abundance of the chemisorption complex II is negligibly small (see Fig.2), none of the two O_2^- species observed here can be directly related to the surface geometry related to reversible H_2 splitting.

The signal attributed to O_2^- species A is obviously saturation limited with respect to the H_2 pressure. It should, therefore, be localized within or close to the local geometry where the chemisorption complex I is hosted, i.e., a 3- or 4-coordinated anion vacancy [21,27]. On the other hand, the signal components attributed to O_2^- species B grow — as the negative "colour centre" IR band at 3528 cm^{-1} (Fig. 5a,b, right), the EPR signal of the F_S^+(H) centre ([22], not shown here) and the OH band at 3692 cm^{-1} (Fig.5a and b, left) — with increasing H_2 pressure. These trends indicate that the formation of the O_2^- species B marks the last step in a whole set of consecutive reactions. The first one is given by UV-induced homolytic H_2 splitting which is intimately related to the production of travelling H• radicals (eq.3). It provides the electrons for the formation of the colour centres $F_S{}^+{}_B$(H). One should, however, not forget that H• radicals resulting out of the UV-induced oxidation of hydride I (eq.4) and hydride II contribute to the $F_S{}^+{}_B$(H) formation, too. The above mentioned last reaction step is then the electron transfer from the colour centre $F_S{}^+{}_B$ (H) to O_2 providing the O_2^- species B paralleled by the corresponding process from $F_S{}^+{}_A$(H) to O_2^- species A. — Trace c in

Fig.5 finally shows that the same surface complexes $OH_{(3702)}...O_2^-[A]$ and $OH_{(3692)}...O_2^-[B]$ also occur in the dark by direct oxidation of the irreversibly formed respective hydride [21].

The major issue of the foregoing discussions is the fact that H_2 splitting on MgO surfaces necessarily involves the presence of low coordinated O^{2-} anions — irrespective of whether the MgO sample is exposed to UV irradiation or not. Additional UV diffuse reflectance spectroscopic studies enable one to get a more detailed insight in these parallel surface reactions. In Fig.6 the dotted trace (a) is related to the same type of MgO sample as those used in the IR and EPR studies. It exhibits two resonance absorptions at 230 and 270 nm superimposed on the continuous diffuse reflection background. They have previously been attributed to electronic excitations of 4- and 3-coordinated surface oxygen anions, respectively [12,33,34]. If the MgO sample is exposed simultaneously to H_2 gas and UV irradiation prior to recording the UV spectra, the intensity of the absorption band at 270 nm is significantly reduced (Figs. 6b and c). Obviously the 3-coordinated O^{2-} species that were transformed into electron hole centres according to eq. 2 have reacted with H_2 in terms of a homolytic splitting process (eq 3) which provides a H· radical for colour centre formation (eq.5) and an OH group absorbing in the IR at 3698 cm^{-1} (Fig.3). Curve c (100 mbar H_2) exhibits a stronger reduction of the absorption at 270 nm than curve b (1 mbar H_2) which agrees with the trends observed for all H· radical related reaction products (OH out of homolytic H_2 splitting, $F_S{}^+{}_A(H)$, $O_2^-[A]$) on raising the H_2 pressure.

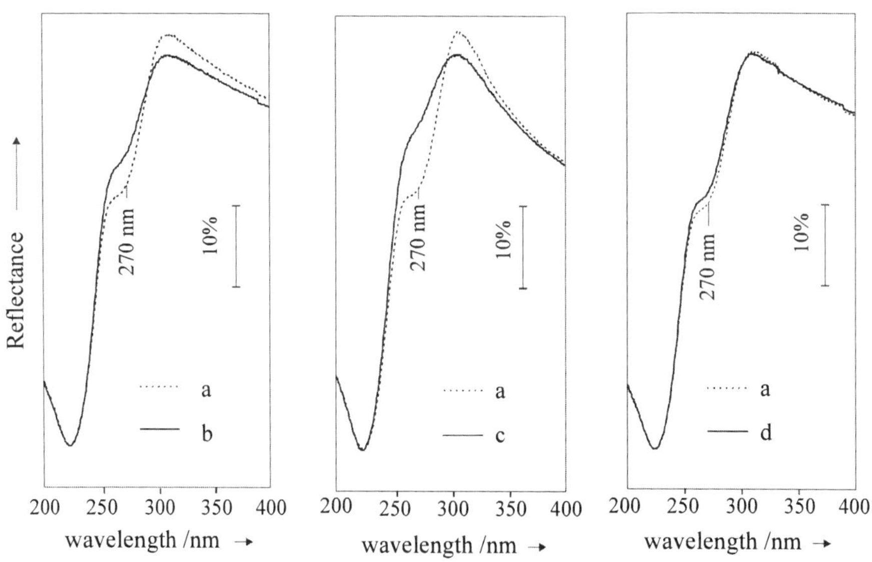

Fig. 6: UV diffuse reflectance spectra of MgO nanoparticles (a) before and after exposition to UV light in the presence of (b) 1 mbar H_2 and (c) 100 mbar H_2. For comparison (d) the spectrum related to the reaction of 1 mbar H_2 in the dark (formation of H_2 chemisorption complex I) is also shown. In order to avoid interference of luminescence in the reflectance measurements, 10 mbar O_2 were added in all experiments (a-d).

If the experimental conditions prior to recording the UV diffuse reflectance spectra are such that exclusively the saturation limited irreversible heterolytic H_2 chemisorption (I) may occur, the intensity loss of the absorption band at 270 nm is small (Fig.6d) as compared to the pretreatment in terms of UV induced homolytic H_2 splitting (Figs. 6b and c). This shows that the majority of the 3-coordinated O^{2-} anions which exhibit a uniform UV excitation at 270 nm, are not able to split H_2 heterolytically in the dark. For this purpose the specific site must — as discussed above — fulfill an additional requirement, namely, of being constituent of a low coordinated anion vacancy.

Fig.6 also provides unambiguous evidence that 4-coordinated O^{2-} anions which give rise to the electronic excitation at 230 nm do not contribute to any irreversible reactions with H_2 — neither under UV irradiation (Fig. 6b and c) nor in the dark (Fig. 6d). The diverse experimental conditions to which the MgO sample is subjected do obviously have no influence on the contour and/ or the position of the band at 230 nm.

All spectra of Fig.6 were recorded after exposing the MgO samples in their specific states to 10 mbar O_2. This is necessary in order to avoid the spectral interference of luminescence of MgO which is efficiently suppressed by O_2 gas. On the other hand, the O_2 addition gives rise to a parallel surface reaction on MgO, namely, the direct oxidation of hydride groups with O_2 (Fig.5c), which leads to the same O_2^- species A and B that were obtained by colour centre oxidation (Fig.5a and b). The relevant issues of the present paper are, however, not affected by hydride oxidation in the dark to any significant extent: the abundance of the OH groups which probe the O_2^- species A and B in the IR spectrum after direct oxidation (Fig.5c, 3702 and 3692 cm^{-1}) is only a fraction of that observed after oxidation of the colour centres (Figs 5a and b). — There is experimental evidence that preadsorption of pure H_2 (no O_2) under conditions where only the chemisorption complex I appears, does not at all affect the UV absorption at 270 nm of MgO under high vacuum. (The minute loss of absorption intensity observed at 270 nm (Fig.6d) should then be explained by the O_2 induced charge transfer from a hydride group to O_2 which also provides an $H^\bullet$ radical for OH formation with 3C O^{2-} anions.)

4. Conclusion

Fig. 7 summarizes the essentials of the different parallel and consecutive reaction steps occurring when MgO nanoparticles are exposed simultaneously to UV light and H_2 gas. There are two different sites involved in heterolytic H_2 splitting, only one site enables homolytic H_2 splitting to occur. Since the hydride groups of both chemisorption complex I and II undergo a UV-induced oxidation, all three reaction steps end up with the production of $H^\bullet$ radicals. These then act as reducing agent for one dominating type of surface anion vacancy to be transformed into colour centres. There is conclusive theoretical [29] and experimental evidence [19,35] that this site has to be described as a so-called double vacancy. As compared to single vacancies, its generation out of an ideal (1,0,0) plane or out of an edge of an ideal MgO cube requires significantly less work against Coulombic attraction. Double vacancies on MgO surfaces are thus likely to be considerably more stable than any other comparable type of site. Therefore, they survive as dominating species the annealing procedure at 1173 K. One type of single anion vacancy also survives, even though less abundant: it is low coordinated and its transformation into colour centre mainly occurs via direct UV induced reduction of an attached hydride group.

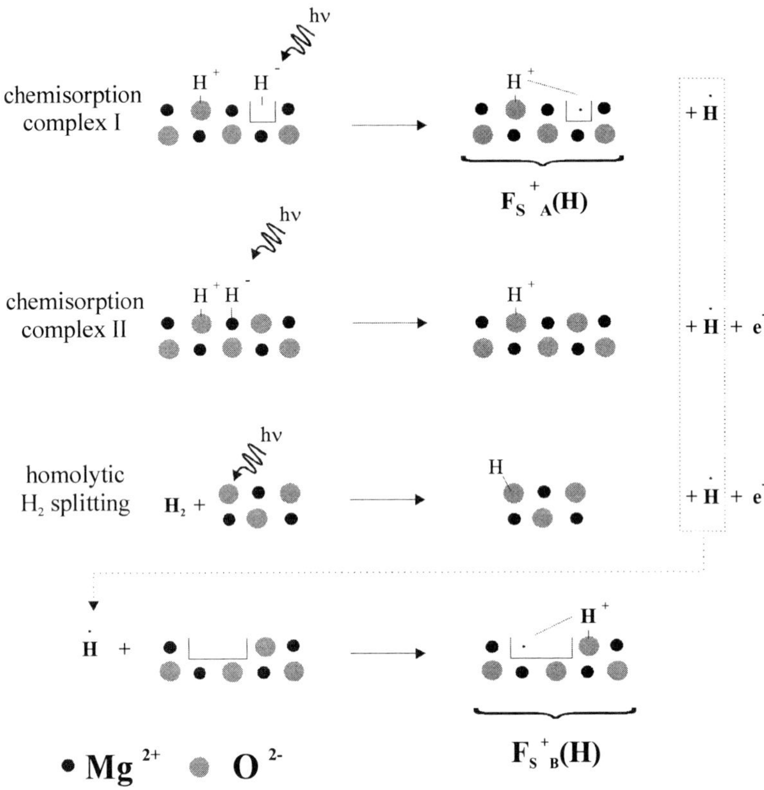

Fig. 7: Reaction scheme for MgO surfaces exposed simultaneously to H_2 gas (>5mbar) and UV light. There are three channels which lead to two different types of colour centres ($F_S{}^+{}_A(H)$ and $F_S{}^+{}_B(H)$).

Fig. 7 does not include the last reaction step in the present report, the bleaching process of the two types of colour centres by O_2 gas. $F_S{}^+{}_A(H)$ is spanned by low coordinated Mg^{2+} cations of a single anion vacancy [10,21]. Different from what an intuitive electrostatic consideration would recommend, the O_2^- is complexed by a vicinal 5C Mg^{2+} [21,36]. This may be explained in terms of a more effective interaction with the closely spaced OH group. On the other hand, the O_2^- produced by bleaching the colour centre in the double vacancy is complexed by one of the 4C or 3C Mg^{2+} cations available in the respective vacancy framework.

MgO and other alkaline earth oxides in pure and doped form play an important role in the catalytic activation of methane [37]. It is likely that one or several of the metal oxide surface sites involved in the H_2 splitting at room temperature are also relevant candidates for methane activation as introductory step for partial oxidation reactions such as oxidative coupling of methane (OCM). Particularly encouraging is the fact that H· radical production occurs on MgO surfaces during H_2 chemisorption under UV irradiation already at room temperature. Similar reaction steps with methane as reactant could provide $CH_3^{\bullet}$ radicals which play a key role in the OCM reaction mechanism [37]. In this case UV activation should considerably reduce the temperature required for the OCM process in the dark.

Acknowledgments

Considerable financial support from the Fonds zur Förderung der wissenschaftlichen Forschung (contract No. P 14731-CHE) and the Max-Buchner-Stiftung (contract No. H-00027/95) is gratefully acknowledged. Furthermore the authors are indebted to Elio Giamello and Salvatore Coluccia for many fascinating discussions on the topic of this paper and other important aspects of life.

References

1. S. Coluccia, M. Baricco, L. Marchese, G, Martra, A. Zecchina, *Spectr. Acta A,* **49**, (1993) 1289
2. E. Knözinger, K.-H. Jacob, S. Singh and P. Hofmann, *Surf. Sci.*, **290** (1993) 388
3. D. Murphy, E. Giamello, *Mol. Engineering* **4** (1994) 147
4. C.G. Harkins, W.W. Shang, T.W. Leland, *J. Phys. Chem.*, **73** (1969) 100
5. S. Coluccia, A.J. Tench, in *Proc. 7th Int.Congr. Catal.*, Tokyo, **1980**, ed.T. Seyama, K. Tanabe, Kodansha, Tokyo, (1981), **B**, 1154.
6. G. Martra, T. Cacciatori, L. Marchese, J.S.J Hargreaves, I.M. Mellor. R.W. Joyner, S. Coluccia, *Catalysis Today*, (2000) in press
7. T. Ito, A. Kawanami, K. Toi, T. Shirakawa, T. Tokuda, *J. Phys. Chem.*, **92** (1988) 3910
8. V. Indovina, D. Cordischi, *Chem. Phys. Lett.*, **43** (1976) 485
9. S. Coluccia, F. Boccuzzi, G. Ghiotti, C. Morterra, *J. Chem. Soc. Faraday Trans.1*, **78**, (1982) 2111
10. E. Knözinger, K.H. Jacob, P. Hofmann, *J.Chem. Soc. Faraday Trans.I*, **89** (1993) 1101
11. R.L. Nelson, J.W. Hale, *Trans. Faraday Soc.*, **52** (1971) 77
12. E.Garrone, A, Zechina, F.S Stone, *Phil.Mag.B*, **42** (1980) 683
13. A. Zecchina, F.S. Stone, *J.Chem. Soc. Faraday Trans.I*, **71**, (1975) 1476
14. S. Coluccia, A.M. Deane, A.J. Tench, *J.Chem. Soc. Faraday Trans.I*, **74** (1978) 2913
15. M. Anpo, Y. Yamada, Y. Kubokawa, S. Coluccia, A. Zecchina, M. Che, J. *Chem. Soc. Faraday Trans.*, **84** (1988) 751
16. V.A. Svets, A.V. Kuznetsov, V.A. Fenin, V.B. Kazansky, *J.Chem. Soc. Faraday Trans.I*, **81** (1985) 2913
17. M. Iwamoto, J.H. Lunsford, *Chem. Phys. Lett.* **1979**, 66, 48, (1979)
18. T. Ito, M. Kato, K. Toi, T. Shirakawa, I. Ikemoto, T. Tokuda, *J. Chem. Soc., Faraday Trans. 1*, **81**, 2835, (1985)
19. M.Sterrer, O. Diwald, E. Knözinger, *J.Phys.Chem.B*, **104** (2000) 3601
20. M.Sterrer, O. Diwald, E. Knözinger, to be published

21. E. Giamello, M.C. Paganini, M. Chiesa, S. Coluccia, G. Martra, D. Murphy, G.Pacchioni, *Surf. Sci.*, **3** (1999) 246
22. O. Diwald, P. Hofmann, E. Knözinger, *Phys.Chem.Chem.Phys,* **1** (1999) 713
23. A. D'Ercole, E. Giamello, C. Pisani, L. Ojamäe, *J. Phys. Chem. B*, **103** (1999) 3872
24. G. Pacchioni, P. Pescarmona, *Surf. Sci.*, **412-413**, (1998) 657
25. O. Diwald, E. Knözinger, G. Martra, *J. Chem. Phys.*, **111** (1999)
26. E. Giamello, M.C. Paganini, D. M. Murphy, A. M. Ferrari, G. Pacchioni, *J. Phys.Chem.*B ,**101** (1997) 971
27. D. Murphy, R.D. Farley, I.J. Purnell, C.C. Rowlands, A. R. Yacob, M.C. Paganini, E. Giamello, *J. Phys. Chem. B*, **103** (1999) 1944
28. D.R. Smith, A.J. Tench, *Chem. Com.*, 1113 (1968)
29. L. Ojamäe, C. Pisani, *J. Chem. Phys.*, **109** (1998) 10984
30. O. Diwald, E. Knözinger, *J. Phys. Chem. B.*, submitted (2001)
31. W. Känzig, M.H. Cohen, *Phys. Rev. Lett.*, **3** (1959) 509
32. M. Che, A.J. Tench, Adv. Catal. **32** (1983) 1
33. P.V. Sushko, A.L. Shluger, *Surf. Sci.*, **421** (1999) 157
34. A.L. Shluger, J.L. Gavartin, M. A. Szymanski, A. M. Stoneham, *Nucl. Instr. and Meth. In Phys. Res. B,* **166-167** (2000) 1
35. J.H. Lunsford, J.P. Jayne, *J. Phys. Chem.* **70** (1966) 3463
36. E. Giamello, D. Murphy, E. Garrone, A. Zecchina, *Spectrochimica Acta*, **49A** (1993) 1323
37. J.H. Lunsford in *Handbook of Heterogenous Catalysis*, (ED. G.Ertl, H. Knözinger, J. Weitkampp, VCH-Weinheim, **1997**), Vol 4, 1843.

Studies in Surface Science and Catalysis 140
A. Gamba, C. Colella and S. Coluccia (Editors)

Intracage chemistry: nitrite to nitrate oxidation via molecular oxygen. A Car Parrinello study.

Ettore Fois[a)], Aldo Gamba[a)] and Gloria Tabacchi[b)]

a) Dipartimento di Scienze Chimiche Fisiche e Matematiche,

Universita' dell'Insubria at Como, Via Lucini 3, I-22100 Como, Italy

b) Max-Planck-Institut für Festkörperforschung

Heisenbergstrasse 1. D-70569 Stuttgart, Germany.

1 Abstract

We present a theoretical study of the oxidation reaction of NO_2^- to NO_3^- by dioxygen in the cages of sodalite. The combined Blue Moon Ensemble and Car Parrinello Molecular Dynamics approaches were used.

Our results indicate the active participation of the zeolite framework in the reaction mechanism via the formation of peroxy-like defects. Moreover a molecular level explanation of the experimentally found first order kinetics is given. A spin-unpolarized Density Functional approach has been adopted for the electron-electron interactions, notwithstanding the triplet state is the ground state for molecular oxygen. However our results suggest that interactions in the zeolite cage may reduce the O_2 triplet-singlet energy gap, therefore justifying the adopted approximation.

2 Introduction

The microscopic description of a chemical reaction in condensed phases is hindered by a series of difficulties that locate the problem at the edge of nowadays physical chemistry.

Many are the issues that challenge researchers in the field: a proper thermal averaging and an accurate description of the quantum many body problem are only two facets of the challenge.

A naive description of the main steps of a reactive event can be sketched: two reactants approach each other in the reaction medium, and the result of such approach (collision) may or may not lead to the products, depending both on the molecular states and reciprocal orientation and the interactions with the medium. While liquid phase reactions are favourite in laboratories, solid matrices are also used as reaction media. Among solids, zeolites, with their arrays of molecular sized cages and channels, are used as *reaction pots* in many synthetic processes both in research laboratories and in large scale industrial plants.

With respect to liquids, reactions in matrices may be easier to study as diffusional and rotational motions of solvent molecules are missing in solids, and moreover, due to their crystalline order, zeolites may be viewed as *simple* environments where a chemical reaction can be investigated microscopically.

In this perspective we have afforded the study, via computer simulation, of an oxidative reaction, $NO_2^- + \frac{1}{2} O_2 \rightarrow NO_3^-$, inside the sodalite cages. Such intracage reaction has been experimentally studied by IR and *uv*-visible spectroscopies, powder X-rays diffraction and thermogravimetry in the temperature range 900-1000 K [1, 2, 3]. It has been proved that (*i*) the reaction occurs in air, (*ii*) the oxidizing agent is molecular oxygen, (*iii*) the reaction is first order with respect to NO_3^- and is completed in 30 hours. However, the reported experiments do not allow a detailed microscopic analysis of the reaction mechanism.

We have studied the intracage oxidation of nitrite sodalite to nitrate sodalite by using the Constrained Molecular Dynamics (Blue Moon Ensemble) method [4, 5] together with the Car Parrinello *ab initio* Molecular Dynamics approach [6]. The application of these two combined techniques has been proved useful when applied to both liquid phase reactions [7, 8] and heterogeneous catalysis [9, 10].

The choice of investigating such reaction stems from two main considerations. Most zeolites-based industrial reactions occur at high temperature in oxidizing conditions such as calcination processes. Therefore the nitrite to nitrate intracage oxidation may give useful indications of general relevance for zeolites technologies in these conditions. The second point is that sodalites with their relatively small number of atoms in the unitary cell can be theoretically studied at high accuracy with an easy computational effort.

We have chosen to adopt perfect pairing of spin in our calculations, therefore one of the reactants, the molecular oxygen, is in the singlet excited state. Such a state is about 100 $KJmol^{-1}$ above the triplet ground state in gas phase. However, in condensed phases, in particular in a highly ionic environment like the studied nitrite sodalite, such approximation may be less drastic with respect to gas phase. We anticipate here that actually interactions with ionic species can significantly reduce the O_2 triplet-singlet gap.

3 Model and Method

It has been found experimentally that both nitrite sodalite $Na_8[Al_6Si_6O_{24}](NO_2)_2$ and nitrate sodalite $Na_8[Al_6Si_6O_{24}](NO_3)_2$ have cubic cells, with lattice parameter larger in the case of the nitrate specimen [1, 3, 11]. Sodalites are porous aluminosilicates characterized by a bcc lattice of cubo-octahedral cavities, the β-cages of many zeolites. Each cage contains four Na^+ ions and one anion, located just in the center of the cage and surrounded by the four cations. The unitary cell contains two of such cages.

In natural sodalite the anion is Cl^-, however many sodalites have been synthesized with different cations and anions. It is known that anions with symmetry lower than spherical, like NO_2^- and NO_3^-, show orientational disorder inside the cavity, namely the crystallographic positions of the anions are not unique. This may be due to either different energy minima or rotational motion.

The reaction occurs in air at high temperature: molecular oxygen diffuses inside nitrite sodalite causing an expansion of the crystals and then giving rise to the oxidized nitrate sodalite (first order with respect to NO_3^-). The same sodalite in N_2 atmosphere does not react, and only the cell expansion is achieved. However the reaction mechanism is not known, indeed, it is quite difficult to devise a simple reaction path. Each NO_2^- is confined in the center of each cage and adjacent cages' centers are separated by more than 7 Å, a distance that corresponds to the separation between nearest NO_2^-'s as well. In order that NO_2^- becomes NO_3^-, the O_2 molecule has to dissociate in atoms and, formally, at the end of the reaction each O atom should be found in different cages to get the NO_3^-'s products. As the dissociation energy of O_2 is 493.7 KJmol^{-1} [12] the simple homolytic event is extremely unlike. Moreover, due to the high ionic character of the species inside the cages, a different reaction path should be favoured with respect to the simple homolytic dissociation. These considerations led us to study the reverse reaction, *i.e.* the $NO_3^- \rightarrow \frac{1}{2}\ O_2 + NO_2^-$ intracage reduction. This choice makes the Blue Moon Ensemble (BME) technique easier to apply. Indeed such approach consists in performing a series of *constrained* Molecular Dynamics simulations, in which a degree of freedom is kept constant (a holonomic constraint) and the constraint force is averaged over each simulation. The degree of freedom is constrained to different values from an initial state to a final state. The integral of the constraint forces corresponds to minus the difference in free energy from the initial to the final state.

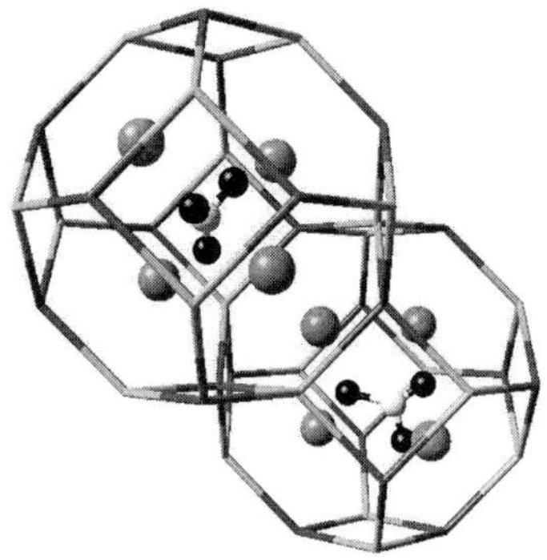

Figure 1: Ball-and-stick reprentation of nitrate sodalite. Black spheres represent oxygens, white spheres represent nitrogens, dark gray spheres represent sodium. The sodalite cages are represented by sticks. Only silicon and aluminum atoms are shown.

Examples of applications of BME have appeared in the literature, both using classical Lagrangean and the the Car Parrinello Lagrangean, such as in the study presented here. We selected the nitrate sodalite as the initial state for the reduction reaction (Figure 1) as it allows a straightforward definition of the reaction coordinate, *i.e.* the constrained degree of freedom, namely one of the NO bond lenghts of NO_3^- in one cage. We performed a series of *ab-initio* Molecular Dynamics simulations with different values of such a distance, constrained by using the RATTLE algorithm [13], starting from its equilibrium value. Moreover other "unconstrained" trajectories were calculated as well: we have simulated the nitrite sodalite, the nitrate sodalite, and the nitrite sodalite with one oxygen molecule in the simulation cell. All simulations were performed with at constant temperature of 1000 K [14] with the Car Parrinello method. Norm-conserving pseudopotentials were used to describe the valence electrons ionic core interactions [15, 16]; the Becke-Perdew gradient corrections to LDA was chosen for the electron-electron interactions [17, 18]. Wavefunctions were expanded in plane waves up to a kinetic energy cutoff of 60 Ry. The Γ point only was used in the Brillouin zone. Experimental lattice parameters were adopted and kept fixed for all simulations. The timestep for the integration of the equations of motion was 0.242 fs and a fictitious mass of 2000 a.u. for the electronic degrees of freedom in the Car Parrinello equations was used. Unconstrained systems were studied for about 4.5 ps, while constrained simulations were continued till convergence in the constraint force was reached, namely 3 to 5 ps depending on the value of the constraint. Ten constrained simulations were perfomed with different values of the constraint.

Notwithstanding the triplet state is the ground state for O_2, we have used a spin-unpolarized functional for a series of reasons, the first and more compelling one being due to the computational cost of using spin-polarized functionals in the case of systems with so many electrons (240 electrons per cell); another reason is that both nitrite and nitrate sodalite (reactant and product) are singlet states. In the gas phase the ground state of molecular oxygen is a $^3\Sigma_g^-$ triplet and the first excited state is a singlet 94.3 KJmol^{-1} higher [19]. Such a difference amounts to 156 KJmol^{-1} with the DFT approximation used here, and to 241 KJmol^{-1} by using the Hartree-Fock approximation, with a triple-zeta + polarization basis set.

The dissociation energy of O_2 is 554 KJmol^{-1} for the triplet and 398 KJmol^{-1} for the singlet dioxygen with the DFT approximation (the experimental value is 493.7 KJmol^{-1}). The calculated gas phase ΔE for the reaction $2NO_2^- + O_2 = 2NO_3$ is -402.2 KJmol^{-1} using a spin-unpolarized functional, rising to -246.2 KJmol^{-1} when a spin-polarized (triplet) functional is used.

However, interactions may contribute to reduce the triplet-singlet energy gap *e.g.* by removing the degeneracy of the highest singly occupied molecular orbitals of ground state oxygen. Test calculations have been carried out in order to substantiate the above statement, and are discussed below.

The nitrite sodalite system was modelled by one cell (lattice parameter 8.923 Å), so that the chemical formula $Na_8[Al_6Si_6O_{24}](NO_2)_2$ corresponds to the atoms in the simulation cell. This choice was adopted also for nitrate sodalite (constrained and unconstrained). The unconstrained nitrite sodalite + O_2 system was simulated by using the cell parameter of nitrate sodalite (lattice parameter 8.996 Å) [3], locating in the starting configuration the O_2 molecule midway between two NO_2^- in the simulation cell.

4 Results

The unconstrained simulation of nitrite sodalite indicates that NO_2^- anions in the β-cages present dynamical disorder, in agreement with experimental data. In particular the simulation showed that the NO_2^- is rotating inside the cavity and this rotational motion is also present at room temperature [20].

The NO_3^- anions, in the unconstrained simulation, show rotational motion as well. However, due to the larger size of NO_3^- anions with respect to NO_2^-, the rotational motion results more hindered. Unfortunately, while for nitrite sodalite all atomic positions have been experimentally resolved, the structure of nitrate sodalite was not completely refined yet, so that no direct comparison with experimental data is possible. However, in this case too, experimental data suggest orientational disorder for the NO_3^- anions inside the β-cages [1].

Interestingly, the unconstrained simulation of nitrite sodalite + O_2 suggested a possible mechanism for the intracage oxidation. The O_2 molecule left soon its starting position (midway between adjacent NO_2^-'s), and diffused towards one of the two NO_2^-,

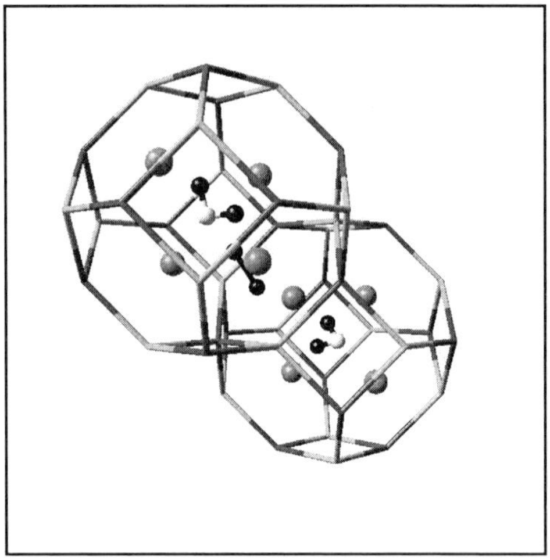

Figure 2: Ball-and-stick snapshot of nitrite sodalite + O_2 from an unconstrained trajectory at 1000 K. Black spheres represent oxygens, white spheres represent nitrogens, dark gray spheres represent sodium. The sodalite cages are represented by sticks. Only silicon and aluminum atoms are shown.

forming a labile complex, namely $[NO_2\cdots O_2]^-$ in equilibrium with the NO_2^- and O_2 species (Figure 2). Such a complex, however, never evolved to the reaction product, *i.e.* one NO_3^- in each cavity, because O_2 never dissociated in the sampled phase space (4.5 ps). But this does not imply that in the simulation conditions the reaction could not occur. The reaction event was not observed during the elapsed time of our simulation due to its low probability: enormously long simulations are needed in order to observe a rare event like a reactive step. However, this simulation is relevant, because it provides information about the structure of the "final state" of the reverse reaction studied by BME. Before examining the constrained simulations, it is worth noticing that the anions in the unconstrained simulations never diffuse from the center of the respective cages. In other words, anions never approach closely to each other. The motion of the nitrate anions in the β-cages can be qualitatively described as a "frictioned" rotational motion, as we observed only two complete rotations of NO_3^- in the elapsed simulation time. This behavior is different from the one observed in nitrite sodalite, where NO_2^- anions rotate around their cage centers on a time scale of about 100 fs. As we are studying the reverse reaction, our "final" NO_2^- should behave in a similar fashion, namely it has to rotate faster than the "initial" NO_3^-.

The initial configuration for the constrained set of simulations was taken from an equilibrated structure of the unconstrained nitrate sodalite. This configuration was slightly modified by increasing the separation of one oxygen atom from the nitrogen in one of the two NO_3^-'s in the cell. In particular, such a distance, from its equilibrium value 1.29 Å, was fixed to 1.5 Å. After equilibration, a trajectory was integrated till the average constraint force did not converge. The final point of this trajectory was used as starting point for a second trajectory with the same NO bond now constrained at 1.75 Å. Such a procedure has been followed for the other constrained simulations. We stopped the BME sampling at a value of the constraint of 4.0 Å, when the reaction was found to occur.

The average constraint forces calculated along the path are reported in table 1. The integral of the constraint forces (with changed sign) is plotted vs the constrained NO distance in Figure 3. Firstly, it should be stressed that the calculated free energy shows unambiguously that the nitrite sodalite + O_2 system is less stable than nitrate sodalite, as shown by experiments. Such a finding make us confident in our computational approach and it is an *a posteriori* validation of the general idea of studying the inverse reaction whenever such a choice is easier to be implemented by means of the BME sampling, with respect to the more conventional approach of following the direct reaction.

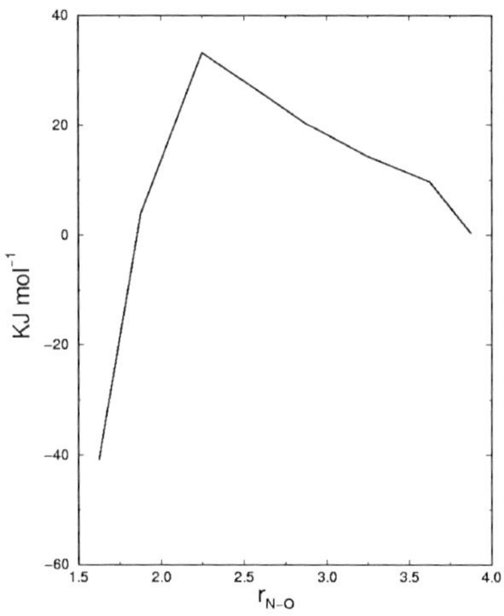

Figure 3: Free energy profile for the simulated reaction *vs* the constrained NO distance. Energy in $KJmol^{-1}$, distances in Å. The free energy zero has been arbitrarily set at a value of the constraint of 4.0 Å. Chemical formula in the panel represent the content of each β-cage at the beginning and at the end of the reaction. The free energy profile refers to a unitary cell (two β-cages).

Table 1: Calculated values of the constraint force f for each constrained distance r. Distances in Å, forces in a.u.

r	1.5	1.75	2.0	2.5	2.75	3.0	3.5	3.75	3.9	4.0
f	-0.0795	-0.0770	-0.0590	0.0142	0.0095	0.0064	0.0027	0.0035	0.011	0.017

Moreover it was possible to calculate from the free energy curve an activation free energy for the oxidative process of the order of 30 $KJmol^{-1}$ and a reaction free energy of about -75 $KJmol^{-1}$.

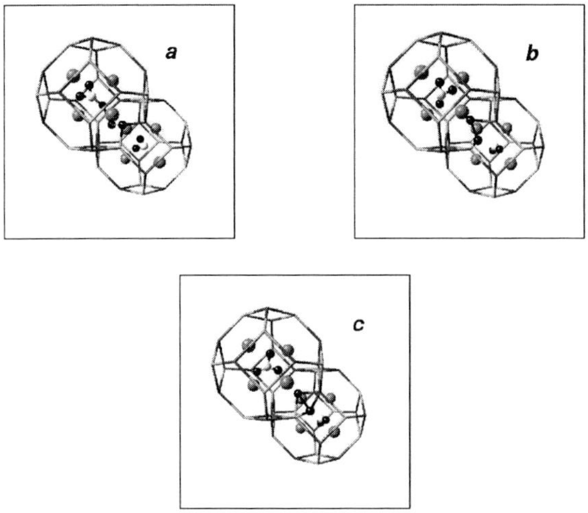

Figure 4: Three ball-and-stick representations of peroxy-like defects. The geometries are taken from the trajectory with the NO constraint fixed at 2.5 Å. Black spheres represent oxygens, white spheres represent nitrogens, dark gray spheres represent sodium. The sodalite cages are represented by sticks. Only silicon and aluminum atoms are shown.

In the first three simulations, with the NO bond constrained at 1.5, 1.75 and 2.0 Å respectively, the constrained NO_3^- behaved quite normally, slowly rotating like a nitrate anion, in line with the other unconstrained anion in the adjacent cavity: at this point the oxygen atom should be still bound to the nitrogen atom. At a value of the constraint of 2.5 Å, the situation changed. First of all, the constraint force changed the sign. This indicates that the forces exerted by the whole system on the constraint have changed sign along the N-O distance, suggesting that the constrained oxygen and nitrogen were no more chemically bonded. In fact, the NO_2^- fraction of the constrained nitrate anion showed a rotational motion faster than that typical of nitrate and more similar to that found for "free" NO_2^- in nitrite sodalite, showing that its motion was quite independent on that of the constrained oxygen.

At this point the constrained oxygen atom should become quite reactive, and as it was still far away from the NO_3^- in the adjacent cage, it could interact strongly with other species, *i.e.* the zeolite framework. Now indeed, the constrained oxygen stays approximately midway between two β-cages, close to the oxygen atom forming the ring shared by two cavities. Actually the constrained oxygen "links" to the framework and forms various kinds of structural defects easily interchanging from one to another. Figure 4 shows some characteristic structures found at this step of the simulation (with the constraint at 2.5 Å). The first defect that we have monitored was a peroxy-defect (Figure 4(a)) where the constrained O is linked to a framework oxygen. This structure then evolves and the constrained oxygen "enters" the sodalite framework. In Figure 4(b), a peroxy-bridge defect is shown, where one O_2 species is placed between an Al and a Si of the frame, each oxygen of the defect being linked to only one of the tetrahedral cations (Si or Al). Figure 4(c) shows another defect, namely an O_2 species that is now shared between two tetrahedral cations, and each oxygen in the defect is linked to both the Al and Si cations.

Care must be taken in naming such defects as *peroxy*. The O_2 species is formed by an oxygen coming from the NO_3^- and one belonging to the framework. During the simulations the total charge of the constrained NO_3^- anion was constantly monitored in order to verify whether the initial negative unitary charge on the nitrate is conserved in the NO_2^- product: this is the case, namely the oxygen that "leaves" the nitrate is neutral, so that the O_2 defect could be considered a real peroxy-species O_2^{-2} formally considering the oxygen coming from the framework to be a closed shell anion O^{2-}.

Recently, some work appeared, where the energetic and geometries of defect centers in zeolites in oxidative conditions were reported [21]. In particular, we quote that the sodalite four-ring system, like the one where we found the peroxy-defects, was studied.

The investigation proved that peroxy-like defects in aluminosilicate structures may have formation energy in the range 40-150 $KJmol^{-1}$. Moreover, their presence is supported by several experimental studies [22, 23, 24, 25].

In the other constrained simulations (constraint respectively at 2.75, 3.0, 3.5, 3.75, 3.9 Å) the formation of this kind of defects has been observed in the whole simulation times, and defects transform each other in a very short time (~100 fs). However for such values of the constraint, the reacting O atom was still bound to the framework. The reactive event occurred only when the constraint was set to an NO distance of 4.0 Å . After few fs, the oxygen previously trapped in the framework defects, left the four ring region and diffused in the adjacent cage colliding with the second NO_3^-. Such collision first led to the transient species $[NO_2 \cdots O_2]^-$, that appeared in the unconstrained nitrite sodalite + O_2 simulation. Then, such species reached rapid equilibrium with the separated NO_2^- and O_2 compounds.

Now the intracage oxidation mechanism can be analyzed. First of all, the finding that the final state of the inverse reduction reaction is similar to the one found in the unconstrained simulation of nitrite sodalite + O_2 strongly supports the idea that the $[NO_2 \cdots O_2]^-$ complex is the first step in the direct reaction (after the O_2 diffusion inside the cavities). Such a complex may activate the molecular oxygen, and actually when the O_2 was complexed to the NO_2^-, its bond length increased and its oscillation became wider, showing a weaker O-O bond (see Figure 5).

Once the O-O bond is weakened one of the oxygen atoms of O_2 molecule can interact with the framework forming the series of defects described above. Such a defect further evolves and the reactive oxygen moves to the adjacent cage where there is still a NO_2^- anion. At this point the atomic oxygen reacts and form a second NO_3^-. This proposed mechanism seems to fit to the first order kinetic with respect to NO_3^-, as described

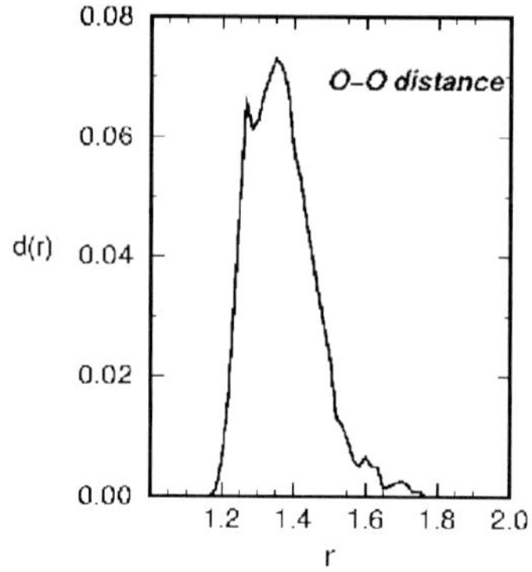

Figure 5: Normalized distribution d(r) of the dioxygen O-O distance r (in Å) in the unconstrained simulation of nitrite-sodalite + O_2 at 1000 K.

in the literature: the formation of the first NO_3^- appears to be the slow step, the trasformation of the labile $[NO_2 \cdots O_2]^-$ complex in a NO_3^- and one O atom linked to the frame should be the rate determining step.

The results presented so far were obtained by using a spin-unpolarized DFT, implying that the O_2 is in the excited singlet state, which is experimentally 94.3 KJmol^{-1} above the triplet ground state (156 KJmol^{-1} within our computational scheme). Were this gas phase energy gap directly transferable to condensed phase systems, the presented results could be biased by the adopted approximation. However, in a highly ionic system like nitrite sodalite, isolated system data may not be pertinent. We have calculated the triplet-singlet energy difference for the isolated $[NO_2 \cdots O_2]^-$ complex,

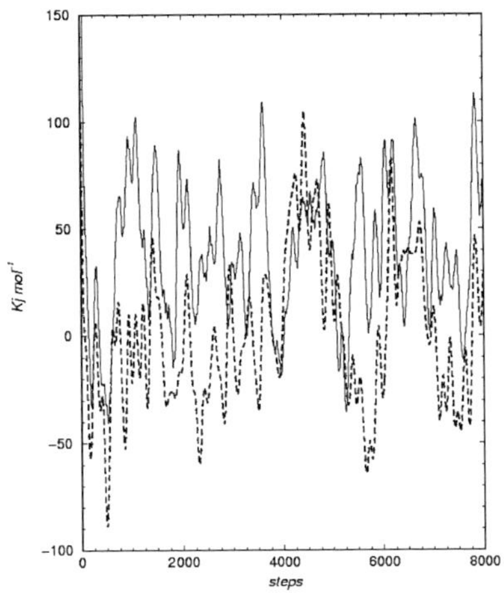

Figure 6: Relative energy of the Triplet state (continous line) vs the Singlet state (dotted line) for the nitrite sodalite + O_2 systems at room Temperature.

by using the same computational scheme for the bulk calculations, for a series of geometries taken from the zeolite trajectories. The differences in energy for such a system reduce to 1-10 $KJmol^{-1}$, always favouring the triplet state. Moreover, we have performed some calculations on the nitrite sodalite + O_2 system by using the spin polarized DFT. In Figure 6 the relative energies for the Singlet and Triplet states at room temperature are reported. The triplet-singlet ΔE's reduce to few $KJmol^{-1}$, but now favouring the singlet state.

Such results indicate that the adopted approximation is a good one for the study of the presented intracage oxidation, and suggest a possible stabilization of singlet oxygen in the nitrite-sodalite system.

5 Conclusions

We have studied the intracage oxidation $NO_2^- + \frac{1}{2} O_2 \rightarrow NO_3^-$ in sodalite by means of the Blue Moon Ensemble and Car Parrinello Molecular Dynamics combined methods. The reaction was simulated by following the inverse reduction process. Such an approach has allowed an easier application of the BME sampling and could be of general interest.

We have found a free energy for the oxidation at 1000 K of the order of -75 $KJmol^{-1}$ and an activation free energy of about 30 $KJmol^{-1}$. A plausible mechanism for this chemical reaction was given, that may explain at microscopic level the phenomenological first order kinetics with respect to NO_3^- found experimentally.

Furthermore, our simulations predict that the sodalite framework is directly involved in the reaction via the formation of defect centers after reacting with the dioxygen.

Beyond the relevance in this particular study, the peroxy-like defects can be reaction intermediates in other zeolites-based oxidations in industrial applications. On the basis of present results, it can be argued that more effective oxidizing media can be obtained by modifying zeolites and mesoporous aluminosilicates in order to allow an easier formation of peroxy-like structures. In this respect, the presented data may suggest a possible activation mechanism of the inert triplet state of dioxygen in the cavities of nitrite sodalite to the more reactive singlet O_2 [26].

In conclusion a microscopic description of a relatively complex chemical reaction "in laboratory conditions" was given. Despite such a description is presented in naive terms, it makes clear, were it necessary, the potentiality of the adopted methods.

References

[1] M.T. Weller, G. Wong, C.L. Adamson, S.M. Dodd, and J.J.B. Roe. *J. Chem. Soc. Dalton Trans.*, page 593, 1990.

[2] J.-C. Buhl. *React. Kinet. Catal. Lett.*, 43:577, 1991.

[3] J.-C. Buhl. *J. Solid State Chem.*, 91:16, 1991.

[4] G. Ciccotti, M. Ferrario, J.T. Hynes, and R. Kapral. *Chem. Phys.*, 129:241, 1989.

[5] E. Paci, G. Ciccotti, M. Ferrario, and R. Kapral. *Chem. Phys. Lett.*, 176:581, 1991.

[6] R. Car and M. Parrinello. *Phys.Rev.Lett.*, 55:2471, 1985.

[7] A. Curioni, M. Sprik, W. Andreoni, H. Schiffer, J. Hutter, and M. Parrinello. *J. Am. Chem. Soc.*, 119:7218, 1997.

[8] E.J. Meijer and M. Sprik. *J. Am. Chem. Soc.*, 120:6345, 1998.

[9] M. Boero, M. Parrinello, and K. Terakura. *J. Am. Chem. Soc.*, 120:2746, 1998.

[10] M. Boero, M. Parrinello, S. Hüffer, and H. Weiss. *J. Am. Chem. Soc.*, 122:501, 2000.

[11] P. Sieger, M. Wiebcke, J. Felsche, and J-C. Buhl. *Acta Cryst.*, C, 47:498, 1991.

[12] D.R. Lide, editor. *CRC Handbook of Chemistry and Physics.* CRC Press, Boca Raron, Florida, 74 edition, 1993-1994.

[13] H.C. Andersen. *J. Comput. Phys.*, 52:24, 1982.

[14] J.Hutter, P.Ballone, M. Bernasconi, P.Focher, E. Fois, S.Goedecker, M. Parrinello, and M. Tuckerman. *CPMD code, version 3.0.* MPI für Festkörperforschung, Stuttgart, and IBM Zürich Research Laboratory, 1990-1996.

[15] L. Kleinman and D.M. Bylander. *Phys.Rev.Lett*, 48:1425, 1982.

[16] N. Troullier and J.L.Martins. *Solid State Commun.*, 74:613, 1990.

[17] A.D. Becke. *Phys. Rev. A*, 38:3098, 1988.

[18] J.P. Perdew. *Phys.Rev. B*, 33:8822, 1986.

[19] G. Herzberg. *Molecular Spectra and Molecular Structure. Vol I - Spectra of Diatomic Molecules. II Ed.* Krieger Publishing Company, Malabar - Florida, 1989.

[20] E. Fois, A. Gamba, and D. Maric. *Il Nuovo Cimento*, 19 D:1679, 1997.

[21] A.A. Sokol, C.R.A. Catlow, J.M. Garces, and A. Kuperman. *J. Phys. Chem. B*, 102:10647, 1998.

[22] K. Yamagishi, S. Namba, and T. Yashima. *J. Phys. Chem.*, 95:872, 1991.

[23] P. Wu, T. Komatsu, and T. Yashima. *J. Phys. Chem.*, 99:10983, 1995.

[24] M. Hunger, J. Käger, H. Pfeifer, J. Caro, B. Zibrowius, M. Bülow, and R. Mostowicz. *J. Chem Soc. Faraday Trans. I*, 83:3459, 1987.

[25] A.R. Leheny, N.J. Turro, and J.M. Drake. *J. Chem. Phys.*, 97:3736, 1992.

[26] C.W. Jefford. In I. Simandi, editor, *Dioxygen Activation and Homogeneous Catalytic Oxidation*, page 555, Amsterdam, NL, 1991. Elsevier Science Publisher.

Studies in Surface Science and Catalysis 140
A. Gamba, C. Colella and S. Coluccia (Editors)

Synthesis, Spectroscopic and Catalytic Properties of Cobalt and Copper Ions in Aluminophosphates with Chabasite-Like Structure. Studies of the NO Reactivity

A. Frache[a], L. Marchese*[a], M. Cadoni[a], S. Coluccia[b], B. Palella[c], R. Pirone[c], P. Ciambelli[d]

[a] *Dipartimento di Scienze e Tecnologie Avanzate, Università del Piemonte Orientale "A. Avogadro", c.so Borsalino 54, I-15100 Alessandria, Italy.*
[b] *Dipartimento di Chimica IFM, Università di Torino, via P. Giuria 7, I-10125 Torino, Italy*
[c] *Istituto di Ricerche sulla Combustione, CNR, P.le Tecchio, 80 Napoli I-80125 Italy*
[d] *Dipartimento di Ingegneria Chimica e Alimentare, Università di Salerno, 84084 Fisciano (SA) Italy*

A combined spectroscopic and catalytic study of the NO reactivity on microporous cobalt-aluminophosphates, copper-aluminophosphates and copper-silicoaluminophosphates with chabasite-related structure, CoAPO-18, CoAPO-34, CuAPO-34 and CuAPSO-34 is reported. CoAPO-18 catalyst was synthesised using N,N-diiso-propylethylamine (DIPE), whereas CoAPO-34, CuAPO-34 and CuAPSO-34 were prepared using morpholine as structure-directing template. NO and CO adsorption on both oxidised and reduced catalysts was monitored by FTIR spectroscopy. These studies revealed that Co^{2+}/Co^{3+} and Cu^{+}/Cu^{2+} redox couples, which should be the locus of the catalytic activity, are present respectively in the cobalt- and copper-containing catalysts. Catalytic tests of both NO oxidation to NO_2 by oxygen and NO reduction by CO were performed. CoAPO-34 catalysts showed exceptionally high performances in the NO oxidation, which were superior of the well-known Cu-ZSM-5, and a very poor activity in the NO reduction. A different behaviour was found for CuAPO-34 catalysts which, beside a good activity in the NO oxidation to NO_2, showed a very high activity in the NO reduction by CO when oxygen was absent in the reaction feed.

1. Introduction

The NO_x abatement in gas emissions from both fixed and mobile sources is a fundamental, and urgent, ecological need, which may be tackled by means of catalytic processes [1]. Owing to such a relevance, an expanding amount of new materials are now-a-days continuously prepared, and among these, zeolite-based ones are of growing interest [1-3]. In this context, metal-containing aluminophosphates and silicon-aluminophosphates with redox properties [4-5], appear attractive, although they have not been used extensively for reactions involving NOx. Good catalytic performances of Cu-SAPO-34 in selective reduction (SCR) of NO with propene [6], Cu-Me-AlPO-11 (Me = Mg^{2+} and Zn^{2+}) in NO decomposition [7] and of CoAPO-34 in NO oxidation to NO_2 [8], were reported. However, the examples of catalytic studies concerning the use of these new zeotype aluminophosphate-based materials in NOx reactions are limited although spectroscopic studies of the NO adsorption on CoAPO-18 and CoAPO-5 [9], CoAPO-34 [8] and Cu-SAPO-34 [10] enlightened the potential of these catalysts.

CoAPO-18 catalysts are microcrystalline aluminophosphates with AEI framework structure [11,12] which have been synthesised by using N,N-diiso-propylethylamine (DIPE) as structure-directing template. CoAPO-34 catalysts have a framework structure which is related to that of the natural chabasite (CHA structural code), and have been synthesised by

* to whom correspondence should be addressed (Dipartimento di Chimica IFM quoted above) – marchese@ch.unito.it

using several types of amines [13] including morpholine, a molecule which addresses very selectively the synthesis to CHA-type structures [8,14].

CoAPO-18 catalysts were first well characterised by spectroscopic tools and it was found that Co^{2+}/Co^{3+} redox couples are present in their zeolite-type framework [12,15-17]. Subsequently, these catalysts were tested in NO oxidation to NO_2 by oxygen and found that they have good catalytic performances [8,9]. Similar FTIR and catalytic results were also obtained for CoAPO-34 catalysts [8]. However, when probed in NO reduction by CO, both these catalysts revealed a surprisingly poor activity, and this prompted us to shift our attention to Cu-containing aluminophosphates.

In this study we report on a comparison between catalytic and spectroscopic properties of cobalt-containing (CoAPO-34 and CoAPO-18) and copper-containing (CuAPO-34 and CuAPSO-34) catalysts. Whilst copper-containing aluminophosphates monitored so far in NOx reactions were prepared via ion-exchange procedure [6,7], in the present work CuAPO-34 and CuAPSO-34 were prepared via one-pot synthesis in which copper salts were added directly in the synthesis gels of the zeolite-type materials. We shall demonstrate that this methodology is more convenient and leads to very active catalysts.

2. Experimental

Synthesis

CoAPO-18 materials were synthesised by using N,N-diiso-propylethylamine (DIPE) as structure-directing template and following the receipt reported in ref. [16]. CoAPO-34 catalysts were prepared by using morpholine and following the procedure reported in ref. [8]. CuAPO-34 and CuAPSO-34 were synthesised by adding CuO at a diluted solution of ortophosphoric acid and, after stirring at 80°C to a complete dissolution, $Al(OH)_3$ was added at room temperature with vigorous stirring until a homogeneous gel was obtained. Subsequently, SiO_2 (only for CuAPSO-34), morpholine and, in some cases, HF were added always under stirring. The resulting gels were crystallized in a teflon-lined autoclave under autogeneous pressure at 190°C for 7-10 days. The crystalline products were filtered, washed with water and dried in open air. Pure, and highly crystalline, phases of CuAPO-34 with ALPO-34-type structure [18] and CuAPSO-34 with SAPO-34-type structure [14,19] were obtained. Table 1 shows the gel compositions used for the crystallisation.

The metal loading in the final products was determined by using an atomic emission spectrometer with an ICP source (JY - mod. 23): Co 1.84 wt %; Cu 2.07 wt % in CuAPO-34 and 2.14 wt % in CuAPSO-34.

Spectroscopic measurements

FTIR experiments on pelletised samples were recorded with a Bruker IFS88 spectrometer at a resolution of 4 cm^{-1}, and by means of specially designed cells which were permanently connected to a vacuum line (ultimate pressure $\leq 10^{-5}$ Torr) to make adsorption-desorption experiments. NO and CO adsorptions were performed at room temperature (RT). The pre-treatments to activate the catalysts were: a) heating at 5°C/min under vacuum up to 550°C; b) heating at 500°C for 1-2 hours; c) heating at 550°C for 20-25 hours under oxygen (100 Torr), and by changing the oxygen two-three times; d) outgassing at 550°C for 1-2 hours. These treatments leaded to cobalt-containing catalysts, in which both Co^{2+} and Co^{3+} ions were contemporarily present within the zeolite framework [9,12,15]; hereafter named *oxidised catalysts*. However, it is known that by outgassing at high temperature copper-containing zeolites, a significant fraction of copper ions are in reduced, Cu^{+}, state, whereas heating at

400-500°C in hydrogen was needed to get all cobalt ions of CoAPOs in divalent state: hereafter named *reduced catalysts*.

Catalytic tests

Catalytic activity measurements were carried out in a flow apparatus by using a quartz fixed bed micro-reactor. The standard conditions adopted for the feed of the NO oxidation were: NO (2000 ppm); O_2 (5% vol.) and balance He; Weight/Feed ratio (W/F) = 0.042 g.s.Ncm^{-3}. The conditions for the NO reduction were: NO (5000 ppm), CO (6000 ppm) and balance He; W/F = 0.12 g·s·Ncm^{-3}. NO, NO_2, N_2O, CO, CO_2 and O_2 were continuously monitored. All as-synthesised products were treated at 550°C heating the samples at 5°C/min under helium flow, and then after 5 hours, switching the gas flow to oxygen. After a further 60 hours at 550°C, the samples were allowed to cool to room temperature (*Oxidised catalysts*). The cobalt-containing catalysts were also treated at 600°C under H_2 flow (heating rate 5°C/min) for 12 hours, whereas the copper-containing catalysts at 600°C under He flow (*Reduced catalysts*).

Table 1: Molar composition of the gels used for the crystallization.

Samples	Me	Al	P	Si	Templ.	HF	H_2O
CoAPO-34	0.04	0.96	1.00	—	1.25[a]	0.35	50
CuAPO-34	0.04	0.96	1.00	—	1.25[a]	0.35	50
CoAPSO-34	0.04	0.96	1.00	0.20	1.35[a]	—	30
ALPO-34	—	1.00	1.00	—	1.25[a]	0.35	50
CoAPO-18	0.04	0.96	1.00	—	0.85[b]	—	25

[a] Morpholine; [b] N,N-diiso-propylethylamine (DIPE)

3. Results and Discussion

Spectroscopic measurements

CO adsorption on CoAPO-18 and CoAPO-34 catalysts were followed and the fraction of surface Co^{2+} ions which can be oxidised to Co^{3+} ions under thermal treatments defined [8,12]. CO linearly adsorbed on Co^{2+} ions [Co^{2+}(CO) complexes], mainly located in framework positions [12], were obtained and their absorption fell at 2080-2085 cm^{-1}. These bands decreased 60 to 40% of the original values after oxidation meaning that a significant fraction of Co^{2+} could be oxidised to Co^{3+} ions. It was proposed that the active species in the catalytic oxidation of NO by oxygen were the oxidable Co^{2+} ions [8]. Fig. 1 shows the adsorption upon different CO dosages at RT on reduced CoAPO-34 catalysts. A band at 2080 cm^{-1}, which gradually decreased with the CO pressure and completely disappeared upon evacuation at room temperature, was observed. At lower CO dosages the band shifted to higher wavenumbers and became broader, signifying that various type of Co^{2+} Lewis acid centres are present in the catalysts [12]. These centres may be either in different framework positions, and this well agrees with the fact that Co^{2+} ions can substitute for various Al ions in the AlPO-34 structure [18], or in extra-lattice defects.

NO adsorption at RT on both reduced (Fig. 2a,c) and oxidised (Fig. 2b,d) CoAPO-18 and CoAPO-34 materials produced, though of different amount, dinitrosyl complexes on Co^{2+} ions (bands in the range 1950-1750 cm^{-1} in FTIR spectra). Particularly on the oxidised catalyst

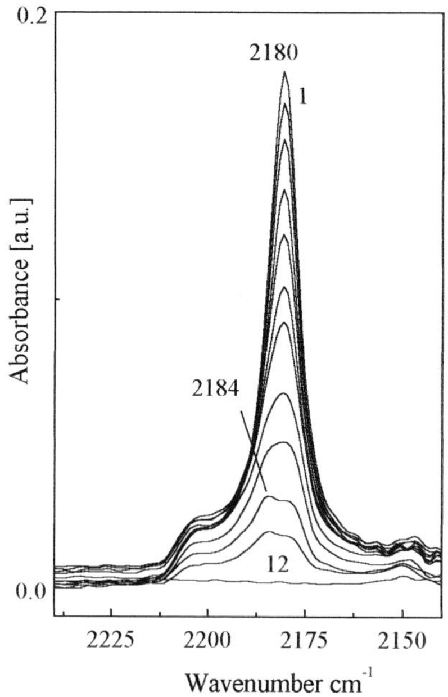

Fig. 1 - FTIR spectra of CO adsorbed at RT on reduced CoAPO-34. Curves 1-11: decreasing CO pressure from 100 to 0.5 Torr. Spectrum 12 was recorded after evacuation at RT for 5 min.

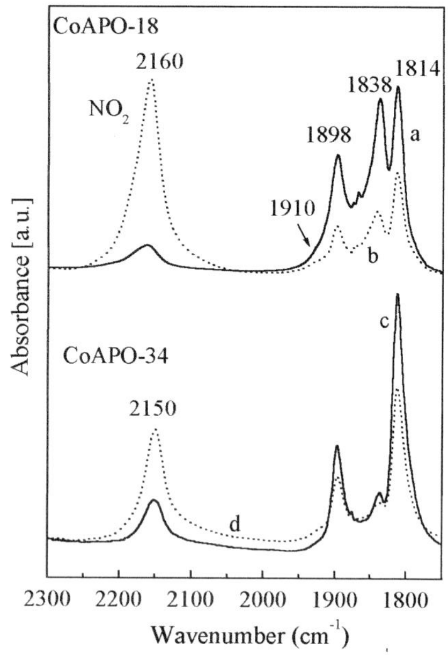

Fig. 2 - FTIR spectra of NO adsorbed at RT on reduced (a,c) and oxidised (b,d) CoAPO-18 and CoAPO-34 catalysts.

where, beside Co^{2+}, a significant fraction of Co^{3+} ions were present [15-17], NO_2 was also formed (bands at 2160 and 2150 cm^{-1}), and some Co^{3+} ions were contemporarily reduced to Co^{2+} [9]. The oxidation of NO by oxygen at RT on reduced CoAPO-18 was also followed by FTIR spectroscopy, and it was found that dinitrosyl complexes were efficiently transformed into NO_2. Similar results were also obtained for CoAPO-34. CO adsorption at room temperature on CuAPO-34 (Fig. 3) produced two sharp bands at 2175 and 2147 cm^{-1} (Fig. 3, curve 1) which gradually decreased in intensity with the CO pressure until, after evacuating the sample at RT (curve 5), a new sharp band at 2152 cm^{-1} was formed. Similar results were obtained for Cu-ZSM-5 and suggested that $Cu^{+}(CO)_2$ dicarbonyl species were formed at higher CO dosages [20-22], and that the dicarbonyls converted into monocarbonyl $Cu^{+}(CO)$ complexes at lower CO dosages [20-24]. When compared to the $Co^{2+}(CO)$ complexes formed on cobalt-containing aluminophosphates, the $Cu^{+}(CO)$ species are much more stable upon evacuation at RT, since in these latter complexes back π-donation is more significant. CO adsorption on CuAPSO-34 led to similar complexes (Fig. 3, top curves), however, significant differences were observed for the two samples. Bands of dicarbonyls and monocarbonyls, whose wavenumbers are reported in table 2, are much broader and more intense in case of CuAPSO-34 (the integrated intensity being around 4 times larger). This signifies that in CuAPSO-34 samples there is a higher concentration of accessible Cu^{+} centres, which are likely to be located in a larger variety of sites [25].

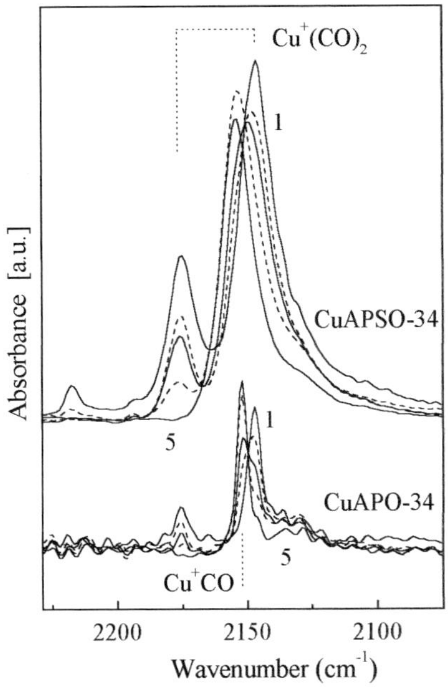

Fig. 3 - CO adsorbed at RT on CuAPO-34 and CuAPSO-34. Spectra 1-4 were recorded after decreasing the pressure from 100 to 1 Torr. Spectra 5 for both samples were recorded after evacuation at RT for 20 minutes.

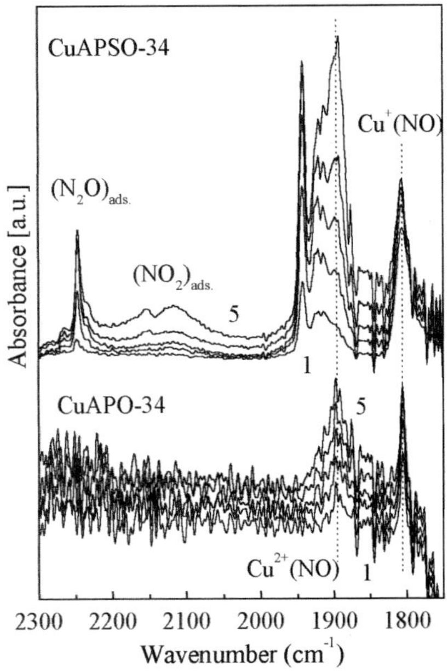

Fig. 4 - NO adsorbed at RT on CuAPO-34 and CuAPSO-34. Spectra 1-5 were recorded after increasing gradually the pressure from 0.5 to 50 Torr.

This is not a surprising result as the presence of Si in the SAPO-34-type structure led to negative charges in the framework [14,17] which could be compensated by copper ions during the zeolite synthesis, similarly to what was obtained by ion-exchange procedure [6,10,25]. By considering that the copper loading is similar in the two samples, it may be additionally inferred from the spectroscopic results that in CuAPO-34 a more significant fraction of copper ions is not accessible to CO adsorption. This is an indirect evidence of the abundance of Cu^{2+} buried in the zeolite framework.

Table 2: Surface species formed by CO adsorption on copper-containing aluminophosphates, and their stretching frequencies [cm^{-1}].

samples	$Cu^+(CO)_2$ [$\nu_{sym.}$]	$Cu^+(CO)_2$ [$\nu_{asym.}$]	$Cu^+(CO)$
CuAPO-34	2175	2146	2154
CuAPSO-34	2175	2147	2152

NO adsorption on CuAPO-34 (fig. 4) led mainly to two weak bands at 1805 cm^{-1}, which is sharper and did not change significantly with the pressure, and at 1895 cm^{-1}, which is broader and has an intensity which changed with both NO pressure and time of contact. Bands in similar positions, and with a similar behaviour, have been observed frequently in Cu-containing zeolites and assigned to mono-nitrosyls adsorbed on Cu^{+} and Cu^{2+} (see table 3) respectively [10,22-24,26,27]. Besides the mono-nitrosyls absorbing at 1805 and 1895 cm^{-1}, new species are formed upon NO adsorption on CuAPSO-34 (Fig. 4). The band intensities of the new species depended by both the NO dosage and time of contact at a fixed NO pressure. The latter effect is not shown in Fig. 4 for sake of brevity. Some of these bands could be assigned with precision on the basis of the data reported for Cu-ZSM-5 catalysts [22,24,26,27] (see details in Fig. 4 and table 3). However, IR features at 1942 and 1920 cm^{-1} seem to be peculiar of the aluminophosphate-based catalysts as, to our knowledge, they have not been observed on other systems. Owing to their position, these new bands should be tentatively assigned to nitrosylic species bonded to extra-framework Cu^{2+} ions in low coordination. More systematic experiments are in progress to define the precise nature of these bands. Whilst the bands in the region 2200-2100 cm^{-1}, due to linearly adsorbed NO_2 species [26], increased constantly with the NO dosage, in the last stages the band at 2247 cm^{-1}, due to adsorbed N_2O [26,27], started to decrease. This behaviour supports the proposal that both this species are intermediates of the NO decomposition to N_2 and O_2 [26,27].

Finally, it may be noted that the band at 1805 cm^{-1} of the mono-nitrosyl complexes, $Cu^{+}(NO)$, is more intense and broader for the CuAPSO-34 catalyst and these are further evidence that Cu^{+} sites in this sample are more abundant and more heterogeneously distributed.

Table 3: Surface species formed by NO adsorption on copper-containing aluminophosphates, and their stretching frequencies [cm^{-1}].

Samples	$(N_2O)_{ads.}$	$(NO_2)_{ads.}$	$Cu^{x+}(NO)_y$ ([a])	$Cu^{2+}(NO)$	$Cu^{+}(NO)$
CuAPO-34	—	—	—	1895	1805
CuAPSO-34	2247	2154 2117	1942, 1920	1895	1805

([a]) Nitrosyl complexes of un-known nature

Catalytic results

The activity of the CoAPOs catalysts in the oxidation of NO to NO_2 [$2\ NO + O_2 \leftrightarrow 2\ NO_2$] was primarily evaluated in view of the possible applications in the reduction of NO_x with hydrocarbons or CO [28-30]. NO_2 is known, in fact, to be intermediate in the NO selective reduction with hydrocarbons on copper-exchanged zeolites [31-33].

Fig. 5 shows that AlPO-34 is not active at all in NO oxidation in the experimental conditions investigated. The very low conversion obtained should be attributed to the homogeneous reaction. A low activity was also observed for the pre-oxidised CoAPO-34 catalyst, the maximum NO conversion being only above 450°C. This suggested that Co^{2+} sites were the main active species in NO oxidation and, in fact, the NO oxidation was much higher for both pre-reduced CoAPO-18 and CoAPO-34 catalysts. The equilibrium value was reached, for both catalysts, at temperatures lower than 400°C, and this represents a very good performance. The activity of the reduced CoAPOs is even more surprising by considering that in comparison to the most active Cu-ZSM-5 catalysts, lower NO inlet concentration (2000 ppm instead of 5000 ppm) and W/F ratio (0.042 instead of 0.33 g·s·Ncm^{-3}) were used [34].

Fig. 6 shows the results of catalytic tests in NO oxidation exhibited by reduced MeAPO-34 (Me = Co and Cu) and CuAPSO-34 catalysts. It is evident that both copper- and cobalt-based aluminophosphates are very active in NO oxidation to NO_2 although CoAPO-34 catalyst is more active than the Cu-containing ones with similar metal loading. In addition, the greater activity exhibited by CuAPSO-34 catalyst in comparison to the CuAPO-34 one can be explained assuming that the insertion of silicon ions into the aluminophosphate framework favours the formation of species more active towards the NO oxidation. This thesis is in good agreement with the spectroscopic data of CO and NO adsorption on CuAPSO-34 which, in this case, are characterised by more intense bands of carbonyls and nitrosyl complexes (Figs. 3-4).

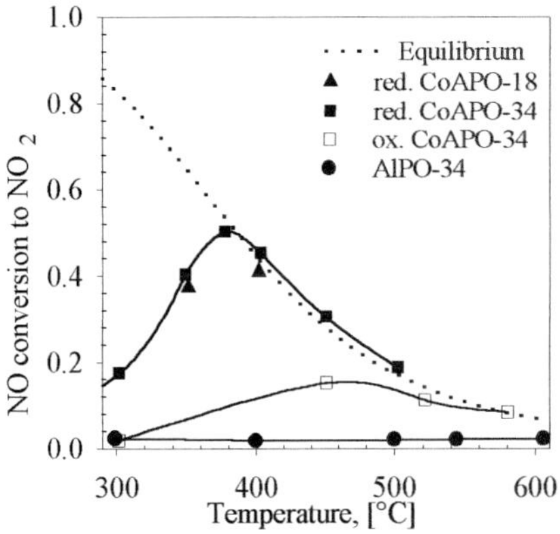

Fig. 5 - NO conversion to NO_2 on CoAPOs catalysts after different pre-treatments. Dotted line represents the e-quilibrium conversion. Feed: NO (2000 ppm), O_2 (5% vol.) and balance He. W/F= 0.042 g·s·Ncm^{-3}.

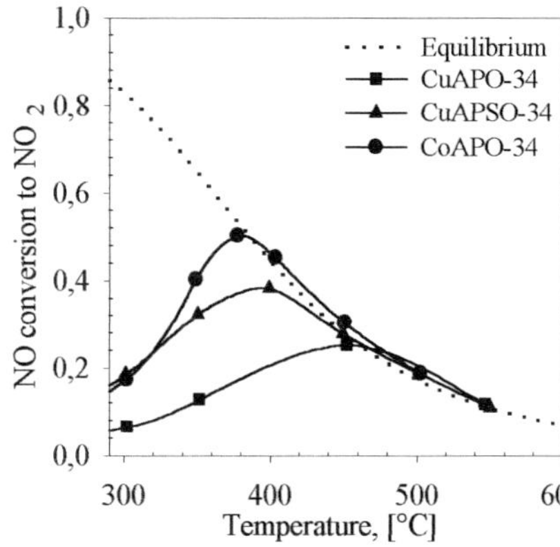

Fig. 6 - NO conversion to NO_2 on MeAPO-34 (Me=Co, Cu) and CuAPSO-34 catalysts. Feed: NO (2000 ppm), O_2 (5% vol.) and balance He. W/F= 0.042 g·s·Ncm^{-3}.

Fig. 7 shows the results of catalytic tests in the reduction of NO with CO, in the absence of O_2, for reduced CuAPO-34 catalyst and suggests that this is a very active catalyst with a high selectivity (>80%) at all temperatures in the 300-550°C range. This is a very interesting result especially if the activity of CuAPO-34 is compared with that of a copper exchanged ZSM-5 catalysts. In fact, the reaction rate measured at 400°C on CuAPO-34 was 1.16 μmol·s^{-1}·g^{-1}, a value about 20 times greater than that observed on a Cu-ZSM5-40 catalyst (40 indicates the SI/Al ratio) with a similar Cu content (1.72% wt.) in the same experimental conditions.

These promising results were not confirmed when an even little amount of oxygen (5000 ppm) was used in the reaction feed. In these conditions,in fact, the only reaction was the oxidation of CO to CO_2 with a total conversion of CO beginning from 450°C. Catalytic tests on CuAPSO-34 systems are in progress.

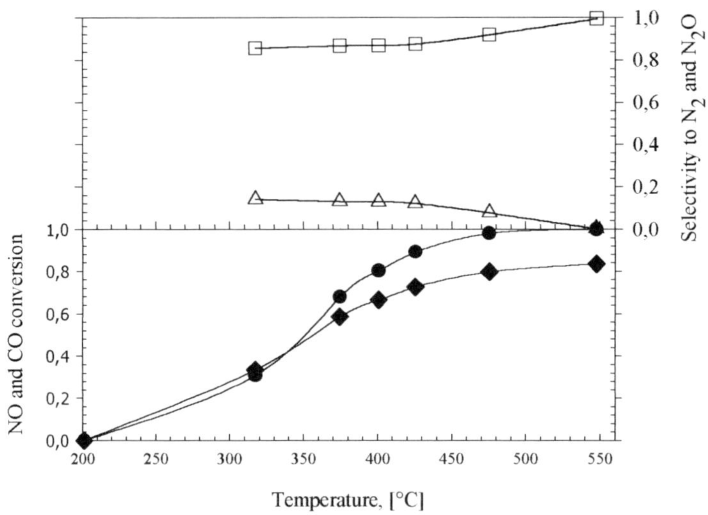

Fig. 7 - NO (●) and CO (◆) conversions in NO+CO reaction over CuAPO34 catalyst. The selectivity to N_2 (□) and N_2O (△) is also reported on the top of the figure. Feed: NO (5000 ppm) and CO (6000 ppm), and balance He. W/F= 0.12 g·s·Ncm^{-3}.

Acknowledgements:
This work has been realised with the contribution "Cofinanziamento 1998" provided by the Italian MURST, "Ministero dell'Università e della Ricerca Scientifica e Tecnologica". The authors thank Dr. F. Soggia for the elemental analysis.

References

1. V.I. Pârvulescu, P. Grange and B. Delmon, Catal. Today, 46 (1998) 233.
2. J.N. Armor, Micropor. Mesopor. Mater., 22 (1998) 451.
3. Y. Traa, B. Burger and J. Weitkamp, Micropor. Mesopor. Mater., 30 (1999) 3.
4. M. Hartman and L. Kevan, Chem. Rev., 99 (1999) 635.
5. J.M. Thomas, Angew. Chem. Int. Ed. Engl., in press.
6. T. Ishihara, M. Kagawa, Y. Mizuhara and Y. Takita, Chem. Lett., (1992) 2119.
7. J. Dedecek, J. CejKa and B. Wichterlová, Appl. Catal. B, Environm., 15 (1998) 233.
8. L. Marchese, E. Gianotti, G. Martra, S. Coluccia, B. Palella, R. Pirone and P. Ciambelli, contribution to the 12th ICC meeting, Granada, Spain, 9-14 July, 2000.
9. E. Gianotti, L. Marchese, G. Martra, S. Coluccia, Catal. Today , 54 (1999) 495.
10. D. B. Akolekar, S.K. Bhargava and K. Foger, J. Chem. Soc., Farad. Trans., 94 (1998) 155.
11. A. Simmens, L.B. McCusker, C. Baerlocher and W.M. Meier, Zeolites 11 (1991) 654.
12. L. Marchese, G.Martra, N. Damilano, S. Coluccia, J.M. Thomas, Stud. Surf. Sci. Catal., 101 (1996) 861.
13. S.T. Wilson, S.O. Oak, E.M. Flanigen and W. Plains, U.S. Pat. 4,567,029, (1986).
14. Marchese, L., Frache, A., Gianotti, E., Martra, G., Causà, M., Coluccia, S., Micropor. Mesopor Mater., 30 (1999) 145.

15. J.M. Thomas, G.N. Greaves, G. Sankar, P.A. Wright, J. Chen, A.J. Dent, L. Marchese, Angew. Chem. Int. Ed. Engl., 33 (1994) 1871.
16. L. Marchese, J. Chen, J.M. Thomas, S. Coluccia, A. Zecchina, J. Phys. Chem., 98 (1994) 13350.
17. S. Coluccia, L. Marchese and G. Martra, Micropor. Mesopor. Mater., 30 (1999) 43.
18. M.M. Hardening and B.M. Kariuki, Acta Cryst., C50 (1994) 852.
19. M. Ito, Y. Shimoyama, Y. Saito, Y. Tsurita, M. Otake, Acta Cryst., C41 (1985) 1698.
20. G. Spoto, A. Zecchina, S. Bordiga, G. Ricchiardi, G. Martra, G. Leofanti and G. Petrini, Appl. Catal. B: Environm., 3 (1994) 151.
21. A. Zecchina, D. Scarano, G. Spoto, S. Bordiga, C. Lamberti and G. Bellussi, Stud. Surf. Sci. Catal., 117 (1998) 343.
22. M.V. Konduru and S.S.C. Chuang, J. Phys. Chem. B, 103 (1999) 5802.
23. K. Hadjiivanov, M. Kantcheva, D. Klissurski, J. Chem. Soc. Farad. Trans., 92 (1996) 4595.
24. Y. Kuroda, R. Kumashiro, T. Yoshimoto and M. Nagao, Phys. Chem. Chem. Phys., 1 (1999) 649.
25. J. Dedecek, B. Wichterlová, Kubát, Micropor. Mesopor. Mater., 32 (1999) 63.
26. E. Giamello, D. Murphy, G. Magnacca, C. Morterra, Y. Shioya, T. Nomura and M. Anpo, J. Catal., 136 (1992) 510.
27. G. Spoto, S. Bordiga, D. Scarano and A. Zecchina, Catal. Letters, 13 (1992) 39.
28. Y.Li, J.N. Armor, J. Catal., 150 (1994) 376.
29. W-X Zhang, H. Yashiro, M. Iwamoto, J. Chem.Soc., Faraday Trans., 91 (1995) 767.
30. D.B. Lukyanov, E.A. Lombardo, G.A. Sill, J.L. d'Itri, W.K. Hall, J. Catal., 163 (1996) 447.
31. P. Ciambelli, P. Corbo, M. Gambino, F. Migliardini, Stud. Surf. Sci. Catal., 116 (1998) 307.
32. M. Iwamoto and H. Hamad, Catal. Today, 10 (1991) 57.
33. M. Guyon, V. Le Chanu, P. Gilot, H. Kessler and G. Prado, Appl.Catal. B : Environmental, 8 (1996) 183.
34. R. Pirone, P. Ciambelli, G. Moretti and G. Russo, Appl. Catal. B: Environmental, 8 (1996), 197.

Studies in Surface Science and Catalysis 140
A. Gamba, C. Colella and S. Coluccia (Editors)

Highly dispersed CaO in mesoporous silica as efficient trap for stabilizing azo dyes

Z. Gabelica[a], M. Shibata[b], H. Hotta[b], T. Suzuki[b] and S. Valange[a,c]

[a] Université de Haute Alsace & ENSCMu, Groupe Sécurité et Ecologie Chimiques, 3, Rue A. Werner, F-68094, Mulhouse Cedex, France
[b] Kao Corp., Tokyo Research Lab., Sumidaku, Tokyo 131-8501, Japan
[c] Present address : LACCO, UMR CNRS 6503, ESIP, 40, Av. du Recteur Pineau, F-86022 Poitiers, France

High surface area MCM-41 mesophases synthesized in the presence of Ca^{2+} ions proved efficient hosts for bulky dye molecules like Lithol Rubine B (LRB). The resulting mesophase-dye composites showed remarkably high light fastness qualities, when compared to Ca-exchanged Al-MCM-41 aluminosilicates, to non composite pigments or to smectite-dye composites. The marked resistance to sunlight fading of such systems was attributed to an optimized Ca-dye interaction favored by a controlled heating of the $Ca(OH)_2$-loaded mesoporous precursors. It was shown that a complete interaction, as found in commercial pigments (aggregates of the Ca salt of LRB), is not appropriate to generate an efficient composite and that a good dispersion and retention of Ca^{2+} ions onto a large surface of a mesophase is a key condition for the composite to show the highest anti photo fading properties. A mechanism involving the dual interaction of CaO particles with the acidic sites located on both the substrate mesophase and the dye molecule was suggested. Any high temperature annealing of the CaO-MCM-41 substrate generates Ca silicate islands where Ca^{2+} ions are more completely surrounded by oxygen ligands and therefore less available for any particular interaction with the dye molecules.

1. INTRODUCTION

Important research was devoted to find appropriate conditions for an efficient encapsulation/anchoring of azo dyes in/on various inorganic substrates in order to improve their light-fastness qualities and water or oil resisting properties. The most widely used inorganic materials to elaborate such dye composites were clay minerals, e.g. smectites [1] or hydrotalcite [2], but also mesoporous silica based materials [3, 4].

We have shown previously [3] that various mesoporous aluminosilicate MCM-41 type materials can behave as efficient supports for bulky acid dye molecules, like Lithol Rubine B (LRB) (3-hydroxy-4-((4-methyl-2-sulphophenyl)azo)-2-naphthalenecarboxylic acid, see Fig. 1).

It was observed that the mesoporous materials synthesized in the presence of Ca^{2+} salts under appropriate pH conditions, yielded composites that showed the most efficient resistance to fading. On the other hand, when Ca^{2+} cations were introduced by classical ion exchange of mesoporous (Na,H)-aluminosilicates, the resulting composites were less resistant to sunlight. Ca^{2+} cations were known to behave as "connectors" between the acid dye and the inorganic substrate [2] but a mechanism describing such interaction was never proposed. Our further research showed that a one-step type preparation of mixed oxidic $CaO-SiO_2-Al_2O_3$

mesostructures yielded more efficient composites [4]. We have suggested a new way for Ca^{2+} ions to interact with the dye molecule and emphasized the actual role of the substrate. It was in particular proposed that a complete Ca-dye interaction was not sufficient to generate an appropriate composite and that probably a good dispersion and retention of Ca^{2+} ions onto a large surface is also a key condition for the composite to show a marked resistance to fading. It was also concluded that when Ca^{2+} ions do incorporate the mesophase framework upon extensive heating, the CaO-MCM-41 system, less efficient composites were obtained.

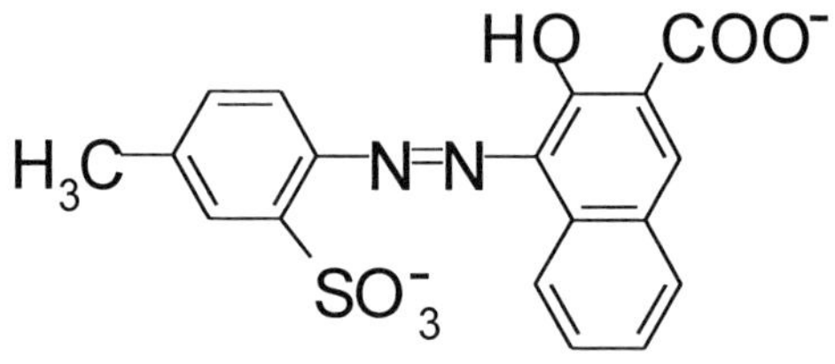

Fig.1. Chemical formula of Lithol Rubine B.

Here we report a more in depth study of the influence of the mesophase surface on the structural state of Ca^{2+} cations under various experimental conditions. The possibility to stabilize small CaO particle and optimize their dual interaction with both the acidic surface of the substrate and the acid functions of the dye molecule is more specifically questioned.

2. EXPERIMENTAL

Al-MCM-41 or (Al,Ca)-MCM-41 mesoporous materials have been prepared in alkaline media (pH 10.5) using Al or Al-Ca nitrates respectively, tetramethylorthosilicate (TMOS) as the silica source, cetyltrimethylammonium bromide (CTMABr) and NaOH, according to previously published recipes [4-5]. The aqueous solution containing Al or Ca nitrates was added to a solution containing NaOH and CTMABr. TMOS was further added to that mixture. The as-synthesized phases were emptied from their surfactant molecules by calcination in static air at a heating rate of 5°C/min., then either cooled down to 30°C at a cooling rate of 10°C/min.

Sample A (coded "Al-MCM-41" in Table 1) is a calcined mesoporous (Na,H)-aluminosilicate further ion-exchanged with calcium chloride by stirring 5 g of calcined sample in 50 g of a 10 wt.% $CaCl_2$ solution for 4 h at 20°C. After exchange, the sample was washed out with distilled water until no more chloride ions were detected. It was expected that the Ca^{2+} ions would neutralize some of the framework negative charges due to the presence of Al^{3+} ions by replacing residual Na^+ or heat generated protons, and stay well dispersed within the silica walls.

Samples B to D (coded "(Al,Ca)-MCM-41" in Table 1) were prepared using the same experimental conditions, from a solution containing both Ca and Al nitrates and by heating the same precursor respectively at 550°C, 750°C and 1000°C.

Table 1
Some physico-chemical characteristics of Ca^{2+}-exchanged Al-MCM-41 and various (Al,Ca)-MCM-41 samples

Sample code	Type	Calcination temp. (°C)	Final ratio Si/Ca	Si/Al	Surface area (m^2/g)
A	Al-MCM-41*	750	86	25	1093
B	(Al,Ca)-MCM-41	550	42	53	871
C	(Al,Ca)-MCM-41	750	44	50	1061
D	(Al,Ca)-MCM-41	1000	46	53	149

* Ca^{2+} ion-exchanged.

All the samples were characterized by XRD, sorption data and XPS, as described elsewhere [4]. Some of these characteristics are included in Table 1.

Dye-MCM-41 composites were prepared by stirring the Ca containing samples in a 0.5% LRB (from Kishi Kasei Co.) aqueous solution for 6 h at 20°C. They are coded as sample N°/LRB (Table 2). The red-pink composites were dried at 20°C and thoroughly washed with water. The amount of dye in each sample and the dye/Ca molar ratios were evaluated by determining the total weight loss upon re-calcining at 550°C the LRB-impregnated samples (Table 2). The dye-MCM-41 composites were deposited onto the siloxane-based paper surface with a paint brush by using a home elaborated method, as described elsewhere [4]. The amount (thickness) of composite applied to the paper was carefully controlled so that all the painted papers exhibited the same color strength, measured by the constant Godlove's value of 55 [6].

The papers were then exposed to the sunlight for 14 h. The color changes (fading) of the papers after exposure were recorded on a Minolta CR-300 colorimeter and the corresponding ΔE values were compared for each sample (Table 3). To check the influence of the way the dye is attached to the MCM-41 surface, sample C was also checked for its anti fading properties in its unwashed state (sample LRB/C_2) and compared to the washed analog (sample LRB/C_1)(Table 3).

3. RESULTS AND DISCUSSION

3.1. Structural state of Ca in different Al-MCM-41 materials

When calcined, both Al-MCM-41 and (Al,Ca)-MCM-41 phases exhibited XRD patterns characterizing pure hexagonal MCM-41 materials [5] and surface areas larger than 800 m^2/g.

The Al bearing phase, when heated up to 750°C prior to direct cooling (sample A), showed a surface area of up to 1093 m^2/g (Table 1) and a substantial pore volume of 0.95 cm^3/g [4], suggesting a better structuration of the mesophase upon a more complete healing of internal or surface defects. XPS data proved undoubtedly that the Ca^{2+} ions in sample A were on the exchange positions of the Al-bearing silica walls [4].

In the case of (Al,Ca)-MCM-41 type samples, because Ca nitrate was initially added into a basic solution (pH around 11), it was speculated that $Ca(OH)_2$ probably precipitated during the formation of the mesophase as a separate phase [4]. Upon the first calcination aiming at removing the surfactant molecules, $Ca(OH)_2$, species present within the mesoporous volume, will undergo dehydration, yielding ultimately isolated CaO. The calcination temperature is also expected to markedly influence the final aggregation of such CaO particles and their consequent interaction within the mesoporous support, as recently observed in the case of other types of oxydic particles (CuO or Cu_2O) on an alumina based mesophase [7]. Their aggregation would, in turn, affect their efficiency in interacting with the dye.

We have therefore checked the effect of calcination temperature on the (Al,Ca)-MCM-41 samples by heating the same precursor at 550°C (sample B), 750°C (sample C) and 1000°C (sample D) (Table 1).

Sample B proved a high surface mesoporous MCM-41 material (BET surface area: 871 m^2/g and pore volume: 0.87 cm^3/g). Its heating to 750°C (sample C) results in a marked improvement of its mesoporous characteristics (BET surface area: 1061 m^2/g and pore volume: 0.99 cm^3/g) while its further annealing up to 1000°C (sample D) resulted in its quasi total collapse (Table 1).

These features obviously reflect the interaction changes between the base particle (CaO) and the acidic aluminosilicate surface that progressively occur during the removal of the surfactant molecules by calcination and the further conditioning of the mesoporous walls. The higher the temperature, the stronger the CaO-wall interaction. As already observed between Cs_2O and mesoporous silicas [8], at very high temperature, the dispersed basic oxide CaO will eventually start to react with the acidic silica walls and form the corresponding Ca silicate phase, independently on the stability of the mesoporous structure of the so formed material, as already observed in the case of bulky Ca silicates [9]. At such temperatures, an independent phenomenon can also occur, namely the collapse of the mesophase, yielding a more bulky aluminosilicate material [10]. So far, XPS could not help to discriminate between the presence of dispersed CaO particles weakly interacting with the aluminosilicate walls at lower temperatures and that of a true Ca (alumino)silicate phase formed at high temperature, because both phases have a similar ionicity [11].

As a conclusion, even if it is clear that the structural state of calcium is different in the Ca^{2+}-exchanged sample A and in samples synthesized in the presence of Ca nitrate (B,C,D), independently on the calcination temperature, we were not able to obtain a straightforward experimental evidence for the progressive reaction of CaO with silica walls at increasing temperatures (both the CaO particles and the Ca silicate domains are far too small to be detected by XRD). However, at that stage, we may assume that, at an adequate temperature, the isolated CaO particles will interact with the substrate strongly enough so as to stay dispersed, but weakly enough so as to prevent the formation of a mesoporous (or bulky) calcium silicate phase. Such particles, partly in interaction with the silica acid sites but also partly available for a further interaction with other acid centers such as the $-COO^-$ or $-SO_3^-$ ends of the dye molecule, could then act as potentially efficient traps and retain the dye strongly enough so as to prevent its reorganization upon a further exposure to light. This hypothesis, as proposed during our previous investigation of similar systems, is here substantiated by the following experiments.

3.2. Use of Ca bearing MCM-41 materials as dye supports

The performance of the various Ca containing MCM-41 samples in behaving as dye supports was evaluated be measuring the color strength (Godlove's A values) of the dye impregnated samples and the strength decrease upon washing (Table 2).

Table 2
Some color properties of the various dye-Ca mesophase composites

Sample code	MCM-41 substrate	Color strength[a]	Dye/100 g calc. sample	Dye/Ca molar ratio
LRB/A	Sample A	58 (92)[b]	5.5	1.4
LRB/B	Sample B	73 (92)[b]	7.3	2.0
LRB/C	Sample C	82 (103)[b]	8.8	1.8
LRB/D	Sample D	75 (82)[b]	7.0	1.9

[a] Godlove's A value. [b] Before washing.

Roughly, the color strength of the washed samples appears logically proportional to the amount of dye left in the pore volume of the mesophase after washing (third and fourth columns in Table 2). So far, sample LRB/C, that contains the highest amount of dye (after washing out the LRB weakly retained on the substrate external surface), also shows the highest color strength. This composite was also shown to possess the highest water-resistance (non bleeding) properties, when it was compared to a large series of similar compounds annealed at different temperatures [4].

The dramatic decrease of the A value for sample LRB/A upon washing indicates that significant amounts of adsorbed dye are easily removed when the sample contains Ca^{2+} ions on the exchange positions of the framework. Oppositely, samples synthesized in the presence of Ca^{2+} ions, especially the one calcined at 750°C (LRB/C), exhibit relatively high A values after washing, suggesting that the dye is more strongly retained on such Ca sites.

The last column of Table 2 clearly confirms our previous findings, namely that the dye/Ca ratios are close to 2 for the MCM-41 materials prepared from gels containing Ca nitrate (samples B, C and D) and is significantly lower than 2 for sample A that involves Ca^{2+} ions on exchange positions. This trend could be generalized for a larger series of samples of each type [4]. Moreover, we have demonstrated that the amount of dye was not dependent on the total amount of Ca present in each type of compound but rather on the structural state of the Ca bearing phases (see below).

One should assume that one dye molecule involving two different acid centers (Fig. 1) can not be neutralized by the same Ca^{2+} ion that is retained on the substrate surface, for sterical reasons. Indeed, the two negative ends of the molecule are too distant from each other so as to interact with the same Ca^{2+} ion to form a chelate type complex. In contrast, such a complex is probably formed in the commercial pigment, that is basically the calcium salt of LRB.

As a result, the dye/Ca ratio of 2 suggests that the ideal neutralization of one CaO molecule (base) is achieved by two different dye molecules through only one of their acid centers, either $-SO_3^-$ or $-CO_2^-$. Because of the potentially stronger acidity of a sulfonate group with respect to a carboxylate, we may speculate that two dye molecules are associated to a CaO particle through their sulfonate ends. Such isolated CaO species are probably well dispersed in the substrate. This dispersion obviously also allows enough room for two bulky dye molecules to approach the basic site (CaO particles).

On the other hand, for CaO to stay well dispersed, the CaO-silica interaction should be strong enough. But this interaction must also be weak enough so as to allow the particles to interact with two acid functions of two dye molecules and to form the corresponding $(LRB)_2Ca$ type salt.

Oppositely, a ratio smaller than 2 means that less than two dye molecules are neutralizing one CaO. Indeed, when the Ca^{2+} ions are present on the exchange positions, they will less readily interact with the dye because they are also partly neutralizing the negative charges of the mesopore walls generated by the presence of framework aluminum.

Finally, one might expect that sample D could already have started to react with CaO at 1000°C and make the corresponding Ca aluminosilicate, even if the mesoporous structure is partly destroyed. However, the dye/Ca ratio in this sample is still close to 2, which suggests that when the Ca silicate islands are generated within the collapsed structure, they stay well dispersed so as to be able to still interact with 2 dye molecules. This can occur not only because their basic character is maintained (although a Ca silicate type phase should have a different basic strength than CaO), but mainly because such interaction is probably still possible on a sterical point of view.

Nevertheless, while the above findings explain fairly well the extent of the Ca-dye interaction (A values) in the different samples, they are not directly correlated to the light fastness properties (resistance to fading) of the composites

3.3. Role of Ca structural state on the photo fading properties of Ca-MCM-41-dye composites

In our previous approach, we have suggested that the structural and/or textural properties of Ca bearing phases, that are directly related to their dispersion, are among the important factors affecting not only the amount of dye in each type of composite, but more particularly their light fastness properties.

In order to shed more light on the role of Ca^{2+} ions on the photo fading properties of the dye composites, we have measured color changes of some selected samples by evaluating the corresponding ΔE values after 14 h of sunlight exposure (Table 3), being understood that a sample showing a small ΔE value has a high stability against fading.

Table 3
Color fading of the various composites after 14 h sunlight exposure

Sample	Substrate	Ca (at.%)[a]	Post ads. treatment	Dye amt (wt. %)	Color change (fading) (ΔE)[b]
LRB/A	Sample A	1.1 (exch.)	washed	5.2	5.9
LRB/B	Sample B	2.1 (synth)	washed	6.8	5.7
LRB/C_1	Sample C	2.2 (synth.)	washed	8.1	3.7
LRB/C_2	Sample C	2.2 (synth.)	unwashed	>10	7.9
LRB/D	Sample D	2.1 (synth.)	washed	6.5	6.1
Commercial pigment[c]		-	-	-	7.7
Ca-exchanged smectite		N.D.	washed	N.D.	6.9

[a] ICP analysis, based on (Ca + Al + Si) = 100%.
[b] $\Delta E = [(\Delta L^*)^2 + (\Delta a^*)^2 + (\Delta b^*)^2]^{1/2}$.
[c] Calcium salt of Lithol Rubine B from Kishi Kasei Co.

Figure 2 compares the color changes upon light exposure of the best of our composite samples (LRB/C_1) and of the commercial pigment (Ca salt of LRB). The new composite proved far more efficient in showing quasi no color change after 14 h.

Among all the composites, sample LRB/C_1 shows the smallest ΔE value, at least smaller than the one measured on sample LRB/A that involves a similar surface area but in which Ca^{2+} ions were introduced through ionic exchange, and also smaller than in its analog that was not heated to the temperature required to generate CaO particles in their optimal structural state (sample LRB/B). In this latter case indeed, the freshly generated CaO particles ($Ca(OH)_2$ dehydrates completely at a temperature close to 500°C), possibly have not yet achieved their optimal dispersion at 550°C.

In sample LRB/C_1, CaO particles are probably optimally retained and dispersed on the substrate surface. The CaO (base) – SiO_2 (acid) interaction is however still strong enough for CaO to stay well dispersed but sufficiently weak to also allow the particles to partly neutralize one acid function of a dye molecule and to form the corresponding Ca salt. In this case, the CaO particles probably do not interact as strongly with the framework as in the case of exchanged sample, in which the interaction might be too strong and impede an optimal reactivity of the Ca^{2+} ions with another (acidic) molecule (dye).

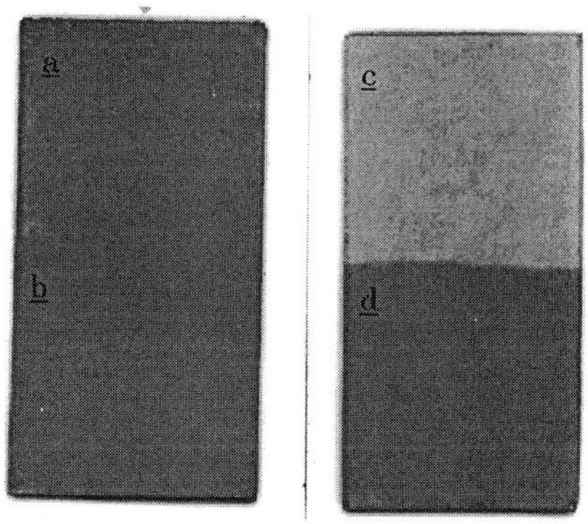

Fig. 2. Color changes of the dye composite sample LRB/C_1 (a: exposed and b: not exposed to light), and of the commercial pigment (c: exposed and d: not exposed to light), after being applied onto the paper, as described previously [4].

3.4. Simplified model explaining the optimal light fastness properties of the novel composites

The above observations suggest a mechanism explaining the optimal resistance of the composites to fading. It is probable that the anti fading properties of LRB dyes are related to their appropriate interaction with Ca^{2+} ions that involve a particular structural state so that one (or even two) dye molecule(s) can interact with them through only one of their acid groups (possibly SO_3^-) and not through both, as in the commercial pigment. For this being possible, the Ca^{2+} ions should be ideally involved in very small CaO entities that are optimally dispersed on a substrate (mesoporous aluminosilicate surface) through an appropriate particle-surface interaction, so that the particles can undergo a dual interaction with the acidic sites located on both the substrate mesophase and the dye molecule.

It is important to note that sample LRB/D, involving a low surface area (and probably also a very small pore volume that could not be evaluated) due to the substrate collapse, also shows a higher ΔE value than its well mesoporous analog LRB/C_1. This experiment strongly confirms that, despite the negligible amount of porous volume left in sample LRB/D, it is the structural state of the Ca^{2+} ions (that here form Ca silicate domains) that plays a major role in conditioning the dye molecules towards the sunlight fading, and neither its total amount, as previously evaluated in each sample [4], nor the total amount of the adsorbed dye, as earlier suggested [12].

Finally, the unwashed sample LRB/C_2 shows a lower stability than its washed analog, although it contains a larger amount of dye. This confirms that most of the dye was left weekly adsorbed (impregnated) on the sample surface and hence more easily decomposed by sunlight than the more strongly bound dye interacting with the superficial Ca^{2+} ions in sample LRB/C1 or even in LRB/D.

Interestingly, the commercial pigment, which is basically the aggregated Ca salt of Lithol Rubine B, also shows a very large ΔE value, similar to the one recorded on sample LRB/C_2, which is in line with the above considerations. On the other hand, LRB/smectite, which is one of the typical dye composites known to possess high anti photo fading properties, is still less efficient than any of our Ca^{2+} based mesoporous compounds. These latter proved better supports than any clay type material for an efficient dye fixing.

The above series of experiments demonstrates that both a high surface area (compare samples LRB/C_1 and LRB/D) and the introduction of Ca^{2+} ions into the MCM-41 substrate so as to generate, through a controlled calcination, small CaO particles highly dispersed within the pore volume during synthesis, are the pre-requisite conditions for MCM-41-Lithol Rubine B composites to show efficient resistance to fading.

REFERENCES

1. Kubo and Y. Nakayama, Shikizai Kyokaishi, 69 (1996) 578.
2. Y. Kubo and M. Tsuji, Shikizai Kyokaishi, 69 (1996) 667.
3. M. Shibata, H. Hotta, T. Suzuki, S. Valange and Z. Gabelica, Chem. Letters, (1999) 1291.
4. Z. Gabelica, S. Valange, M. Shibata, H. Hotta and T. Suzuki, Micr. Mes. Mater., in press.
5. Z Gabelica, J.M. Clacens, R. Sobry and G. Van den Bossche, Stud. Surf. Sci. Catal., 97 (1995) 143.
6. I.H. Godlove, Am. Dyestuff Report, 43 (1954) 685.
7. S. Valange, J. Barrault, A. Derouault, Z. Gabelica, Micr. Mes. Mater., in press.
8. C. Noda Perez, E. Moreno, C.A. Henriques, S. Valange, Z. Gabelica and J.L.F. Monteiro, Microporous and Mesoporous Mater., 41 (2000) 137.
9. P. Tarte and M. Liegeois-Duyckaerts, Spectrochim. Acta, 28A (1972) 2029.
10. S.C. Shen and S. Kawi, J. Phys. Chem. B, 103 (1999) 8870.
11. F. Liebau, Structural Chemistry of Silicates, Springer Verlag, Berlin, 1985.
12. H. Zollinger, Color Chemistry, VCH Verlag, Weinheim, 1987.

Studies in Surface Science and Catalysis 140
A. Gamba, C. Colella and S. Coluccia (Editors)

Preparation and characterisation of WO_x/SnO_2 nanosized powders for thick films gas sensors

G.Ghiotti[a], A. Chiorino[a], F.Prinetto[a], M.C. Carotta[b] and C. Malagù[b]

[a] Dip. di Chimica IFM, Università di Torino, Via Pietro Giuria 7, 10125 Torino, Italy.

[b] INFM-Dipartimento di Fisica. Università di Ferrara, Via Paradiso 12, 44100 Ferrara,Italy.

This work gives results about the characterisation of SnO_2 materials, pure and W^{6+}-added, prepared via a sol-gel route. The materials were characterised as powders or thick films using a variety of techniques. The powder and thick film morphology was analysed by TEM, HRTEM and SEM, the powder texture by volumetric measurements. Powders and films made by regularly shaped and nano-sized particles, even after thermal treatments at 850 °C were obtained. Electrical and FT-IR spectroscopic measurements were employed on films and powders, respectively. Electrical and FT-IR measurements show that W lowers the response of the tin oxide towards CO, but enhances its ability to sense NO_2. The surface species formed by CO and NO_2 interaction were carefully investigated by FT-IR spectroscopy.

1. INTRODUCTION

A significant part of the current research on gas sensors is aimed at obtaining semiconducting materials to achieve highly sensitive and selective long term operating sensors. One route is to expertise new preparation methods of well known materials leading to nanocrystalline powders. Indeed, grain size reduction is one of the main factors enhancing the gas sensing properties. Another route is to text new materials. Recently WO_3 doped SnO_2 [1] or Sn_xWO_{3+x} mixed oxides [2] have been proposed as gas sensors, on the basis of the well known outstanding sensitivity of WO_3 to NO_x [3]. The aim of this work is to obtain SnO_2 and WO_x/SnO_2 powders showing very regularly shaped and nano-sized particles, resistant to thermal treatments generally used in the thick film preparation. Both SnO_2 and WO_x/SnO_2 materials were characterised as powders or thick films using a variety of techniques.

2. EXPERIMENTAL.

2.1. Powders.

SnO_2 powder was prepared by the following sol-gel process: a given amount of deionised water was added drop-wise to a n-butanol solution 0.7 M of tin(II)2-ethylexanoate (Strem

Chemicals) and the mixture was stirred at room temperature (RT) during 3h [4]. The molar ratio of water to Sn was 4 and the pH of the solution was adjusted to be 1 with HNO_3. The resulting gel was dried overnight at 95° C and calcined at 550°C, the powder obtained was named SN. The W^{6+} added materials were prepared by impregnation at incipient wetness of the SN powder with given amounts of aqueous solutions of ammonium tungstate to obtain samples containing 1 and 5 W mol %. The impregnated powders, dried 3h at 120° C, are thereafter labelled W1 and W5. TEM and HRTEM images of the powders were obtained with a Jeol 2000 EX electron microscope equipped with a top entry stage. The powders were compressed in self supporting discs of about 50 mg cm^{-2} and put in an IR cell, which allowed thermal treatments in *vacuum* and controlled atmospheres. Samples were initially submitted to alternate outgassing-oxidizing cycles at 650 °C or 850 °C (the temperature appears in the sample label, e.g.: SN-650, W1-850 etc). Before running the spectra of the samples in CO, or NO_2, oxygen was pumped out briefly to 10^{-3} mbar. High purity gases (from Praxair) were used: O_2 and CO without further purification. NO_2 was prepared in laboratory by contacting freshly distilled NO with O_2, in excess of the 2:1 reaction stoichiometry, during 3 weeks at RT and then was purified from O_2 by freezing. BET surface areas of SN-650, W1-650 and W5-650 pellets were determined by N_2 physisorption using a Micromeritics ASAP (10^{-4} Pa) apparatus.

2.2. Films.

The thick films were obtained starting from miniaturised laser pre-cut 96% alumina substrates (2.5 x 2.5 mm^2 and 0.25 mm thick for each device) provided with a heater element on the backside, a Pt-100 resistor for the control of the sensor operating temperature and a gold front interdigitated contacts. The thick films were prepared starting from pastes obtained by adding to the above powders an organic vehicle together with a small percentage of glass fritt for improving the adhesion of the layers to the substrates. The sizes of the deposited layers were 2 by 2 mm and approximately 30 μm thick. The samples were then fired for 1 h at 650 °C or 850 °C in air. All the conductance measurements were performed in a closed test chamber at a flow rate of 0.5 lt min^{-1}, at temperature up to 450°C in wet (50% R.H.) air and in 100 ppm of CO or in 10 ppm of NO_2 in air. SEM images of the films were done with a 360-Cambridge scanning electron microscope.

3. RESULTS AND DISCUSSION.

3.1. Morphology and texture.

TEM images of SN powder showed aggregates of particles, rounded shaped and homogeneous in size (10-20 nm). W1 and W5 samples showed more densely packed aggregates of particles of homogeneous size (20 nm) with very indented borders. HRTEM images of the three powders showed particles with fringe distances of the cassiterite (110) and (101) planes. No crystalline particles distinct from SnO_2 could be put in evidence. However,

Figure 1. SEM images of: a) SN-850, b) W5-850 films

small spots, absent in the case of SN sample appeared on some particles of W5 sample. The small spots, 0.5-1 nm in size, can be due to the presence of small aggregates of tungsten-oxide units. SEM images of SN-850 films (see Fig.1) showed densely packed aggregates of rounded particles homogeneous in size (40-60 nm), while W1-850 and W5-850 films showed aggregates of particles with less regular shape and less homogeneous in size (40-90 nm). SN-650 , W1-650 and W5-650 pellets showed surface area of 31, 25 and 27 m^2 g^{-1} respectively.

3.2. FT-IR characterization.

FT-IR spectra in O_2. The SN-650/850 spectra were those expected for a nano-dispersed phase, with a stoichiometry "seriously perturbed" mainly by oxygen vacancies (V_O), as illustrated in a previous paper [5]. Two main differences were evident after W addition: i) the overall IR transmittance was markedly higher; ii) new peaks, of medium and low intensity, in the low frequency side of the spectrum (1150-800 cm^{-1}) appeared, as shown in Figure 2 for SN-650, W1-650 and W5-650 samples (curves 1, 2 and 3, respectively). Curves 2 and 3 have been translated along the absorbance axis to allow a better comparison with curve 1, owing to the different sample transmittance. The increase in transmittance can be interpreted as due to a decreased concentration of free electrons and of electrons trapped in

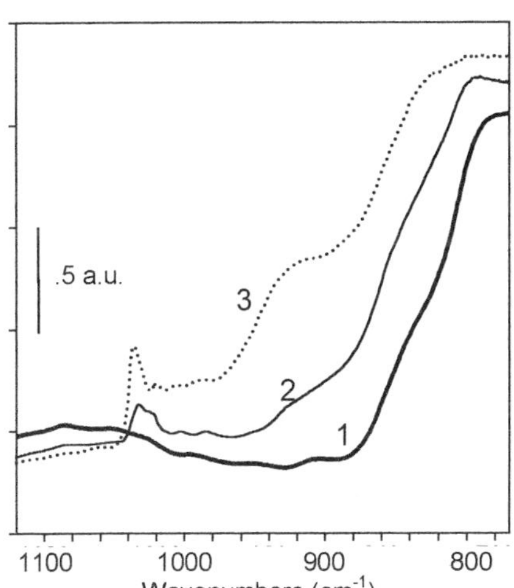

Figure 2. FT-IR spectra of: 1, SN-650; 2, W1-650 and 3, W5-650 samples in the 1200-780 cm^{-1} region.

oxygen vacancies, the W^{6+} ions acting as acceptor levels deeper than oxygen vacancies. If so, we will expect a decrease of the conductance in air of the materials by W addition. The new peaks observed fall in a region where the vibration modes of W=O, $(W\text{-}O\text{-}W)_n$ and $(Sn\text{-}O\text{-}W)_n$ are expected to fall [6, 7].

FT-IR spectra in CO. The response to pure CO at RT of the SN-650/850 pellets was the following one: CO reacted with surface oxygen ions (O_2^-, O^-, O^{2-}) giving carbonate-like species, which partially decomposed to CO_2 releasing electrons to the tin oxide; the electrons released repopulated the conduction band (c.b.) and the oxygen vacancies (V_O^{++}), markedly decreasing the sample transmittance. In particular, the electronic repopulation of the V_O^{++} could be revealed by the increase of a very broad absorption extending all over near and medium IR regions, previously assigned to the photoionization of the monoionized oxygen vacancies, $V_O^{+} + h\nu \rightarrow V_O^{++} + e^-(c.b.)$ [8-10]. The surface reduction increased by increasing the contact times and/or the equilibrium pressure. The behaviour above described is illustrated in Fig. 3a for SN-650 sample (taken as an example): the reported curves are the differences between absorption spectra after and before the interaction with CO at different contact times and/or equilibrium pressures and they clearly show the increase of the very broad absorption previously discussed. The species formed at the surface were easily detected, the sharp peaks superimposed to the broad electronic absorption are, indeed, the vibration modes of the surface species formed by CO interaction. The peaks observed in the 1800-1000 cm^{-1} spectral region could be assigned to different surface carbonate-like species formed by the reductive adsorption of CO on different sites.

The main species formed were:

i) chelating and bridging bidentate carbonates ($\nu(C{=}O)$ at 1650-1550 cm^{-1}, $\nu_{asym}(OCO)$ and $\nu_{sym}(OCO)$ at 1270 and 1020 cm^{-1}, respectively);

ii) "strongly perturbed CO_2" or "carboxylate species" ($\nu_{asym}(OCO)$ and $\nu_{sym}(OCO)$ at 1730-50 and 1270 cm^{-1}, respectively);

iii) CO_2 linearly coordinated to unsaturated surface Sn^{4+} ions ($\nu_{asym}(OCO)$ at 2330 cm^{-1}).

At high CO pressure the CO_2 linearly coordinated was displaced by CO as shown by the disappearance of the peak at 2330 cm^{-1} and the increase of a band at 2203 cm^{-1} assignable to CO linearly bound to unsaturated surface Sn^{4+} ions. It is worth noting that after the CO removal and the subsequent O_2 admission at RT the sample transmittance was almost completely restored. So we can conclude that O_2 was adsorbed giving surface species O_2^- and possibly O^- able to trap the electrons previously transferred to the materials.

Similar results were shown by SN-850 sample, but the surface reactivity was lower. Lower amounts of species were formed and lower amounts of electrons were released to the SnO_2.

As for the W added materials, W1-650/850 and W5-650/850 showed a behaviour similar to that of SN-650/850 ones. However, within the same temperature pretreatment, SN appeared as the most sensitive material. The comparison is illustrated in Fig. 3b for the materials pretreated at 650 °C. Taking the integrated intensity of the broad electronic absorption as a measure of the response to CO, the sensitivity is markedly and progressively lowered by

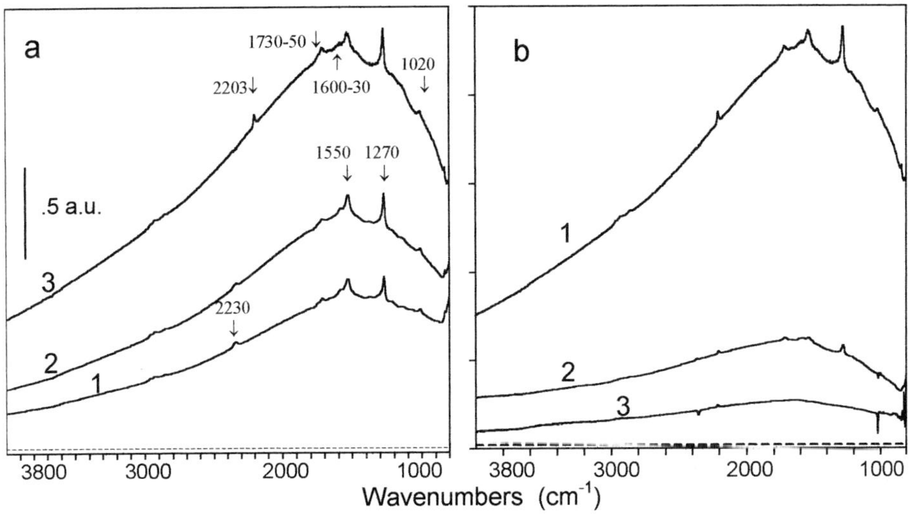

Figure 3. a) Effects of CO at RT on FT-IR spectra of SN–650 sample: curve1, $1x10^{-1}$ mbar; curve 2, 2 mbar; and curve 3, 7 mbar. b) Comparison between the effects of CO (7 mbar) on FT-IR spectra of SN-650 (curve 1), W1-650 (curve 2) and of W5-650 (curve 3). Dashed lines: zero of absorbance.

increasing both the temperature pretreatment and the W loading. Actually, these results cannot be taken as a measure of the sensitivity of the three materials to respond to the CO gas; indeed the more or less good response to CO at a fixed temperature of all the materials has been measured by the more or less marked increase in their conductance when passing from wet air to 100 ppm of CO in wet air, i.e. when CO is not pure but diluted in air. However, the reported FT-IR results can be taken as a preliminary and coarse test of the sample sensitivity towards CO. Indeed the previous results are in qualitative agreement with the electrical ones as it will be shown in the paragraph **3.3**. No measurements have been done at higher temperature in pure CO to avoid the reduction of the WOx-phase and possible irreversible phenomena. However, we will develop our research studying the interaction of CO/O_2 mixtures in the temperature range of the electrical measurements.

FT-IR spectra in NO_2. As for NO_2, we only examined the interaction at RT of 5 mbar of the gas with SN-650, W1-650 and W5-650 pellets, and followed the spectra evolution with the contact time. The results can be resumed as follows. By interaction with SN-650 pellet, a quick reduction process at the surface immediately occurred with electrons release to the c.b. and the subsequent V_O^{++} repopulation to V_O^{+}, revealed by the quick increase of the very broad absorption extending all over near and medium IR regions, already described (see Fig. 4), who reached its maximum intensity in 2 sec. Sharp vibrational bands appeared, but of very weak intensity, none of them assignable to surface species formed in a reaction involving a total electron transfer from NO_2 to the material. Actually, the most intense peak is one at 1178 cm^{-1}, assignable to surface $\nu_{asym}(ONO)$ mode of chelating ionic nitrites (NO_2^-). The

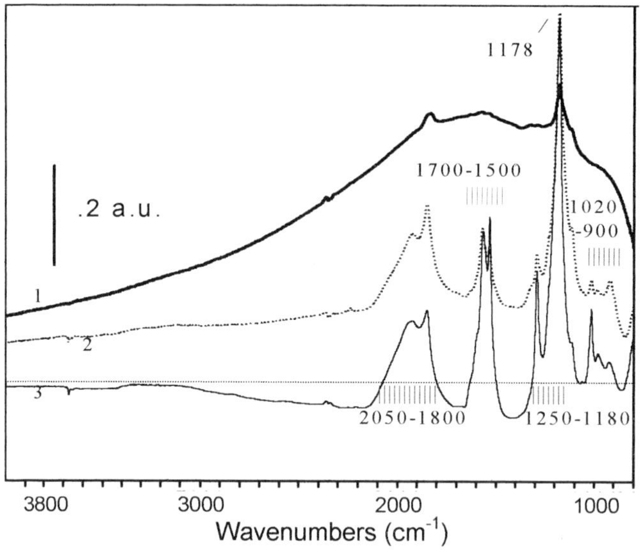

Figure 4. Time resolved effects of 5 mbar of NO_2 at RT on SN-650 FT-IR spectra . Spectra after 2 sec, 10 sec and 1min (curves 1, 2 and 3 respectively). The dotted line represents the zero of absorbance.

surface reductive mechanism that we propose is the following: $NO_2 + V_O^{2+} \rightarrow NO_2^+ + V_O^+$. The NO_2^+ species, that should have a stretching mode at 2400 cm^{-1}, was probably formed in too small amounts to be detected. At higher contact time, the electron population of the V_O^+ began to decrease revealing an oxidation process with a kinetic slower than the reductive one. The electron population decrease was paralleled by the growth of sharp peaks related to the formation of a variety of NO_x surface species:

i) chelating nitrites (ν_{asym}(ONO) mode at 1178 cm^{-1});

ii) several types of chelating and bridging nitrates (ν(N=O) modes in the range 1700-1500 cm^{-1}, ν_{asym}(ONO) modes in the range 1250-1180 cm^{-1} and ν_{sym}(ONO) modes in the range 1000-900 cm^{-1});

iii) $NO^{\delta-}$ and $NO_2^{\delta+}$ weakly adsorbed species (bands in the range 1800-2050 cm^{-1}).

When the reactions stopped after 1 min, the sample was slightly oxidized, that is the electron concentration in the V_O^+ was lower than before the NO_2 admission: in fact, taking as reference the dotted line representing the line of absorbance zero in Fig. 4, a broad, weak, and negative absorption is evident, showing the same shape of the electronic absorption assigned to $V_O^+ + h\nu \rightarrow V_O^{++} + e^-$(c.b.).

At variance, W1-650 pellet was immediately oxidized by NO_2 (see Fig. 5a), with decrease of the electron population of the V_O^+: a broad, weak and negative absorption was already evident after a contact time of 2 sec, at the same time weak vibrational peaks related to the formation of a variety of NO_x surface species appeared. At higher contact times the electronic

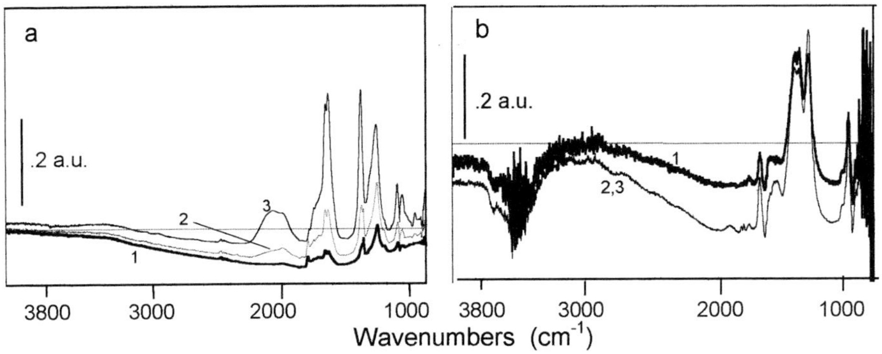

Figure 5. Time resolved effects of 5 mbar of NO_2 at RT on FT-IR spectra of: a) W1-650 and b) W5-650 samples. Curves 1, 2 and 3, spectra after 2 sec, 10 sec and 1min respectively. The dotted line represents the zero of absorbance.

population remained unchanged, while the vibrational peaks related to the NO_x surface species continued to increase during 1 min. The surface species formed were similar to those present at the surface of the SN-650 material, but in different relative amounts. In particular chelating nitrites, $NO^{\delta-}$ and $NO_2^{\delta+}$ weakly adsorbed species were markedly decreased. Chelating and bridging nitrates were the species predominant at the surface, their vibration frequencies being slightly different from those present on SN-650, indicating that tungsten ions were involved in the coordination of these species. Concerning W5-650 sample (see Fig.5b), again the electron population of the V_O^+ began to decrease immediately after NO_2 interaction, revealing oxidation processes, which stopped after 10 sec. Also in this case the electron concentration decrease was paralleled by the growth of sharp peaks related to the formation of NO_x surface species. However, chelating nitrites and $NO^{\delta-}$ or $NO_2^{\delta+}$ weakly adsorbed species were completely absent, only one type of bridging nitrate was still present (ν(N=O) mode at 1700 cm^{-1}, ν_{asym}(ONO) modes at 1250-1180 cm^{-1} and ν_{sym}(ONO) at 1000 cm^{-1} respectively), while the main surface species were nitrate of ionic type (ν_{asym}(ONO) mode at 1450 cm^{-1} and ν_{sym}(ONO) mode at 1370 cm^{-1}) not detected on the other two materials. This indicates that the surface chemistry was markedly changed. At the equilibrium (after 1 min), the values of the integrated negative intensity of the broad electronic absorption were in the following order: W5-650< W1-650< SN-650 sample (compare the intensities of the broad negative absorption of the curves 3 in Figs. 4 and 5). Taking this integrated negative intensity as a measure of the electronic response to NO_2, W5 appeared the most sensitive one, the response decreasing with the decrease of W loading.

3.3. Electrical characterisation.

In order to investigate the influence of the tungsten addition onto the electrical properties of tin oxide as host material, we performed conductance measurements as a function of temperature in wet air (40% relative humidity). Fig. 6a shows the Arrhenius plots of pure and

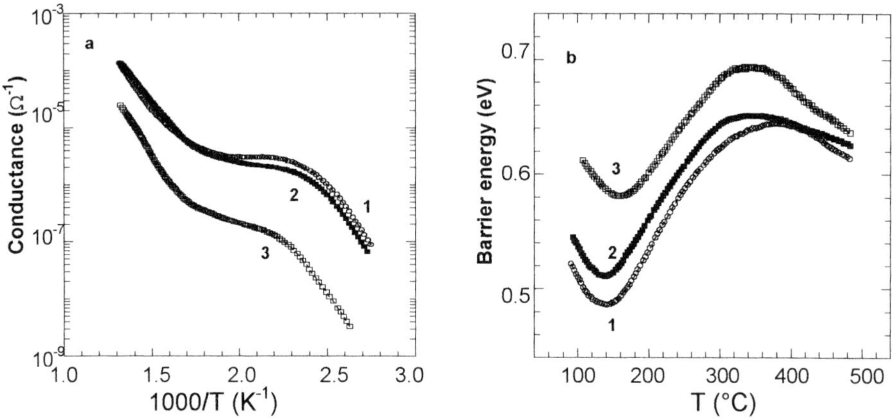

Figure 6. Temperature dependence of the: (a) conductance and (b) energy barrier in wet (40% RH) air of 1, SN-850, 2, W1-850 and 3, W5-850 thick films heated to 850°C.

tungsten-added SnO_2 thick films fired at 850 °C (SN-850, W1-850 and W5-850) obtained changing the temperature between 100 °C and 500 °C at the heating rate of 3 K/min. The samples exhibited three regions of conductivity as usual in tin oxide sensors and previously reported [11, 12] for thick films of SnO_2 differently synthesised. Tungsten addition caused a significant decrease in conductance (about one order of magnitude at the temperature of 300°C) when the W loading was of 5%. At the same time also the intergrain barrier energy values at the various temperatures increased, as it is shown in Fig. 6b (the experimental method to determine them is described in [13]).

Since the difference between the minimum and the maximum values of the energy barrier (ΔE) is lower in the case of W added materials than on pure SnO_2, the above result can not be associated with a greater amount of oxygen ions adsorbed which is at the contrary expected to increase ΔE [14]. In fact, from the Fig. 6b, ΔE is 0.15 eV in the case of SN-850, 0.13 eV for W1-850 and finally 0.11 eV for W5-850. Therefore the decreased conductance has to be attributed to the increased number of holes in the valence band injected by the acceptor-like W ions. The SnO_2 and W-added SnO_2 based films were tested as gas sensors towards carbon monoxide and nitrogen dioxide.

Concerning the SN-850, W1-850, W5-850 samples, the tests showed that the ability of the thick films to sense CO was lowered increasing the W loading, in agreement with the barrier energy behaviour. In fact it has been demonstrated [14] that ΔE is strongly related to the sensitivity to reducing gases, like carbon monoxide. The W content also affected the temperature of the maximum response. In particular W addition stabilised the response at the different working temperatures: W5 thick films showed a weak response almost unaltered in all the examined range 200-350°C. Concerning NO_2, conductance measurements showed that the ability of the materials to sense the gas was enhanced increasing the W loading.

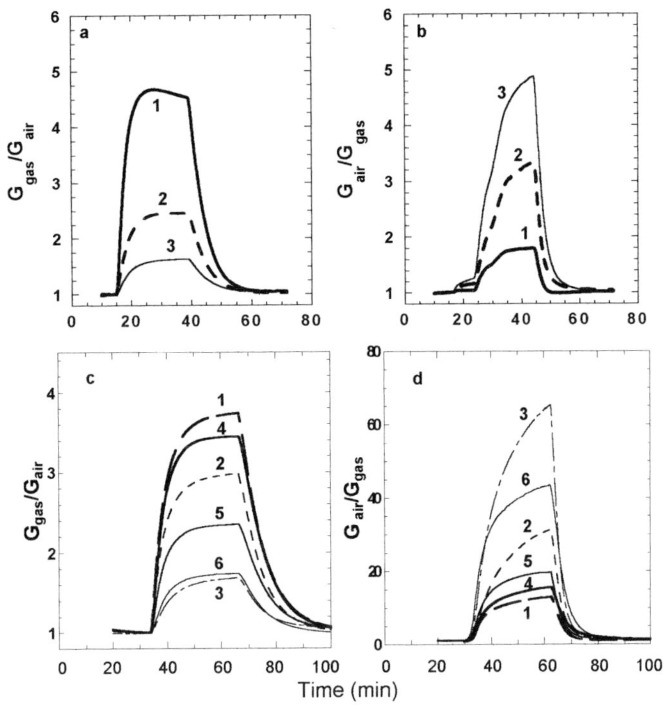

Figure 7. Electrical response to: (a) 100 ppm of CO (G_{gas}/G_{air}), (b) 10 ppm of NO_2 (G_{air}/G_{gas}) in wet air (40% RH) of Sn-850 (curves 1), W1-850 (curves 2) and W5-850 (curves 3) thick films, measured at 250°C; (c) CO (100 ppm) in wet air (40%RH) of all samples measured at 300 °C; (d) NO_2 (10 ppm) in wet air (40%RH) of all samples measured at 200 °C. In sections (c) and (d) curves 1 to 6 refer respectively to: Sn-650, W1-650, W5-650, Sn-850, W1-850 and W5-850 thick films.

The enhancement was particularly marked at 250°C (about three times passing from SN-850 to W5-850). It is worthy to note that at this working temperature SN-850 showed the highest response to CO, its response being more than three times higher than that of the W5 film (see Fig. 7a and b). Furthermore, all the three materials showed the maximum response to NO_2 at 150°C, temperature at which the response of the three films to CO was very low.

The SN-650, W1-650, W5-650 samples behaved in the same way as the samples fired at 850 °C; the only difference was the temperature of the maximum response to CO, 300 °C instead of 250 °C. In Fig. 7c, the response to 100 ppm of carbon monoxide in wet air (40% of relative humidity) of a complete series of samples at the working temperature of 300 °C is reported. Moreover the response to 10 ppm of NO_2 in wet air of the same series of samples at the working temperature of 200 °C is shown in Fig. 7 d. More generally the films fired at 650 °C showed a higher response to the gases probably because of a greater specific surface area. It is worthy to note that at all the working temperatures the films showed by repeated measurements a steady response, both to CO and to NO_2 (not reported in figure).

4. CONCLUSIONS.

SEM, TEM and HRTEM microscopies show that the goal of obtaining powders and films made by nanosized particles is attained. Electrical measurements show that W lowers the conductance of the films in air, in good agreement with the FT-IR transmittance increase attributed to a decreased concentration of free electrons and of electrons trapped in oxygen vacancies by W^{6+} addition. Electrical measurements show that W lowers the response of SnO_2 to CO and also affects the temperature of the maximum response. At variance the W addition enhances its ability to sense NO_2, leaving unaltered the temperature of the maximum response. FT-IR transmission changes of the three materials, pretreated either at 650°C or at 850°C, contacted with CO and NO_2 at RT, are qualitatively in good agreement with the electrical measurements. FT-IR spectroscopy also provided information on the nature of the surface species formed by interaction with the two examined gases.

ACKNOWLEDGEMENTS

Financial support was partially provided by the Italian CNR (Progetti Finalizzati MADESS II)

REFERENCES.

1. D. H Yun., C. H. Kwon and K. Lee, Sensors and Actuators B, 35 (1996) 48.
2. J. L. Solis and V. Lannto, Sensors and Actuators B, 48 (1998) 322.
3. D-S. Lee, S. Han, J. Huh and D-D.Lee, Sensors and Actuators B, 60 (1999) 57.
4. C. D. Feng, Y. Shimizu, M. Egashira, J. Electrochemical Soc., 141 (1994) 220.
5. A. Chiorino, G. Ghiotti, F. Prinetto, M. C. Carotta and G.Martinelli, Sensors and Actuators B, 44 (1997) 474.
6. J. Leyer, R. Margraf, E. Taglauer and H. Knözinger, Surface Science, 201 (1988) 603.
7. J. A. Horsley, I. E. Wachs, J. M. Brown, J. H. Via and F.D. Hardcastle, J. Phys. Chem., 91 (1987) 4014.
8. I. E. Wachs, F.D. Hardcastle and S.S. Chan, Spectroscopy, 1 (1986) 5
9. G. Ghiotti, A. Chiorino, and F. Boccuzzi, Sensors and Actuators B,19 (1989)151.
10. A. Chiorino, G.Ghiotti, F. Prinetto, M. C. Carotta and G.Martinelli, Sensors and Actuators B, 44 (1997) 474.
11. M.C. Carotta, C. Dallara, G. Martinelli, L. Passari and A. Camanzi, Sensors and Actuators B, 3 (1991) 191.
12. G. Ghiotti, A. Chiorino, G. Martinelli and M.C. Carotta, Sensors and Actuators B, 24-25 (1995) 520.
13. V. Lantto, P. Romppainen and S. Leppävuori, Sensors and Actuators, 14 (1988) 149.
14. G. Martinelli and M.C. Carotta, Sensors and Actuators, 7 (1-3) (1992) 717.

Studies in Surface Science and Catalysis 140
A. Gamba, C. Colella and S. Coluccia (Editors)

Factors affecting the isomorphous substitution of Al with Fe in the MFI-type zeolite in the presence of TPABr and ethylene glycol

G. Giordano and A. Katovic

Dipartimento di Ingegneria Chimica e dei Materiali,
Università della Calabria, Via Bucci, 87030 Rende (CS), Italy

This paper describes the isomorphous substitution of aluminium by iron in the MFI type zeolite by direct synthesis in presence of oxalate or phosphate iron complexes, starting from a system containing TPABr and ethylene glycol (EG). The effect of Si/Al, Si/Fe, TPABr/Si and EG/Si ratios was investigated in order to understand the influence of these parameters on the nature of the products as well as their properties.

1. INTRODUCTION

There is a growing interest in the preparation of Fe- and Fe-Al-MFI type catalysts due to their capability in the N_2O reduction followed by the direct oxidation of benzene to phenol [1-3]. According to the literature, the first reaction step is the decomposition of N_2O with the formation of an "α-oxygen", that is able to oxidize different hydrocarbons, especially benzene to phenol. Pirutko *et al.* [1] studied the oxidation of benzene over Fe—MFI type catalyst, Centi *et al.* [2] showed that an "α-oxygen" is formed during the reduction of N_2O over Fe-Al-MFI type catalyst, and finally Pirutko *et al.* [3] observed the direct formation of phenol over an Fe-Ti-silicalite catalyst.

Different ways have been proposed in the "open" and patent literature for the incorporation of metals into zeolites and for the partial or total substitution of aluminum. Ion exchange methods are very frequently used for the incorporation of mono and bivalent metals, e.g., in the preparation of Cu-MFI type zeolites, extensively studied for the SCR reaction [4, 5]. Solid state reactions are also used for the introduction of copper, iron and other metals [6, 7]. The TVD technique is another interesting method for the preparation of zeolites containing transition metals [8], and finally the direct synthesis in presence of metal salts [9, 10] or metal complexes is also used.

In this paper the last mentioned technique was chosen. Accordingly, two different complexes of iron with oxalic acid or with ortophosphoric acid were prepared, in order to obtain Fe-MFI and Al-Fe-MFI zeolites starting from a hydrogel containing tetrapropylammonium bromide (TPABr), ethylene glycol (EG) or both of them. The most expensive reactant in the preparation of MFI type zeolite is the organic compound containing nitrogen, normally a TPA salt or tetraprolylammonium hydroxide. One of the problems connected with the use of this type of organic compound is that the after-synthesis thermal treatment results in the formation of nitrogen oxides or ammonia, two compounds with heavy environmental impact. To overcome this problem, a nitrogen-free organic molecule (ethylene glycol) has been used [11] instead of TPA compounds.

The role played by (a) the aluminum and iron amount in the starting hydrogel, (b) the $TPABr/SiO_2$ ratio and (c) EG/SiO_2 ratio on the characteristics of the products obtained is one of the main purpose of the present investigation.

2. EXPERIMENTAL

Two different hydrogel systems were studied. The first system was prepared from the following initial reaction mixture:

A $$x\ Na_2O - y\ TPABr - z\ Al_2O_3 - SiO_2 - q\ Fe_2O_3/pHA - 20\ H_2O,$$

where $x = 0.1\text{-}0.32$; $y = 0.02\text{–}0.08$; $z = 0\text{–}0.05$; $q = 0.005\text{-}0.025$; the ratio $p/q = 3$ and HA stands for $H_2C_2O_4$ or H_3PO_4.

The second system was prepared from the following initial reaction mixture:

B $$0.16\ Na_2O - x\ TPABr - y\ EG - z\ Al_2O_3 - SiO_2 - q\ Fe_2O_3/p\ HA - 10\ H_2O,$$

where $x = 0\text{– }0.08$; $y = 0\text{–}6.0$; $z = 0\text{–}0.05$; $q = 0.005\text{–}0.1$; the ratio $p/q = 3$ and HA stays for $H_2C_2O_4$ or H_3PO_4.

The synthesis procedure was the following: sodium aluminate (Carlo Erba) was added in a sodium-hydroxide (Carlo Erba) solution and after the homogenization the organic compound(s) (TPABr from Fluka and EG from Carlo Erba) and the silica source (silica-gel BDH) were added. In another beaker a solution of iron complex with oxalic or ortophosphoric acid, made from iron nitrate (Carlo Erba) and the specific acid (Carlo Erba), was prepared. This solution was slowly added to the aluminum silicate hydrogel. For the synthesis of the silica form, the procedure was identical but without introduction of the aluminum source. In some experiments related to the system B an amount of 3–10 wt. % of MFI (Al and Fe free) calcined seeds was added.

Syntheses were performed at 170 °C in Teflon lined autoclaves in static conditions under autogenous pressure. The products were cooled down to room temperature, washed and dried at 105 °C overnight. The characterization of the samples was performed by the following techniques: powder X-ray diffraction (Philips PW 1730/10, Cu $K\alpha_1$ radiation), thermal analysis, combined TG-DTG-DSC (Netzsch STA

409), scanning electron microscopy equipped with a EDX instrument (Steroescan 360 Cambridge Instruments) and atomic absorption spectrophotometry (GBC 932 AA).

3. RESULTS AND DISCUSSION

In Table 1 the products obtained for both systems A and B in the presence of oxalic acid or phosphoric acid are reported.

Table 1

The nature of the sinthesized products varying the composition of the starting hydrogel in the type A and B systems

Sample	Si/Fe_{gel}	Si/Al_{gel}	TPABr	EG	Time (d.)	Product
1 (A)	100	∞	0.08	0.0	1	MFI
2 (A)	10	∞	0.08	0.0	3	MFI
3 (A)	100	100	0.08	0.0	1	MFI
4 (A)	10	∞	0.02	0.0	3	MFI
5(A)	10	10	0.02	0.0	2	amorphous
6 (A)	10	10	0.02	0.0	4	MFI
7 (A)	∞	25	0.02	0.0	2	MFI
8 (B)	20	∞	0.08	3	1	MFI
9 (B)	6.67	∞	0.04	3	2	MFI
10 (B)	6.67	∞	0.04	3	5	MFI
11 (B)	20	10	0.04	3	2	MFI
12 (B)	10	10	0.04	3	2	MFI
13 (B)	6.67	10	0.04	3	6	MFI
14 (B)	20	∞	0.02	3	2	MFI
15 (B)	6.67	∞	0.02	3	2	MFI + [Q]
16 (B)	6.67	∞	0.02	3	4	[Q]+ MFI
17 (B)	20	20	0.02	3	2	MFI
18 (B)	20	10	0.02	3	5	MFI
19 (B)	10	10	0.02	3	5	MFI
20 (B)	6.67	10	0.02	3	11	amorphous
21 (B)	∞	20	0.0	3	3	MFI
22 (B)	20	20	0.0	3	2	MFI
23 (B)	20	10	0.0	3	5	MFI

The secondary crystalline phases are reported in brackets [ ]. Q stands for quartz.

First of all it can be observed that, in the presence of only TPABr (runs in the system A), it is possible to obtain Fe-MFI type zeolite in a large Si/Fe range and in a short time, using both oxalic or phosphoric acid with or without the presence of aluminum. The products obtained are stable either from a chemical or from a thermal standpoints. In fact, if the reaction time is prolonged, the transformation of MFI phase in other phases was not observed, and also, at the end of a thermal treatment at high temperature (850 °C) the MFI is the only phase detected.

The crystallization time is a function of the initial iron and aluminium content in the starting hydrogel (see samples 1 (A) and 2 (A) and samples 4 (A), 5 (A) and 6 (A), Table 1).

In the type B system, the formation of the Al-MFI type (see sample 21 (B) in Table 1) is very easy in a wide range of Si/Al ratios in the presence of only EG molecules, what confirms the results reported in a previous paper [11]. Having in mind the results of the paper [11], in which it was demonstrated that silica-rich systems did not give any pure MFI type zeolite, in this study we tested only the Al-rich EG systems. Of course, the presence of TPA ions increases the crystallization rate (see samples 11(B) and 18 (B), and 12 (B) and 19 (B), Table 1). When the TPABr amount in the initial hydrogel changes, not only the crystallization rate varies, but also the stability of the product. As a matter of fact, when the content of TPABr decreases from 0.08 to 0.02, the stability of the MFI product decreases as well. The iron-rich samples with the TPABr content equal to 0.08, (sample 8 (B) in Table 1), are very stable after 6 days, on the contrary the sample 14 (B) (Table 1), after a crystallization time of 4 days, shows the co-crystallization of MFI with quartz and cristobalite. Also the samples, obtained in the presence of 5 or 10 wt. % of seeds and after a 1 day aging at room temperature (such as sample 14 (B) and 16 (B) in Table 1) after 4 days for crystallization show the co-cristallization of MFI and quartz. Even for the system B the amount of iron, aluminium and TPABr in the initial reaction mixture influence the reaction time: when the metal content increases and the TPABr amount decreases, the crystallization time is prolonged, see samples 8(B) and 14 (B) in Table 1 for the influence of TPABr, samples 9 (B) and 13 (B) in Table 1 for the influenece of aluminium, and finally samples 11 (B) and 13 (B) in Table 1 for the iron effect. Finally, a large amount of iron with a small content of TPABr do not allow any crystallization even after longer reaction times (see sample 20 (B) in Table1).

The presence of EG is able to address the formation of MFI type zeolite in the presence of iron and aluminium (see samples 22 (B) and 23 (B) in Table 1) for the Si/Al and Si/Fe ratios indicated in the table. In the absence of EG we do not detect any crystalline phases in the systems investigated.

For both systems the pH value plays an important role in the crystallization process. As a matter of fact, the pure MFI type zeolite formed when the pH value of the initial hydrogel ranged between 9 to 12.5.

In Table 2 the chemical characterization of the samples obtained from systems A and B is reported.

It can be observed that for type A synthesis system the amount of iron incorporated into the zeolite is related to its content in the TPABr-rich hydrogel systems (see samples 1 (A) and 2 (A) in Table 2). For a higher TPABr amount, equal to 0.08, even in the presence of aluminium, iron is incorporated preferentially respect to aluminium (see samples 3 (A), 24 (A) and 25 (A) in Table 2). When the TPABr content in the initial reaction mixture decreases down to 0.04, the amount of iron found in the zeolite decreases (see sample 26 (A) in Table 2). Finally, for a TPABr content equal to 0.02, the aluminium atoms are preferentially incorporated in the MFI framework, so much so that only a little amount of iron is detected in the chemical composition.

Table 2
Chemical analyses of the samples obtained from system A and B

Sample	$TPABr_{gel}$	Si/Al_{gel}	Si/Fe_{gel}	Si/Al_{cryst}	Si/Fe_{cryst}
1 (A)	0.08	∞	100	-	85.5
2 (A)	0.08	∞	10	-	15.4
3 (A)	0.08	100	100	97.7	73.8
24 (A)	0.08	50	50	51.6	37.3
25 (A)	0.08	25	25	42.3	29.6
26 (A)	0.04	∞	25	-	48.0
6 (A)	0.02	10	10	25.0	307
7(A)	0.02	25	∞	16.5	-
8 (B)	0.08	∞	20	-	258
9 (B)	0.04	∞	6.67	-	252
11 (B)	0.04	10	20	23.5	418
12 (B)	0.04	10	20	23.0	459
13 (B)	0.04	10	6.67	24.1	442
14 (B)	0.02	∞	20	-	504
19 (B)	0.02	10	10	29.1	403
21 (B)	0.0	20	∞	13.7	-
22 (B)	0.0	20	20	35.0	423
23 (B)	0.0	10	20	18.1	369

This behaviour can be explained by the different preference exhibited by the solution cations for the iron- and aluminium- containing anions. It has been demonstrated by different authors in the synthesis of B-MFI type zeolite [12, 13] that Na^+ interacts better with $[Si\text{-}O\text{-}Al]^-$, whereas TPA^+ prefers $[Si\text{-}O\text{-}Fe]^-$ groups.

In the case of the type-B system (with EG) a low iron content in the MFI samples has been constantly found, even in the presence of a high TPABr content (see sample 8 (B) in Table 2). On the contrary, the amount of aluminium in the zeolitic framework is related to its content in the starting hydrogel and is almost constantly a high amount. This confirms that the hydrogel systems that contain both sodium and alcohol, such as EG, lead to a high amount of sodium in the crystals [11, 14], which means also that they favour the Al incorporation, because of the favourable interaction between Na^+ and the $[Si\text{-}O\text{-}Al]^-$ groups. Probably the EG molecules compete with the TPA^+ cations and fill the zeolitic channels, as demonstrated by thermal analysis. In fact intact EG molecules were found in the MFI channels.

The size of the crystals obtained from system A decreases when the amount of iron incorporated into the zeolite increases, and changes from 10 μm (samples 1 (A) in Tables 1 and 2) to 2.3 μm (sample 2 (A) in Tables 1 and 2). Accordingly, also the morphology changes from brick-like to sphere (see Figs. 1 and 2).

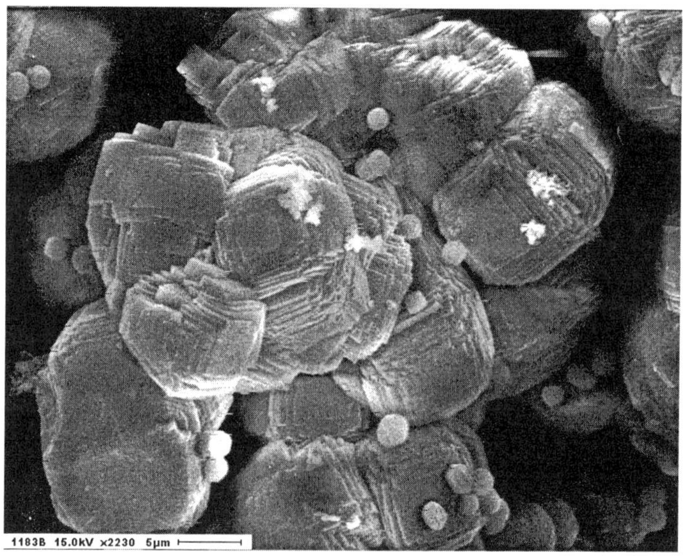

Fig. 1. Scanning electron micrograph of Fe-MFI sample 1 (A).

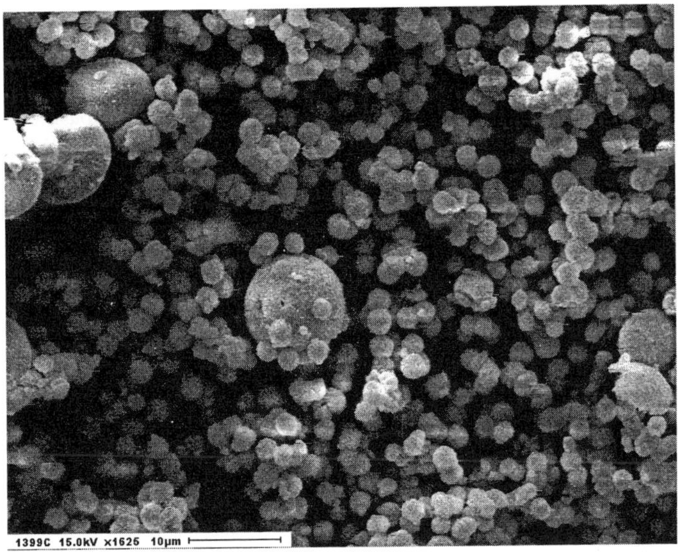

Fig. 2. Scanning electron micrograph of Fe-MFI sample 2 (A).

These observations give an indirect confirmation of the iron incorporation into the MFI framework. The presence of aluminium in the zeolite allows a reduction in crystal size compared to that of the Al-free samples. In fact the sample 3 (A) shows a spherical habitus with a diameter of 5 µm, whereas the sample 24 (A) evidences the same morphology with a smaller size (2.7 µm) due to the high amount of aluminium and iron incorporated into the zeolitic framework.

The morphology of the samples obtained from the system B shows a brick-like shape for both iron and iron-aluminium systems. Also in this case the crystal size decreases when the amount of Al incorporated into the zeolite increases and changes from 12 to 1 µm.

4. CONCLUSIONS

It is impossible to obtain Fe-MFI type zeolite from hydrogel systems containing EG in the absence of TPA ions, but only Al-Fe-MFI zeolite and in specific Si/Al and SI/Fe ranges. On the contrary, the formation of Al-MFI type zeolite is very easy in a wide range of Si/Al ratios in presence of only EG molecules.

When the amount of TPABr in the initial hydrogel decreases, the stability of the MFI product decreases too and the co-cristallization of MFI and quartz is observed after short periods of time. It is possible to obtain MFI products starting from various ratios of Al/Fe in the hydrogel. The crystallization time is a function of the initial TPABr, aluminium and iron content. The presence of Al in the hydrogel stabilizes the MFI structure, so it is possible to obtain MFI in absence of TPABr, but after prolonged crystallization times.

The amount of iron incorporated into the zeolitic framework is a function of its content in the starting hydrogel in the case of the TPABr-rich system (type-A system), and it is preferentially introduced into the zeolite if compared to aluminium. If the amount of TPABr in the initial reaction mixture decreases, the iron content in the zeolite crystals also decreases and for the samples synthesized in presence of aluminium, this latter metal is incorporated into the MFI framework.

The systems containing EG in the hydrogel do not allow a high iron incorporation. Aluminium is readily incorporated in the framework, iron is detected only in traces. This behaviour can be explained with the preferential interaction of $[Si\text{-}O\text{-}Fe]^-$ with TPA^+ and $[Si\text{-}O\text{-}Fe]^-$ with Na^+.

The crystal size of the MFI products depends on the amount of iron and/or aluminium in the initial hydrogel and on the TPABr content as well.

REFERENCES

1 L.V. Pirutko, D.P. Ivanov, K.A. Dubkov, V.V. Terskikh, A.S. Kharitonov and G.I. Panov, Proc. 12th Int. Zeolite Conf., M.M.J. Treacy et al. (eds.), Materials Research Society, Warrendale, Vol II, 1999, p. 1245.

2 G. Centi, G. Grasso, F. Vazzana and F. Arena, Stud. Surf. Sci. Catal., 130 (2000) 635.

3 L.V. Pirutko, V. Chernyavsky, K.A. Dubkov, A. Uriarte, M. Rodkin and G.I. Panov, Ext. Abstr. ZMPC 2000, Sendai (Japan) August 2000, p. 90 (in press in Microporous and Mesoporous Materials).

4 M. Shelef, Chem. Rev., 95 (1995) 209.

5 P. Ciambelli, P. Corbo, M. Gambino, S. Iacoponi, G. Minelli, G. Moretti and P. Porta, Stud. Surf. Sci. Catal., 97 (1995) 295.

6 H.G. Karge and H.K. Beyer, Stud. Surf. Sci. Catal., 69 (1991) 43.

7 F. Cosentino, A. Katovic, G. Giordano, P. Lentz and J. B. Nagy, Stud. Surf. Sci. Catal., 125 (1999) 109.

8 F. Vazzana and G. Centi, Catal. Today, 53 (1999) 683.

9 P. Ratnasamy and R. Kumar, Catal. Today, 9 (1991) 329.

10 R. Kumar, A. Raj, S. Baran Kumar and P. Ratnasamy, Stud. Surf. Sci. Catal., 84 (1994) 109.

11 G. Giordano, F. Di Renzo and F. Fajula, Stud. Surf. Sci. Catal., 83 (1994) 41.

12 R. de Ruiter, J.C. Jansen and H. van Bekkum, Zeolites, 12 (1992) 56.

13 F. Testa, R. Chiappetta, F. Crea, R. Aiello, A. Fonseca and J. B.Nagy, Stud. Surf. Sci. Catal., 94 (1995) 349.
14 G. Giordano, F. Di Renzo, F. Remoué, F. Fajula, D. Plee and P. Schulz, Stud. Surf. Sci. Catal., 84 (1994) 141.

Studies in Surface Science and Catalysis 140
A. Gamba, C. Colella and S. Coluccia (Editors)

Synthesis and adsorption properties of iron containing BEA and MOR type zeolites

G. Giordano[a], A. Katovic[a] and D. Caputo[b]

[a]Dipartimento di Ingegneria Chimica e dei Materiali, Università della Calabria, Via P. Bucci, 87030 Rende (CS), Italy

[b]Dipartimento di Ingegneria dei Materiali e della Produzione, Università Federico II di Napoli, Piazzale V. Tecchio 80, 80125 Napoli, Italy

Due to their particular catalytic properties, iron- and iron-aluminium-substituted zeolites are especially interesting for application in oxidation processes. In this paper the preparation procedure and the adsorption properties of Fe-Al-BEA and Fe-Al-MOR zeolites are described and the results are compared with the data obtained for the corresponding aluminium forms of both zeolites.

1. INTRODUCTION

The present effort in many acid-catalyzed reactions with zeolites is to substitute inorganic Brønsted and Lewis acids by solid acids. The first step in the fine tuning of zeolite properties is usually the isomorphous substitution of metals other than Al into the zeolite framework. The modified product selectivity and conversion are due to the induced changes in the zeolite acidity.

In the last years a great interest was paid to the catalytic properties of iron-containing zeolites that show interesting activities in different industrial reactions. The Fe-BEA zeolite is reported to be a good catalyst in the vapour phase alkylation processes [1], the Fe-TON zeolite shows very high activity and selectivity in the olefin isomerization [2, 3]. Finally, new applications of zeolitic catalysts in the partial oxidation catalysis, such as the Solutia Inc. processes for benzene hydroxylation to phenol using Fe-MFI, open a novel route for the use of zeolites in oxidation processes [4, 5]. On the other hand, the catalytic properties of the metal-modified MOR type zeolite in the isomerization process are well known.

In the present work a series of synthesis was carried out in order to obtain iron-containing zeolites with a pore channel system larger than that of the MFI type zeolite. Attention was paid in particular on the BEA and MOR type zeolites.

This paper deals with the preparation and the adsorption characterization of two different BEA and MOR type, Al- and Fe, Al-containing, zeolites. Iron is incorporated during the zeolite synthesis from a Fe-oxalate complex added to the starting hydrogel [6, 7]. The influence of the Si/(Al+Fe) ratio on the crystallization field of both zeolite types is also investigated.

2. EXPERIMENTAL

The zeolite samples of Al-MOR and Fe,Al-MOR were prepared starting from the following molar composition:

A $x\ Na_2O - 0.2\ TEABr - y\ Al_2O_3 - z\ Fe_2O_3/w\ C_2H_2O_4 - SiO_2 - 20\ H_2O$,

where TEABr stands for tetraethylammonium bromide, $C_2H_2O_4$ is oxalic acid; $0.1 \leq x \leq 0.27$; $0.015 \leq y \leq 0.025$; $0 \leq z \leq 0.025$, w/z was constantly equal to 3.

The zeolite samples of Al-BEA and Fe,Al-BEA were prepared starting from the following molar composition:

B $x\ Na_2O - 0.2\ TEAOH - y\ Al_2O_3 - z\ Fe_2O_3/w\ C_2H_2O_4 - SiO_2 - 20\ H_2O$,

where TEAOH stands for tetraethylammonium hydroxide, $C_2H_2O_4$ is oxalic acid; $0.1 \leq x \leq 0.15$; $0.015 \leq y \leq 0.02$; $0 \leq z \leq 0.015$, w/z was constantly equal to 3.

The hydrogels were obtained by adding precipitated silica (BDH) to a previously prepared homogeneous solution consisting of the organic compound (Fluka), sodium hydroxide (Carlo Erba), sodium aluminate (Carlo Erba) and distilled water. In another beaker, a solution of iron complex with oxalic acid was prepared, starting from iron nitrate (Carlo Erba) and oxalic acid (Carlo Erba). This solution was slowly added to the hydrogel, which, after 1 hour of homogenization, was transferred into the autoclaves.

The hydrothermal syntheses were perfomed in static conditions under autogeneous pressure at 140 and 170 °C for BEA and MOR type respectively, while the crystallization time varied as a function of the inital hydrogel composition and depended mainly on the aluminium and iron content.

The solid products were recovered by filtration, dried and checked by powder X-ray diffraction (XRD). The zeolite samples were further characterized by scanning electron microscopy (SEM), energy dispersive X-ray analysis (EDX), thermal analysis, chemical analysis (by atomic absorption spectrophotometry) and adsorption measurements.

The adsorbed amount of toluene was measured under equilibrium conditions by a gravimetric technique using a McBain-type adsorption balance [8]. The equilibrium gas pressure was measured through electronic devices (Edwards Datametrics 1500), while a Heto thermostating unit allowed a temperature control within the range of ± 0.1°C. Before each measurement the samples were thermally activated at 650 °C in air for 2 hours (for 12 hours only in the case of the sample MOR 756) and afterwards at 350°C under high vacuum ($P<10^{-5}$ mbar) for 3 hours (Edwards Diffstak MK2 diffusion pump). After cooling to 25°C, aliquots of toluene vapour were allowed, from time to time, to enter the balance chamber and a sufficient time was awaited to attain equilibrium. Adsorption isotherms were outlined by plotting the adsorbed toluene amount (on dry basis) versus gas partial pressures at equilibrium.

3. RESULTS AND DISCUSSION

First of all it is useful to underline that an unsuccessful attempt to obtain a pure iron form of BEA and MOR type zeolite was undertaken, but evidently the absence of aluminium in the initial hydrogel does not allow the formation of crystalline phases even for long reaction times.

Table 1
Products obtained starting from the hydrogel A (see experimental)*

Sample	Na_2O	Al_2O_3	Fe_2O_3	Time, days	Products
M1	0.15	0.015	-	7	Q
M2	0.10	0.02	-	9	BEA+Q+L
M3	0.13	0.02	-	7	MOR+Q
M4	0.15	0.02	-	2.5	MOR
M5	0.27	0.02	-	2.5	MOR
M6	0.15	0.025	-	4	MOR
M7	0.15	0.02	0.005	5	MOR
M8	0.15	0.02	0.010	8	MOR
M9	0.17	0.02	0.015	11	MOR
M10	0.21	0.02	0.020	34	MOR+Am
M11	0.23	0.02	0.025	40	Am

* Moles of reactants per mol of SiO_2; Q = quartz; L = layered phase; Am = amorphous.

In Table 1 the results on the synthesis of the Al-MOR and Al-Fe-MOR zeolites are shown. As expected, the amount of Al in the starting hydrogel is the crucial parameter during the crystallization process. Indeed the zeolite MOR was obtained from the mixtures with a Si/Al ratio lower than 33.3. Alkali content influences the nature of the products. Starting from the same Al content, the dense phase (quartz) co-crystallizes together with a zeolitic phase which changes into zeolite BEA when the amount of sodium hydroxide in the starting hydrogel decreased, and a layered phase appeared in the case of the lowest alkali content tested. This confirms that a great amount of sodium hydroxide is necessary for the synthesis of MOR type zeolite [9]. In any case for the formation of Al or Fe-Al-MOR zeolite the pH value of the starting hydrogel must be higher than 13. In the presence of only iron, this reaction system leads to the formation of quartz and in presence of aluminium content ranging from 0.005 to 0.015 no crystalline phases were detected also for long reaction times. Since with this hydrogel system it was impossible to obtain mordenite in the absence of aluminium, we tested the iron incorporation in the system that contains the lowest amount of Al (equal to 0.02). The maximum amount of iron that could be introduced into the zeolite was equal to 0.015 (Si/Fe = 33.3); in the case of higher iron contents in the hydrogel, the system did not evolve versus mordenite formation. The crystallization time was prolonged when iron content in the starting hydrogel increases.

In the system without iron (sample M4), MOR crystallize in 2.5 days, on the contrary for a higher Si/Fe ratio (equal to 33.3, sample M9) 11 days are required. Only for the highest iron loading it was necessary to increase the amount of sodium hydroxide in order to maintain the pH of the hydrogel higher than 13, the value which was previously found to be essential to obtain pure MOR type zeolite.

In Table 2 the results on the synthesis of BEA zeolite type are reported. First of all, when the reaction temperature increased from 140 to 170 °C the crystallization time was shortened, but the thermal stability of the product decreased. As a matter of fact, from the reaction system B2 the BEA type zeolite was obtained after 7 days at 140 °C and after 4 days at 170 °C, but if the reaction time was prolonged up to 8 days at 170 °C, the products were BEA and quartz. On the contrary, at 140 °C only the BEA phase formed even after 18 days. The BEA type zeolite was obtained also in presence of a mixture of TEAOH and TEABr, but the preliminary experiments carried out with this mixture showed that for a complete crystallization it is necessary to prolong twice the crystallization time.

Table 2
Products obtained starting from the hydrogel B (see experimental)*

Sample	Na_2O	Al_2O_3	Fe_2O_3	Time, days	Products
B1	0.10	0.015	-	17	BEA+cr
B2	0.10	0.02	-	7	BEA
B3	0.10	0.02	0.005	10	BEA
B4	0.10	0.02	0.010	18	BEA
B5	0.12	0.02	0.015	21	MOR

* Moles of reactants per mol of SiO_2; cr = cristobalite.

As in the case of the MOR type zeolite, it was impossible to obtain the Al-free (or Fe-BEA), starting from the studied hydrogel system. Consequently, the experiments for the iron introduction were carried out at the lowest Al content in the hydrogel (equal to 0.02).

First of all, it must be noted that the pH values play an important role in the formation of BEA type zeolite. As in the case of the MOR type zeolite the pH value of the starting hydrogel must be higher than 13. The crystallization field of Fe,Al-BEA is more limited than that of the corresponding form of the MOR zeolite. In fact, the iron Al-BEA zeolite crystallized from the hydrogels having a Si/Fe ratio ranging from 100 to 50. Higher amounts of iron in the initial reaction mixture (Si/Fe = 33.3) did not lead to the BEA crystallization, in fact, after long reaction times, the only crystalline phase detected was mordenite (see sample B5). The crystallization time is a function of the iron content in the starting hydrogel. As a matter of fact, it increases with the iron content, e.g., the reaction time was 7 days in the absence of iron (sample B2) and 18 days with a Si/Fe = 50 (sample B4).

Table 3
Chemical and morphological characterization of BEA and MOR type zeolites*

Sample	Si/Al_{gel}	Si/Fe_{gel}	$Si/Al_{zeolite}$	$Si/Fe_{zeolite}$	Morphology	Cryst. size
M4	25	-	24.1	-	rod-like	5x2 μm
M7	25	100	24.7	59.1	rod-like	9x4 μm
M8	25	50	24.9	37.5	butt.-like	12x6μm
M9	25	33.3	23.9	26.6	butt.-like	13x4μm
B2	25	-	23.7	-	spheres	0.2-0.6μm
B3	25	100	28,1	59.4	spheres irr	1-2μm
B4	25	50	28.5	38.9	spheres irr	1-3μm

*"butt." stands for butterfly and "irr." for irregular.

The results on the aluminium and iron incorporation into the crystalline solid are presented in Table 3. Bearing in mind that the Si/Al ratio is constant for the examined range of the Si/Fe ratios, it can be noticed that the incorporation of Al does not depend on the amount of iron present in the starting hydrogel for the MOR type zeolite (samples M4 - M7). No differences were observed in absence of iron (see samples M4 and M9). Secondly, the Fe content in the zeolite increases proportionally with its initial value. The morphology and the crystal size of MOR samples change with the iron content in the crystals. The iron free sample and the sample with a smaller amount of iron in the zeolite (samples M4 and M7) show crystals with a rod-like morphology, having length and width that increase from 5 x 2 μm up to 9 x 4 μm. This difference is probably due to the longer reaction time (induction and growth) required for sample M7 respect to the sample M4 (see Table 1). The iron-rich samples (M8 and M9) present a butterfly-like morphology with greater dimensions then M4 and M7 samples. Probably, also this increase in the crystal size is due to the crystallization time required for the formation of the zeolite (see Table 1).

Similarly to the MOR zeolite, also the BEA type zeolite presents an aluminium incorporation not depending on the iron amount in the hydrogel. A decrease in Al incorporation was observed for the iron samples compared to the iron-free sample (samples B2–B4). The iron content in the BEA crystals increases with the amount of iron present in the initial reaction mixture. The morphology of the BEA crystals is spherical and changes from regular to irregular with the introduction of iron. The crystal size also increases with the presence of iron and this is probably due to the longer time required for the crystallization (see Table 1).

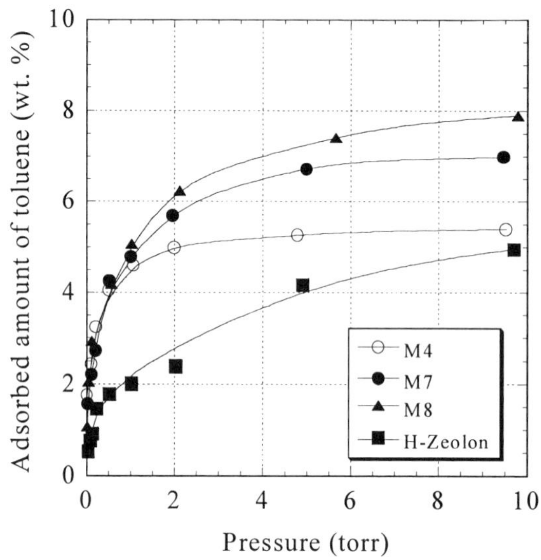

Fig. 1. Adsorption isotherms of toluene at 25 °C for the zeolite MOR (samples M4, M7, M8), synthesized in this study, compared to that of the commercial H-Zeolon.

The results obtained from the toluene adsorption measurements on the BEA and MOR type zeolites are shown in Figs. 1 and 2, respectively. The results of Fig. 1 show that all MOR type zeolites obtained with the procedure described in the experimental part present a channel systems with large pores. As a matter of fact, the commercial sample H-Zeolon, taken as a large pore reference mordenite, showed an adsorption capacity lower than that of the Al-MOR and Al-Fe-MOR samples. The adsorption capacity of the iron-aluminium form was greater than that of the aluminium form and increased with the amount of iron incorporated in the structure. This indicates that iron introduction into the zeolitic framework is not competitive with the aluminium's incorporation. The proof of the iron incorporation into the MOR framework is the modification of the crystal morphology (see Table 3, sample M4 and M8, M9) of the Al-MOR and Al-Fe-MOR samples. In the absence of iron, rod-like crystals of 5 x 2 μm in size are formed. Starting from the Al-Fe-MOR zeolite the crystal morphology changes drastically and assumes a butterfly-like shape (4 x 13 μm for sample M9).

Further evidence about the iron incorporation was given by thermal analysis (not presented here). In fact, only in the DSC curves of Al-Fe-MOR samples there was a shoulder at high temperature.

The unchanging amount of Al in the zeolite MOR samples indicates that the isomorphous substitution of the Fe^{3+} cations takes place at the expenses of the framework Si. Most likely, the observed pore expansion of the Al-Fe-MOR zeolite, demonstrated from the data of Fig. 1, is due by the substitution of the longer Fe-O bond (1.84 Å) in the place of the Si-O bond (1.61 Å).

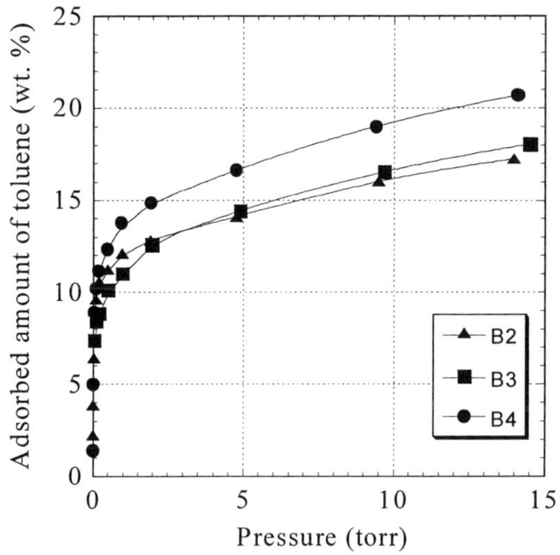

Fig. 2. Adsorption isotherms of toluene at 25 °C for the zeolite BEA, synthesized in the present study (samples B2, B3 and B4 in Table 2).

A similar behaviour was also shown by the BEA type zeolite, Fig. 2. The results show that the iron-poor sample (B3) present an adsorption capacity equal to the iron-free form (sample B2). Only at high pressure the adsorption capacity of the sample B3 is greater than that of the sample B2. On the contrary the iron-rich form (sample B4) presented, in the entire pressure range, a higher adsorption capacity. As in the case of the MOR type zeolite, we can assume that also in the BEA zeolite iron is incorporated in the framework. As a support to this hypothesis there are the differences in the morphology and crystal size and the the increased adsorption capacities interpreted in terms of pore expansion due to the formation of Fe-O bond in the BEA structure.

A further indirect support to Fe incorporation in both BEA and MOR zeolites is that after calcination at 750 °C the relevant samples exhibited a white colour. If iron had not been in the zeolitic structure the samples would have become brown after thermal treatment.

4. CONCLUSIONS

The use of iron oxalate complex is a good way for the preparation of Al-Fe-BEA and Al-Fe-MOR zeolites. The limit of the described procedure is that it is impossible to obtain pure iron forms, because in the absence of aluminium in the starting hydrogel no BEA or MOR type zeolitic phases formed. The incorporation of iron in the zeolitic framework is proportional to its amount in the initial reaction mixture. The maximum iron incorporation in the two zeolites corresponds to a Si/Fe ratio in the hydrogel equal to 33 and 50 for MOR and BEA zeolites, respectively. The iron incorporation modifies the morphology and the crystal size of the products.

The presence of iron increases the adsorption capacity of both BEA and MOR zeolites. In the case of MOR type zeolite all samples obtained with this procedure show a large pore channel system, larger than the commercial H-Zeolon, used as a reference mordenite.

REFERENCES

1. A.V. Smirnov, F. Di Renzo, O.E. Lebedeva, D. Brunel, B. Chiche, A. Tavolaro, B.V. Romanovsky, G. Giordano, F. Fajula and I. Ivanova, Stud. Surf. Sci. Catal., 105 (1997) 1325.
2. M.A. Asensi, A. Corma and A. Martinez, Journ. Catal., 158 (1996) 561.
3. M.A. Asensi, A. Corma, A. Martinez, M. Derewinski, J. Krysciak and S.S. Tamhankar, Appl. Catal (A), 174 (1998) 163.
4. L.V. Pirutko, D.P. Ivanov, K.A. Dubkov, V.V. Terskikh, A.S. Kharitonov and G.I. Panov, Proc. 12th Int. Zeolite Conf., Baltimore, 1998, p. 1245.
5. G. Centi, G. Grasso, F. Vazzana and F. Arena, Stud. Surf. Sci. Catal., 130, 2000, p. 635.
6. R. Kumar, A. Raj, S. Baran Kumar and P. Ratnasamy, Stud. Surf. Sci. Catal., 84 (1994) 109.
7. G. Giordano, A. Katovic, A. Fonseca and J. B.Nagy, Stud. Surf. Sci. Catal., 2001, in press.
8. D. Caputo, B. de Gennaro, M. Pansini, C. Colella, In "Natural Microporous Materials in Environmental Technology", P. Misaelides, F. Macasek, T.J. Pinnavaia and C. Colella (eds.), NATO Science Series, Series E: Applied Sciences, Vol. 362, Kluwer Academic Publishers, Dordrecht, 1998, p. 225.
9. P.A. Jacobs and J.A. Martens (eds.), Synthesis of High-silica Aluminosilicate Zeolites, Elsevier, Amsterdam, 1984.

Studies in Surface Science and Catalysis 140
A. Gamba, C. Colella and S. Coluccia (Editors)

Local Structures of Active Sites on Mo-MCM-41 Mesoporous Molecular Sieves and their Photocatalytic Reactivity for the Decomposition of NO_X

Shinya Higashimoto[a], Rie Tsumura[a], Masaya Matsuoka[a], Hiromi Yamashita[a], Michel Che[b], and Masakazu Anpo[a]*
[a]Department of Applied Chemistry, Graduate School of Engineering
Osaka Prefecture University
1-1 Gakuen-cho, Sakai, Osaka 599-8531, Japan
anpo@ok.chem.osakafu-u.ac.jp
[b]Laboratoire de Réactivité de Surface, Université P. et M. Curie, UMR7609, CNRS
4 Place Jussieu, 75252 Paris Cedex 05, France

Investigations of mesoporous Mo-MCM-41 molecular sieves have shown that they include the tetrahedrally coordinated Mo-oxides, which exhibits a photoluminescence spectrum attributed to the radiative decay from the charge transfer excited triplet state. Also, the photocatalytic decomposition reaction of NO into N_2 was found to proceed effectively and efficiently by the coexistence of CO. The dynamic quenching of the photoluminescence spectrum by the addition of NO or CO indicates that the charge transfer excited triplet state of the Mo-oxides having a tetrahedral coordination plays a significant role in this reaction.

1. INTRODUCTION

Zeolites involving transition metal ions within their framework and cavities are known to exhibit unique and fascinating properties for applications not only in catalysis but also for various photochemical processes [1,2]. Zeolites incorporated with transition metal ions in a highly dispersed state were found to exhibit photocatalytic reactivity for such reactions as oxidation using NO or O_2 as an oxidant and the metathesis reaction of olefins. Also, the urgency of the reduction of global air pollution makes the need to address the elimination of CO as well as NO_x of great concern. Since CO poisons the catalysts and causes a decline in reactivity, it is vital that the reaction can proceed even in the presence of CO.

The present study deals with a characterization of the local structure of the Mo-oxides of Mo-MCM-41 and their photocatalytic reactivity, significantly for the decomposition of NO into N_2 in the presence of CO by means of XAFS (XANES and EXAFS), XRD, ESR and in-situ photoluminescence measurements.

2. EXPERIMENTAL

The Mo-MCM-41 mesoporous molecular sieves (0.5, 1.0, 2.0, 4.0 Mo wt%) were synthesized using tetraethylorthosilicate (TEOS) and $(NH_4)_6Mo_7O_{24}{\cdot}4H_2O$ as the starting materials and cetyl trimethylammonium bromide (CTMABr) as the template in accordance with previous literature [3]. After the products were recovered by filtration, washed with distilled water several times and dried at 373 K for 12 h, calcination of the samples were carried out under a dry flow of air at 773 K for 8 h. Prior to spectroscopic measurements and photocatalytic reactions, the catalysts were degassed at 773 K for 2 h and calcined in O_2 (> 20 Torr) at 773 K for 2 h, then degassed at 473 K for 2 h. The photoluminescence and lifetimes were measured at 77 K with a Shimadzu RF-501 spectrofluorophotometer and an apparatus for lifetime measurements, respectively. XAFS (XANES and EXAFS) spectra were obtained at the BL-10B facility of the High Energy Acceleration Research Organization (KEK) in Tsukuba. The XAFS spectra of the dehydrated samples were recorded at the Mo K-edge absorption in the transmittance mode at 293 K. The EXAFS data were examined by the EXAFS analysis program Rigaku EXAFS (REX). The photocatalytic reactions of NO in the absence and presence of CO molecules were carried out at 298 K with a high pressure mercury lamp through a UV cut filter (λ > 270 nm). The products were analyzed by online gas chromatography.

3. RESULTS AND DISCUSSION

3.1. Local Structure of Mo-oxides in the Mo-MCM-41 Catalyst

The results of the XRD patterns and the BET surface area of the Mo-MCM-41 catalyst indicated that they have a hexagonal lattice having mesopores larger than 20 Å and that they possess a high BET surface area (≒1000 m^2/g) as compared with amorphous silica (300 m^2/g), so that they can be considered effective photocatalysts [3].

Figure 1 shows XAFS (XANES and EXAFS) spectra of the Mo K-edge spectra of Mo-MCM-41 to be: 1.0, 2.0 and 4.0 Mo wt%. A characteristic feature of the XANES spectra of Mo-MCM-41 is the appearance of a pre-edge peak due to the so-called 1s-4d transition of the Mo atoms, suggesting the presence of a terminal Mo-oxonium group (Mo=O). It was also found that the shape of the XANES

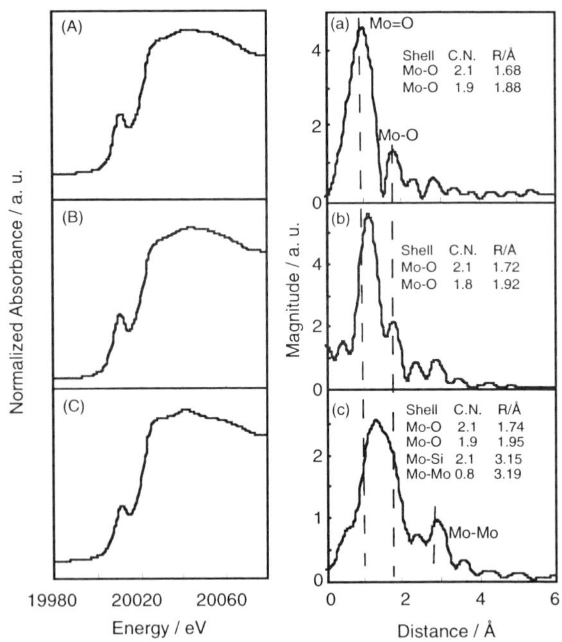

Fig. 1. XANES and FT-EXAFS spectra of the Mo-MCM-41 with various Mo loadings of (A, a) 1.0, (B, b) 2.0, and (C, c) 4.0 Mo wt%. C.N.;coordination number, R/Å; interatomic distance

spectra are quite similar to that of the K_2MoO_4 compound with four-fold coordinated molybdenum atoms, indicating that Mo-MCM-41 consists of Mo-oxides having a tetrahedral coordination. The pre-edge peak intensity is the highest in Mo-MCM-41 (1.0 Mo wt%), however, it decreases slightly and the shape of the XANES spectra changes when the Mo content is increased from 1.0 to 4.0 Mo wt%. In the EXAFS radial structure function of Mo-MCM-41 (1.0 Mo wt%), there are well-resolved peaks due to the presence of neighboring oxygen atoms (Mo-O) which can be observed at ca. 0.8-2.0 Å (without phase-shift correction). On the other hand, an additional peak can be observed at ca. 3.0 Å due to the Mo-O-Mo bond or Mo-O-Si bond in Mo-MCM-41 (4.0 Mo wt%). The curve fitting analysis showed that highly dispersed tetrahedrally coordinated Mo-oxides are formed having two shorter Mo-O bonds (1.68 Å) and two longer ones (1.88 Å) in Mo-MCM-41 (1.0 Mo wt%), while in Mo-MCM-41 (4.0 Mo wt%), oligomeric tetrahedral Mo-oxides $(MoO_4^{2-})_n$ with an additional Mo-O-Mo bonds (3.19 Å) and Mo-O-Si (3.15 Å) could be observed.

Mo-MCM-41 (1.0 Mo wt%) exhibits a photoluminescence spectrum at around 400-600 nm upon excitation at around 295 nm (defined as X), which coincides with the photoluminescence spectrum of the tetrahedrally coordinated Mo-oxide species highly dispersed on SiO_2, as shown in Fig. 2 [1,4]. After the addition of NO molecules on the Mo-MCM-41 catalyst at the room temperature, the efficient quenching of the photoluminescence spectrum was found at 77 K, their intensity depending on the amount of added NO molecules accompanied by the shortening of the emission lifetime of the excited triplet state. According to the following equation: $\tau = \Phi / k_q$, in the given constant k_q, if the photoluminescence yields are quenched, the lifetimes of the charge transfer excited triplet state become short in the presence of the quenching molecules. (k_q, τ, Φ are the absolute reaction rate constant, the lifetime and the photoluminescence yields of the charge transfer excited triplet state in the presence of quenching molecules, respectively.) The excitation and emission spectra are attributed to the following charge transfer processes on the Mo-O moieties of the tetrahedral molybdate ions (MoO_4^{2-}), involving an electron transfer from the O^{2-} to Mo^{6+} ions and a reverse radiative decay from the charge transfer excited triplet state. The width and the wavelength at the maximum intensity of the emission band do not change upon varying the excitation wavelength, indicating that there is only one luminescent with a

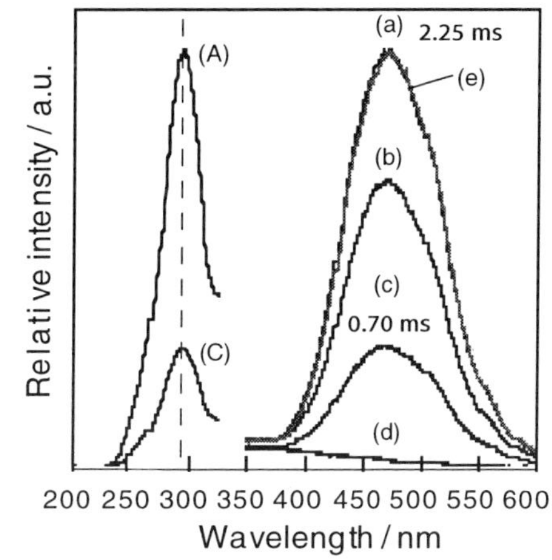

Fig. 2. Effect of the addition of NO on the photoluminescence spectrum and its excitation spectrum of the Mo-MCM-41 with a Si/Mo ratio of 160. Pressure of added NO : (A, a) 0, (b) 0.07, (C, c) 0.4 Torr, (d) excess, (e) degassed after (d).

$$[\,Mo^{6+} = O^{2-}\,] \xrightarrow{h\nu} [\,Mo^{5+} = O^{-}\,] \xrightarrow{h\nu'} [\,Mo^{6+} = O^{2-}\,]$$

photoluminescence lifetime of 2.25 ms at 77 K. These results indicate that the Mo-oxides showing absorption in the region of 295 nm, form a tetrahedral coordination in a highly dispersed state.

On the other hand, as shown in Fig. 3, there are at least two luminescent species (the absorption spectrum can be deconvoluted into two components having wavelength regions of X and Y in Fig. 3: 295 and 310 nm, respectively) on Mo-MCM-41 (4.0 Mo wt%) [5]. The increase of Mo content, from 1.0 to 4.0 Mo wt%, leads to the formation of not only the emitted X site with a photoluminescence lifetime of 2.25 ms but also another emitting site (in the region of Y) probably induced by Mo-Mo interactions via oxygen ions (Mo-O-Mo), which cause a more efficient radiationless energy leading to a decrease in the photoluminescence lifetime (0.91 ms). Taking into account the XAFS data, only the isolated tetrahedrally coordinated Mo-oxides are formed in lower Mo loadings, while two types of tetrahedrally coordinated Mo-oxides at highly dispersed levels and $(MoO_4^{2-})_n$ are formed in higher Mo loadings, as shown in Fig. 4.

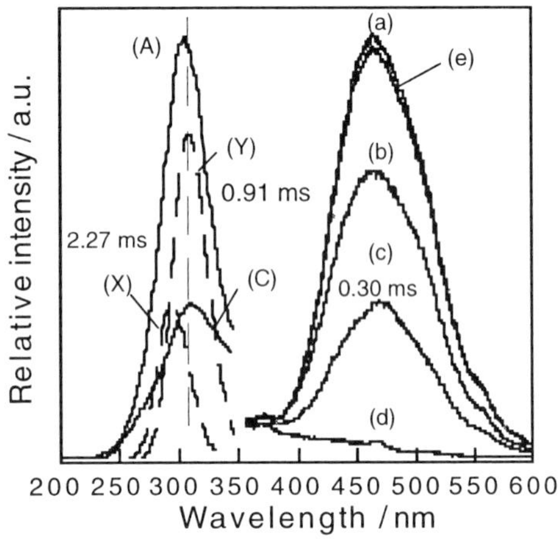

Fig. 3. Effect of the addition of NO on the photoluminescence spectrum and its excitation spectrum of the Mo-MCM-41 with a Si/Mo ratio of 40. Pressure of added NO : (A, a) 0, (b) 0.6, (C, c) 4 Torr, (d) excess, (e) degassed after (d). Spectrum (A) can be deconvoluted into (X) and (Y).

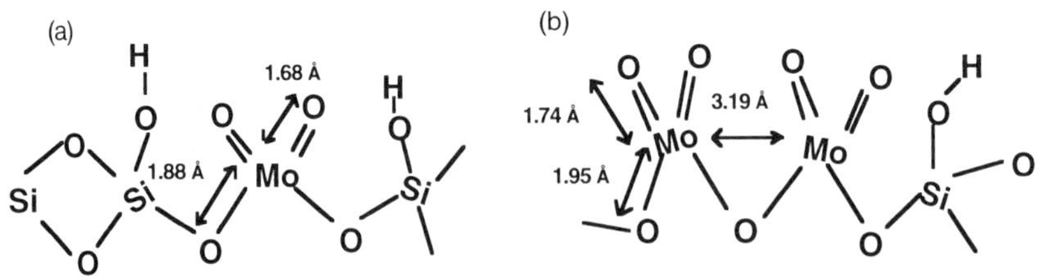

Fig. 4. Proposed local structures of the isolated (a) as well as oligomeric (b) Mo-oxide sites in the Mo-MCM-41 catalysts.

3.2. Photocatalytic Decomposition Reaction of NO by the coexistence of CO on the Mo-MCM-41 Catalyst and its Photocatalytic Reactivity

In blank tests, silicious MCM-41 or bulk MoO_3 did not exhibit photocatalytic reactivities in the presence of a mixture of NO and CO under UV-irradiation. The photocatalytic decomposition reactions of NO in the absence and presence of CO were performed on Mo-MCM-41 mesoporous molecular sieves. UV-irradiation of the Mo-MCM-41 in the presence of NO led to the evolution of N_2 as well as N_2O and NO_2. Moreover, it was found that the photocatalytic decomposition reactions of NO was dramatically enhanced by the coexistence of CO, leading to the formation of N_2 and CO_2. UV-irradiation of Mo-MCM-41 in the presence of a mixture of NO and CO leads to the formation of N_2 and CO_2 with a good linearity against the UV-irradiation time, while the turnover number (TON), (defined as the value of the number of photo-formed N_2 molecules divided by the total number of Mo species in the catalyst) exceeded unity after irradiation for 2 h, as shown in Fig. 5. These results clearly indicate that the reaction proceeds photocatalytically [6]. After UV-irradiation for 3 h, NO conversion and selectivity for the formation of N_2 reached close to 100%, leading to the formation of small amounts of N_2O during this reaction in gas. Figure 6 shows the relationship between the yields of N_2 formation for the decomposition of NO on Mo-MCM-41 (0.5-4.0 Mo wt%) by the coexistence of CO and the relative intensity of the absorption spectra observed in the total region (X and Y) of the catalyst.

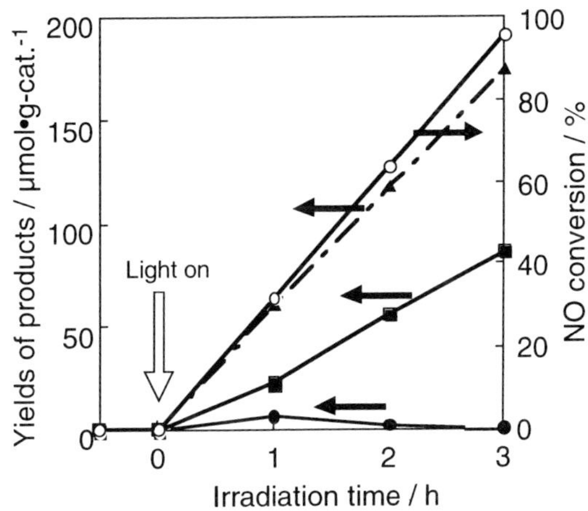

Fig. 5. Reaction time profiles of the photocatalytic decomposition reaction of NO by the coexistence of CO on Mo-MCM-41 (1.0 Mo wt%). Yields of CO_2 (▲), N_2 (■), N_2O (●), conversion of NO (○). Amount of added NO or CO; 180 μmol•g-cat^{-1}

The yields of N_2 formation for the photocatalytic reactions have a good relationship with the intensities of the photoluminescence as well as with the amount of generated Mo^{4+} ions through the photo-reduction of Mo^{6+} with CO (the number of Mo^{4+} ions are estimated by the number of photo-formed CO_2 molecules) [6]. Moreover, the yields of N_2 per Mo atoms of 1 g in the photocatalyst are almost constant in lower loadings of Mo (0.5, 1.0 Mo wt%), however, they dramatically decrease in higher Mo loadings (2.0, 4.0 Mo wt%). These results indicate that the photo-reduced Mo^{4+} ions by the co-existence of CO through the charge transfer excited triplet state of both tetrahedral Mo-oxides at the highly dispersed level as well as $(MoO_4^{2-})_n$ play a significant role in the photocatalytic decomposition of NO, leading to the formation of N_2 and CO_2. In particular, this reaction proceeds effectively and

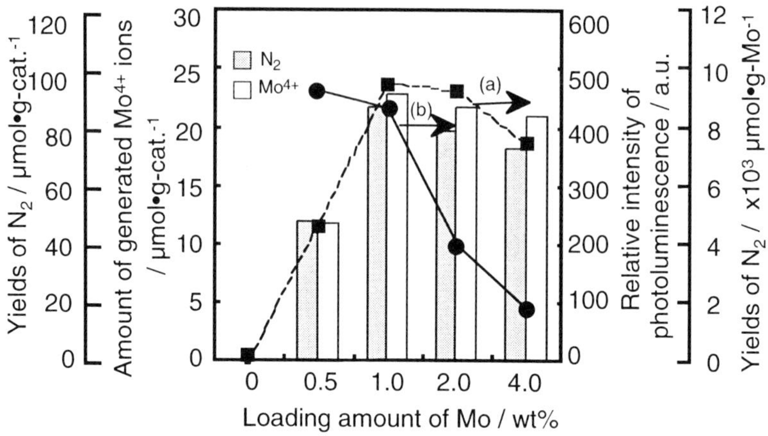

Fig. 6. Relationships between the yields of N_2 formation for the photocatalytic decomposition reactions of NO by the coexistence of CO for 3 h (left), amount of the generated Mo^{4+} ions in the photoreduction of Mo^{6+} with CO under UV-irradiation for 0.5 h, the relative intensity of the photoluminescence (dotted line; a) and the yields of N_2 per Mo atoms of 1g (solid line; b) of Mo-MCM-41 (0.5, 1.0, 2.0, and 4.0 Mo wt%). Added NO or CO: 180 μmol•g-cat.$^{-1}$

efficiently on the Mo-MCM-41 catalysts having highly dispersed tetrahedrally coordinated Mo-oxides.

UV-irradiation of Mo-MCM-41 in the presence of CO alone and its subsequent evacuation at 293 K led to an efficient quenching of the photoluminescence. Moreover, no ESR signals due to the Mo^{5+} ions were detected under UV-irradiation, suggesting that the charge transfer excited triplet state of the $[Mo^{5+} - O^{-}]^*$ complex reacts with CO, leading to the formation of Mo^{4+} ions as well as CO_2. Exposure of the photo-reduced Mo^{4+} ions into NO molecules led to the formation of N_2O under dark conditions [7]. In fact, after subsequent evacuation, the photoluminescence intensity recovered, although not to its original intensity due to the formation of such carbonyl species as $[Mo^{4+} - CO]$, which decompose only upon heating at temperatures above 373 K [6-10]. In fact, the exposure of NO or N_2O onto Mo^{4+} ions under dark conditions led to the formation of N_2O and N_2, respectively. After the subsequent evacuation of the catalyst, the photoluminescence intensity recovered by the oxidation of Mo^{4+} to Mo^{6+} ions. Previously, Kazansky et al. proposed the reaction mechanism for the photocatalytic decomposition of NO by the coexistence of CO in the MoO_3/SiO_2 system [7]. These results support that the following reactions play a role in the photocatalytic decomposition reaction of NO in the presence of CO. In-situ photoluminescence and ESR measurements demonstrated that this reaction proceeds in a redox cycle between

$$[Mo^{6+} - O^{2-}] \xrightarrow{h\nu} [Mo^{5+} - O^{-}]^*$$

$$[Mo^{5+} - O^{-}]^* + CO \longrightarrow Mo^{4+} + CO_2$$

$$Mo^{4+} + NO \longrightarrow Mo^{6+} \cdots NO^{2-}$$

$$Mo^{6+} \cdots NO^{2-} + NO \longrightarrow Mo^{6+} = O^{2-} + N_2O$$

$$Mo^{4+} + N_2O \longrightarrow Mo^{6+} = O^{2-} + N_2$$

alternating Mo^{6+} and Mo^{4+} ions, *i.e.*, the photo-formed Mo^{4+} ions through the reaction of the charge transfer excited triplet state of Mo-oxides with CO are oxidized to the original $Mo^{6+} = O^{2-}$ species in the presence of NO or N_2O.

4. CONCLUSIONS

Investigations of the local structures of mesoporous Mo-MCM-41 molecular sieves using various spectroscopic methods showed that in such systems the highly dispersed tetrahedrally coordinated Mo-oxides are formed in lower Mo loadings, while tetrahedrally coordinated Mo-oxides $(MoO_4^{2-})_n$ with an additional Mo-O-Mo bond are formed in higher Mo loadings. The photocatalytic decomposition reactions of NO into N_2 and CO_2 products were found to proceed efficiently even in the coexistence of CO. In-situ photoluminescence studies in these systems clearly indicated that the charge transfer excited triplet state of the tetrahedrally coordinated Mo-oxides plays a significant role in these reactions and this reaction proceeds in a redox cycle between the Mo^{6+} and Mo^{4+} ions.

RERERENCES

1. M. Anpo, and M. Che, Adv. Catal., 44 (1999) 119, and references therein.
2. M. Anpo ed., Photofunctional Zeolites, NOVA Science Publishers Inc. (2000).
3. W. Zhang, J. Wang, P. T. Tanev, and T. J. Pinnavaia, J. Chem. Soc., Chem. Commun. (1996) 979.
4. M. Anpo, M. Kondo, S. Coluccia, C. Louis, and M. Che, J. Am. Chem. Soc., 111 (1989) 8791.
5. S. Higashimoto, R. Tsumura, S. G. Zhang, M. Matsuoka, H. Yamashita, C. Louis, M. Che, and M. Anpo, Chem. Lett., (2000) 408.
6. R. Tsumura, S. Higashimoto, M. Matsuoka, H. Yamashita, M. Che, M. Anpo, Catal. Lett. 68 (2000) 101.
7. I. R. Subbotina, B. N. Shelimov, V. B. Kazansky, A. A. Lisachenko, M. Che, and S. Coluccia, J. Catal., 184 (1999) 390.
8. C. C. Williams, and J. G. Ekerdt, J. Phys. Chem., 97 (1993) 6843.
9. B. N. Shelimov, A. N. Pershin, and V. B. Kazansky, J. Catal., 64 (1980) 426.
10. A. N. Pershin, B. N. Shelimov, and V. B. Kazansky, Kinet. Katal., 21 (1980) 494.

Studies in Surface Science and Catalysis 140
A. Gamba, C. Colella and S. Coluccia (Editors)

The factorial experimental design applied to the zeolite synthesis

A. Katovic[a], M. Cosco[b], P. Cozzucoli[b] and G. Giordano[a]

[a]Dipartimento di Ingegneria Chimica e dei Materiali e [b]Dipartimento di Economia e Statistica, Università della Calabria, Via P. Bucci, 87030 Rende (CS), Italy

The experimental design is an efficient methodology that can be applied to the preparation of zeolites, with the advantage of reducing the number of experiments. In this work the factorial experimental design has been applied to the preparation of the FAU type zeolite.

1. INTRODUCTION

The synthesis of zeolites is influenced by many chemical and physical factors. The first step in the zeolite synthesis is the determination of the crystallization fields. There are always a lot of inevitable events that can result in an unwilling surprise during a synthesis of a chosen zeolite type, even when the procedure is well-known and studied by other research groups. This is due to the fact that the exact synthesis conditions are hardly reproduced according to the specifications found in the literature because of the variability of the sources of materials and circumstances of each laboratory. Considering that a variation of only one parameter, i.e., silica source, hydrogel preparation mode, etc, can change the nature of the final product, the synthesis procedure is usually strictly planned. This requires a great deal of experimental work that proceeds usually by changing one of the factors at a time. The experimental design is a tool that can be successfully applied in order to reduce the number of experiments and to evaluate the main and the joint effects of the various factors on the response [1-3].

Recently the combinatorial chemistry was proposed and applied in the field of inorganic chemistry for applications concerning thin film preparation [4], luminous bodies [5], heterogeneous catalysts [6]. The use of this methodology requires a large number of experiments, whereas the approach based on computer simulation (such as Molecular Dynamics) is relatively time consuming. On the contrary, experimental design requires few experimental data and the mathematical model is relatively simple.

The aim of this work was to apply of the well-known experimental plans used in various areas of the engineering planning in the field of zeolite synthesis. The scope was to understand how the particular aspects of zeolite chemistry can be fitted by the methodology of the full factorial experimental design in order to obtain accurate and reproducible responses in agreement with the empirically acquired results.

Based on the different Si/Al ratios, two faujasite-type zeolites are known: zeolite X (Si/Al = 1–1.5) and zeolite Y (Si/Al = 2-3), both presenting a three dimensional channel system with large pore openings (ca. 7.4 Å) and large cavities (13 Å). The aluminum content in the zeolite crystals determines the characteristics and therefore their application: zeolite X is used for ion exchange, adsorption and drying processes, wheres zeolite Y is utilized in heterogeneous catalysis [7]. In this work the zeolite X was chosen for the investigation.

In this work a 2^3 factorial experimental design was used to examine the synthesis of zeolite X (FAU).

2. EXPERIMENTAL

In a first set of experiments the following molar composition of the hydrogel was chosen for the synthesis of the FAU-type zeolite:

I $$2\ Na_2O - 0.5\ Al_2O_3 - A\ SiO_2 - B\ H_2O.$$

The second set of experiments was undertaken from the following starting mixture:

II $$B\ Na_2O - 0.5\ Al_2O_3 - A\ SiO_2 - 100\ H_2O.$$

In both cases two chemical factors were considered variable (A and B), while the crystallization time was considered as a third factor (C). The other factors were maintained constant: temperature (70 °C), silica (water-glass, Merck) and alumina source (sodium aluminate, Carlo Erba) and the preparation mode of the hydrogel. The crystallization runs were performed in polypropylene bottles in static conditions.

Using the 2^3 factorial experimental design plan, the influence of the chosen factors on the finally obtained product was determined. The content of the zeolite (response) found in the samples was quantitatively determined by X-ray diffraction technique. The yield of the zeolite X in the solid product was determined by confronting each XRD pattern with the XRD patterns of the commercial zeolite 13X (UOP standard) and mixtures prepared with the same and amorphous phase. The intensities of the five most intense peaks were selected for the quantitative measurements.

3. RESULTS AND DISCUSSION

According to he 2^3 factorial experimental design, 11 runs were performed in the first set, including three experiments with central values of A, B and C factors (see Table 1). The experimental factors chosen were the molar content of silica (A), the molar content of water (B) and the crystallization time. This first set of experiments resulted useful as a screening test for the verification of the choice of the synthesis parameters.

Table 1

Factors considered in the first set of runs

Factors	Low level (-1)	High level (+1)	Central point
A = SiO_2	1 mole	2 moles	1.5 moles
B = H_2O	50 moles	100 moles	75 moles
C = cryst. time	6 h	18 h	12 h

The results obtained are shown in Fig. 1 as a cube-plot representation. The zeolite FAU was obtained only in four runs and in the case of the three central points (yield: 0.4/0.43/0.45 not presented in the figure).

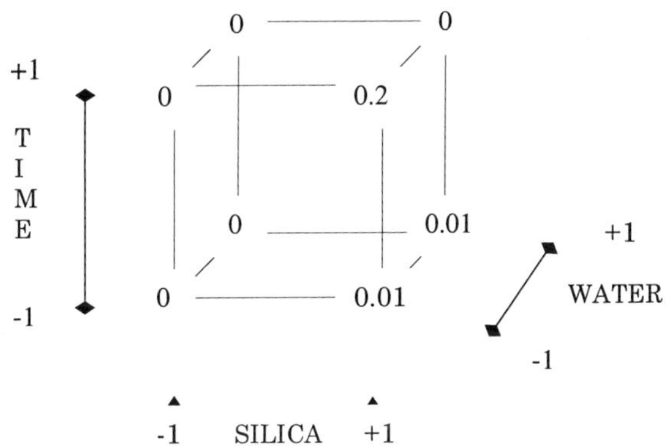

Fig. 1. Cube-plot representation of the experimental results in the first set of experiments.

Table 2

Factors considered in the second set of runs

Factors	Low level (-1)	High level (+1)	Center point
A = SiO_2	2 moles	4 moles	3 moles
B = Na_2O	1.5 moles	2.5 moles	2 moles
C = cryst. time	18 h	30 h	24 h

The amount of the FAU crystalline phase never exceeded 50 wt % of the solid product. As expected, the other crystalline phase present in the final product was zeolite LTA. Accordingly, the levels of the experimental factors are in the range of the overlapping crystallization fields of both zeolites. From the estimates of factor effects and the corresponding confidence intervals (not presented), indications emerged for a larger level of factor A (SiO_2 content) and C (crystallization time). The factor B (H_2O content) resulted statistically not significant, so the value of 100 moles was saved for the further set of experiments.

Table 3
The 2^3 factorial design containing the main factors, the interactions and the response.

	A	B	C	AB	AC	BC	ABC	Response
(1)	-	-	-	+	+	+	-	0.01
a	+	-	-	-	-	+	+	0.2
b	-	+	-	-	+	-	+	0.9
ab	+	+	-	+	-	-	-	0.01
c	-	-	+	+	-	-	+	0.15
ac	+	-	+	-	+	-	-	0.5
bc	-	+	+	-	-	+	-	0.95
abc	+	+	+	+	+	+	+	0.015
	0	0	0	0	0	0	0	0.5
	0	0	0	0	0	0	0	0.55
	0	0	0	0	0	0	0	0.6

The second series of experiments was run with the values of the experimental factors reported in Table 2, whereas the data concerning the factorial design and the corresponding responses are reported in Table 3.

It can be seen that the FAU zeolite is obtained from the hydrogels having a low Si/Al ratio and a high sodium hydroxide contents. The co-crystallization of LTA and FAU zeolites is observed in the case of synthesis procedures (1) and c (the content of LTA is 0.4 and 0.1, respectively, not reported in Table 3). In the first case the level of all experimental factors is low, whereas in the second the silica level is low, so the presence of the LTA zeolite was expectable. The unexpected fact was that the zeolite FAU was not obtained in a higher amount from the hydrogels with the highest silica content. Probably the high content of sodium hydroxide prevents the formation of the zeolite FAU, by affecting the nucleation step of the crystallization process.

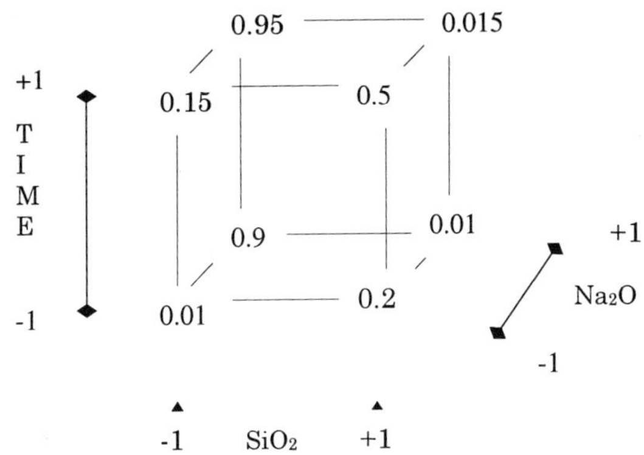

Figure 2. Cube-plot representation of experimental results for the second set of experiments.

It is difficult to draw definite indications from the results of the second set of experiments reported in Fig. 2, because there is a negative effect of factor A and the response (the yield of the zeolite FAU in the final product) changes in a different way when the factor B passes from the lowest to the highest level, suggesting an antagonistic effect between these two factors.

Table 4
The factor effect estimates

Effects	A	B	AB	C	AC	BC	ABC
Estimates	-0.321	0.253	-0.591	0.123	0.028	-0.096	-0.051

Table 5
The confidence intervals

Effects	Lower confidence limit	Upper confidence limit
A	-0.473	-0.169
B	0.101	0.409
C	-0.028	0.275
AB	-0.743	-0.439
AC	-0.123	0.180
BC	-0.248	0.052
ABC	-0.203	0.100

The first statistical confirmation of the above observations comes from the values of the estimates of factor effects reported in Table 4 and the related 95% confidence intervals (see Table 5).

The only intervals, that do not contain zero, i.e., those concerning the main effects of silica (A), sodium hydroxide (B) and their interaction (AB), must be considered statistically significant. From the practical point of view, the statistically non significant effect of the crystallization times tested means that once the zeolite FAU starts to crystallize alone it is thermodynamically stable for the given experimental conditions.

It is useful to find a model that describes, in the field of the experimental conditions under observation, the relation between the response and the factors proved to be significant. The model might be useful for interpolation, forecast, as well as verification of the real significance of the effects.

From the obtained results, the hypothetical model can be written in the following way:

$$y = \beta_0 + \beta_1 x_1 + \beta_2 x_2 + \beta_{12} x_1 x_2 + \varepsilon$$

where y defines the response, x_1 and x_2 the A and B factors, respectively, expressed in coded units, that assume values −1 and +1 for the low and high factor level. The term x_1x_2 stands for the interaction between the two mentioned factors A and B. ε defines the casual error, while β_0, β_1, β_2 and β_{12} are the unknown coefficients that have to be estimated.

The model can be expressed in the matrix form, becoming:

$$y = X\beta + \varepsilon,$$

where y stands for the observation vector of the response variable having a 11 x 1 dimension, while X is the observation matrix of the coefficients of the 11 x 4 dimension.

In the case of the experimental conditions chosen in the second series, the estimate regression model can be written in following way:

$$y = (0.398636) - (0.161625)x_1 + (0.126875)x_2 - (0.295625)x_1x_2.$$

It can be seen that every estimated coefficient is exactly equal to one half of the corresponding effect. In the case of the effect of factor A, this means that if x_1 increases for a unity (coded) the response should decrease for 0.160625 in percentage points. All this confirms in terms of numerical values the observations discussed above. To test the adequacy of the proposed estimated model, it is necessary to analyze the variance, the significance of every single parameter and at the end, the test lack of fitting [1]. All these procedures were done and the estimated model was found to predict the relation existing between the factors and the response for the chosen experimental conditions. The further step is to utilize the estimated model for the forecast in order to extend the investigation to other experimental conditions. This is possible by finding the response surfaces from the proposed estimated model that give the indications for the direction to follow when choosing the new experimental factors and their levels.

At the end an experiment which embraces the results obtained from applying the 2^3 factorial design and the practical laboratory experience was undertaken. The initial hydrogel was the reaction mixture prepared as a center point in the second series of experiments. The solid product obtained after 24 h at 70 °C contained roughly 55 wt.% of zeolite FAU. Increasing the crystallization temperature by 10 °C and leaving unchanged the other parameters, the zeolite FAU without amorphous phase was formed, whose morphology (cube-like) and crystal size (<2 μm) are shown in Fig. 3.

Fig. 3. SEM picture of zeolite FAU obtained from the following hydrogel after 24 h at 80 °C: 2 Na_2O – 0.5 Al_2O_3 – 3 SiO_2 – 100 H_2O.

4. CONCLUSIONS

The effectiveness of the experimental design applied to the zeolite synthesis has been successfully tested. The advantages of the experimental design are the following: reducing the number of experiments results in a decrease in the investigation time and the reduction of its costs. Moreover, it is possible to separate the individual effects from the interaction of the same. There is also a possibility to determine the direction of the crystallization process due to expression of all effects in terms of numerical values.

In the case of the zeolite-type taken for the investigation, the crystallization field of the zeolite FAU has been determined by fulfilling the previously achieved results during the experimental work regarding a zeolite-type (LTA) whose synthesis conditions in part overlap, as they usually co-crystallize. The way to obtain pure FAU starting from hydrogels with different Si/Al ratios was successfully validated through a small number of runs. Very interesting observations regarding the role of water and sodium emerged from these experiments that are taken for further examination.

REFERENCES

1. D.C. Montgomery, Design and Analysis of Experiments, John Wiley & Sons, New York, 1997.
2. G.E.P. Box, W.G. Hunter and J.S. Hunter, Statistics for Experimenters, John Wiley & Sons, New York, 1978.
3. D. M. Himmelblau, Process Analysis by Statistical Methods, John Wiley & Sons, New York, 1970.
4. R.B. van Dover, L.F. Schneemeyer and R.M. Fleming, Nature 392 (1998) 162.
5. E. Danielson, J.H. Golden, E.W. McFarland, C.M. Reaves, W.H. Weinberg and X.D. Wu, Nature 389 (1997) 944.
6. K. Yajima, Y. Ueda, H. Tsuruya, T. Kanougi, Y. Oumi, S.S.C. Ammal, S. Takami, M. Kubo and A. Miyamoto, Stud. Surf. Sci. Catal. 130A (2000) 401.

R. Szostak, Handbook of Molecular Sieves, Van Nostrand Reinhold, New York, 1992.

Studies in Surface Science and Catalysis 140
A. Gamba, C. Colella and S. Coluccia (Editors)

Spectroscopic characterisation and photocatalytic properties of Mg^{2+}-modified MCM-41

Gianmario Martra[a,*], Hisao Yoshida[a,b], Valeria Dellarocca[a], Chizu Murata[b], Tadashi Hattori[b], Salvatore Coluccia[a] and Leonardo Marchese[a,c]

[a]Dipartimento di Chimica IFM, Università di Torino, Via P. Giuria 7, I-10125 Torino, Italy

[b]Department of Applied Chemistry, Graduate School of Engineering, Nagoya University, Nagoya 464-8603, Japan

[c]Dipartimento di Scienze e Tecnologie Avanzate, Università del Piemonte Orientale «A. Avogadro», C.so Borsalino 54, I-15100 Alessandria, Italy

Two Mg^{2+}-modified mesoporous silica systems (Mg loading: 15 wt% as MgO) were prepared by impregnation of a purely siliceous MCM-41 with $Mg(OCH_3)_2$ methanol solution and $Mg(NO_3)_2$ aqueous solution followed by calcination. They were characterised using XRD and diffuse-reflectance UV-Vis spectroscopy. Adsorption properties towards CO_2 (monitored by FTIR spectroscopy) and catalytic activities for propene photooxidation were also investigated.

1. INTRODUCTION

Heterogeneous basic catalysts have received growing attention in the last decade [1]. The increasing awareness of environmental protection and sustainable industrial and technological growth are stimulating researches for new heterogeneous basic catalysts which may avoid large volumes of aqueous basic solutions used in several industrial processes. Heterogeneous basic catalysts appear quite attractive to envisage new and more convenient synthesis routes, in particular in the field of fine chemistry [2].

Among the various types of basic solids, magnesium oxide plays a central role in the investigations of the properties of surface centers of basic heterogeneous catalysts and on their interaction with several types of substrates, both for the model character of this material and its activity as an actual catalyst. Particular attention has been devoted to the properties of magnesium oxide powders with high specific surface area, which can be produced by controlled decomposition of the parent hydroxide [3]. These samples exhibit a significant fraction of sites in low coordination (in edge and corner positions of the MgO microparticles), which have been recognized as the active centers in the reactivity towards simple molecules [4,5]. Many efforts to increase the dispersion of MgO have been made, since the number of the surface centers in the lowest coordination increases as the size of MgO particle decreases.

* Author to whom correspondence should be addressed; e-mail: martra@ch.unito.it

Related to this aim is the preparation of MgO powders by chemical vapor deposition [6], and the dispersion of MgO on SiO_2 [7] and within the cavities of faujasite zeolites [8]. Noticeably, in the last case, the location of the basic oxide inside a porous matrix was intended for the introduction of an additional shape selectivity effect related to the host structure. However, the access to the faujasite cavities with a diameter of *ca.* 7 Å is limited, and bulky reactant cannot reach the inner catalytic active centers.

The recent discovery of mesoporous silicas, exhibiting channels with diameter in the range 15 to 100 Å, resulted in the availability of a support with very attractive features [9,10]. Among such materials, the MCM-41, developed by the Mobil Research Laboratories in 1992 [9], is the most studied one. MCM-41 has very high specific surface area, typically around 1000 $m^2 \cdot g^{-1}$, and a regular porous structure suitable for shape selectivity effects, which are useful for preparing new heterogeneous catalysts for the synthesis of fine chemicals. Thus, many efforts have been made to prepare active supported catalysts by the insertion of redox and acid-base centers [11 and references therein].

Furthermore, silica-based systems received attention as effective catalysts for several photochemical reactions, e.g olefin photo-isomerisation [12], olefin photo-oxidation [13-15], photomethathesis [16] and methane coupling [17]. Recently, it was also reported that silica mesoporous materials are more active photocatalysts than amorphous silica [18]. Among them, Mg/SiO_2 systems [13] were found to act as catalysts for the photo-oxidation of propene to propene oxide in the presence of molecular oxygen, which is an attractive path for the production of this industrially important chemical.

On these bases, we carried out a research devoted to the preparation of new catalytic materials, trying to disperse MgO on a purely siliceous MCM-41 support by impregnation with different precursors, and investigating their structural, electronic and chemical properties by spectroscopic techniques and photocatalytic tests.

2. EXPERIMENTAL

2.1. Materials

The MCM-41 silica support was synthesized according to the method described in the literature [19]. The as-synthesized product was calcined at 823 K (elevating rate 5 $K \cdot min^{-1}$) for 1 hour under N_2 flow, followed by successive heating at 823 K under O_2 flow for 6-8 hours. After the calcination, the MCM-41 was impregnated with two different solutions: i) methanol solution of $Mg(OCH_3)_2$ and ii) aqueous solution of $Mg(NO_3)_2$. Commercially obtained methanol (Carlo Erba), $Mg(OCH_3)_2$ and $Mg(NO_3)_2$ (Merck) were used. In the former case, after mixing the MCM-41 with a solution of $Mg(OCH_3)_2$ in methanol for 1 hour, the mixture was filtered at room temperature. By determination of the amount of Mg in the filtrate, it was confirmed that all $Mg(OCH_3)_2$ was loaded on MCM-41. In the latter case, after impregnating the MCM-41 with an aqueous $Mg(NO_3)_2$ solution, water was evaporated to dryness on heater. Then, both were dried at 323 K for 12 hours and calcined under O_2 flow at 823 K for 5 hours. The MgO loading calculated from the amount of the starting materials was *ca.* 15 wt% for both samples. The sample prepared from methanol solution will be hereafter referred to as Mg/MCM-41(15M), and that from aqueous solution as Mg/MCM-41(15W). High purity CO_2 (Matheson), for the spectroscopic measurements of adsorbed species, and high purity propene (Takachiho) and oxygen (Nagoya Nissan), for the photo-oxidation tests, where used after through several freeze/liquefy-pump-thaw cycles.

2.2. Methods

X-ray diffraction spectra of the samples were recorded with a PW 1050 Philips spectrometer with Co Kα incident radiation. Transmission electron micrographs and electron dispersion spectroscopy (EDS) analysis were obtained with a Philips CM 12 instrument equipped with Philips 9100 attachment.

UV-Vis diffuse reflectance (DR UV-Vis) and FTIR spectroscopic measurements were carried out on the samples put in suitably designed cells permanently attached to conventional vacuum lines (residual pressure: 1×10^{-6} Torr; 1 Torr= 133.33 Pa) allowing all thermal and adsorption-desorption experiments to be carried out in situ. For DR UV-Vis measurements the powder sample was put in an optical quartz cell. The spectra were recorded by using a Perkin-Elmer Lambda 19 spectrophotometer equipped with an integrating sphere and using $BaSO_4$ as reference. For FTIR measurements, the samples, in form of self-supporting pellets, were placed in a quartz cell with KBr windows. The spectra were collected with a Bruker IFS 88 instrument (resolution: 4 cm^{-1}), and were reported in absorbance, having subtracted the spectrum of the samples before the adsorption. Their intensity was normalized with respect to the "density" ($mg \cdot cm^{-2}$) of the pellets.

Photo-oxidation tests were carried out as described in the literature [14]. Before the photoreaction, the samples were heated in air up to 673 K, and then evacuated. Subsequently the samples were treated in oxygen (100 Torr) at 673 K for 1 h, followed by evacuation at 673 K for 1 h. The photo-oxidation of propene was carried out for 2 h in a closed quartz reaction vessel (123.6 cm^3). The temperature of the catalyst bed was elevated up to 313-323 K by photoirradiation. The reactants were propene (100 μmol, 15 Torr) and oxygen (200 μmol, 30 Torr). The catalyst (200 mg) was spread on the flat bottom (12.6 cm^2) of the vessel. A 200 W Xe lamp was used as a light source. Products in gas phase and products desorbed by photoirradiation for 10 min were analyzed separately by gas chromatography, followed by analysis of desorbed products by heating at 573 K.

3. RESULTS AND DISCUSSION

3.1. Structural features

The XRD pattern of the calcined MCM-41 (Fig. 1, a) exhibited the main (100) diffraction peak at 2.86° (2θ), d_{100} = 3.58 nm, corresponding to a regular hexagonal arrangement of channels with unit cell parameter (a) of 4.14 nm [9]. The presence of (110), (200) and (210) diffractions at 4.91, 5.64 and 7.32° suggested that the hexagonal structure of the siliceous matrix was very regular and that the material had good crystallinity [20].

The intensity of XRD pattern of uncalcined Mg/MCM-41(15M) sample (Fig.1, b) was lower than the bare MCM-41 (Fig. 1a). After calcination the (100) diffraction peak appeared slightly decreased in intensity and shifted to 30.6° (d_{100}= 3.35 nm; unit cell parameter (a)= 3.87 nm), while the weaker pattern at higher angle disappeared (Fig. 1, c). This indicates that the structure of the MCM-41 support was affected by the loading with magnesium species, resulting in a loss of regularity of the mesoporous structure. The reduction of the d-space, which is the distance between the pores, would suggest a decrease of the channel size. A similar effect was reported also for V/MCM-48 systems [21].

No peaks appeared in the 2-8° 2θ range of the XRD profile of the uncalcined Mg/MCM-41(15W) (Fig. 1, d), whilst this sample showed a clear diffraction pattern in the higher angle region (not shown) assigned to $Mg(NO_3)_2$, which would distort the regularity of

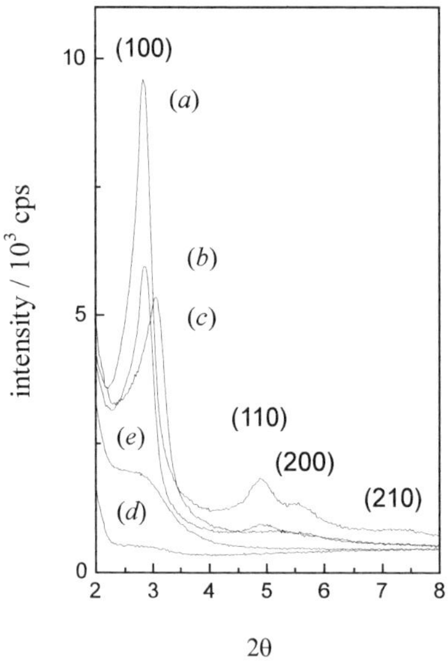

Fig. 1. XRD patterns of MCM-41 (*a*), uncalcined (*b*) and calcined (*c*) Mg/MCM-41(15M), uncalcined (*d*) and calcined (*e*) Mg/MCM-41(15W).

the array of channels. After calcination, and the consequent decomposition of the $Mg(NO_3)_2$ phase, the (100) diffraction was restored to a very limited extent (Fig. 1, e). This indicates that the functionalisation of MCM-41 with $Mg(NO_3)_2$ as source of Mg^{2+} resulted in the ruin of the regular hexagonal arrangement of channels of the support. Inspection by transmission electron microscopy did not find out significant differences in the morphology of the particles of the bare support and of the two Mg^{2+}-modifided MCM-41 samples, which in all cases exhibited a globular shape with irregular borders (images not reported). Noticeably, for both the Mg/MCM-41 systems no MgO microcrystals on the external surface of the grains of the support or as separate phase were observed, and in both cases EDS analysis revealed that Mg^{2+} was homogeneously distributed within the siliceous matrix.

3.2. Electronic state of the supported species

UV-Vis diffuse reflectance spectroscopy revealed that MCM-41, Mg/MCM-41(15W) and Mg/MCM-41(15M) have different electronic states (Fig. 2). Whilst the bare MCM-41 support did not absorb in the 200-400 nm range (Fig. 2, a), as expected for pure siliceous materials [22], a sharp band at 210 nm in the case of the Mg/MCM-41(15W) sample (Fig. 2, b) and two bands at 220 nm (main component) and 280 nm (heavy shoulder) in the spectrum of the Mg/MCM-41(15M) system (Fig. 2, c) were observed.

Pure MgO powders with very high surface area exhibit two bands due to electronic surface states in the 220-280 nm range [5,23]. The UV spectrum of the Mg/MCM-41(15M) sample appeared quite similar to that of highly dispersed MgO powders. In fact, these materials exhibit a band at *ca.* 220 nm, assigned to charge transfer transition involving Mg^{2+}-O^{2-} pairs, located on the edges of MgO crystallites, and a second component in the 260-280 nm range, attributed to sites on corner positions. For MgO powders outgassed at temperature lower than 973 K, charge transfer transitions involving OH groups may contribute to the band at 260-280 nm [24]. Therefore, it is proposed that a MgO-like phase, highly dispersed on the silica support, exists in the Mg/MCM-41(15M).

On the contrary, the presence of small MgO particles in the Mg/MCM-41(15W) sample can be ruled out, and its single band at 210 nm may tentatively be assigned to charge transfer transitions involving Mg^{2+} anchored to the walls of the siliceous matrix.

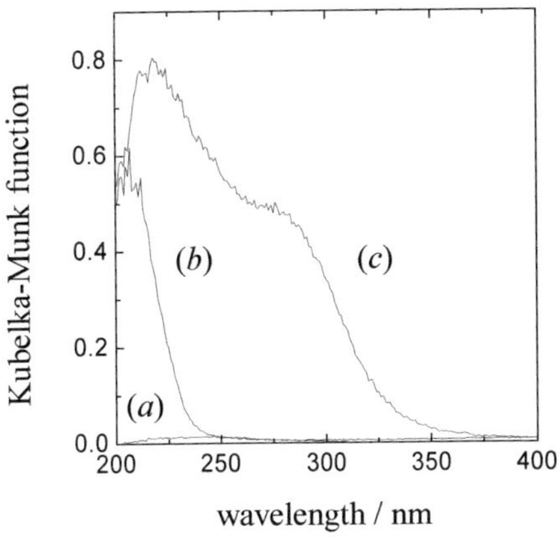

Fig. 2. DR UV-Vis spectra of MCM-41 (*a*), Mg/MCM-41(15W) (*b*), and Mg/MCM-41(15M) (*c*). All samples were outgassed at 873 K for 1 h and the spectra recorded *in situ.*

3.3. Reactivity toward CO_2 monitored by FTIR spectroscopy

CO_2 is widely employed as a probe molecule to monitor the properties of basic centers of metal oxides, but it is also useful to probe the Lewis acid features of such materials, as it can interact with surface cations via linear end-on complexes [25 and references therein]. Fig. 3 shows the spectra of carbon dioxide adsorbed on the three samples, in the presence of gas phase CO_2 in the range where the asymmetric stretching mode of CO_2 molecules absorbs. In the case of the bare MCM-41, a main band at 2340 cm^{-1} was observed (Fig. 3, a), which is assigned to carbon dioxide molecules weakly interacting with the pore wall of the mesoporous structure. Two weaker shoulders at *ca.* 2360 and 2330 cm^{-1} were also present, due to the branches of the roto-vibrational profile of CO_2 in gas phase. Similar results for CO_2 adsorbed on silicalite were reported [26]. It is of interest that CO_2 adsorption on MCM-41 resulted in a very weak perturbation, possibly through hydrogen bonding, of a small fraction of the hydroxyl groups (silanols) located on the silica walls, whose frequency was shifted downward of ca. 50 cm^{-1} [27].

The adsorption of CO_2 on the Mg/MCM-41(15M) sample produced a broader and more complex band, exhibiting a maximum at 2344 cm^{-1} and a prominent shoulder at 2355 cm^{-1} (Fig. 3, b). The latter component is in a range where the asymmetric stretching band of CO_2 linearly adsorbed on cations in low coordination are observed [25-28], and can be assigned to carbon dioxide molecules end-on coordinated to Mg^{2+} cations. The maximum at 2344 cm^{-1} was close to the frequency of CO_2 molecules interacting with silanols of the bare support (Fig. 3, a), and exhibited essentially a similar dependence on the CO_2 pressure (not shown). This band, also in this case, can be assigned to CO_2 weakly interacting with the walls the siliceous pores.

Essentially the same spectral features, just slightly reduced in intensity, were observed in the case of the Mg/MCM-41(15W) sample (Fig. 3, c). Although XRD data showed that the regular mesoporous structure of MCM-41 was not maintained for this system, this sample

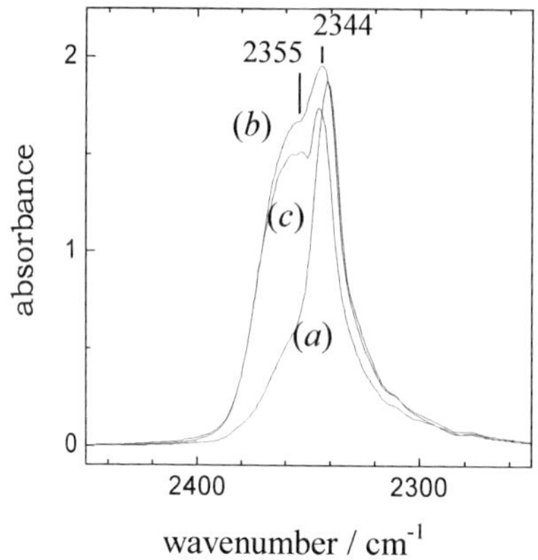

Fig. 3. IR spectra of CO_2 (100 Torr) adsorbed at room temperature on MCM-41 (*a*), Mg/MCM-41(15M) (*b*), and Mg/MCM-41(15W) (*c*). The samples were outgassed at 823 K for 45 min before CO_2 adsorption.

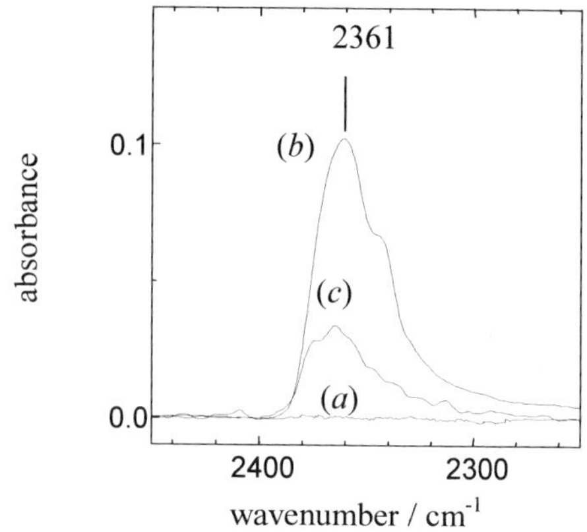

Fig. 4. IR spectra of MCM-41 (*a*), Mg/MCM-41(15M) (*b*), and Mg/MCM-41(15W) (*c*) outgassed at 823 K for 45 min, contacted with 100 Torr CO_2 and then outgassed at room temperature for 30 min.

exhibited almost the same amount of adsorbed CO_2. It means that this sample maintains the adsorption capability even without the regular hexagonal structure.

Fig. 4 shows the spectra after evacuation at room temperature, and indicates that the molecules in interaction with siliceous frame are fully desorbed (Fig. 4, a), whilst those adsorbed on Mg^{2+} cations are partially retained. The amount of CO_2 irreversibly adsorbed onto Mg^{2+} is larger for Mg/MCM-41(15M) (Fig. 4, b) than Mg/MCM-41(15W) (Fig. 4, c). This suggests the presence of stronger Mg^{2+} Lewis acid sites on MgO particles in Mg/MCM41 (15M). In figure 5 the IR spectra of the same samples in the presence of CO_2 in the low frequency range are reported. Whilst no bands were observed in the case of the bare MCM-41 (Fig. 5, a), a series of absorptions were present in the spectra of CO_2 adsorbed on the two Mg/MCM-41 samples (Fig. 5, b, c), with a main band at 1666 cm^{-1}, shoulders at 1730 (weak) and 1622 (strong) cm^{-1}, and bands at 1440 and 1380 cm^{-1}.

The narrow peak at 1380 cm^{-1} is assigned to the ν_1 mode of CO_2 molecules linearly adsorbed on Mg^{2+} ions. This mode, which corresponds to the CO_2 symmetric vibration, is IR inactive (and Raman active) for the molecule in the gas-phase. However, it becomes IR active for the molecule in the adsorbed state [26,28]. The 1380 cm^{-1} peak is absent in the spectra of CO_2 adsorbed onto the bare MCM-41, and this strengthens the assignment proposed, as the interaction of CO_2 with pure siliceous materials is extremely weak so that the molecule is hardly perturbed as compared with the gas phase.

The components at 1730, 1666, 1622 and 1440 cm^{-1} are due to various carbonate-like species [25], which are produced by reaction of CO_2 with basic centers. On the Mg/MCM-41(15M) sample, the bands due to carbonate-like groups appeared more intense than Mg/MCM-41(15W) and the distribution of adsorbed carbonate species was similar but not identical.

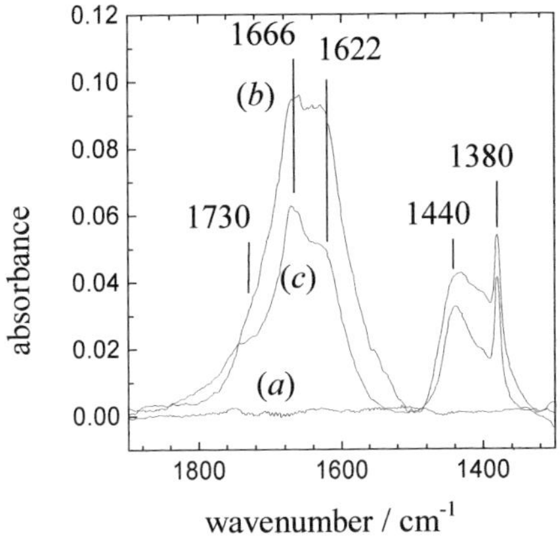

Fig. 5. IR spectra of CO_2 (10 Torr) adsorbed at room temperature on MCM-41 (*a*), Mg/MCM-41(15M) (*b*), and Mg/MCM-41(15W) (*c*). The samples were outgassed at 823 K for 45 min before CO_2 adsorption.

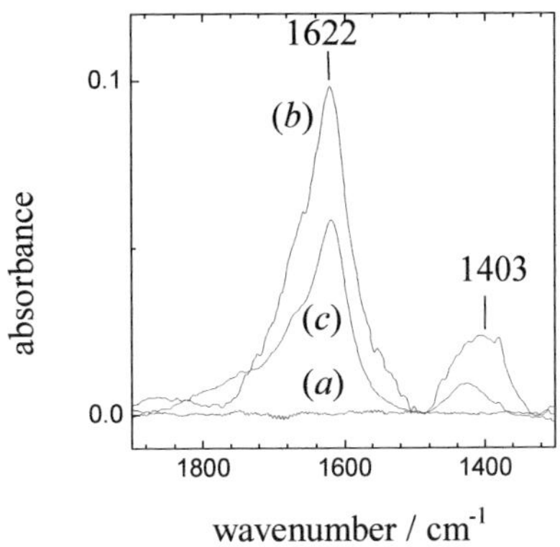

Fig. 6. IR spectra of MCM-41 (*a*), Mg/MCM-41(15M) (*b*), and Mg/MCM-41(15W) (*c*) outgassed at 823 K for 45 min, contacted with 100 Torr CO_2 and then outgassed at room temperature for 30 min.

It should be noticed that no MgO particles are present in the Mg/MCM-41(15W), as indicated by the DR UV-Vis spectra (Fig. 2, b). Consequently, it can be proposed that basic centers able to react with CO_2 to produce carbonate-like species are the oxygen atoms bridging the supported Mg^{2+} ions with the silica walls of the support.

By decreasing the amount of adsorbed CO_2, the components at 1730, 1666 were progressively depleted, the band at 1622 cm^{-1} slightly increased in intensity and the sharp peak at 1381 cm^{-1}, due to CO_2 linearly adsorbed on Mg^{2+} cations, almost disappeared (Fig. 6). This different reversibility of the various bands indicated both the presence of basic sites interacting with CO_2 molecules with different strength and the occurrence of some changes in the structure of the adsorbed carbonate-like species during the desorption process [25,29]. As shown in Fig. 6, the amount of irreversibly adsorbed species was larger on the Mg/MCM-41(15M) sample (Fig. 6, b) than on the Mg/MCM-41(15W) one (Fig. 6 c).

3.4. Photocatalytic activity

The photocatalytic activity of the pure MCM-41 and of the two Mg/MCM-41 systems was studied for the photo-oxidation of propene with molecular oxygen. In order to obtain reliable insight on the selectivity of the process, propene conversion was kept low. The products observed were propene oxide (PO), ethanal (acetaldehyde, AA), propanone (acetone, AC), propanal (propionaldehyde, PA), prop-2-enal (acrolein, AL), alcohols (methanol, ethanol and propan-2-ol), hydrocarbons (ethene and butenes) CO and CO_2.

Bare MCM-41 was found to behave as effective catalyst for the partial photo-oxidation of propene, with a good selectivity towards PO and a low selectivity towards CO_2 (Table 1, entry 1).

Table 1 Catalytic activity of the bare MCM-41 support and of the two Mg/MCM-41 systems in the photo-oxidation of propene.

Catalyst	Conversion (%)	Selectivity (%)					
		CO_2	PO[a]	AA[b]	AC[c]	PA[d]	others[e]
1 MCM-41	1.3	10.9	28.0	23.5	16.0	1.1	20.5
2 Mg/MCM-41(15M)	2.7	5.3	17.3	26.1	18.3	18.2	14.8
3 Mg/MCM-41(15W)	1.3	6.9	13.8	12.4	29.3	10.5	27.1

[a] PO = propene oxide; [b]AA = acetaldehyde; [c]AC = acetone; [d]PA = propionaldehyde; [e]others = acrolein, alcohols, hydrocarbons and carbon monoxide.

Both Mg/MCM-41 systems exhibited further lower selectivity to CO_2, and a distribution of the partial photo-oxidation products different from the pure MCM-41. The Mg/MCM-41(15M) sample showed a good selectivity to AA (Table 1, entry 2), while the Mg/MCM-41(15W) sample exhibited an improved selectivity to AC (Table 1, entry 3). When compared to the bare MCM-41, a significant increase in the selectivity towards PA was obtained in the case of Mg/MCM-41(15M) (Table 1, entry 2), while the Mg/MCM-41(15W) sample shows a significant increase of the selectivity towards both AC and PA (Table 1, entry 3).

These results clearly indicated that the presence of magnesium ions resulted in a modification of the photocatalytic performance of the bare MCM-41 support, and that the preparation method has great influence on the products selectivity. Differences in the product distribution between the two Mg/MCM-41 systems may be ascribed to the fact that Mg/MCM-41(15M) sample contains MgO-like particles, while in the Mg/MCM-41(15W) sample anchored Mg^{2+} ions are present.

4. CONCLUSIONS

The impregnation of MCM-41 with a methanol solution of $Mg(OCH_3)_2$ and subsequent calcination [sample Mg/MCM-41(15M)] produced dispersed MgO-like particles on MCM-41 and the regularity of the structure of the MCM-41 matrix was partly maintained. By contrast, highly dispersed Mg^{2+} ions anchored to the siliceous MCM-41 surface are produced when an aqueous solution of $Mg(NO_3)_2$ was used as a precursor [sample Mg/MCM-41(15W)], and the regularity of the structure of the mesoporous host was not preserved.

The Mg^{2+} loading successfully modified the acid-base properties of MCM-41 towards CO_2 adsorption, and the two preparation methods produced slightly different distributions of adsorbed species. As for the photooxidation activity, the two Mg/MCM-41 samples exhibited high selectivity for partial oxidation of propene by molecular oxygen instead of CO_2 formation, and the product selectivity was much dependent on the preparation route.

ACKNOWLEDGEMENTS

Dr. Enrica Gianotti (Dipartimento di Chimica IFM, University of Turin) is gratefully acknowledged for assistance in several experimental measurements. Dr. Alberto Frache (Dipartimento di Scienze e Tecnologie Avanzate, University of East Piedmont "A. Avogadro") is acknowledged for helping in the synthesis of the MCM-41 used as support.

REFERENCES

1. (a) H. Hattori, M. Misono and Y. Ono, Acid-Base Catalysis II, Kodansha (Tokyo) - Elsevier (Amsterdam), 1994; (b) H. Hattori, Chem. Rev. 95 (1995) 537; (c) D. Barthomeuf, Cat. Rev.-Sci. Eng. 38 (1996) 521.
2. (a) P.E. Hathaway and M.E. Davis, J. Catal., 119 (1989) 497; (b) G. Suzukamo, M. Fukao, T. Hibi and K. Chikaishi, in K. Tanabe, H. Hattori, T. Yamaguchi and T. Tanaka (Editors), Proc. Int. Symp. Acid-Base Catalysis, Kodansha (Tokyo) - VCH (Basel), 1989, p. 405; Y. Kakuno and H. Hattori, J. Catal., 85 (1984) 509.
3. (a) J. Tench and G.T. Pott, Chem. Phys. Lett., 26 (1974) 590; (b) S. Coluccia, M. Deane and A.J. Tench, J. Chem Soc. Faraday Trans. 1, 74 (1978) 2913; (c) E. Garrone, A. Zecchina and F. S. Stone, Phil. Mag. B, 42 (1980) 683; (d) A. Zecchina, S. Coluccia, G. Spoto, D. Scarano and L. Marchese, J. Chem. Soc. Faraday Trans., 86 (1990) 703; (e) S. Coluccia, M. Baricco, L. Marchese, G. Martra and A. Zecchina, Spectrochimica Acta, 49A (1993) 1289; (f) Y. Yanagisawa, K. Takaoka and S. Yamabe, in H. Hattori, M. Misono and Y. Ono (Editors), Proc. Int. Symp. Acid-Base Catalysis II, Kodansha (Tokyo) – Elsevier (Amsterdam), 1994, p. 201; (g) T. Yoshida, T. Tanaka, H. Yoshida, T. Funabiki, S. Yoshida and T. Murata, J. Phys. Chem., 99 (1995) 10890.
4. (a) S. Coluccia and A.J. Tench, Proc. 7th Int. Congr. on Catal., Tokyo, Japan, 1980, p. 1160; (b) G. Martra, E. Borello, E. Giamello and S. Coluccia, in H. Hattori, M. Misono, Y. Ono (Editors), Proc. Int. Symp. Acid-Base Catalysis II, Kodansha (Tokyo) – Elsevier (Amsterdam), 1994, p. 169; (c) E. Garrone, D. Bartalini, S. Coluccia, G. Martra, D. Tichit and F. Figueras, in H. Hattori, M. Misono, Y. Ono (Editors), Proc. Int. Symp. Acid-Base Catalysis II, Kodansha (Tokyo) – Elsevier (Amsterdam), 1994, p. 183.
5. S. Coluccia and L. Marchese, in K. Tanabe, H. Hattori, T. Yamaguchi and T. Tanaka (Editors), Proc. Int. Symp. Acid-Base Catalysis, Kodansha (Tokyo) – VCH (Basel), 1989, p. 207.
6. E. Knoezinger, K.H. Jacob, S. Singh and P. Hofmann, Surf. Sci., 290 (1993) 388.
7. (a) H. Yoshida, T. Tanaka, K. Nakatsuka, T. Funabiki and S. Yoshida, in H. Hattori, M. Misono and Y. Ono (Editors), Proc. Int. Symp. Acid-Base Catalysis II, Kodansha (Tokyo) – Elsevier (Amsterdam), 1994, p. 473; (b) H. Yoshida, T. Yoshida, T. Tanaka, T. Funabiki, S. Yoshida, T. Abe, K. Kimura, T. Hattori, J. Phys. IV France, 1997, 7, Colloque C2, 991.
8. H. Tsuji, F. Yagi, H. Hattori and H. Kita, in L. Guczi, F. Solymosy and P. Tetenyi (Editors), New Frontiers in Catalysis, Proc. 10th Int. Congr. Catal., Elsevier and Akademiai Kiadò, Budapest, 1993, p, 1171.

9. J. S. Beck, J. C. Vartuli, W. J. Roth, M. E. Leonowicz, C. T. Kresge, K. D. Schmitt, C. T.-W. Chu, D. H. Olson, E. W. Sheppard, S. B. McCullen, J. B. Higgins and J. L. Schlenker, J. Am. Chem. Soc., 114 (1992) 10834.
10. (a) T. Yanagisawa, T. Shimizu, K. Kuroda, and C. Kato, Bull. Chem. Soc. Jpn., 63 (1990) 988; (b) S. Inagaki, Y. Fukushima, and K. Kuroda, J. Chem. Soc., Chem.Commun., (1993) 680.
11. A. Corma, Chem. Rev., 97 (1997) 2373.
12. A. Morikawa, M. Hattori, K. Yagi and K. Otsuka, Z. Phys. Chem. N.F., 104 (1977) 309.
13. (a) H. Yoshida, T. Tanaka, M. Yamamoto, T. Funabiki, S. Yoshida, Chem. Commun., (1996) 2125; (b) H. Yoshida, T. Tanaka, M. Yamamoto, T. Yoshida, T. Funabiki, S. Yoshida, J. Catal. 171 (1997) 351.
14. (a) H. Yoshida, C. Murata and T. Hattori, Chem. Commun., (1999) 1551; (b) H. Yoshida, C. Murata, and T. Hattori, J. Catal., 194 (2000) 364.
15. (a) Y. Kubokawa, M. Anpo and C. Yun, Proc. 7th Int. Congress on Catalysis, Tokyo, 1980, T. Seiyama and K. Tanabe (Editors), Kodansha, Tokyo, 1980, Vol. B, p.1170; (b) A. Ogata, A. Kazusaka and M. Enyo, J. Phys. Chem., 90 (1986) 5201.
16. (a) H. Yoshida, T. Tanaka, S. Matsuo, T. Funabiki and S. Yoshida, J. Chem. Soc., Chem. Commun., (1995) 761; (b) T. Tanaka, S. Matsuo, T. Maeda, H. Yoshida, T. Funabiki, S. Yoshida, Appl. Surf. Sci., 121/122 (1997) 296.
17. (a) Y. Kato, H. Yoshida, T. Hattori, Chem. Commun., (1998) 2389. (b) H. Yoshida, Y. Kato and T. Hattori, Stud. Surf. Sci. Catal., 130 (2000) 659.
18. (a) H. Yoshida, K. Kimura, Y. Inaki, T. Hattori, Chem. Commun., (1997) 129; (b) H. Yoshida, C. Murata, Y. Inaki, T. Hattori, Chem.Lett., (1998) 1121; (c) Y. Inaki, H. Yoshida, K. Kimura, S. Inagaki, Y. Fukushima and T. Hattori, Phys. Chem. Chem. Phys., 2 (2000) 5293.
19. F. Rey, G. Sankar, T. Maschmeyer, J. M. Thomas, R. G. Bell and G.N. Greaves, Topics in Catal., 3 (1996) 121.
20. (a) Q. Huo, D.I. Margolese and G.D. Stucky, Chem. Mat., 8 (1996) 1147; (b) R. Mokaya, W. Zhou and W. Jones, Chem. Commun., (1999) 51.
21. M. Morey, A. Davidson, H. Eckert and G. Stucky, Chem. Mat., 8 (1996) 486.
22. L. Marchese, T. Maschmeyer, E. Gianotti, S. Coluccia and J.M. Thomas, J. Phys. Chem. B, 101 (1997) 8836.
23. A. Zecchina, M. G. Lofthouse and F. S. Stone, J. Chem. Soc., Faraday Trans.1, 71 (1975) 1476.
24. (a) M. Anpo, K. Chiba, S. Coluccia and G. Martra. Res. Chem. Intermed., 19 (1993) 495; (b) N. De Leo, Thesis, University of Turin, Dipartimento di Chimica IFM, 1997.
25. G. Busca and V. Lorenzelli, Mater. Chem., 7 (1982) 89.
26. T. Yamazaki, M. Katoh, S. Ozawa and Y. Ogino, Mol. Phys., 80 (1993) 313.
27. V. Dellarocca, G. Martra, L. Marchese et al., unpublished results.
28. H. Förster and M. Schumann, J. Chem. Soc., Trans. 1, 85 (1989) 1149.
29. G. Martra, S. Coluccia, P. Davit, E. Gianotti, L. Marchese, H. Tsuji, H. Hattori, Res. Chem. Intermed., 25 (1999) 77.

Studies in Surface Science and Catalysis 140
A. Gamba, C. Colella and S. Coluccia (Editors)

Electroceramics in solid oxide fuel cell technology

A. J. McEvoy

Laboratory for Photonics and Interfaces, Department of Chemistry,
Ecole Polytechnique Fédérale de Lausanne, CH-1015 Lausanne, Switzerland

Energy technology is the key to sustaining the standard of living generally acquired, at least in the industrialised world, over the past century. However, the domination by fossil-fuel fired thermomechanical systems will be challenged under the coinciding pressures of resource limitation, demand expansion and environmental factors. A re-examination of the role of electrochemical systems, both for the more efficient utilisation of resources and as energy storage devices, is now taking place. Battery and fuel cell technology, having stagnated for most of the last century, are set for a more prominent role in the coming years. Oxide ceramic materials are already of increasing significance in fuel processing; in high temperature fuel cells and ionic transport membranes they can have similar significance in end use of those fuels.

1. ORIGINS AND PRINCIPLES

Until recently the received opinion had it that the energy technology of the 21st century would evolve in response to the exhaustion of high quality deposits of primary fossil fuels. Now, in contrast, the critical challenge is seen to be a sink, rather than a source, problem given the inability of the environment to accommodate the dispersal of combustion products without significant negative consequences. Particularly ominous are the climatic implications of atmospheric carbon dioxide accumulation. In either scenario, however, limitation of resources or restrictions on disposal, the imperative of conservation presents itself nonetheless, and the technical response will find increasing use for ceramic and oxide materials. Much of the present volume deals with the preparation and properties of zeolites and their particular significance for fuel processing in the petroleum industry. After combustion of the fuels also, catalysts supported on ceramics are now a standard component for emissions control. In the nuclear energy sector, oxide fuels have established their role. The refractory properties of ceramic oxides have even led to their evaluation as components in engines, not just as thermal barriers but also, when selected for high-temperature mechanical properties, as turbine blades and turbocharger rotors. In these cases they can increase the efficiency of thermomechanical systems, which however still remain subject to Carnot cycle limitations. In the solid oxide fuel cell (SOFC), where the electrochemical behaviour of certain oxide ceramics is exploited, even this fundamental restriction on all thermomechanical systems is circumvented, since although operating at a high temperature it is not a heat engine but a direct generator of electricity through a chemical oxidation reaction.

The ionic nature of conduction in certain solids is known since the 19th century, and the first solid electrolyte fuel cell was probably the glass or porcelain device with platinum electrodes and alcohol as fuel, reported by Gaugain in 1853 (1). Later there was considerable interest in metal oxides as phosphors for gas-lighting appliances, the "lime-light", and by 1899 Nernst had identified a suitable oxide formulation for an electrically-heated analogue (2). It is anomalous that the centenary of what is now our standard solid oxide electrolyte, yttria-stabilised zirconia (YSZ), could therefore have passed practically unnoticed. The advanced understanding of conduction in solid solutions of oxide ceramics a century ago is clearly demonstrated in the thesis of Reynolds in 1902 (3). The early history of this technology has recently been reviewed by Möbius (4).

In the fuel cell application, a difference of chemical potential of oxygen is established by admission of a fuel to one surface of the ceramic electrolyte membrane, the other being exposed to air. Charge exchange is permitted through conducting electrodes attached to each surface, that on the fuel side acting as the anodic site for oxidation of the fuel, thereby acquiring a negative potential with respect to the cathode where oxygen from the air is reduced to ions. These oxygen ions in turn, with a thermally activated mobility, provide the charge carriers for current through the electrolyte. The operating principle of a solid oxide fuel cell is presented schematically in Fig. 1 As will be seen, ceramics are also essential for the electrodes as well as the electrolyte.

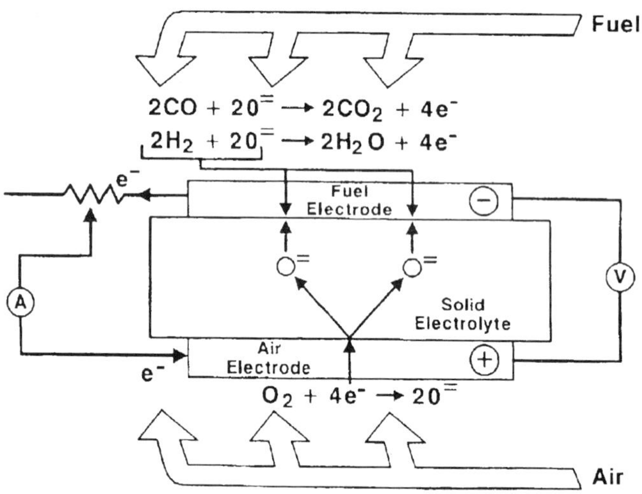

Fig. 1. Operating principle of a fuel cell using an oxide electrolyte with anion (O^{2-}) mobility (after Ref. 5).

2. CERAMIC ELECTROLYTE MATERIALS

As a wide band gap oxide, the ideal solid state electrolyte has an extremely low electronic conductivity; charge transport through the material is purely ionic. This ionic mobility requires an open crystal structure, typically fluorite for simple oxides. While several binary oxides with electrolytic behaviour occur in this structure in the pure state, zirconia and hafnia require stabilisation as solid solutions with an admixture of elements of lower valence, typically yttrium, giving yttria-stabilised cubic zirconia (YSZ). Associated is an added advantage, with charge compensation for the lower valency cations by the appearance of one vacancy in the oxygen sublattice for every two yttrium atoms on zirconium sites. With thermal activation at elevated temperatures, significant ionic conductivity is achieved due to the mobility of these anion vacancies. In Fig. 2 the half cell of a stabilised cubic structure is shown, with a substituting cation and compensating anion site vacancy.

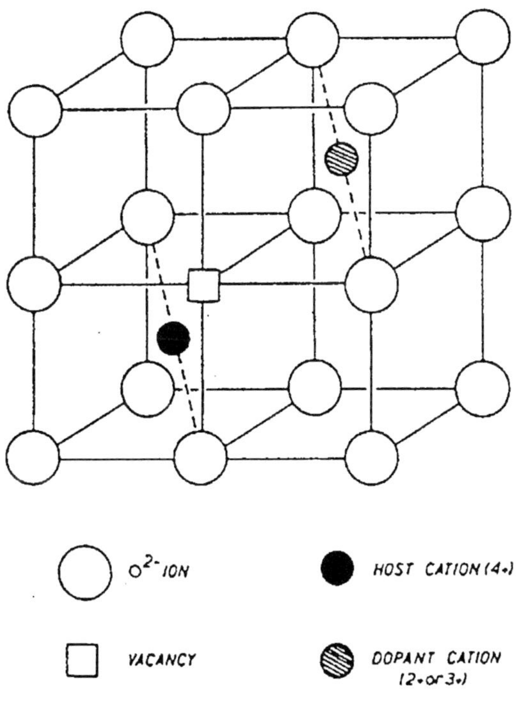

Fig. 2. Stabilised cubic structure of oxide electrolyte, showing a substituting cation and compensating anion site vacancy.

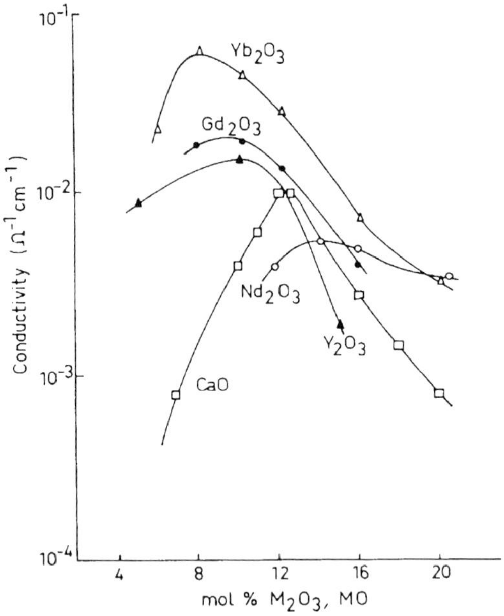

Fig. 3. Effect on the ionic conductivity of stabilised zirconia at 800°C of choice of solute oxide and its concentration (after Ref. 6).

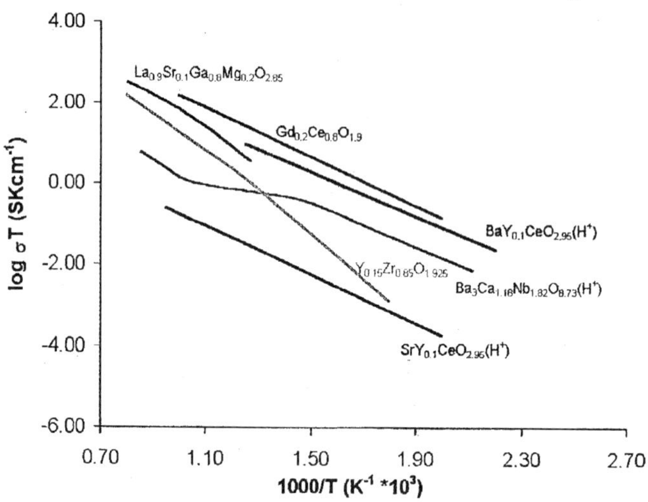

Fig. 4. Conductivity of several oxide ceramic anion and cation conductors suitable for fuel cell and selective membrane applications (after Ref. 8).

Selection of an appropriate solute is important for the formulation of an effective electrolyte. Maximum conductivity, for example, seems to be associated with a size homogeneity between the substituting species and the majority cation in the cubic structure, as well as its concentration in solid solution. Figure 3 presents the effects on the ionic conductivity of stabilised zirconia at a fixed temperature, on variation of the cationic substituting species. It is evident that the optimised yttrium solid solution has a conductivity of about 0.015 S cm^{-1} at 800°C, so that only a very thin electrolyte membrane can provide a technically acceptable current density at that temperature. The well-established Westinghouse SOFC system therefore operates closer to 1000°C to take advantage of the rapid increase of electrolyte conductivity with temperature (7) (see also Fig. 7). This dependance, particularly steep for YSZ, is presented for several solid ionic conducting materials in Fig. 4.

3. CATHODES

The selection of a material suitable as a fuel cell cathode for reduction of oxygen is not trivial. The choice by Baur et al. (9) for the first modern solid oxide fuel cell confronted all the principal requirements in their selection of iron oxide. With an oxide, electrode stability in a high-temperature oxidising environment was secured even when the air (with entrained atmospheric water vapour) is admitted to the cathode at temperatures close to 1000°C. The contact to the electrolyte shows only a limited reactivity and the oxide is an adequate electronic conductor. In addition, iron being a transition metal, some degree of redox activity could be expected to catalyse the cathodic reaction, and given the non-stoichiometry of iron oxide structures, there could also be expected at least a limited oxygen ion mobility and therefore some degree of mixed electronic and ionic conduction, delocalising the charge exchange reaction from the three-phase boundary where electrode, electrolyte and air coincide. Unfortunately, however, the mixed valence condition gives way to poorly conducting Fe_2O_3 in real cell conditions. Other oxides were therefore evaluated, such as indium-tin mixed oxide (ITO), today well known as a transparent electrode in optoelectronics. However, as a glassy substance, Tannenberger and van den Berghe (10) had to insist in their patent on depositing it as a dendrite structure to facilitate oxygen migration along the grain boundaries to reach the electrochemically active zone at the interface with the electrolyte. Interestingly enough, their patent mentioned cobalt and chromium oxides in solid solution with lanthana and strontia, which adopt the perovskite structure. At the same time the Rohr group with the Brown Boveri Company in Germany had identified the promise of lanthanum manganite (11) and the manganese, nickel and chromium perovskites were patented. The preferred materials selection now established for SOFC systems was essentially defined in the 1980's, with a lanthanum-strontium manganite cathode (LSM), a YSZ electrolyte and as anode a YSZ-transition metal (Ni or Co) cermet composite (12,13). The chromium compound finds application as a high-temperature electronic conductor because of its unique stability under an extreme range of oxygen activity, including the anode, fuel, environment in SOFC service, whereas the cobaltite and ferrite perovskites are marked by their mixed conductivity with both electron and ion mobility.

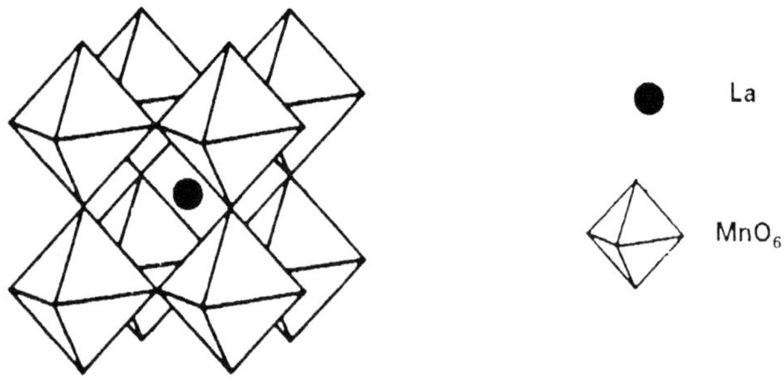

Fig. 5. The ABO_3 perovskite structure. The lanthanum ion (A) takes position within the octahedra representing the oxygen anion sublattice. These octahedra in turn centre on the manganese (B) sites

The perovskite ABO_3 structure is ideally cubic, but easily distortable to orthorhombic, and adaptable to nonstoichiometries associated with vacancies and multiple cation valences (Fig. 5). These are precisely the properties which relate to oxygen adsorption and charge exchange processes. These materials also readily form solid solutions, and in particular strontium on lanthanum (A) lattice sites increases the conductivity of lanthanum manganite, adapts its thermal expansion coefficient to match more closely that of zirconia electrolyte, and suppresses phase changes. In a perovskite structure, each cation, A and B, sublattice interpenetrates the anion sublattice of corner-sharing octahedra. In the solid solution LSM, where the substituting strontium occupies sites in the lanthanum sublattice, charge equilibrium is maintained by the oxidation of a corresponding proportion of Mn(III) ions to oxidation state IV. Exposed to air, it has an apparent excess of oxygen anions, but excluding the possibility of interstitial oxygen, this is in fact a lanthanum deficiency, again compensated by a higher oxidation state of Mn. As a consequence anion mobility is low in air, and remains so until sufficiently reducing conditions are established to induce oxygen defects in the lattice. As shown in Fig. 6 this occurs only under low oxygen partial pressure or under cathodic polarisation which is electrochemically equivalent. The corresponding cobaltite consistently has an oxygen substoichiometry however, and in principle therefore would form a more effective mixed conduction cathode material. However for high temperature cells it is unsuitable due to more rapid interfacial reaction with YSZ, forming a $La_2Zr_2O_7$ insulating interphase. It can therefore be used only in the more recent intermediate temperature cells.

4. ANODES

The function of the anode is to provide the site for oxidation of the fuel. Immediately several issues of compatibility are raised, both in regard to the chemistry of the solid state components and with the fuels and reaction products, as well as the relevant electrical and catalytic properties for effective charge exchange, and a structure compatible with fuel access to the reaction site and gaseous oxidation product discharge. It is an unfortunate fact at this time that no single material offers adequate performance in all these aspects. The required electronic conductivity would imply the use of a metal. In general, however, porous metals tend to densify at the operating temperature of SOFC systems, thereby reducing the effective surface area for contact with the reagent gases and gradually eliminating the required porosity. The consequence is a highly polarised, and ultimately ineffective, interface.

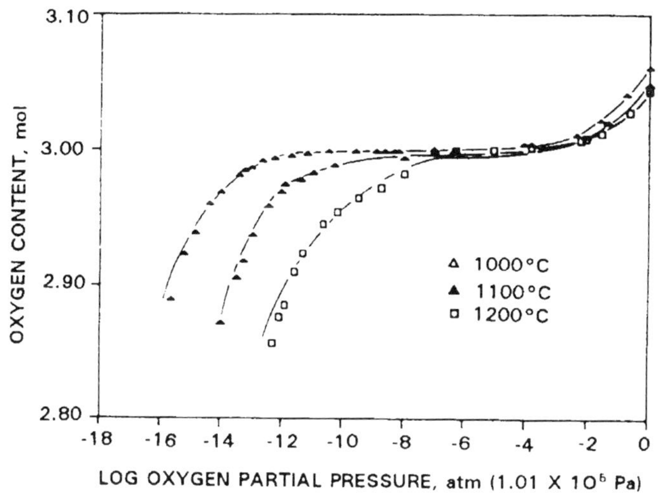

Fig. 6. Variation of stoichiometry of $La_{0.9}Sr_{0.1}Mn\ O_{(3\pm\delta)}$ with temperature and oxygen partial pressure (after Ref. 14).

The use of a ceramic oxide could also be evaluated; ceria, CeO_2, for example, in solid solution with the oxide of a trivalent metal is, like zirconia, an oxygen ionic conductor, with the additional advantage that under conditions such as the environment of the anode it acquires also electronic conductivity due to partial reduction of Ce(IV) to Ce(III). Given this mixed conductivity, there is the apparent advantage that charge exchange and evolution of the oxidation product would be delocalised from the three-phase boundary of coincidence of electrolyte, electrode and gas. However, this simple solution is excluded by the significant difference of thermal expansivity between the two ceramics, ceria and zirconia, leading to interfacial stress and delamination under temperature variation, and also by the limited conductivity of the reduced material. The possibility of use of ceria as electrolyte with integral anode is also restricted since under open-circuit conditions the reduction effect propagates as a front into the interior of the material, so that the effective electrolyte zone is extremely thin, some degree of electronic conductivity exists through the ceramic and constitutes an internal short-circuit, reducing the cell voltage and representing an energy loss (15). In addition the partial reduction induces a lattice expansion, again with associated mechanical stress. Therefore even though ceria-based cells may eventually be significant for lower- or intermediate-temperature solid oxide devices, they are not presently favoured.

The standard compromise at this time is a composite or cermet made up of distinct grains of ceramic and metal. For compatibility with the electrolyte zirconia is the ceramic of choice, a considerable fraction being sub-micron to facilitate sintering during anode formation. Some admixture of larger grains can be helpful to produce an interconnected YSZ structure, within which the metal provides electronically conductive continuity while being stabilised against grain growth and densification. The standard cermet uses nickel for its chemical stability under anode conditions, combined with electrochemical activity and low price in comparison with cobalt or noble metals. Metals such as iron are not acceptable, as despite the presence of hydrogen they can corrode in the steam evolved as a combustion product during fuel cell operation. Normally the cermet is formed from a mixture of NiO and YSZ in approximately equal proportions, the reduction of the oxide to metallic nickel on first contact with the fuel providing the required porosity for gas diffusion and exchange. Use of nickel however imposed some restrictions on fuel gas chemistry. With hydrocarbon vapours there is a tendency to pyrolyse on nickel at elevated temperatures, with the deposition of vitreous carbon. Therefore the most convenient fuels for commercial fuel cell operation, natural gas or other hydrocarbons such as LPG, require pretreatment, typically reforming with steam or carbon dioxide, both available by partially recycling exhaust gas if necessary, in order to provide the mixture of hydrogen and carbon monoxide which actually fuel the cell. The reforming catalysts are a further example of use of oxides in fuel cell technology; there is the more distant prospect also of oxide anodes, probably conducting perovskites, which do not induce fuel pyrolysis and may therefore accommodate to direct oxidation of hydrocarbons (16).

5. INTERCONNECTIONS AND CELL CONFIGURATIONS

The oxide materials and their functionality within an individual cell or electrode-electrolyte assembly have been presented. However, as the Nernst potential, the maximum voltage obtainable from the difference of oxygen partial pressure on anode and cathode, is little over 1 V at the operating temperature of a SOFC, provision for series connection is necessary. Three system concepts have been developed in this connection, individual tube or planar cells, or the as yet largely unproven integrated interconnection device (Fig. 7). In all cases an interconnect material is required, with a lanthanum cobaltite solid solution being widely favoured. In the tubular system a restricted amount of this material is required as a strip-type interconnect, and is well proven in service. It is more questionable in planar systems, where oxygen loss at the contact to an anode causes lattice expansion and stress. Additionally there is a minor ionic conductive effect, representing an efficiency loss. However the principal objection is that the chromite interconnector/separator, of the same area as the individual planar cells, is of comparable cost and weight. One of the principal motivations for the present trend towards lower temperature operation of solid oxide fuel cells is the replacement of the ceramic interconnect by a metallic component. At a temperature near 800°C oxide-dispersion stabilised chromium interconnects formed by powder metallurgy

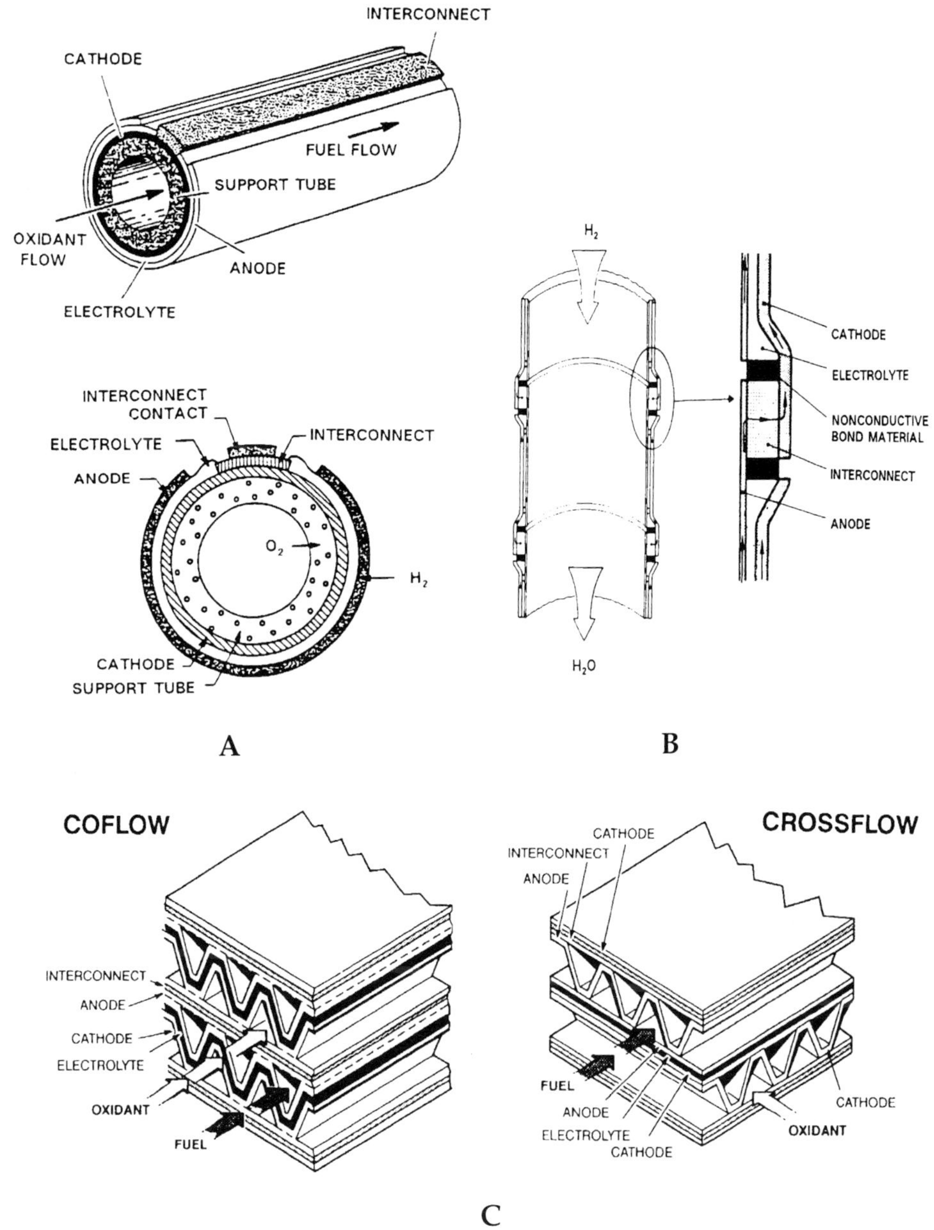

Fig. 7. Solid oxide fuel cell configurations. A: Siemens-Westinghouse tubular cell; B: Tubular integrated interconnector concept. Similar interconnected systems exist in planar geometry; C: Planar SOFC designs, differing only in gas flow manifolding.

are serviceable, but conventional engineering steels require even lower temperatures. This in turn imposes a thickness limitation on the electrolyte and high-performance electrode-electrolyte interfaces where the catalytic properties compensate for the decreasing thermal activation of the charge exchange reactions.

6. CONCLUSIONS

The technology of fuel cells is subject to intense research and evaluation, both for fixed and mobile applications, given the prospect of higher efficiency in fuel utilisation associated with lower emissions, both motivations being particularly significant for our objective of a sustainable society. The role of oxide materials in one fuel cell system, that based on the use of an electroceramic oxide ion conductor as electrolyte, has been reviewed in detail. Space does not permit the presentation of the synthesis of the oxide compounds required in SOFC practice, or the fabrication procedures adopted. Suffice it to remark that every established procedure for preparation of ceramic structures has been evaluated, from tape casting and powder pressing to sputtering and vacuum plasma spraying, for the manufacture of SOFC components, and the search remains open. Neither has there been a reference to other applications of solid state oxide electroceramic devices, such as the oxygen-selective membrane in iron-cobalt perovskite and its possible applications in selective oxidation reactions in chemical engineering; solid electrolytes for high-temperature electrolysis; life-support oxygen supplies in aerospace and even medical applications; or the whole area of materials conductive by hydrogen or hydroxyl ion mobility (Fig. 4) which opens a new class of fuel cells as well as applications in hydrogen isotope separation and dehydrogenation reactions.

The choice of materials dictated by the required electrical, thermomechanical and chemical properties has however been presented, with the consequences for fuel processing and systems operation. It is evident that the considerable advantages expected from high temperature SOFC technology are obtained only as a result of the availability of adequate materials and the development of appropriate processing techniques subsequently incorporated in a carefully engineered system. In its development and future applications therefore the fuel cell is clearly a crossroads of chemistry, energy technology and society.

7. ACKNOWLEDGEMENTS

Studies in solid state ionics, high temperature electrochemistry and fuel cells have been financed at EPFL by the Swiss Federal Office of Energy, by the Federal Office of Education and Science for participation in European Union research projects, and by the National Priority Programme for Materials (now terminated). The work continues in the context of the International Energy Agency programme for research, development and demonstration of advanced fuel cells, and of the European Science Foundation consortium OSSEP.

REFERENCES

1. J.-M. Gauguin, C. R. Acad. Sci., 37 (1853) 584.
2. W. Nernst, U.S. Patent 685 730, filed 24 August 1899.

3. H. Reynolds, Uber die Leitfähigkeit fester Mischungen bei hohen Temperaturen, Thesis, Univ. Göttingen, Germany, 1902.
4. H.-H. Möbius, J. Solid State Electrochem., 1 (1997) 2.
5. Fuel Cell Handbook, 4th. Edition, 1998, U.S. Dept. of Energy, p. 5.2.
6. J.F. Baumard and P. Abelard, in Sci. Technol. Zirconia ll, N. Claussen, M. Rühle and A. H.Heuer (Eds.), Ad. Ceram. 12 (1984) 555.
7. S.C. Singhal, in Solid Oxide Fuel Cells V, U. Stimming, S.C. Singhal, H. Tagawa and W. Lehnert (Eds.), Electrochem. Soc. Proc. Vol. PV97-40 (1997) 37.
8. J.T.S. Irvine, OSSEP brochure, European Science Foundation, Strasbourg, 2000.
9. E. Baur and R. Brunner, Z. Elektrochem. 43 (1937) 725 ; E. Baur and H. Preis, *ibid.*, 43 (1937) 727.
10. P. van den Berghe and H. Tannenberger, U.S. Patent 3,834,943 (1974); Swiss Patent 594292 (1977).
11. F.-J. Rohr, German Patent Demand P 26 14 728.4 (1976).
12. G.A. Louis, J.M. Lee, D.L. Maricle and J.C. Trocciola, U.S. Patent 4,248,941 (1981).
13. P. Reichner, Eur. Patent EP O 285 727 (1987) (Westinghouse Corp.).
14. J.H. Kuo, H.U. Anderson and D.M. Sparlin, J. Solid State Chem., 83 (1989) 52.
15. M. Gödickemeier, K. Sasaki and L.J. Gauckler, in 2nd European Solid Oxide Fuel Cell Forum, B. Thorstensen (Ed.), p. 717.
16. J.T. Irvine, P.R. Slater, A. Kaiser, J.C. Bradley, P. Holtappels and M. Mogensen, in 4th European Solid Oxide Fuel Cell Forum, A.J. McEvoy (Ed.), p. 471; A.-L. Sauvet, J. Guindet and J. Fouletier, *ibid.*, p. 567.

Studies in Surface Science and Catalysis 140
A. Gamba, C. Colella and S. Coluccia (Editors)

Synthesis and characterization of Co-containing zeolites of MFI structure

E. Nigro[a], F. Testa[a], R. Aiello[a], P.Lentz[b], A. Fonseca[b], A. Oszko[c], P. Fejes[d], A. Kukovecz[d], I. Kiricsi[d], and J. B.Nagy[b]

[a]Dipartimento di Ing. Chimica e dei Materiali, Università della Calabria, 87030 Rende (Cs), Italy
[b]Laboratoire de RMN, Facultés Universitaires Notre-Dame de la Paix, 5000 Namur, Belgium
[c]Institute of Solid State and Radiochemistry, University of Szeged, 6720 Szeged Hungary
[d]Applied Chemistry Department, University of Szeged, 6720 Szeged Hungary

Co-containing zeolites of MFI structure were synthesized using alkaline media. The orthorhombic-monoclinic symmetry transition suggests that at least the Co(II) ions also occupy tetrahedral framework positions. The XPS data clearly show that the samples contain both framework tetrahedral and extraframework octahedral Co(II) ions at ion exchange positions. The diffuse reflectance UV-visible spectra show unambiguously the presence of tetrahedral Co(II) ions in the structure.

1. INTRODUCTION

The isomorphous substitution of silicon within the zeolitic framework is an important problem and a challenge for elements different from aluminium. Although the introduction of boron, gallium, or iron is relatively easy and well documented [1], few studies are devoted to the introduction of Co(II) into the framework of zeolites [2]. As both the framework and the extraframework Co-species seem to be active in catalysis [3], it is of paramount importance to synthesize and well characterize Co-containing zeolites [4].

We report in this study the hydrothermal synthesis of Co-silicate of MFI structure in alkaline media. The samples were characterized by XRD, chemical analysis, thermal analysis, ^{27}Al-NMR, diffuse reflectance UV-visible spectroscopy and XPS spectroscopy.

2. EXPERIMENTAL

Two different gels were prepared. For the first series gels (series A) using sodium silicate the final composition were: $35SiO_2$-xNa_2O-$yCo(CH_3COO)_2 4H_2O$-3.4TPABr-$8.4H_2SO_4$-$808H_2O$ with x=11, 12.9 and 15 and y=0.5, 1, 1.5 and 2. The second type of gels B were prepared using also $Al(OH)_3$ (Pfaltz & Bauer) and their global compositions were $35SiO_2$-xNa_2O-$0.5Co(CH_3COO)_2$-$0.5Al_2O_3$-3.4TPABr-$808H_2O$ with x=3, 6, 9 and 12. The gels, after complete homogenization, were put in PTFE-lined 25 cm^3 stainless-steel auoclaves. The Co-containing samples were obtained by hydrothermal synthesis at 170°C after two days. After quenching the autoclaves, the products were recovered, filtered, washed with distilled water and finally dried a 80°C overnight. The samples were characterized by various physicochemical techniques such as XRD, chemical and thermal analysis, XPS, diffuse reflectance UV-visible spectroscopy and ^{27}Al-NMR spectroscopy.

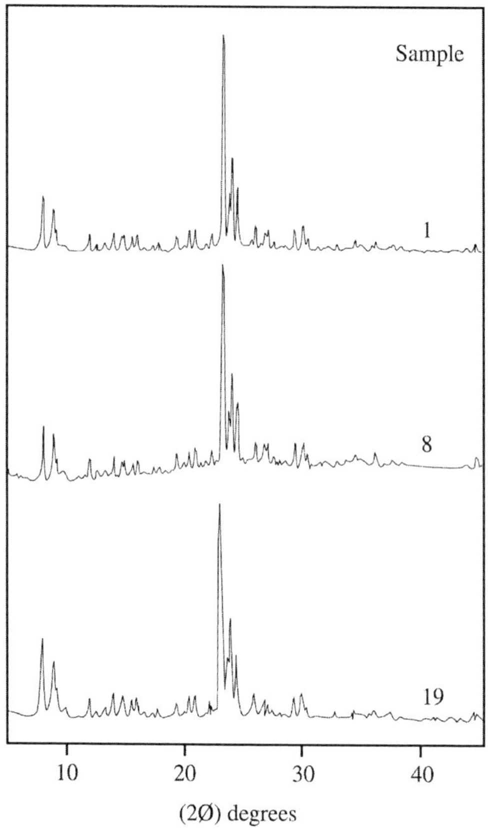

Fig. 1. XRD patterns of samples 1, 8 and 19 (see Table 1).

3. RESULTS AND DISCUSSION

Fig. 1 illustrates some XRD patterns at various Co content. It can be seen that the material of sample n°1 has a well crystalline MFI structure of orthorhombic symmetry. Sample n° 8 shows the presence of some amorphous phase as well. Indeed, these samples contain a rather high Co-content where most of the Co(II) ions are certainly extra-framework ions.

In the first series A, where the Co- MFI samples were obtained from sodium silicate source the crystallinity of the samples decreases both with the increase of alkalinity (x) and with the increase of Co-content (y) in the gel. The decrease of the crystallinity with increasing alkalinity is also observed in series B in presence of aluminium. However, sample n° 19 shows a high crystallinity (Fig. 1). Note, that high alkalinity is required for the Co-ZSM5 synthesis. Indeed, when no NaOH was added to the reaction mixture, no crystalline material was obtained after two days of synthesis. In addition, the initial gel was pink showing the presence of octahedrally coordinated Co, while for the gels prepared with high Na_2O content, the gels were all blue having the charecteristic colour of $Co(OH)_4^-$ ions.

Table 1
Chemical composition of the as-made Co-containing MFI samples
a) Samples obtained from gels A of composition $35SiO_2$-xNa_2O-$yCo(CH_3COO)_2$-$3.4TPABr$-$8.4H_2SO_4$-$808H_2O$ at 170°C after two days of synthesis

Sample	x	y	Co/u.c.	Na/u.c.	H_2O/u.c	TPA/u.c.	TPA/u.c. T(°C)	
							LT[a]	HT[b]
1	11	0.5	2.7	8.3	0	3.6	1.8-402	1.8-456
2	11	1	3.8	7.3	7.3	3.6	2.2-395	1.4-456
7	12.9	1.5	6.6	9	9.3	3.5	1.9-408	1.6-454
8	12.9	2	12.7	16.5	7.0	3.4	1.8-398	1.6-456
11	15	1.5	5.6	10.5	7.4	3.8	406(450)[c]	
12	15	2	10	15.6	8.3	3.1	406(450)[c]	

b) Samples obtained from gels B of composition $35SiO_2$-$0.5Co(CH_3COO)_2$-$0.5Al_2O_3$-xNa_2O-$3.4TPABr$-$808H_2O$ at 170°C after two days of synthesis.

Sample	x	Co/u.c.	Al/u.c.	Na/u.c.	H_2O/u.c.	TPA/u.c.	T(°C)
18	3	2.0	2.4	5.6	8.5	3.4	461
19	6	2.0	3.1	2.8	9.2	3.4	470
20	9	1.8	3.2	2.0	10.3	3.2	471
21	12	1.6	4.2	1.6	9.8	3.0	469

a.) LT:low temperature peak; b.) HT: high temperature peak; c.) shoulder

The chemical analysis by Atomic Absorption yielded the values of Co/u.c. and Na/u.c., while the thermal analysis led to the TPA/u.c. and H_2O/u.c. (Table 1).

The Co/u.c. values are rather high and suggest that the greater part of the Co is extra-framework. These values increase with increasing Co-content of the gel (Table 1a). The Na/u.c. values are also rather high and suggest that most of the Co (II) ions are extra-framework. Indeed, in the case of Al-ZSM-5 samples, only some 6.9 Na/u.c. was found for an Al content of 8.3 Al/u.c. [5] (Table 1a and b). The H_2O/u.c. varies in a random manner and does not seem to be linked to any Co/u.c. or Na/u.c. variation.

In the (Co, Al)-ZSM-5 samples (series B), the Co/u.c. values are rather low with respect to the Co-ZSM-5 samples (Table 1b). In addition, the Co/u.c. values are decreasing with increasing Al/u.c. values of the samples. It is rather well known that Al is more easily incorporated into the MFI lattice and its presence disfavours the incorporation of Co in the samples. Note that the ^{27}Al-NMR spectra show the presence of both framework tetrahedral (in large amount) and extra-framework octahedral (in small amount) aluminium species. The Na/u.c. values decrease with increasing Al/u.c. showing that in these samples Na^+cations are not preferentially linked to the negative charge created by the Al in the tetrahedral lattice. The H_2O/u.c. values are close to 9.5 and do not vary much as a function of the Al- or the Co-content.

The DSC results are very revealing on the possible incorporation of Co in the tetrahedral framework. Fig. 2 illustrates the DSC curves of three samples. When the framework Co content is low (samples of series A) two peaks characterize the decomposition of the occluded TPA^+ ions. The low temperature peak (LT) at ca 400°C and the high temperature peak (HT) at ca 455°C. The temperature of decomposition is not influenced by the formal Co/u.c values.

The TPA/u.c. does not vary much as a function of Co/u.c. and is close to an average value of 3.7/u.c. Approximately half of it is decomposed at low temperature and the other half at high temperature (Table 1a).

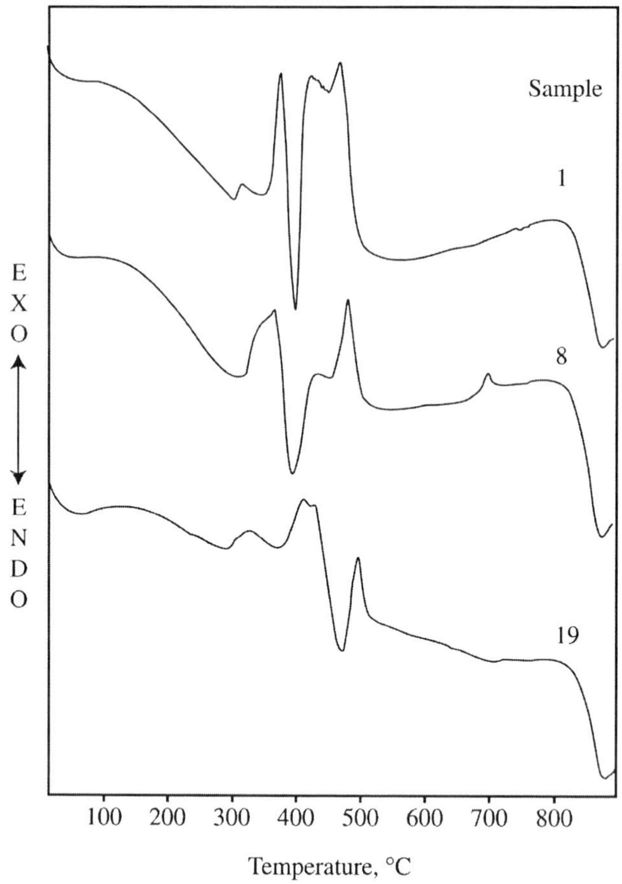

Fig. 2. DSC curves of samples 1, 8, and 19 (see Table 1).

Only at high Co content (samples 8 and 12) there is some decrease in the TPA/u.c. values suggesting the incorporation of some of Co in the samples.

The (Co,Al)-ZSM-5 samples are once again more revealing (Table 1b). The TPA/u.c. decreases with increasing Al/u.c. as it was previously reported [5, 6]. In addition, only one d.t.a. peak was observed at already 2.4 Al/u.c. suggesting that some of the Co is also incorporated in the structure.

Diffuse reflectance (DR) spectroscopy of Co^{2+} permits the observation of d-d transitions in the near infrared and visible region and charge transfer (CT) transitions in the ultraviolet region. Co^{2+} is the only common d^7 ion and because of its stereochemistry the respective spectra have been extensively studied [4]. The two representative cases: Co^{2+} ion in octahedral (O_h) and tetrahedral (T_h) [4] crystal field can be interpreted in the same way as the octahedral d^2 ion (e.g. V^{3+}) and octahedral d^3 ion (e.g. Cr^{3+}), respectively. (For the corresponding Tanabe-Sugano diagrams see in [4]). As all d-d transitions for octahedral complexes possessing symmetry centre are symmetry-forbidden, these bands are in the spectra exceptionally weak. Thus, the DR spectra of zeolites, ZSM-5 zeolites in particular, with framework (FW) and extra-framework (EFW) Co^{2+} content are dominated by

- intense $O^{2-} \Rightarrow Co^{2+}$ charge transfer transitions, characteristic essentially for Co^{2+} in T_h coordination around 200-230 nm;
- the most intense T_h transition, ν_3 [corresponding to $^4A_2(e^4t_2^3) \Rightarrow {}^4T_1(P)$] around 590-630 nm and a weaker one, ν_2, which is already in the near infrared (1400-1500 nm). (The third d-d transition, ν_1, is also in the infrared region [2000-3300 nm], but its examination is hindered by overlying vibrational bands).

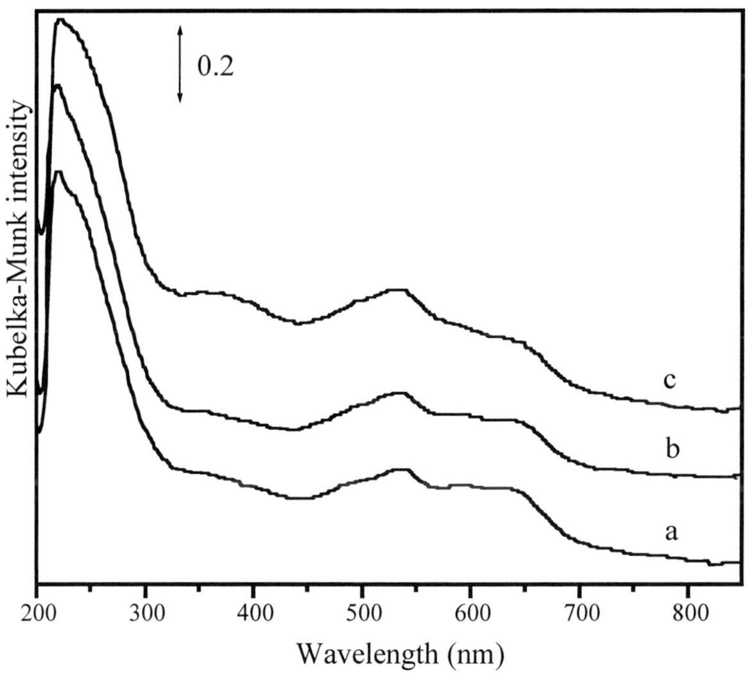

Fig. 3. Diffuse reflectance UV-VIS spectra of samples No. 1 (a), 18 (b) and 21 (c) recorded at a resolution of 2 nm.

Table 2
Visible diffuse reflectance spectra of selected as-made Co-ZSM-5 samples

Sample	Band Intensities (a.u.)		
	480 nm Co Octa	530nm+590nm+640nm Co Tetra	340nm Co(III)
1	11	21	25
18	8	17	13
20	0	19	46
21	17	20	26

Fig. 3 exhibits in the 460-700 nm visible region (from blue to red) at least 4 badly resolved bands (at 486, 530, 584 and 643 nm), of which the last three are T_h d-d transitions. This triplet is indicative for high-spin (d^7) Co^{2+} in tetrahedral crystal field.

The nearly invisible 357 nm band is supposed to be due to high-spin Co^{3+} in unknown coordination [3].

Table 2 compares the relative intensities of the 480 nm band (octahedral Co(II)), the sum of 530, 590 and 640 nm bands (tetrahedral Co(II)) and the 340 nm band (Co(III)). The values are in arbitrary units and only their variations have a meaning for the state of Co ions in the ZSM-5 samples. The octahedral Co(II) ions should be extraframework ions, the tetrahedral Co(II) ions could be due to both framework and extraframework species and the Co(II) ions are considered essentially extra-framework secies.

It can be seen that tetrahedral Co(II) ions are present in all studied samples in roughly similar amounts. The amount of extraframework Co(II) ions are similar in samples 1 and 21, while sample 18 and 20 show only a rather small amount. The amount of extraframework Co(III) ions is low in sample 18, somewhat higher in samples 1 and 21 and rather high in sample 20. No systemic variations could be detected as a function of the formal Co/u.c. content (Table 1).

XPS or electron spectroscopy for chemical analysis (ESCA) is used for characterizing surface species.

In the case of Co^{2+} the 2p electron transitions are observed which, after ionisation, become separated into two levels: $2p_{3/2}$ and $2p_{1/2}$. The $2p_{1/2}$ peak is shifted upwards in energy by about 15.7 eV. Each peak has one, more or less intense, shoulder, a so called satellite, thus, we see not two, but rather four peaks (3 peaks and 1 strong shoulder) in the XP spectrum. The appearance of these "excess" peaks is due to multiplet splitting of the respective electron orbitals.

Table 3
XPS data of selected Co-ZSM-5 samples

Sample	Binding energies (eV)		Satellites (eV)	
	Co ($2p_{3/2}$)	Co ($2p_{1/2}$)		
Reference Co/ZSM-5	780.9	796.8	785.9	802.85
1	780.95	796.95	785.95	803.05
8	781.4	797.25	786.75	803.35
12	781.05	797.1	786.6	803.35
21	781.05	797.1	786.0	803.2

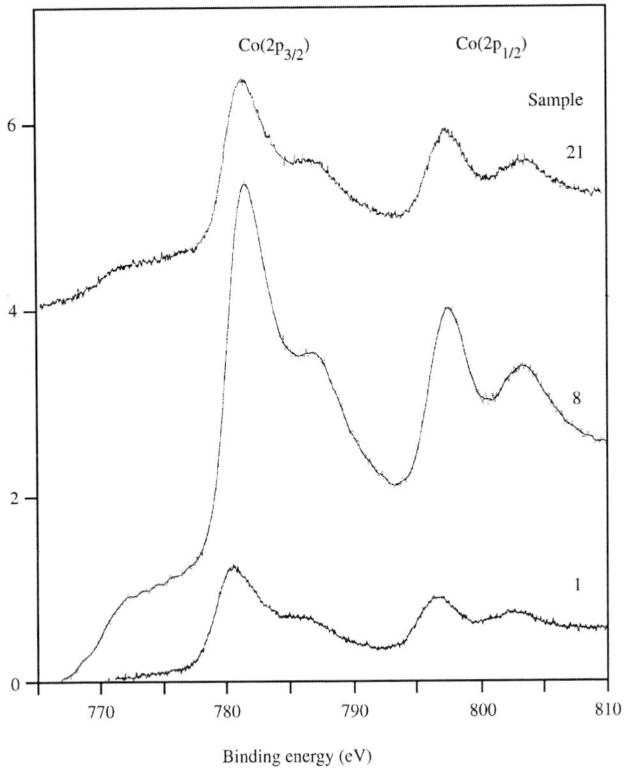

Figure 4. XPS spectra of $Co(2p_{3/2})$ and $Co(2p_{1/2})$ of samples n° 1, 8 and 21.

The paper deals with two series of Co-ZSM-5 samples: a). (n ° 1 to 12, cf. Table 1) is characterised by variable x/y ratios in the slurry. The 2nd series, b)., contained both Co^{2+} and Al^{3+} and they were charged in a constant Co/Al = 0.5 ratio. It would have been very informative to take the XP spectra of each sample, however, the well known general trend of such syntheses in alkaline media (resulting only in partial substitution) and the registered UV-VIS spectra (see above) revealed unequivocally that it is impossible to attain complete incorporation of Co^{2+}, thus, only three samples (n° 1,8 and 21) have been chosen for closer examination.

In a previous study we succeeded to detect a slight, but unmistakable 1.9 eV difference in the locations of $2p_{3/2}$ XP signals of FW and EFW Fe^{3+} ions in heat-treated Fe-ZSM-5 samples [7]. Even though the theme was not treated in the literature so far, it was deemed to be interesting to try to find a small shift for FW and EFW Co^{2+} too. This would have been indicative of mixed coordination, i.e. the presence of both tetrahedral and octahedral Co ions.

The XP spectra of the Co^{2+} photoelectron region for the Co-ZSM-5 samples n° 1, 8, 12 and 21 are given in Fig. 4. Table 3 provides information on the binding energies of the $Co(2p_{3/2})$, $Co(2p_{1/2})$ (with respect to Si(2p)) and of the satellites. If we add a fourth Co-ZSM-5 sample to the selected ones (synthesized from magadiite in the presence of Co-pyrocatecholate by P. Fejes) (in Table 3 marked as reference, maybe it is not an overstatement that the increase of Co^{2+} content (cf. Table 1.) brings about a little upward shift in the location of the $2p_{3/2}$ peaks which do not attain 781.5 eV, typical for free Co^{2+}, even at sample n° 8, bearing the largest Co^{2+} content (12 Co^{2+}/u.c. corresponding to Si/Co = 6.56). It is believed that the binding energies of core level electrons for Co^{2+} in tetrahedral coordination (similar to the respective Fe^{3+}levels) are smaller than those in octahedral symmetry and in mixed coordination the spectrometer, being not able to resolve the two peaks (separated by a few tenths of eV), registers only one peak at the weighted average of the two binding energies. This manifests itself as an upward energy shift when the percentage of octahedral Co^{2+} increases with the cobalt content.

4. CONCLUSIONS

The various physicochemical techniques used showed unambiguously that part of the Co(II) ions is incorporated into the tetrahedral framework positions. However, despite the joint effort of the different techniques no quantitation of the tetrahedral Co(II) ions was possible.

5. ACKNOWLEDGMENT

The present work is a part of a project coordinated by A. Zecchina and cofinanced by the Italian MURST (Cofin 98, Area 03). This work was carried out with the financial support of the Regione Calabria (POP 97/99). The authors thank the Belgian Program on Inter University Poles of Attraction initiated by the Belgian State, Prime Minister's Office for Scientific Technical and Cultural Affairs (OSTC-PAI-IUAP No. 4/10 or Reduced Dimensionality Systems) for financial support.

REFERENCES

1. R. Szostak, Molecular Sieves. Principles of Synthesis and Identification, Van Nostrand Reinhold, New York, 1989, p. 205.
2. T. Inui, J.-B. Kim and T. Takeguchi, Zeolites, 17 (1997) 354.
3. K. Kagawa, Y. Ichikawa, S. Iwamoto and T. Inui, Microporous and Mesoporous Mater., 25(1998) 15.
4. A. Verberckmoes, B. M. Weckhuysen and R.A. Schoonheydt, Microporous Mesoporous Mater., 22 (1998) 165.
5. G. Debras, A. Gourgue, J. B.Nagy and G. De Clippeleir, Zeolites, 5 (1985) 377.
6. R. Aiello, F. Crea, E. Nigro, F. Testa, R. Mostowicz, A. Fonseca and J. B.Nagy, Microporous Mesoporous Mater. 28 (1999) 241.
7. P. Fejes, J. B.Nagy, J. Halász and A. Oszkó, Appl. Catal. A: General, 175 (1998) 89.

Studies in Surface Science and Catalysis 140
A. Gamba, C. Colella and S. Coluccia (Editors)

Synthesis and Characterization of dye-containing MCM-41 materials

B. Onida[1], B. Bonelli[2], M. Lucco-Borlera[1], F. Geobaldo[1], C. Otero Areán[3] and E. Garrone[1]

[1]Dipartimento di Scienza dei Materiali e Ingegneria Chimica, Politecnico di Torino, Corso Duca degli Abruzzi 24, 10129 Turin, Italy.

[2]Dipartimento di Chimica IFM, Università di Torino, Via P. Giuria 7, 10125 Turin, Italy.

[3]Departamento de Química, Universidad de Las Islas Baleares, Palma de Mallorca, Spain.

Congo Red and Curcumin containing MCM-41 materials have been synthesized and characterized by means of X-ray diffraction, UV-Visible and FTIR spectroscopy. Dyes are incorporated in the micellar phase of silica-surfactant mesophase and their spectroscopic properties confirm that they are in a solvated state, where both surfactant and silica wall may act as a solvent. Dyes maintain their pH indicator properties and are accessible to ionic species, such as H_3O^+ and Cu^{2+}.

1. INTRODUCTION

MCM-41 is the hexagonal phase of silica-based mesoporous materials collectively designated as M41S [1].

The incorporation of functional molecules in these systems is a new development of advanced materials research and various examples of inclusion chemistry in ordered mesoporous host materials are known [2].

Recently, successful attempts have been made to prepare both dye-functionalized mesostructured silica [3-5] and dye-containing zeolitic materials [6, 7]. The growing interest for such systems is due to their potential application in optical devices and as chemical sensors.

Congo Red and Curcumin (hereafter designated as CR and CC, respectively) are pH indicators, with pH intervals of 3-5.2 and 7.4-8.6 respectively. Congo Red may also act as chelating agent towards metal ions and it has accordingly been successfully tested in ion-sensing [8] and chromatographic separations of metal traces [9] in solution. The chemical structure of both CR and CC is shown in Scheme 1.

Among the different methods reported in the literature for introducing dye-molecules in ordered mesoporous silica, a simple one consists in adding the dye to the synthesis gel, so obtaining a material where the dye is embedded in the micellar phase [3]. Following this approach, precursors of MCM-41, i. e. micelle-bearing samples containing CR and CC have been prepared and characterized by means of UV-visible spectroscopy. The pH indicator properties and complexing ability of dye-containing materials have been also investigated.

2. EXPERIMENTAL

Following a procedure similar to that reported in [10], several MCM-41 samples have been

prepared, namely a dye-free MCM-41 and several dye-containing MCM-41 (Table 1).
The reactants were tetraethylorthosilicate (TEOS, Aldrich), distilled water, hexadecyltrimethylammonium bromide (CTAB), Congo Red (Merck) or Curcumin (Merck) and NaOH (Carlo Erba).
For the synthesis, performed at room temperature, 2,0 g of CTAB was added to a basic solution (0,96 g of NaOH in 475 ml of H_2O) under stirring. When the solution became homogenous, proper amounts of the corresponding dye were added. To the coloured solution 10 ml of TEOS was added, giving rise to a coloured slurry. After 2 hours the product was filtered, washed first with distilled water and then with ethanol and dried at ambient temperature.
For both CR and CC, two different samples have been prepared, with a Si/dye ratio equal to 100 and 25, respectively. The composition of the synthesis mixtures was 1 TEOS · 0,125 CTABr · 0,400 NaOH · 0,01-0,04 dye · 525 H_2O.

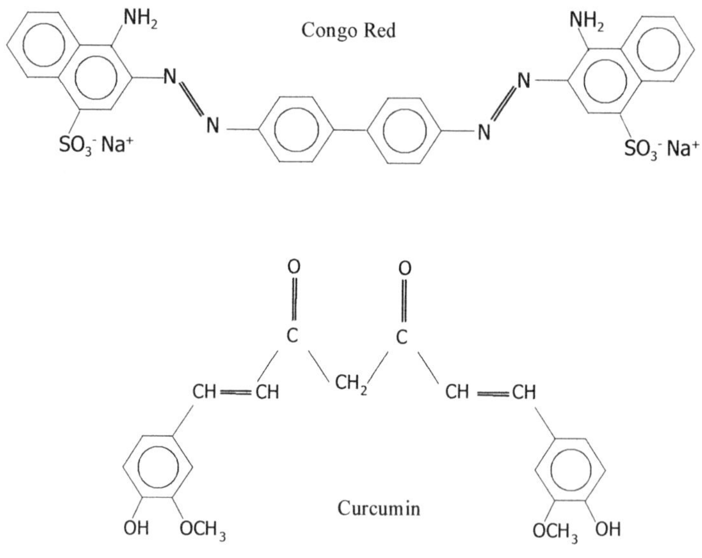

Scheme 1

The as-synthesized materials, *i. e.* micelle-bearing systems, were characterised by XRD and UV-Vis spectroscopy. XRD patterns were recorded using an X'Pert Philips diffractometer, with CuKα radiation. Reflectance UV-visible spectra were recorded using a Varian Cary 5 spectrophotometer.
For comparison, two surfactant-free CR/SiO_2 systems were prepared. One was produced by precipitation of SiO_2 from a basic solution of TEOS in presence of CR. The other was obtained by impregnation of the calcined MCM-41 with a CR solution. For the calcination, MCM-41 was heated in air flow up to 773 K (5K/min) and maintained at the same temperature for 6 hours.
Table 1 reports the samples studied, together with their label and related Si/dye ratio.

Table 1 Nomenclature and composition of the samples studied

Name	Sample	Si/dye
MCM-41	Dye-free MCM-41	∞
CR-MCM-41	CR containing MCM-41	100
CR-MCM-41(4)	CR containing MCM-41	25
CC-MCM-41	CC containing MCM-41	100
CC-MCM-41(4)	CC containing MCM-41	25
I-CR-MCM-41	Calcined MCM-41 impregnated with CR	~ 100
CR-SiO_2	SiO_2 precipitated in presence of CR	~ 100

3. RESULTS AND DISCUSSION

3.1 Characterization of the materials

CR-MCM-41, CR-MCM-41(4), CC-MCM-41 and CC-MCM-41(4) samples showed the typical colours of the basic form of the indicators, i. e. bright red for CR and red-wine for CC, because of the pH of the synthesis mixture. The filtrates were colourless, indicating that the dye was quantitatively incorporated in the material. When washing with water or ethanol the filtrate was colourless, which indicates that no leaching of dye occurred in these conditions. In

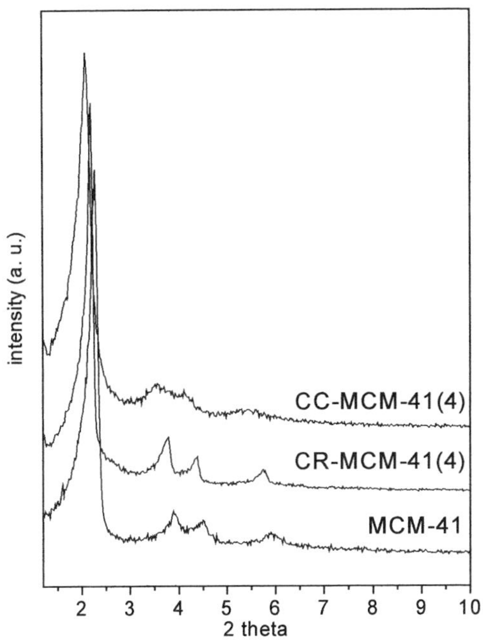

Table 2 Lattice parameter of the different samples

Sample	a (Å)
MCM-41	44.4
CR-MCM-41	46.4
CR-MCM-41(4)	46.4
CC-MCM-41	44.5
CC-MCM-41(4)	48.8
CC-MCM-41(4) w*	44.9

* washed with ethanol/acetone mixture

Figure 1. XRD patterns of dye-free and dye containing MCM-41.

contrast, washing with water both I-CR-MCM-41 and CR-SiO_2 brought about dye leaching, so that a coloured solution and a colourless powder were obtained. Dye leaching from CR-MCM-41, CR-MCM-41(4), CC-MCM-41 and CC-MCM-41(4) samples where obtained washing with an ethanol/acetone mixture.

These data indicate that surfactant micelles are necessary for retaining the dye in the material. Indeed, it is known that ionic organic dyes, such as pH indicators, can be removed from water solution by adding ionic surfactants, which collect the dye in the froth. Also, ionic dyes such as CR have themselves a tendency to form micelles [11].

Figure 1 shows the comparison among XRD patterns of samples MCM-41, CR-MCM-41(4) and CC-MCM-41(4). In all cases, the peaks characteristic of the hexagonal phase are present. A shift to lower angles of all peaks, particularly evident for the d_{100}, is observed in the presence of dye molecules, indicating an increase of d-spacing and thus of the parameter of the hexagonal cell. Table 2 gives the cell parameter *a* of the different MCM-41 samples, calculated as $a = 2d_{100}/\sqrt{3}$ [12, 13] .

The increase of cell parameter indicates that the dye is incorporated into the surfactant micelles, giving rise to some swelling of the micelles and enlargement of the cell [12]. After washing with ethanol/acetone mixture, which causes dye leaching, the XRD spectrum changes, in that peaks become broader and the *a* value decreases (data concerning CC-MCM-41(4) in Table 2). This suggests that together with the dye, a certain amount of surfactant is extracted, causing a shrinkage of the structure. These results support the idea that the dye is present inside the micellar phase of the material.

The enlargement of the cell is higher for CC-MCM-41(4) than for CR-MCM-41(4). This can be due both to the different shape and steric hindrance of the two molecules and/or to a different orientation of the molecules within the micelles.

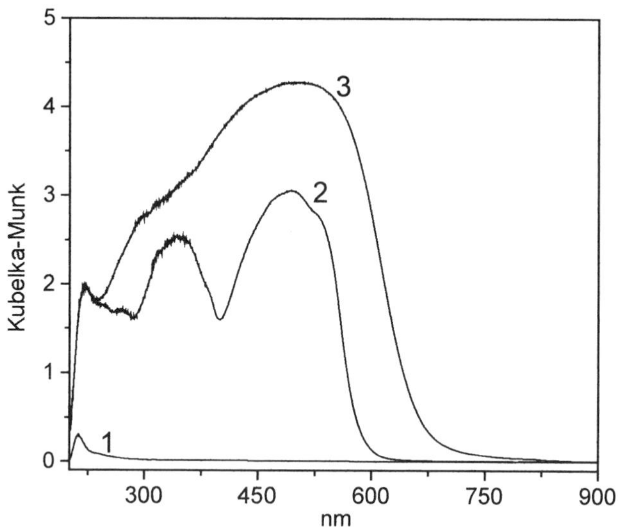

Figure 2. UV-visible spectra of MCM-41 (curve 1), CR-MCM-41 (curve 2) and CC-MCM-41 (curve 3)

Figure 2 shows the UV-visible spectra of MCM-41, CR-MCM-41 and CC-MCM-41 In spectrum 1 no absorption is observed, whereas in curves 2 and 3 intense bands due to chromophores of dye are present. A single broad absorption is observed for CC-MCM-41, whereas two maxima are present in CR-MCM-41. The different absorption tails account for the difference in colour.

Figure 3 compares the spectrum of CR-MCM-41 with that of pure, crystalline CR. In the latter spectrum, a complex absorption band is observed, where two maxima are seen at about 430 nm and 560 nm. The electronic transitions of CR-MCM-41 appear shifted to lower wavelength with respect to the crystalline dye and two better defined maxima are present at about 340 nm and 500 nm, together with two shoulders at 380 nm and 530 nm, respectively (labelled with an asterisk in the figure). The spectrum of CR-MCM-41 resembles that of CR in solution, as shown adjacent to Figure 3. In both water and ethanol solutions two bands are observed at 340 nm and about 500 nm, being the latter assigned to the π-π^* transition of the

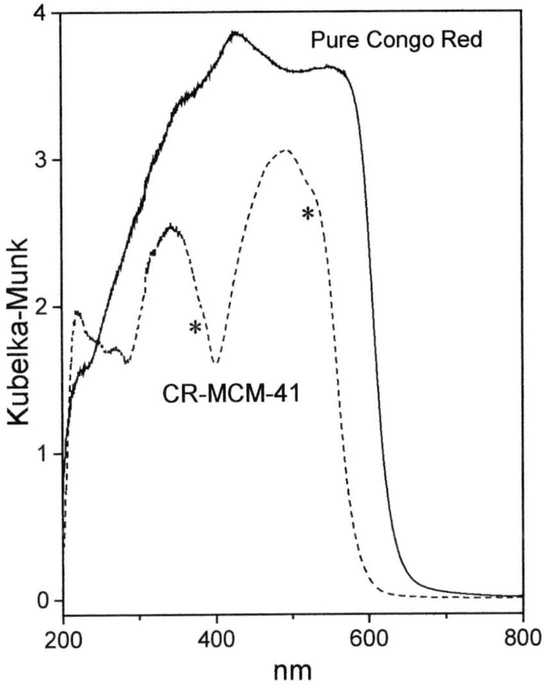

Figure 3 UV-visible spectra of CR-MCM-41 and pure CR

UV-Vis absorptions of CR

System	λ_{max} (nm)
Water	340 500
Ethanol (95%wt)	340 510
CTAB water solution	320; 368 (s) 500; 536 (s)
CR-MCM-41	340; 380 (s) 500; 530 (s)
solid	430 560

(s) shoulder

azo group of the dye [8].

The presence of two shoulders in CR-MCM-41, not seen in aqueous solutions, may be due to the interaction of the dye with surfactant molecules. Indeed, two shoulders are also observed at 368 nm and 536 nm in the spectrum of CR in aqueous solutions of surfactant (spectra not reported), where electrostatic interactions between dye and surfactant occur [11].

In conclusion, the electronic spectra in Figure 3 show that CR in MCM-41 precursor is in a solvated state, where the surfactant and maybe also the silica wall act as a solvent.

Most probably the sulfonic groups interact with the cationic heads of the surfactant molecules *via* electrostatic interactions. When the synthesis is carried out in basic conditions, as in the present case, in the surfactant/silica mesophase the cationic parts of the surfactant molecules interact with the negatively charged silica surface [14]. Thus it can be assumed that in CR-MCM-41 the sulfonic groups of the CR molecule are located at the interface surfactant-silica, whereas the rest of the molecule, i. e. aromatic rings, amino groups and azo groups, is embedded in the micelle.

3.2 Accessibility of dye molecules.
The accessibility of the confined dye to species in solution has been investigated in two ways. In the first experiment, CR-MCM-41 and CC-MCM-41 have been put in contact with acidic solution, to assess whether the dye molecules were reachable by $[H_3O^+]$ species and were able to display their pH-indicator properties.
CR-MCM41 changes colour, from red (basic from) to blue (acidic form), upon contact with an acidic solution at pH<1. This indicates that protonation of CR occurs and that the azo group is accessible to $[H_3O^+]$ ions from aqueous solution. When pH increases because of dilution of the acidic solution, the sample changes from blue to red, indicating that the protonation of embedded CR is reversible.
The pH at which the colour of CR-MCM-41 changes from red to blue is lower than that of CR in aqueous solution. The same behaviour has been observed for CC-MCM-41 and for both dyes in CTAB-containing aqueous solutions. Taking into account that most probably in all these cases the dye is embedded in the micellar phase, this can be explained considering a partition of H_3O^+ species between the hydrophilic phase (water itself) and the hydrophobic phase (micelles). The higher affinity of the hydronium ion for water with respect to the micelle causes a higher $[H_3O^+]$ value in aqueous solution (to which pH is referred) than in the micelles. This means that, in order to observe colour change from the basic form to the acidic form for the micelle-embedded dye, a higher $[H_3O^+]$ in the water solution must be achieved, *i. e.* a lower pH, with respect to the value observed for CR aqueous solution
The second experiment consisted in putting CR-MCM-41 in contact with a solution of Cu^{2+} ions. CR may act as a sensitive photometric reagent toward metal ions, such as Cu^{2+}, giving rise to changes in the electronic spectrum. Chelation involves the azo and the amino groups, causing a red shift of the π-π^* [8]. The aim of the experiment was therefore to see whether complexation of copper ions by the dye may still occur when the dye is immobilised in the MCM-41 material.
CR-MCM-41 was maintained under stirring in $CuSO_4$ solution for 12 hours, then filtered and washed. Two solvents have been considered, *i. e.* water and ethanol (95% wt).
Figure 4 compares the UV-visible spectra in the range 400-900 nm of CR-MCM-41 as synthesised and after contact with water and ethanol solutions of $CuSO_4$. After contact with aqueous solution (curve 2) the band due the π-π^* transition at about 500 nm appears more intense and less structured and no new absorptions are seen. This can be due to a small change in the solvation state of the dye. Instead, after contact with ethanol solution (curve 3), a new absorption appears at about 630 nm and comparing curve 2 and curve 3 a sort of isosbestic point is seen at 570 nm. Absorption at 630 nm is ascribed to π-π^* trasition of molecules complexing Cu^{2+} ions, in agreement with what observed for other metal ions complexed by CR linked to a polymeric matrix [8]. A fraction of this absorption is probably due to the weaker d-d transition of Cu^{2+} ions [15]
The different effect of the two solvents can be explained considering the different hydrophobic properties of the solvated ions. Ethanol-solvated ions have most probably a

higher hydrophobicity, i. e. affinity for the micellar phase, than aquo-ions, so the former diffuse inside the dye-containing mesophases more easily than the latter.
Both experiments shows that in CR-MCM-41 dye molecules are still accessible to species from solution.

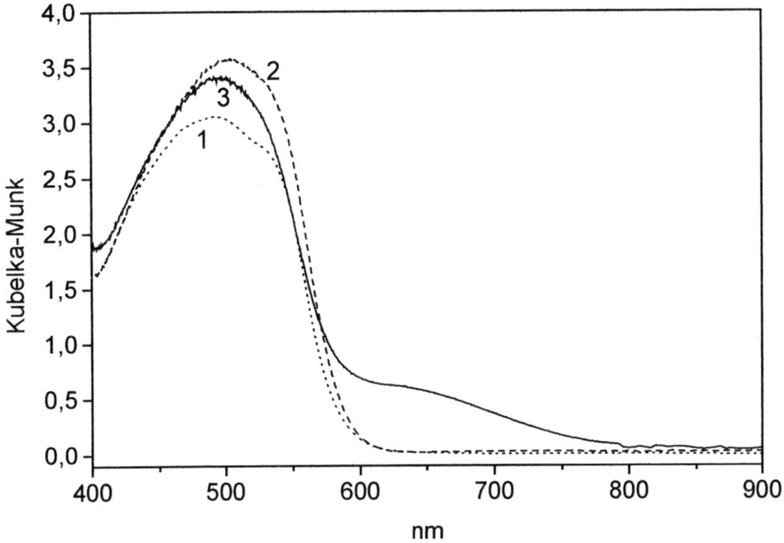

Figure 4 Spectral changes of CR-MCM-41 upon contact with $CuSO_4$ solutions. Curve 1: as-synthesized CR-MCM-41; curve 2: upon contact with $CuSO_4$ water solution; curve 3: upon contact with $CuSO_4$ ethanol solution.

4. CONCLUSIONS

Dye-containing MCM-41 precursors have been prepared adding pH indicators (Congo Red and Curcumin) to the synthesis gel. Dye molecules are incorporated in the micellar phase in their basic form, because of the synthesis conditions. The ionic parts of the molecules are most probably located at the silica/surfactant interface. The electronic spectra of the materials suggest that dyes are in a solvated state, where surfactant and silica wall may act as a solvent. Though embedded in the micelles, dyes are still accessible to ionic species from solution, such as H_3O^+ and Cu^{2+}. They still act as pH indicators in that reversible protonation occurs, with concomitant colour change when the materials are put in contact with solutions of proper pH. The pH at which colour changes take place is lower than that observed for dyes in aqueous solutions. This is ascribed to a H_3O^+ partition equilibrium between water solution and micellar phase in the material, for which $[H_3O^+]$ is higher in the former than in the latter. At least part of the dye molecules in CR-MCM-41 is able to give complexation of metal ions, such as Cu^{2+}, which causes a red shift of the π-π* electronic transition of the azo group.

References

[1] C. T. Kresge, M. E. Leonowicz, W. J. Roth, J. C. Vartuli and J. S. Beck, *Nature (London)*, 1992, 359, 710.

[2] K. Moller and T. Bein, *Chem. Mater.*, 1998, 10, 2950
[3] H. S. Zhou, H. Sasabe and I. Honna, *J. Mater. Chem.*, 1998, 8, 515.
[4] B. Lebeau, C. E. Fowler, S. R. Hall and S. Mann, *J. Mater. Chem.*, 1999, 9, 2279.
[5] Z. Gabelica, S. Valance, M. Shibata, H. Hotta and T. Suzuki, *International Symposium on Mesoporous Molecular Sieves*, Quebec City, Canada, 2000.
[6] S. Wohlrab, R. Hoppe, G. Schulz-Ekloff, D. Wöhrle, *Zeolites*, 1992, 12, 862.
[7] J. L. Meinershagen and T. Bein, *J. Am. Chem. Soc.*, 1999, 121, 448.
[8] T. S. Lee, C. Yang, *Polymer Bulletin*, 1999, 42, 655.
[9] S. A. Nabi, A. Gupta and A. Sikarwar, *Annali di Chimica*, 1999, 89, 419
[10] Q. Cai, W. Y. Lin, F. S. Xiao, W. Q. Pang, X. H. Chen and B. S. Zou, *Microp. Mesoporous Mater.*, 1999, 32, 1.
[11] F. Sebba, *Nature*, 1960, 184, 1062.
[12] J. S. Beck, J. C. Vartuli, W. J. Roth, M. E. Leonowicz, C. T. Kresge, K. D. Schmit, C. T-W. Chu, D. H. Olson, E. W. Sheppard, S. B. McCullen, J. B. Higgins and J. L. Schlenker, *J. Am. Chem. Soc.*, 1992, 114, 10832.
[13] A. Galarneau, D. Desplantier, R. Dutartre, F. Di Renzo, *Microp. Mesoporous Mater.*, 1999, 27, 297.
[14] G. D. Stucky, D. Zhao, P. Yang, W: Lukens, N. Melosh, B. F. Chmelka, *Stud. Surf. Sci. Catal*, 1998, 117, 1.
[15] B. N. Figgis, *Introduction to Ligand Fields*; Interscience: London, 1966.

Studies in Surface Science and Catalysis 140
A. Gamba, C. Colella and S. Coluccia (Editors)

Modeling of breakthrough curves in fixed–bed zeolite columns

F. Pepe[a], D. Caputo[b] and C. Colella[b]

[a]DICMA (Dipartimento di Ingegneria Chimica, Mineraria e delle Tecnologie Ambientali), Università di Bologna, V.le Risorgimento 2, 40136 Bologna, Italy.

[b]Dipartimento di Ingegneria dei Materiali e della Produzione, Università di Napoli Federico II, P.le Tecchio 80, 80125 Napoli, Italy

A diffusional model for the interpretation of breakthrough curves obtained in fixed–bed ion exchange processes was presented. The model, based on the linear driving force (LDF) approximation, takes into account both fluid–particle and intra–particle resistances to diffusion and makes no specific assumption on their relative magnitude or on the form of the ion exchange isotherm. Its only parameters, to be evaluated on the basis of the experimental data, are a solid–liquid and an intra–particle mass transfer coefficient. In order to verify the model, its previsions were checked against some published results, relative to Pb^{2+} removal from water by means of a Na–exchanged phillipsite–rich tuff. The model was capable of satisfactorily reproducing the dependence of the experimental breakthrough curves on solution composition. An analysis of the model results showed that the external mass transfer coefficient does not depend on the composition of the liquid phase. On the other hand it appeared that, in agreement with previous published experimental results, the internal mass transfer coefficient varies with the total normality of the solution.

1. INTRODUCTION

Treatment of liquid wastes with zeolites is a very promising technique, especially for removal of ammonium and heavy metal ions from both municipal and industrial wastewater (Pansini, 1996; Colella, 1999). Indeed, many natural and synthetic zeolites exhibit good ion exchange properties in terms of cation exchange selectivity and capacity for the main toxic and noxious cations. In addition, natural zeolites are readily available in many areas of the world at relatively low costs (Colella, 1999).

Despite the considerable amount of studies on equilibrium and kinetics of ion exchange processes involving zeolites, a lack of data still exists concerning the modeling of zeolite columns for scale up of laboratory data to engineering size equipments. The main problem encountered is in the complex interplay among the different mass transfer resistances involved in the process. Indeed, the overall transfer rate depends on the convective transport from the liquid bulk to the solid–liquid interface, on liquid phase diffusion within the pore of the particles, and on solid phase diffusion within the zeolite crystals. Glueckauf (1955) proposed the linear driving force (LDF) model, according to which both liquid–solid and intra–particle mass transfer rates linearly depend on their respective driving forces. Following this approach, the mass transfer rate r (moles, or equivalents, transferred per unit time and surface) can be evaluated by the following equations:

$$r = k_e\left(c - c^*\right) \tag{1}$$

$$r = k_i\left(q^* - q\right) \tag{2}$$

in which c is the liquid bulk concentration of the ion being transferred, q is its average concentration in the solid particle, c^* and q^* are its interfacial concentrations, in solution and on the solid respectively, and k_e and k_i are an external (solid–liquid) and an internal (intra–particle) mass transfer coefficient, respectively. Equations (1, 2) have to be associated with the condition that equilibrium prevails at the solid–liquid interface, so that c^* and q^* are subject to the equilibrium relationship.

Equations (1, 2) constitute a powerful tool for the analysis of the mass transfer process. Indeed, if a reliable equilibrium relationship is available for the ion exchange reaction and the mass transfer coefficients are known, then it is possible to evaluate the mass transfer rate for given compositions of the liquid and solid phases. On the other hand, experimental data relative to the mass transfer rate can be used to evaluate unknown mass transfer coefficients.

However, it has to be observed that Eq. (2) is rigorously valid only when a linear isotherm exists or, which is the same, when the intra–particle concentration profile of the ion being transferred is parabolic. In order to avoid this limitation, Hall *et al.* (1966) followed a numerical approach, integrating the mass transfer equations to obtain a fairly general representation of the breakthrough curves, valid under many experimental conditions. However, their work concerned adsorption rather than ion exchange, and is only applicable to cases in which the equilibrium conforms to the Langmuir–type isotherm. On the other hand, Zhang and Ritter (1977) and Lai and Tan (1991) proposed different corrections to Eq. (2), with the aim of making it capable of handling non–linear isotherms.

The subject has recently been reviewed by Carta and Cincotti (1998), who numerically integrated the mass transfer equations for intra–particle transport, to obtain the exact concentration profiles within the solid particle in different conditions. In this way they were able to evaluate the relative merits of the different approximations proposed and, from the results of their analysis, suggested an expression for the intra–particle mass transfer rate. Their expression explicitly takes into account both the form of the ion exchange isotherm and the value of the partition ratio $\Lambda=\rho_s Q/\varepsilon N$, in which Q is the ion exchange capacity of the solid considered, ρ_s is its density, ε is the void fraction and N is the total normality of the solution.

Despite the corrections proposed, Ruthven (1984) indicated that in most cases the "original" LDF model, defined by Eqs. (1, 2), gives acceptable results when used for the evaluation of breakthrough curves. Therefore, such model will be used in the following to interpret and model breakthrough curves obtained in a fixed–bed ion exchange process, in which a naturally occurring phillipsite–rich tuff is used to remove Pb^{2+} from water. Experimental data used are those of a previous investigation, in which the Na–exchanged form of the tuff was found to display a great selectivity and a good exchange kinetics for Pb^{2+}, even in presence of considerable amounts of Na^+ as interfering cation (Pansini *et al.*, 1996). The diffusional model which will be proposed considers both equilibrium and mass transfer relationships and it can be used regardless of the form of the equilibrium isotherm.

The aim of the present paper is to test the applicability of the model, and to use it to evaluate the external and internal mass transfer coefficients in the experimental conditions considered by Pansini *et al.* (1996).

2. MODELING OF THE ION EXCHANGE PROCESS

The fixed–bed ion exchange process is described by the material balances on the species being transferred in both the liquid and the solid phase. In particular, when a binary system is considered, only a couple of material balances has to be taken into account, a stoichiometric relationship being sufficient to give the liquid and solid concentrations of the other species. The material balances for the species considered have the following form (McCabe *et al.*, 1985):

$$\varepsilon D \frac{\partial^2 c}{\partial z^2} - u \frac{\partial c}{\partial z} - \varepsilon \frac{\partial c}{\partial t} - (1-\varepsilon) a r = 0 \tag{3}$$

$$\rho_s \frac{\partial q}{\partial t} = a r \tag{4}$$

Equation (3) refers to the liquid phase, and the non–idealities in its flow are taken into account by means of the dispersion model (Levenspiel, 1999). Equation (4), on the other hand, refers to the solid phase. In these equations z is the axial coordinate along the bed, t is the time, D is the axial dispersion coefficient, u is the liquid superficial velocity, and a is the (external) specific surface area of the solid particles. Coupling between these two equations is given by the presence in both of the rate term r, which can be evaluated by means of Eqs. (1, 2).

The boundary conditions for the system of Eqs. (3, 4) depend on the specific problem under consideration. Here, since the removal from the liquid phase of a species that initially is not present in the bed is studied, the following conditions were taken into account:

$$t = 0, \quad 0 \le z \le L: \quad c = 0, \quad q = 0 \tag{5a}$$

$$t > 0, \quad z = 0: \quad \varepsilon D \frac{\partial c}{\partial z} = u(c_o - c) \tag{5b}$$

$$t > 0, \quad z = L: \quad \frac{\partial c}{\partial z} = 0 \tag{5c}$$

These conditions state that at the beginning of each experiment the species being removed from the liquid phase is not present in the column (5a); that its concentration in the solution fed to the fixed bed is constant and equal to c_o (5b); that the fixed bed is a "closed" system, the length of which is L (5c, see Levenspiel, 1999).

The dimensionless equations corresponding to Eqs. (3. 4) are the following:

$$\frac{1}{P\grave{e}} \frac{\partial^2 \gamma}{\partial \zeta^2} - \frac{\partial \gamma}{\partial \zeta} - \frac{1}{\Lambda} \frac{\partial \gamma}{\partial \theta} - \psi = 0 \tag{6}$$

$$\frac{\partial \omega}{\partial \theta} = \psi \tag{7}$$

Here γ is the liquid phase concentration (γ=c/N), ω is the solid phase concentration (ω=q/Q), ζ is the axial coordinate (ζ=z/L), θ is the time (θ=t/τ), ψ is the exchange rate, and eventually $P\grave{e}$ is the Pèclèt number ($P\grave{e}$=uL/D). The characteristic time τ is given by the following expression:

$$\tau = \frac{(1-\varepsilon)\rho_S QL}{Nu} \tag{8}$$

On the other hand, the dimensionless exchange rate ψ, is given by:

$$\psi = \frac{(1-\varepsilon)La}{uN} r \tag{9}$$

Eventually, the dimensionless forms of the boundary conditions (5) are:

$$\theta = 0, \quad 0 \le \zeta \le 1: \quad \gamma = 0, \quad \omega = 0 \tag{10a}$$

$$\theta > 0, \quad \zeta = 0: \quad \frac{\partial \gamma}{\partial \zeta} = P\grave{e}(\gamma_o - \gamma) \tag{10b}$$

$$\theta > 0, \quad \zeta = 1: \quad \frac{\partial \gamma}{\partial \zeta} = 0 \tag{10c}$$

The system of partial differential Eqs. (6, 7), with the boundary conditions (10), was integrated at finite differences, using a Fortran program explicitly written for this purpose.

3. EXPERIMENTAL

Ion–exchange experiments were carried out by contacting a simulated wastewater, containing Pb^{2+} and Na^+ with a Neapolitan yellow tuff coming from Chiaiano (Naples, Italy), containing 61% phillipsite and having a cation exchange capacity Q=2.01 eq/kg. Two kinds of experiments were performed (see Pansini *et al.*, 1996, for details). In static experiments, the Pb–Na–phillipsite exchange isotherm was obtained, at 25°C and total normality N=100 eq/m^3. In dynamic experiments, breakthrough of Pb^{2+} from a fixed–bed column containing the phillipsitic tuff in its Na–form was studied.

Static experiments were carried out using small samples of pure phillipsite, obtained from the original tuff by means of an enrichment process. In order to obtain the exchange isotherm, such samples were allowed to react with solutions containing different amounts of $Pb(NO_3)_2$ and $NaNO_3$ (with the required value of N), for more than three days. The dynamic experiments, on the other hand, were carried out in a glass column, which contained a tuff bed, previously Na–exchanged. The bed was made of 48.0 g of granules of the 24–60 mesh fraction (considered as to spheres with d_p=0.37 mm; ρ_b=810 kg/m^3); its i.d. was 14. 3 mm, its length was L= 0.37 m. Three different solutions were fed to the fixed–bed column, all of them with a flow rate of $1.17\cdot10^{-7}$ m^3/s (corresponding to u=$7.26\cdot10^{-4}$ m/s). In all three solutions the Pb^{2+} concentration was 92 g/m^3 (c_o=0.89 eq/m^3), but the Na^+ concentrations were 0, 500 and 1000 g/m^3 (corresponding to total normalities N=0.89, 22.6 and 44.4 eq/m^3, respectively). Samples of the solution flowing from the column were periodically withdrawn by means of a temporized sample collector and analyzed for Pb^{2+} and Na^+, leading to the breakthrough curves in the three experimental conditions considered: such curves were compared with the model described above.

4. RESULTS AND DISCUSSION

The static experiments allowed to evaluate the Pb–Na equilibrium over phillipsite at N=100 eq/m^3. However, in order to apply the LDF model, it is necessary to know the ion exchange isotherms also in the conditions considered for the dynamic experiments (see Eqs. 1, 2). Therefore, it was necessary to extrapolate the data relative to N=100 eq./m^3 down to N=0.89, 22.6 and 44.4 eq/m^3: this was accomplished using the model proposed by Pabalan (1994), in which thermodynamic quantities were computed as reported by Caputo *et al.* (1995). The results of the extrapolation are reported in Fig. 1 in which the experimental data for N=100 eq/m^3 are reported together with the calculated data for N=44.4, 22.6 and 0.89 eq/m^3.

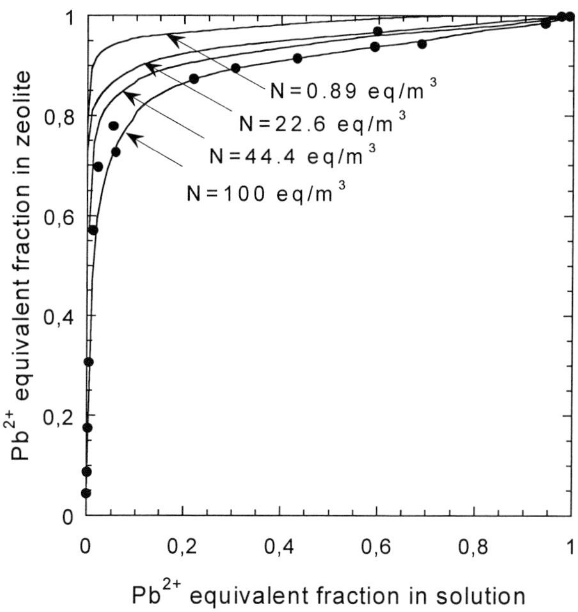

Fig. 1. Pb–Na ion exchange isotherms over phillipsite at T=25°C.
•: experimental data at N=100 eq/m^3 (from Pansini *et al.*, 1996); —: model results (see text).

The dynamic experiments allowed to determine the breakthrough curves for Pb at the three total normalities considered. Both model and experimental results are reported in Fig. 2, in which the ratio between the outlet and the inlet Pb^{2+} ion concentration is reported as a function of time. As expected, the experimental results showed that at the beginning of each experimental run, and for about 100 h, almost no Pb^{2+} could be detected in the effluents. Afterwards, c_{out} gradually increased, asymptotically reaching c_o: in these conditions it can be assumed that the zeolite contained in the tuff is in equilibrium with the aqueous solution. Fig. 2 shows that the higher the Na^+ concentration, the faster the saturation. Furthermore, it appears that, at least in the narrow range investigated here, the Na^+ concentration in the liquid stream does not appreciably affect the slope of the breakthrough curves.

In order to check the validity of the model described in Section 2, integration of Eqs. (6, 7) was carried out, considering the experimental conditions of the three dynamic runs in the boundary conditions (10). As already indicated, once the boundary conditions and the equilibrium isotherms are assigned, the only parameters of the model are the mass transfer coefficients k_e and k_i. A simple "best fit" procedure led to a very satisfactory agreement between model and experimental results. It is important to observe that all three breakthrough curves were obtained using the same value for the external mass transfer coefficient (namely, k_e=5.1·10^{-6} m/s): this is in agreement with the physical nature of k_e, which only depends on fluid dynamic conditions. On the other hand, the internal mass transfer coefficient k_i varied with the total normality of the solution used. In particular, for N=0.89 eq/m^3 it was k_i=1.4·10^{-6} eq/m^3; for N=22.6 eq/m^3 it was k_i=1.5·10^{-6} eq/m^3; for N=44.4 eq/m^3 it was k_i=1.6·10^{-6} eq/m^3. The fact that the internal resistance to diffusion varies with the overall ionic strength of the solution is in agreement with previous published experimental results (Ciambelli *et al.*, 1985).

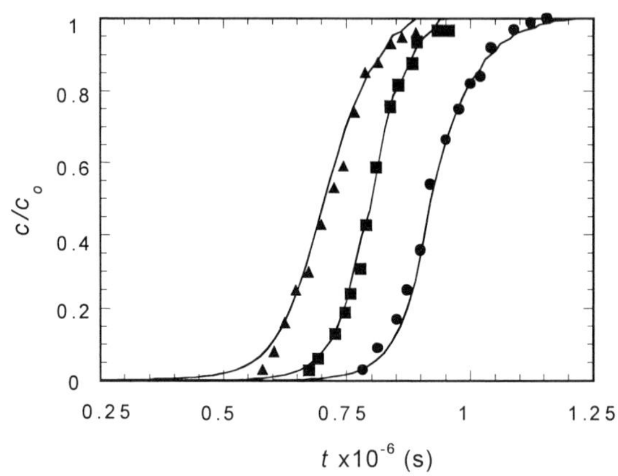

Fig. 2. - Comparison between model (continuous lines) and experimental results.
●: 0 g/m^3 Na^+ in the inlet solution; ■: 500 g/m^3 Na^+; ▲: 1000 g/m^3 Na^+.

5. CONCLUSIONS

Pb^{2+} removal from water by means of a Na–exchanged phillipsite–rich tuff was studied, using a diffusional model. The model is based on the linear driving force (LDF) approximation and takes into account both fluid–particle and intra–particle resistances to diffusion, making no specific assumption on their relative magnitude or on the form of the ion exchange isotherm.

Breakthrough curves obtained in a fixed–bed ion exchange process by Pansini *et al.* (1996) were interpreted by means of the model, the only parameters of which are a solid–liquid and an intra–particle mass transfer coefficient. The experimental results were obtained using a constant superficial velocity and a fixed amount of solid in the bed, but varying the solution composition. In particular, three different Na^+ concentrations (and therefore total normalities) were used, in order to test the tuff capability to remove Pb^{2+} even in presence of high concentrations of interfering ions. A very satisfactory agreement between model and experimental results was found, using the same value for the external mass transfer coefficient for all three runs (namely, k_e=5.1·10^{-6} m/s), and different values for the internal mass transfer coefficient (namely, k_i=1.4·10^{-6} eq/m^3 for N=0.89 eq/m^3; k_i=1.5·10^{-6} eq/m^3 for N=22.6 eq/m^3; k_i=1.6·10^{-6} eq/m^3 for N=44.4 eq/m^3).

LIST OF SYMBOLS

a	specific external surface are of the solid particles, m^{-1}
c	liquid phase concentration, eq./m^3
D	dispersion coefficient, m^2/s
d_p	average particle diameter, m
k_e	external (solid–liquid) mass transfer coefficient, s^{-1}
k_i	internal (intra–particle) mass transfer coefficient, kg/m^{3}s
L	bed length, m
N	total normality, eq./m^3
$Pè$	Pèclèt number, dimensionless
Q	ion exchange capacity, eq./kg
q	solid phase concentration, eq./kg
r	mass transfer rate, eq./m^{2}s
t	time, s
u	superficial velocity of the liquid, m/s

Greek symbols

γ	dimensionless liquid phase concentration
ε	void fraction, dimensionless
ζ	dimensionless axial coordinate
θ	dimensionless time
Λ	partition ratio, dimensionless
ρ_s	solid density, kg/m^3
τ	characteristic time, s
ψ	dimensionless exchange rate
ω	dimensionless solid phase concentration

Subscripts and superscripts

*	equilibrium
out	outlet
o	inlet

REFERENCES

Caputo D., Dattilo R. and Pansini M. (1995) In Proc. 3° Convegno Nazionale Scienza e Tecnologia delle Zeoliti, R. Aiello (ed.), De Rose, Montalto, Italy, p. 143.
Carta G. and Cincotti, A. (1998), Chem. Eng. Sci., 53, 3483.
Ciambelli P., Corbo P., Porcelli C. and Rimoli, A. (1985) Zeolites, 5, 184.

Colella C. (1999) In "Natural Microporous Materials in the Environmental Technology", P. Misaelides, F. Macasek, T.J. Pinnavaia and C. Colella (eds.), NATO Sciences Series, Series E: Applied Sciences, Vol. 362, Kluwer A.P., Dordrecht, The Netherlands, p: 207.

Glueckauf E. (1955) Trans Faraday Soc., 51, 1540.

Hall K.R., Eagleton L.C., Acrivos, A. and Vermeulen, T. (1966) Ind. Eng. Chem. Fundam., 5, 212.

Lai C.C. and Tan C.S. (1991), A. I. Ch. E. J., 37, 461.

Levenspiel O. (1999) Chemical Reaction Engineering, 3rd Edition, Wiley and Sons, NY.

McCabe W.L., Smith J.C. and Harriot P. (1985) Unit Operations of Chemical Engineering, 4th Edition, McGraw Hill, NY.

Pabalan R.T. (1994) Geochim. Cosmochim. Acta, 58, 4573.

Pansini M. (1996) Mineral. Deposita, 31, 563.

Pansini M., Colella C., Caputo D., de'Gennaro M. and Langella, A. (1996) Microporous Materials, 5, 357.

Ruthven D.M. (1984) Principles of Adsorption and Adsorption Processes, Wiley and Sons, NY.

Zhang R. and Ritter J.A. (1977) A. I. Ch. E. J., 54, 376.

Studies in Surface Science and Catalysis 140
A. Gamba, C. Colella and S. Coluccia (Editors)

A kinetic study of NO decomposition on Cu-ZSM5

R. Pirone[a], P. Ciambelli[b], B. Palella[c] and G. Russo[a]

[a]Istituto di Ricerche sulla Combustione, CNR, P.le V. Tecchio 80, 80125 Napoli, Italy

[b]Dipartimento di Ingegneria chimica e alimentare, Università di Salerno, 84084 Fisciano, Salerno, Italy

[c]Dipartimento di Ingegneria chimica, Università di Napoli Federico II, P.le V. Tecchio 80, 80125 Napoli, Italy

The mechanism of NO decomposition has been investigated by carrying out a kinetic study on over-exchanged Cu-ZSM5 zeolite at Si/Al = 80. NO decomposition tests has been performed in experimental conditions close to those interesting in the applications. In order to test several rate equations in fitting the experimental data, it has been developed a specific numerical code based on a derivative-free minimisation algorithm. The kinetic model results in steady-state mass balance equations in the gas and solid phase for an isothermal plug flow reactor. The generation term in the differential mass balance equations is constituted by the decomposition rate of NO, and the contemporary presence of NO oxidation, producing NO_2 in thermodynamic equilibrium with NO and O_2 has been also taken into account. An intrinsic second order dependence on NO partial pressure was found together with an inhibiting effect of O_2. Among different mechanistic rate equations, the best fit was obtained by an overall decomposition kinetics, generated by a five-step reaction mechanism based on the reversible adsorption of NO on the catalyst, the consequent partial reduction to N_2O and copper sites oxidation, N_2O fast decomposition and oxygen reversible desorption. The parametric identification gives a correct dependence of the estimated parameters on temperature: the kinetic constants are monotonously dependent on temperature, even in the presence of a maximum in NO decomposition rate, evidenced by the experimental results and several literature references.

1. INTRODUCTION

NO decomposition is one of the most investigated DeNO$_x$ reactions, since it is thermodynamically favoured up to about 1000°C and could potentially perform nitric oxide abatement without adding any reducing agent. Unfortunately, up to now no catalyst has been found active enough to perform practical applications. The most active catalyst among all those investigated is Cu-exchanged ZSM5 zeolite [1]. Notwithstanding the high hydrothermal instability which strongly limits deNOx performances, Cu-ZSM5 could act as a "model system" in the research of novel catalysts for NO decomposition, since the Cu ions exchanged in the ZSM5 matrix show unique features when compared for example to those of copper ions implanted in mesoporous silico-aluminates or alumino-phosphates [2].

An impressing number of publications has been devoted to investigating Cu-ZSM5 catalyst of NO decomposition or selective reduction, as recently reviewed by Pârvulescu et al. [3]. However, notwithstanding the considerable amount of research work been devoted to this subject, many basic questions seem still unanswered. Several reviews covering this topic [1-4] put into a clear evidence that an unanimous agreement on most significant aspects does not exist yet. Major disputed questions are the attribution of the copper oxidation state of active sites, the role and the state of extra-lattice oxygen species (ELO) in NO decomposition and the mechanism of N-N bond formation. Other aspects seem to be better clarified, as the evidence that over-exchanged copper species are the most active ones [5-7] as well as that at high temperature Cu sites undergo "self-reduction" without any reducing species [8-15]. Most authors propose Cu^+ as active site (although relevant differences between several theories should be pointed out) [8-12,16-21]. On the contrary, Shelef [22] proposed that the reaction entirely proceeds on square-planar Cu^{+2} sites through the formation of *gem*-dinitrosyl adsorbed intermediates.

The structure of ELO species and the related state of copper in the different oxidation states is highly disputed. The group of Sachtler [16-17] proposed the existence of bridged O^{-2} ions $(Cu^{+2}–O^{-2}–Cu^{+2})^{+2}$, while that of Bell [12,18] hypothesised the presence of ELO species associated to isolated copper sites ($Cu^{+2}–O^-$ and $Cu^{+2}–O_2^-$). Consequently, different reaction mechanisms have been hypothesised. Following the two possible concepts on ELO identity, at least two major differences could be depicted among reaction mechanisms proposed. Aylor et al. [18] hypothesised that adjacent pairs of nitrosyl species adsorbed on Cu^+ react to form N_2O which is subsequently converted to N_2. Similar assumptions were made by Giamello et al. [11] and Spoto et al. [19], both hypothesising extra-lattice oxygen species as $Cu^{+2}–O^-$. On the contrary, other groups evidenced the importance of NO_2 as well as of adsorbed nitrate species formation, whose existence could be strictly connected to oxygen bridged ions between copper pairs, as key intermediates in the decomposition of NO [20-21,23].

While most of the investigations on the reaction mechanism are based on spectroscopic measurements performed at low temperature, very rarely these studies focused the attention to the consequent kinetic behaviour of NO decomposition. A power law rate equation has been often used to correlate conversion data, but NO decomposition rates dependent on NO partial pressure with an exponent ranging from 0.9 up to 1.8 have been reported [6,9,24-25].

A specific and still unexplained aspect of the reaction kinetics is the presence of a reversible maximum of NO conversion to N_2 with increasing the reaction temperature [9,14,25-29]. Despite significant improvements in the comprehension of the reaction mechanism, the kinetic equation proposed by Li and Hall in 1991 [25] is still used as a reference. This was also the conclusions from the results of the infrared spectroscopic investigation of Aylor et al. [18], which reported that the rate of NO decomposition is linearly dependent on NO partial pressure and is lowered by O_2 dissociative adsorption on the active sites:

$$r_{NO \to N_2} = \frac{k[NO]}{1 + K_{O_2}\sqrt{[O_2]}} \tag{1}$$

However, the NO first order depending rate equation neither is adequate to describe the overall kinetics from experimental results of integral reactors [7,24], neither is capable to give a maximum value of NO conversion.

Recently, a more complex kinetic study was performed by Tomašić et al. [29-30], who compared three different mechanistic kinetic models on the basis of activity measurements performed in an integral flow reactor. However, the kinetic equation proposed is not able to describe the influence of temperature on NO decomposition rate, even if it satisfactorily fits conversion data obtained in the absence of O_2 in the feed at each temperature investigated.

In this paper, we have studied the kinetics of NO decomposition over Cu-ZSM5 in a range of experimental conditions quite close to those interesting for practical applications. The NO conversion data were modelled with several reaction rate equations, all based on different plausible reaction mechanisms. The ultimate goal of the work is to develop a kinetic model able to represent the reactor behaviour in all experimental conditions investigated.

2. EXPERIMENTAL

H-ZSM5 zeolite (Si/Al = 80), synthesized according to the procedures described in [7,14], was ion exchanged with a copper acetate aqueous solution (60 mM) for two hours at 50°C. Subsequent drying at 110°C followed the exchange procedure. Copper content of the catalyst, as measured by atomic absorption, was 3.24 wt.% (5.1×10^{-4} g·atom/g_{cat}), corresponding to about five times the ion exchange capacity (526% over-exchange), 100% exchange corresponding to the atomic ratio Cu/Al = 0.5. A more detailed investigation of the state of copper in over-exchanged ZSM5 has been already reported [7,31]. In any case, in the present investigation we did not observe the presence of any bulk copper oxide species in the catalyst sample.

Catalytic activity measurements were carried out with a fixed-bed microreactor consisting of a quartz tube (I.D. 1.0 cm) with annular cross section, in which a porous disk supports the catalyst particles. Particle size ranged from 200 to 300 μm. The reactor was inserted in an electrical furnace supplied with three heated and temperature controlled zones. Catalytic bed axial temperature was measured by a Chromel-Alumel thermocouple, placed in a quartz tube, coaxial and internal to the reactor. Three Brooks mass flow controllers were used to measure the flow rates of high purity gases: NO (1% vol.) + He, He (99.998%) and O_2 (99.998%). Contact time in the runs, defined as the ratio between the mass of catalyst loaded (W) on total feeding flow rate (F), ranged from 0.06 to 0.28 g·s·Ncm^{-3}. Before each experiment, the catalyst was treated in He for 2 h at 550°C and quenched down to a given temperature. Gas analysis was performed by Hartmann & Braunn non-dispersive IR continuous analyser for the measurement of NO, NO_2 and N_2O concentrations (URAS 10 E), and paramagnetic analyser for the concentration of O_2. Hewlett Packard HP 5890 gas-chromatograph (Molecular Sieve 5A column) was used for the measurements of N_2 concentration.

3. RESULTS AND DISCUSSION

In order to evidence the specific features of NO decomposition kinetics in a range of experimental conditions as close as possible to those interesting for practical applications the feed concentration of NO has been kept in the range from 600 to 5000 ppm, i.e. at values lower than those typically investigated for this reaction [6,9,25]. O_2 was eventually added to the feed, as it is well known that it inhibits the catalyst activity. This effect had not been very deeply investigated, although sometimes it had been roughly quantified [25]. Moreover, the present study does not deal with the effect of water on the catalyst activity, since the deactivation occurring in the presence of H_2O is so strong to prevent any possible attempt to investigate reaction kinetics. A quite large number of kinetic runs (101) was performed, by varying NO and O_2 feed molar fractions, contact time and temperature. Table 1 summarises the experimental conditions investigated.

Table 1
Number of experiments performed at different temperature as a function of all other conditions

O_2, ppm	0				500				1000				2000				5000				8000			
W/F, 10^{-2} g·s·Ncm^{-3}	6	11	21	28	6	11	21	28	6	11	21	28	6	11	21	28	6	11	21	28	6	11	21	28
NO, ppm																								
600	0	2	0	0	0	0	2	0	0	0	0	0	0	0	2	0	0	0	2	0	0	0	0	0
1000	0	3	4	2	0	0	2	0	0	5	1	2	0	0	2	0	0	0	3	2	0	0	0	0
2000	0	0	4	0	0	0	2	0	0	0	0	0	0	0	2	0	0	0	2	0	0	0	0	0
3000	0	0	4	0	0	0	2	0	0	0	0	0	0	0	2	0	0	0	3	0	0	0	0	0
4000	0	0	4	0	0	0	2	0	0	0	0	0	0	0	2	0	0	0	2	0	0	0	0	0
5000	1	3	4	1	0	0	3	0	1	3	1	0	0	0	3	0	1	5	4	3	0	0	3	0

Experimental results are presented in Figs. 1 and 2. Fig. 1 reports the NO conversion obtained as a function of temperature in NO decomposition tests carried out in the absence of O_2 in the feed. This set of data is representative of the already discussed singular temperature dependence of NO decomposition rate over Cu-ZSM5. Fig. 2 depicts the effect of NO and O_2 concentration in the feed on NO conversion to N_2.

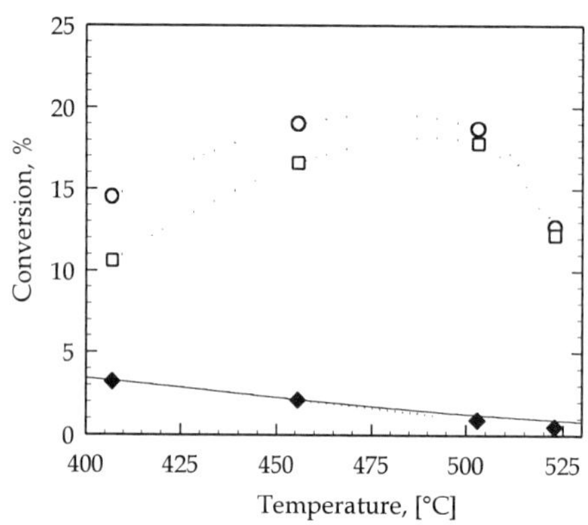

Fig. 1. NO conversion as a function of the reaction temperature: (○) overall, (□) to N_2 and (◆) to NO_2. Continuous line represents the calculated thermodynamic limit for NO conversion to NO_2. Feed: NO (0.5% vol.), balance He; W/F = 0.21 g·s·cm^{-3}

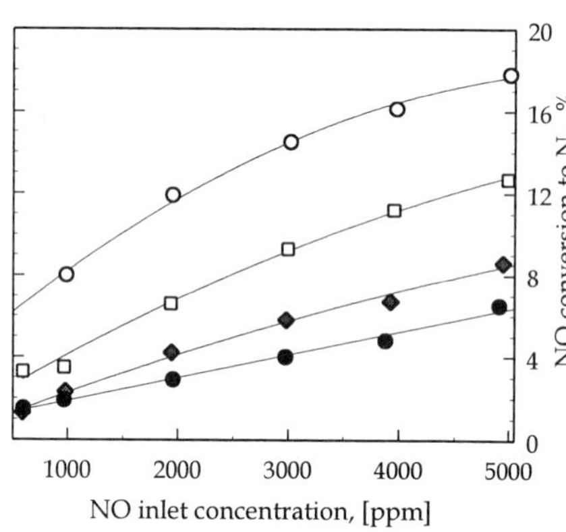

Fig. 2. NO conversion to N_2 as a function of NO inlet concentration at 500°C. Feed containing O_2 at different concentrations: (○) 0, (□) 1000, (◆) 2000, (•) 5000 ppm; W/F = 0.21 g·s·cm^{-3}.

Fig. 1 shows that the formation of N_2 and, consequently, the total NO consumption increase with increasing temperature, reaching a maximum value at about 475°C. The presence of the maximum was also found in previous investigations, but is not clearly explained yet [7,9,25,28]. After several cycles of increasing-decreasing the reaction temperature, the catalyst did not undergo any significant deactivation, indicating that the effect of temperature on the activity is reversible. In the range of temperatures from 400 to 525°C no N_2O was detected in the reaction products, as also reported elsewhere [7,9], while NO_2 was produced via the Cu-catalysed reaction of the undecomposed NO and the oxygen formed by NO decomposition:

$$2\,NO + O_2 \rightleftarrows NO_2 \qquad (2)$$

As we have already demonstrated [7], the yield to NO_2 also decreases with increasing temperature, because of thermodynamic limitations.

The effect of the reactants concentration on NO conversion to N_2 has been much less investigated in comparison with the effect of temperature. Many authors reported a nearly first order reaction kinetics for NO decomposition over Cu-ZSM5, so expecting quite no influence of the inlet concentration on the overall final conversion [6,9,24-25]. We have already reported that the reaction order does differ from the unity [7] and data reported in Fig. 2 also show that the NO yield to N_2 is significantly increasing with increasing nitric oxide concentration in the feed, in good agreement with previous results. This happens also in the presence of O_2, even at lower values of NO conversion.

Fig. 2 also shows that O_2 inhibits the NO decomposition rate. The conversion of NO to N_2 decreased by the presence of O_2 in the feed gas mixture already at very low O_2 concentration and a concurrent greater production of NO_2 (not reported in the figure) was observed. The conversion of NO to NO_2 always approached the thermodynamic equilibrium value, which increases by increasing O_2 concentration.

3.1 Model Reacting System

We have modelled the reacting system as an isothermal plug flow reactor, by neglecting both axial dispersion and internal and external mass transfer resistance. These assumptions are justified by experimental findings as well as by theoretical calculations.

The reactor isothermicity is assured by (i) the very low adiabatic temperature raise (≤ 3°C) and (ii) the use of a micro-reactor with annular cross section (higher external surface per unit volume) heated by a “three zone” electrical furnace. Moreover, an uniform temperature profile has been revealed by the internal thermocouple placed along the reactor axis.

The estimation of the reactor Peclet number ($Pe = v_{gas}L/D_a$) allowed us to neglect the axial back-diffusion (D_a):

$$Pe = Pe_{d_p} \cdot \frac{L}{d_p} \approx 40 \qquad (3)$$

being: $L = 1$–4 cm,
$d_p = 212$–300 μm, and
$Pe_{d_p} = 2$ for catalytic fixed bed reactors for $Re_p > 0.01$–0.1 [32].

The assumption of neglecting the terms related to the intra- and extra-particle diffusion has been checked by carrying out kinetic runs with different catalyst particle size and different reactant flow rate as well as by theoretical estimates.

The measured values of pressure drop in the micro-reactor were significantly variable with the experimental conditions, maximum values reaching about 0.5 atm. The dependence of the reaction kinetics on total pressure changes has been taken into account in the model: Pressure has been considered linearly variable along the reactor axis, as expected by the Ergun Law [33] for low values of Reynolds number ($Re_p \approx 10$).

Consequently, the reactor model is constituted by a system of N+1 equations, where N is the number of chemical species present in the system (NO, NO_2, N_2 and O_2, neglecting the presence of N_2O; N = 4) and another unknown variable is pressure. The equations are one momentum balance (in the form of simplified Ergun Law), and four mass balance relationships. The presence of NO_2 among the reaction products has been related to the catalytic activity of Cu-ZSM5 towards the oxidation of NO to NO_2, as revealed by our previous investigation in similar experimental conditions [7], as well as by the present results (Fig. 1). It has been hypothesised that reaction (2) proceeds in parallel to NO decomposition, having not assumed that NO_2 formation is responsible for copper reduction from Cu^{+2} (inactive in decomposing NO) to Cu^{+} (the active site), as also proposed by some author [20-21,23].

In the model, we assume that the rate of NO catalytic oxidation is so high that equilibrium concentration of NO, NO_2 and O_2 are retained throughout the reactor. This hypothesis is not directly suggested by the experimental findings, because they just prove that equilibrium conditions are established at the reactor exit. However, it is supported by NO oxidation measurements carried out with a contact time about five times lower than the minimum value investigated in the experimental campaign here presented, in which the equilibrium conversion of NO to NO_2 was reached.

Consequently, the model results in the following system of differential equations:

$$\begin{cases} \dfrac{d}{dz} y_{N_2} = \dfrac{v_{gas}}{C^0_{tot}} \cdot r_{N_2} \\ 2\dfrac{d}{dz} y_{N_2} + \dfrac{d}{dz} y_{NO_2} + \dfrac{d}{dz} y_{NO} = 0 \\ 2\dfrac{d}{dz} y_{O_2} + 2\dfrac{d}{dz} y_{NO_2} + \dfrac{d}{dz} y_{NO} = 0 \\ K_p = \dfrac{1}{p} \dfrac{y^2_{NO_2}}{y^2_{NO} y_{O_2}} \\ p(z) = p_{in} - \dfrac{p_{in} - p_{out}}{L} \cdot z \end{cases}$$

The boundary conditions are:

$$\begin{cases} y_{N_2}\big|_{z=0} = y^o_{N_2} = 0 \\ y_{O_2}\big|_{z=0} = y^o_{O_2} \\ y_{NO_2}\big|_{z=0} = y^o_{NO_2} \\ p\big|_{z=0} = p_{in} \end{cases}$$

where $r_{N_2} = r_{N_2}(p, y_{NO}, y_{O_2})$

3.2 Modelling Results

Various rate equations have been considered to fit the experimental data. The kinetic parameters involved in the equations have been estimated by means of a derivative-free minimisation algorithm, in order to compare the different models with respect to the sum of square residuals, $\hat{\sigma}^2$ and the correlation index, R^2, defined as:

$$\sigma^2 = \frac{\Sigma_i \left(x^{exp}_{N_2} - x^{mod}_{N_2}\right)^2}{n_{exp} - N_{par}} \qquad R^2 = 1 - \frac{\Sigma_i \left(x^{exp}_{N_2} - x^{mod}_{N_2}\right)^2}{\Sigma_i \left(x^{exp}_{N_2,i}\right)^2}$$

The preliminary description of reaction kinetics has been performed by involving two power law-type rate equations:

$$1.\ r_{N_2} = k \cdot p_{NO}^{m} \qquad \text{and} \qquad 2.\ r_{N_2} = k \cdot p_{NO}^{m} \cdot p_{O_2}^{n}$$

Table 2
Power law rate equations: estimated values of kinetic parameters

	400°C	450°C	500°C	520°C
	$r_{N_2} = k \cdot p_{NO}^{m}$ *			
k, mol/(g·h·atm^{m})	0.043 ± 0.006	0.38 ± 0.88	1.96 ± 3.6	1.04 ± 2.7
m	1.25 ± 0.47	1.53 ± 0.46	1.80 ± 0.33	1.74 ± 0.46
σ^2	$8.9 \cdot 10^{-4}$	$1.8 \cdot 10^{-3}$	$1.3 \cdot 10^{-3}$	$6.0 \cdot 10^{-4}$
R^2	0.671	0.681	0.835	0.802
	$r_{N_2} = k \cdot p_{NO}^{m} \cdot p_{O_2}^{n}$			
k, mol/(g·h·atm^{m+n})	0.010 ± 0.006	0.452 ± 0.347	0.756 ± 0.47	3.52 ± 0.99
m	1.5 ± 0.11	2.1 ± 0.18	2.0 ± 0.12	2.3 ± 0.22
n	- 0.37 ± 0.046	- 0.38 ± 0.065	- 0.30 ± 0.039	-0.25 ± 0.052
σ^2	$3.2 \cdot 10^{-5}$	$5.5 \cdot 10^{-5}$	$7.2 \cdot 10^{-5}$	$3.0 \cdot 10^{-5}$
R^2	0.988	0.990	0.991	0.990

* Parameters identification carried out over 32 experiments performed in the absence of O_2 in the feed.

As already reported [7], some author found a linear dependence of NO decomposition kinetics on NO partial pressure [6,25] by performing the data fitting with a power law rate equation not involving the dependence of O_2 partial pressure (equation 1). With the same equation, a reaction order slightly higher than unity (1-1.5) was reported by Iwamoto et al. [9], while a value significantly higher (1.6) was already estimated by us [7] and Centi et al. (1.8) [24]. In both latter investigations the range of NO concentration (from 600 to 5000 ppm) was lower than in previous studies (1-14 %vol.).

With reference to the experimental data obtained in the absence of oxygen in the feed, the results of modelling do confirm that the reaction order for NO decomposition, variable with temperature, is significantly higher than unity. At 500°C, the value of exponent 'm' was estimated as 1.8 (Tab. 2), very close to that already estimated in similar conditions [7] and in good agreement with the results of Centi et al. [24] for an over-exchanged Cu-ZSM5 zeolite with Si/Al = 25. It is evident that equation (1) cannot fit the data when O_2 is present in the feed, since it is unable to describe the inhibiting effect of O_2 on NO decomposition. In addition, the correlation index R^2 is very low (in the range 0.71 – 0.83), even taking into account only those runs in which very low O_2 concentration was present in the gas phase (no O_2 in the feed, NO conversion to O_2 lower than 20%), showing that the effect of O_2 partial pressure on reaction kinetics is never negligible.

A meaningful improvement of the fitting is obtained with a power law rate equation with two exponents (equation 2), i.e. including also the dependence on O_2 partial pressure. The reaction order estimated for NO (Table 2) is still changing, but less strongly, with the temperature and is always higher than that estimated with equation (1). Moreover, it is observed that the order rate estimated for NO partial pressure in the temperature range from 450 to 525°C was about 2, while at the lowest temperature investigated it is reduced to 1.5, likely due to the great amount of NO_2 produced by NO oxidation which may influence the fitting in a stronger way than at higher temperatures. The inhibiting effect of O_2 partial pressure results in the negative value for n in equation (2), whose absolute value slightly decreases from 0.37 to 0.25 with increasing temperature. The correlation index for equation (2) is very high (0.988 - 0.991), showing that taking into account the effect of both reactants and using a more complete kinetic model (which also describes the parallel occurrence of NO oxidation to NO_2), all conversion data (101 experiments) can be satisfactorily fitted.

However, power law rate equations could only merely correlate conversion data, quantifying the effect of reactants partial pressure on NO decomposition kinetics, but cannot describe nor give information on the reaction mechanism. On the contrary, a fitting carried out with reaction mechanism based rate equations is potentially more interesting for mechanism comprehension. Consequently, a series of kinetic models, all derived from possible reaction schemes and congruent with the findings of power law equations fitting, have been tested. Table 3 reports the list of the ten kinetic equations tested and the corresponding reaction mechanisms which rate equations could be considered as based on. In particular, Tab.3a refers to the models previously proposed by other authors, i.e. three kinetic equations discussed by Tomašić et al. [29-30] and one tested by Centi et al. [24].

Table 3a
Kinetic models: reaction mechanism, corresponding rate equation, estimated parameters

Reaction mechanism and corresponding rate equation	Estimated values of kinetic constants				
		Temperature, °C			
Model *Centi*		400	450	500	525
$NO + Cu^+ \rightleftarrows [NO^- – Cu^{+2}]$	$k_s K_{NO}[Cu_{tot}]$	23.9 ± 100	22.6 ± 24	7.5 ± 9.2	8.37 ± 20
$NO + [NO^- – Cu^{+2}] \rightarrow [(NO)_2^- – Cu^{+2}]$	K'_{NO} ($\cdot 10^{-2}$)	19 ± 130	3.9 ± 13	0.69 ± 8.8	3.08 ± 27
$[(NO)_2^- – Cu^{+2}] \rightarrow N_2 + O_2 + Cu^+$	K''_{NO} ($\cdot 10^{-3}$)	48.5 ± 1.3	0.97 ± 15	0.13 ± 115	0.409 ± 32
$r_{N_2} = \dfrac{(k_1 \cdot l_{tot}) \cdot K_1 \cdot K_2 \cdot p_{NO}^2}{1 + K_1 \cdot p_{NO} + K_2 \cdot p_{NO}^2}$	σ^2	$9.67 \cdot 10^{-4}$	$3.66 \cdot 10^{-4}$	$1.33 \cdot 10^{-3}$	$4.14 \cdot 10^{-4}$
	R^2	0.658	0.982	0.831	0.910
Model *Tomašić* 1 *		400	450	500	525
$2 NO + S \rightarrow N_2 + 2 O{\cdot}S$	k	27 ± 16	17 ± 3.1	15 ± 1.4	6.9 ± 0.64
$2 O{\cdot}S \rightleftarrows O_2 + 2 S$	K'_D ($\cdot 10^{-2}$)	115 ± 130	9.8 ± 5.3	3.4 ± 1.1	0.97 ± 0.38
$r_{N_2} = \dfrac{k \cdot p_{NO}^2}{\left(1 + \sqrt{K'_D \cdot p_{O_2}}\right)^2}$	σ^2	$1.78 \cdot 10^{-4}$	$1.29 \cdot 10^{-4}$	$8.64 \cdot 10^{-5}$	$2.32 \cdot 10^{-5}$
	R^2	0.943	0.985	0.989	0.994
Model *Tomašić* 2 *		400	450	500	525
$NO + S \rightleftarrows NO{\cdot}S$	k_s	33.1 ± 9.3	18.4 ± 5	13.5 ± 3.0	6.94 ± 2.5
$2 NO{\cdot}S \rightarrow N_2 + O_2{\cdot}S + S$	K_A	334 ± 99	107 ± 52	45 ± 31	33 ± 57
$O_2{\cdot}S \rightleftarrows O_2 + S$	K'_D ($\cdot 10^{-2}$)	16 ± 5.0	3.6 ± 1.3	2.2 ± 0.53	1.1 ± 0.41
$r_{N_2} = \dfrac{k \cdot p_{NO}^2}{(1 + K_A \cdot p_{NO} + K'_D \cdot p_{O_2})^2}$	σ^2	$7.3 \cdot 10^{-5}$	$2.0 \cdot 10^{-4}$	$1.8 \cdot 10^{-4}$	$7.5 \cdot 10^{-5}$
	R^2	0.974	0.978	0.977	0.985
Model *Tomašić* 3 *		400	450	500	525
$NO + S \rightleftarrows NO{\cdot}S$	k	70.0 ± 51	24.1 ± 6.4	15.6 ± 3.5	7.30 ± 2.7
$NO + NO{\cdot}S \rightarrow N_2 + O_2{\cdot}S$	K_A ($\cdot 10^{-2}$)	26 ± 23	2.9 ± 1.8	1.3 ± 2.2	0.76 ± 1.4
$O_2{\cdot}S \rightleftarrows O_2 + S$	K_D ($\cdot 10^{-2}$)	200 ± 150	18 ± 7.6	0.92 ± 0.22	3.4 ± 1.6
$r_{N_2} = \dfrac{k \cdot p_{NO}^2}{1 + K_A \cdot p_{NO} + K'_D \cdot p_{O_2}}$	σ^2	$5.8 \cdot 10^{-5}$	$1.4 \cdot 10^{-4}$	$1.3 \cdot 10^{-4}$	$5.8 \cdot 10^{-5}$
	R^2	0.979	0.985	0.984	0.988

* The symbol 'S' indicates a generic active site, following authors' nomenclature.

Table 3b
Kinetic models tested: reaction mechanism, corresponding rate equation and estimated parameters

Reaction mechanism and corresponding rate equation	Estimated values of kinetic constants	Temperature, °C			
Model *Centi modified*		400	450	500	525
$NO + Cu^{+} \rightleftarrows [NO^{-}-Cu^{+2}]$	$k_s K_{NO}[Cu_{tot}]$	64.0 ± 84	20.5 ± 12	15.7 ± 6.9	7.25 ± 6.6
$NO + [NO^{-}-Cu^{+2}] \rightarrow [(NO)_2^{-}-Cu^{+2}]$	$K'_{NO}\ (\cdot 10^{-2})$	22 ± 44	2.9 ± 7.2	1.3 ± 3.8	0.76 ± 7.3
$[(NO)_2^{-}-Cu^{+2}] \rightarrow N_2 + [O_2-Cu^{+2}]$	$K''_{NO}\ (\cdot 10^{-4})$	264 ± 2900	-0.7 ± 0.04	0.55 ± 500	0.41 ± 1100
$[O_2^{-}-Cu^{+2}] \rightleftarrows O_2 + Cu^{+}$	$K_{O_2}\ (\cdot 10^{-2})$	180 ± 240	18.6 ± 13	9.2 ± 4.4	3.4 ± 4.0
$r_{N_2} = \dfrac{k_s K_{NO}[Cu_{tot}] \cdot p_{NO}^2}{1 + K'_{NO} \cdot p_{NO} + K''_{NO} \cdot p_{NO}^2 + K_{O_2} p_{O_2}}$	σ^2	$6.10 \cdot 10^{-5}$	$1.49 \cdot 10^{-4}$	$1.32 \cdot 10^{-4}$	$6.67 \cdot 10^{-5}$
	R^2	0.979	0.985	0.984	0.988
Model 1		400	450	500	525
$2\,NO + Cu^{+} \rightarrow N_2 + [O_2^{-}-Cu^{+2}]$	$k\,[Cu_{tot}]$	20 ± 11	12 ± 2.1	10 ± 0.88	6.2 ± 0.61
$[O_2^{-}-Cu^{+2}] \rightleftarrows O_2 + Cu^{+}$	$K_{O_2}\ (\cdot 10^{-2})$	200 ± 180	16 ± 8.9	6.3 ± 1.7	346 ± 1.4
$r_{N_2} = \dfrac{k \cdot [Cu_{tot}] \cdot p_{NO}^2}{1 + K_{O_2} \cdot p_{O_2}}$	σ^2	$2.26 \cdot 10^{-4}$	$2.63 \cdot 10^{-4}$	$1.63 \cdot 10^{-4}$	$8.01 \cdot 10^{-5}$
	R^2	0.922	0.972	0.980	0.988
Model 2		400	450	500	525
$NO + Cu^{+} \rightleftarrows [NO^{-}-Cu^{+2}]$	$k_s[Cu_{tot}]^2$	0.14 ± 0.02	0.35 ± 0.17	0.56 ± 0.31	0.47 ± 0.98
$2\,[NO^{-}-Cu^{+2}] \rightarrow N_2 + [O^{-}-Cu^{+2}]$	K_{NO}	338 ± 98	62 ± 41	32 ± 21	17 ± 39
$2\,[O^{-}-Cu^{+2}] \rightleftarrows O_2 + 2\,Cu^{+}$	K_{O_2}	74 ± 19	29 ± 6.7	20 ± 2.8	9.8 ± 2.1
$r_{N_2} = \dfrac{k_s K_{NO}[Cu_{tot}]^2 \cdot p_{NO}^2}{\left(1 + K_{NO} \cdot p_{NO} + K_{O_2} \cdot p_{O_2}^{0.5}\right)^2}$	σ^2	$4.14 \cdot 10^{-5}$	$9.12 \cdot 10^{-5}$	$7.39 \cdot 10^{-5}$	$2.33 \cdot 10^{-5}$
	R^2	0.985	0.990	0.991	0.995
Model 3		400	450	500	525
$2\,NO + Cu^{+} \rightarrow N_2O + [O^{-}-Cu^{+2}]$	$k \cdot [Cu_{tot}]$	7±600 $(\cdot 10^{17})$	28 ± 7.7	19 ± 2.6	8.1 ± 1.1
$N_2O + Cu^{+} \rightarrow N_2 + [O^{-}-Cu^{+2}]$	K_{O_2}	12 ± 9 $(\cdot 10^{18})$	190 ± 73	78 ± 18	33 ± 11
$N_2O + [O^{-}-Cu^{+2}] \rightarrow N_2 + O_2 + Cu^{+}$	σ^2	$1.10 \cdot 10^{-4}$	$7.30 \cdot 10^{-5}$	$6.30 \cdot 10^{-5}$	$4.2 \cdot 10^{-5}$
$2\,[O^{-}-Cu^{+2}] \rightleftarrows O_2 + 2\,Cu^{+}$	R^2	0.956	0.987	0.992	0.986
$r_{N_2} = \dfrac{k \cdot [Cu_{tot}] \cdot p_{NO}^2}{1 + K_{O_2} \cdot p_{O_2}^{0.5}}$					
Model 4		400	450	500	525
$NO + Cu^{+} \rightleftarrows [NO^{-}-Cu^{+2}]$	$k_s[Cu_{tot}]^2$	0.049 ± 0.025	0.16 ± 0.057	0.44 ± 0.31	0.87 ± 0.98
$NO + [NO^{-}-Cu^{+2}] \rightarrow N_2O + [O^{-}-Cu^{+2}]$	$K_{NO}\ (\cdot 10^{2})$	390 ± 19	2.1 ± 1.9	0.49 ± 0.27	0.089 ± 0.04
$N_2O + Cu^{+} \rightarrow N_2 + [O^{-}-Cu^{+2}]$	$K_{O_2}\ (\cdot 10^{2})$	211 ± 50	1.9 ± 0.87	0.85 ± 0.21	0.33 ± 0.01
$N_2O + [O^{-}-Cu^{+2}] \rightarrow N_2 + O_2 + Cu^{+}$	σ^2	$7.60 \cdot 10^{-5}$	$2.45 \cdot 10^{-5}$	$5.97 \cdot 10^{-5}$	$1.73 \cdot 10^{-5}$
$2\,[O^{-}-Cu^{+2}] \rightleftarrows O_2 + 2\,Cu^{+}$	R^2	0.973	0.997	0.993	0.997
$r_{N_2} = \dfrac{k_s K_{NO}[Cu_{tot}] \cdot p_{NO}^2}{1 + K_{NO} \cdot p_{NO} + K_{O_2} \cdot p_{O_2}^{0.5}}$					
Model 5		400	450	500	525
$NO + Cu^{+} \rightleftarrows [NO^{-}-Cu^{+2}]$	$k_s[Cu_{tot}]^2$	3.1 ± 13	0.13 ± 0.003	0.34 ± 0.0004	4 ± 0.0083 $(\cdot 10^{-2})$
$NO + [NO^{-}-Cu^{+2}] \rightarrow [(NO)_2^{-}-Cu^{+2}]$	$K_{NO}\ (\cdot 10^{2})$	3.9 ± 36	21 ± 24	56 ± 84	1.8 ± 130
$[(NO)_2^{-}-Cu^{+2}] \rightarrow N_2 + [O_2-Cu^{+2}]$	$K_{O_2}\ (\cdot 10^{2})$	215 ± 810	1.8 ± 0.78	0.77 ± 0.21	0.29 ± 0.068
$2\,[O^{-}-Cu^{+2}] \rightleftarrows O_2 + 2\,Cu^{+}$	σ^2	$1.18 \cdot 10^{-4}$	$6.84 \cdot 10^{-5}$	$6.57 \cdot 10^{-5}$	$1.80 \cdot 10^{-5}$
$r_{N_2} = \dfrac{k_s K_{NO}[Cu_{tot}] \cdot p_{NO}^2}{1 + K_{NO} \cdot p_{NO}^2 + K_{O_2} \cdot p_{O_2}^{0.5}}$	R^2	0.958	0.988	0.992	0.996

The results of the fitting analysis show that none of the kinetic models previously proposed do satisfactorily fit our experimental data. The model proposed by Centi, in the absence of any O_2-related term in the model, could describe the reaction kinetics only when very little amount of oxygen is present, as revealed by the very low values of R^2 obtained by fitting all conversion data. Neither the models proposed by Tomašić et al. [29-30] give satisfactory fitting of our conversion data. The equation named by us *Tomašić 1* was reported as that which best fitted the experimental data of the authors. Even our results show that this is the best fitting model (except at 400°C), but it must be noticed that the estimated values for the surface reaction rate constant k are continuously decreasing with increasing the reaction temperature. Since also the authors estimated the parameter k not monotonously dependent on temperature, it may be concluded that this model is not able to describe to the appearance of a maximum of NO conversion and then the overall kinetics in the whole field of experimental conditions investigated.

The other models proposed by Tomašić et al., i.e. those which take into account the adsorption of NO, do not give better performances in fitting our data, in agreement with the findings of the authors.

At this step of the work we evaluated the fitting performances of six further rate equations, derived from assumed reaction mechanisms and congruent with previous findings obtained with power law rate equations. Among them, it has been reported the model named *Centi modified*, which represents our proposal to take into account the effect of O_2 partial pressure on the overall kinetics under the same hypotheses of the model of Centi [24]. The results of parameters identification, carried out for each temperature investigated, are reported in Tab.3b and show different performances of the models. The model *Centi modified* does not produce effective results, since the relevant R^2 values are still very low and the minimisation algorithm estimated some unacceptable parameters value (for example, a negative value for K''_{NO} at 450°C).

The best data fitting was obtained by model (4), not only because of the lower values of $\hat{\sigma}^2$ and higher values of R^2 but also because it is the only model for which a monotonous dependence on temperature was estimated for each parameter. Tab. 3 also reports the reaction scheme which every model involved in the study could be considered as the description of. In particular, the reaction mechanism whose kinetic description is the equation (4) is below reported and discussed. It is constituted by five different reaction step:

NO adsorption:	$NO + Cu^{+} \rightleftarrows [NO^{-}{-}Cu^{+2}]$	(M1)
N_2O formation:	$NO + [NO^{-}{-}Cu^{+2}] \rightarrow N_2O + [O^{-}{-}Cu^{+2}]$	(M2)
N_2O decomposition over reduced sites:	$N_2O + Cu^{+} \rightarrow N_2 + [O^{-}{-}Cu^{+2}]$	(M3)
N_2O decomposition over oxidised sites:	$N_2O + [O^{-}{-}Cu^{+2}] \rightarrow N_2 + Cu^{+}$	(M4)
Catalyst self-reduction:	$2\,[O^{-}{-}Cu^{+2}] \rightleftarrows O_2 + 2\,Cu^{+}$	(M5)

The following assumptions were made to obtain the kinetic model (4) from the reaction mechanism (M1-M5): reaction (M2) was assumed as rate determining step, while the kinetics of N_2O decomposition (step M3-M4) was considered much faster than all other reaction rates. Moreover, the reversibility of steps (M1) and (M4) were assumed and, consequently, the amount of $[NO^{-}{-}Cu^{+2}]$ and $[O^{-}{-}Cu^{+2}]$ drawn by equilibrium relationships.

All these assumptions are congruent either with our previous results either literature data. The hypotheses that reduced copper species are the active sites for NO decomposition as well as that the reaction mechanism proceeds through a redox cycle is widely accepted [8-12,16-

21]. On the other hand, the formation of nitrous oxide as intermediate species in NO conversion to N_2 is still debated, but following the former hypotheses of Giamello et al. [11] and Valyon and Hall [34], Aylor et al. [18] proposed that N_2O is produced by thermal decomposition of $Cu^+(NO)_2$, while the isotopic studies of Chang and McCarty [35] also showed that NO decomposition into N_2 passes through the formation of N_2O, without any contribution of lattice oxygen.

In a previous investigation, we showed that feeding NO at low temperature (80°C) to Cu over-exchanged (640%) ZSM5 after He pre-treatment at elevated temperatures, a transient production of N_2O is observed as result of catalyst re-oxidation [14]. Similar results have been obtained in the present study. Fig.3 shows that, after catalyst pre-treatment of 2 h in He flow at 550°C, causing copper ions reduction from Cu^{+2} to Cu^+, the response to a step change of NO gaseous concentration (0-600 ppm) at 80°C, results in the transient formation of N_2O due to the re-oxidation of copper sites.

The behaviour of the system in the transient experiment at low temperature proves that catalyst Cu^+ sites are able to convert nitric oxide into nitrous oxide at low temperature, until all copper is re-oxidised to Cu^{+2}. The phenomenon is activated by temperature, but at higher temperatures (> 300°C) the formation of gaseous N_2O is less observable, since the kinetics of its decomposition becomes significant, as shown in Fig.4.

We had already proved that in the range of temperature investigated in the present study, N_2O decomposition is very much faster than NO decomposition [36-37], so justifying the assumption that no contribution to the overall kinetics (M1-M5) is given by step (M3-M4). Moreover, we also discussed the assumption that both reduced and oxidised copper sites of Cu-ZSM5 are active in decomposing N_2O [37].

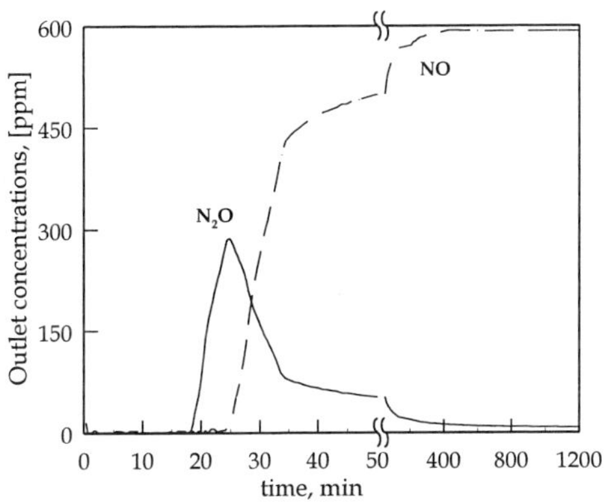

Fig. 3. NO and N_2O outlet concentrations as functions of time after a step change of NO inlet concentration (from 0 to 600 ppm) at 80°C on pre-reduced catalyst.

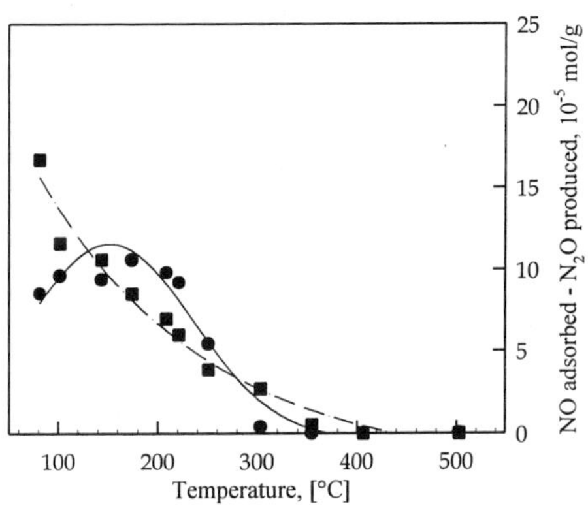

Fig. 4. NO adsorbed (■) and N_2O produced (●) as functions of temperature in NO concentration step change experiments on pre-reduced catalyst.

On the basis of the above hypotheses, the estimated parameters obtained by data fitting with the kinetic model (4) assume a correct chemico-physical meaning. In particular, it is worthwhile to look at Arrhenius-type temperature dependence of the three parameters involved in model (4): k_1, K_{NO} and K_{O_2}.

The values of the activation energy and adsorption heat estimated (reported in Fig. 5) are in good agreement with literature references and physical significance. The activation energy estimated for the kinetic constant k_1, which represents the activation energy of the rate-determining reaction step (M2), i.e. of the oxidation of copper site through the interaction between one gaseous and one adsorbed NO molecules and the formation of gaseous N_2O is 25.7 Kcal/mol, and cannot be compared with any other similar estimation for NO decomposition. Indeed, it has been pointed out that the overall reaction rate is not monotonously increasing with temperature (so the apparent activation energy is strongly dependent on temperature, up to became negative above 500°C).

The adsorption heat of NO on Cu-ZSM5 was evaluated by analysing the dependence of K_{NO} on temperature. The estimated value is 58.3 Kcal/mol, i.e. very close to DFT calculations performed by Trout et al. [12], who reported NO adsorption heat on Cu-ZSM5 in the range 42.8-77.7 Kcal/mol.

The estimated heat of adsorption of O_2 was indeed 43.6 Kcal/mol. The complex interaction of oxygen with Cu-ZSM5 was studied by Valyon and Hall [38-39]. In particular, isotopic studies evidenced the high mobility of oxygen atoms over catalyst surface and allowed them to show that the binding energy of oxygen atoms on copper sites is not constant, but varying with catalyst coverage degree (or catalyst treatment temperature) in the range from 11 to 54 Kcal/mol [38].

Finally, in order to verify the ability of model (4) to expect a maximum value of NO conversion to N_2 in the same experimental conditions of Fig. 1, when expressing all kinetic parameters with monotonous functions of temperature, we have performed numerical simulations by using the Arrhenius parameters. Fig. 6 shows the results of this investigation demonstrating that model (4) well describes the occurrence of a peak of NO conversion to N_2 as a function of the reaction temperature. This happens because both upper and lower sides of the ratio constituting the rate equation (4) are decreasing functions of the increasing temperature. $k_s K_{NO}[Cu_{tot}]$ is proportional to $\exp[-(E_{act} + \Delta H_{NO})/RT]$, being E_{act} + ΔH_{NO} a negative value (-32.6 kcal/mol). At lower temperature, the dependence of the lower side of the ratio is stronger since $K_{NO} \cdot p_{NO} \gg 1 + K_{O_2} \cdot p_{O_2}^{0.5}$ and, hence, proportional to $\exp[-\Delta H_{NO}/RT]$. As a consequence, the whole expression increases with increasing temperature.

At higher temperature, $1 \gg K_{NO} \cdot p_{NO} + K_{O_2} \cdot p_{O_2}^{0.5}$, so the whole expression decreases when temperature

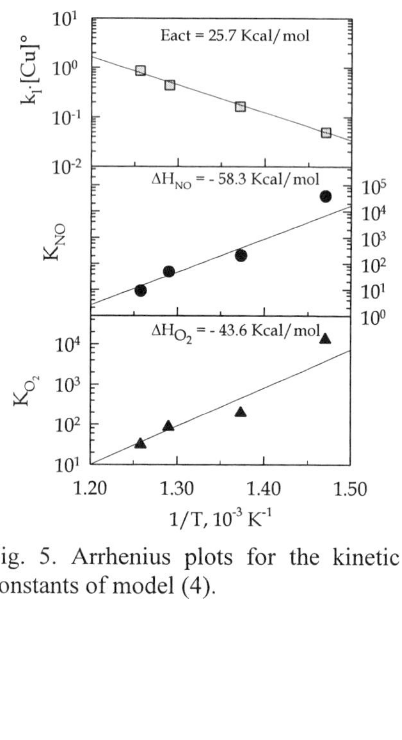

Fig. 5. Arrhenius plots for the kinetic constants of model (4).

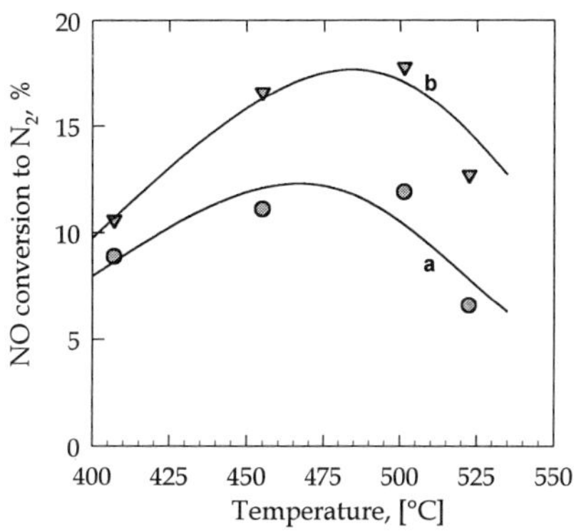

Fig. 6. Effect of temperature on NO conversion to N_2, as simulated by model (4) for two representative values of the inlet NO concentration: a) 2000, b) 5000 ppm. W/F = 0.21 g·s·Ncm^{-3}.

increases.

1. CONCLUSIONS

The kinetics of nitric oxide decomposition over a Cu-ZSM5 catalyst has been investigated by measuring the conversion of NO and analysing the distribution of products in different experimental conditions. By modelling the conversion data with two competitive reactions (NO oxidation and NO decomposition) occurring in a plug flow reactor, we have derived a rate equation based on a five-steps reaction mechanism characterised by the formation of N_2O as key intermediate. The kinetic model developed is capable of well reproducing the presence of a maximum of NO decomposition rate as a function of temperature.

REFERENCES

1. M. Iwamoto and H. Yahiro, Catal. Today, 22 (1994) 5.
2. M. Iwamoto and H. Hamada, Catal. Today, 10 (1991) 57.
3. V.I. Pârvulescu, P. Grange and B. Delmon, Catal. Today, 46 (1998) 233.
4. G. Centi and S. Perathoner, Appl. Catal. A, 132 (1995) 179.
5. J. Valyon and W. K. Hall, Catal. Lett., 19 (1993) 109.
6. M.C. Campa, V. Indovina, G. Minelli, I. Pettiti, P. Porta and A. Riccio, Catal. Lett., 23 (1994) 141.
7. R. Pirone, P. Ciambelli, G. Moretti and G. Russo, Appl. Catal. B, 8 (1996) 197.
8. W. K. Hall and J. Valyon, Catal. Lett., 15 (1992) 311.
9. M. Iwamoto, H. Furukawa and S. Kagawa, In New Dev. Zeo. Sci. Tech., Y. Murikami et al. eds., Kodansha/Elsevier, Tokyo/Amsterdam, 1986, 943.
10. M. Iwamoto, H. Yahiro, N. Mizuno, W.-X. Zhang, Y. Mine, H. Furukawa and S. Kagawa, J. Phys. Chem., 96 (1992) 9360.
11. E. Giamello, D. Murphy, G. Magnacca, C. Morterra, Y. Shioya, T. Nomura and M. Anpo, J. Catal., 136 (1992) 510.
12. B.L. Trout, A.K. Chakraborty and A.T. Bell, J. Phys. Chem., 100 (1996) 17582.
13. J. Dědeček and B. Wichterlová, J. Phys. Chem., 98 (1994) 5721.
14. R. Pirone, E. Garufi, P. Ciambelli, G. Moretti and G. Russo, Catal. Lett., 43 (1997) 255.
15. M.C. Nunes Amorim de Carvalho, F. Barboza Passos and M. Schmal, Appl. Catal. A, 193 (2000) 265.
16. T. Beutel, J. Sárkány, G.-D. Lei, J. Y. Yan and W. M. H. Sachtler, J. Phys. Chem., 100 (1996) 845.
17. G.D. Lei, B.J. Adelman, J. Sarkany and W.M.H. Sachtler, Appl. Catal. B, 5 (1995) 245.

18. A.W. Aylor, S.C. Larsen, J.A. Reimer and A.T. Bell, J. Catal., 157 (1995) 592.
19. G. Spoto, S. Bordiga, D. Scarano and A. Zecchina, Catal. Lett., 13 (1992) 39.
20. M. V. Konduru and S.S.C. Chuang, J. Phys. Chem. B, 103 (1999) 5802.
21. M. V. Konduru and S.S.C. Chuang, J. Catal., 187 (1999) 436.
22. M. Shelef, Catal. Lett., 15 (1992) 305.
23. X. Solans-Monfort, V. Branchadell and M. Sodupe, J. Phys. Chem. A, 104 (2000) 3225.
24. G. Centi, C. Nigro, S. Perathoner and G. Stella, In Environmental Catalysis, J. N. Armor (ed.), ACS Symp. Ser., 552 (1994) 22.
25. Y. Li and W. K. Hall, J. Catal., 129 (1991) 202.
26. B. Ganemi, E. Björnbom and J. Paul, Appl. Catal. B, 17 (1998) 293.
27. Z. Schay, H. Knözinger, L. Guczi, G. Pál-Borbély, Appl. Catal. B, 18 (1998) 263.
28. Z. Schay and L. Guczi, Catal. Today, 17 (1993) 175.
29. V. Tomašić, Z. Gomzi and S. Zrnčević, Appl. Catal. B, 18 (1998) 233.
30. V. Tomašić, Z. Gomzi and S. Zrnčević, React. Kinet. Catal. Lett., 64 (1998) 89.
31. P. Ciambelli, P. Corbo, M. Gambino, G. Minelli, G. Moretti and P. Porta, Catal. Today, 26 (1995) 33.
32. R.H. Perry and D.W. Green, Perry's Chemical Engineers' Handbook, 7th ed., McGraw-Hill, 1997
33. S. Ergun, Chem. Eng. Prog., 48 (1952) 89.
34. J. Valyon and W.K. Hall, J. Phys. Chem., 97 (1993) 1204.
35. Y.-F. Chang and J.G. McCarty, J. Catal. 165 (1997) 1.
36. P. Ciambelli, E. Garufi, R. Pirone, G. Russo and F. Santagata, Appl. Catal. B, 8 (1996) 333.
37. P. Ciambelli, A. Di Benedetto, E. Garufi, R. Pirone and G. Russo, J. Catal., 175 (1998) 161.
38. J. Valyon and W.K. Hall, J. Catal., 143 (1993) 520.
39. J. Valyon, W. S. Millman and W. K. Hall, Catal. Lett., 24 (1994) 215.

Studies in Surface Science and Catalysis 140
A. Gamba, C. Colella and S. Coluccia (Editors)

Characterisation of palladium catalysts supported on hydrotalcite-derived mixed oxides.

Federica Prinetto[a], Giovanna Ghiotti[a], Nigamananda Das[b], Didier Tichit[b], Bernard Coq[b].

[a] Dipartimento di Chimica IFM, Università di Torino, Via P. Giuria 7, 10125 Torino, Italy.

[b] CNRS-ENSCM UMR 5618, 8 Rue de l'Ecole Normale, 34000 Montpellier, France.

A series of dispersed palladium catalysts with Pd loading in the range 0.05-1.5 wt-% have been prepared by impregnation either with Pd acetylacetonate or Pd chloride of Mg/Al mixed oxide supports obtained by thermal decomposition of hydrotalcite compounds. The surface acid-base properties and the metal phase were investigated by CO_2 and CO interaction, respectively, followed by FT-IR spectroscopy. Pd catalysts obtained using the acetylacetonate precursor were active and selective in the "one-pot" synthesis of methyl isobutyl ketone (MIBK) from acetone. In particular, catalysts with Pd loading in the range 0.2-0.5 wt-% showed the highest selectivity toward MIBK, thanks on one hand to the high number and strength of basic sites and on the other hand to the high dispersion of Pd and specific interaction between metal particles and strongly basic O^{2-} sites. At variance, catalysts obtained using the chloride precursor exhibited low activity and selectivity, due to their poor basicity.

1. INTRODUCTION.

Hydrotalcite (HT), $[Mg_6Al_2(OH)_{16}](CO_3) \cdot 4H_2O$, is a naturally occurring anionic clay, which displays a lamellar structure consisting of brucite-like layers and compensating anions situated in the interlayer space [1, 2]. HT-type materials, easily obtained in laboratory by coprecipitation of the M^{2+} and M^{3+} salts in alkaline media, show interesting applications as anionic exchangers and catalyst precursors. The thermal decomposition of HT compounds leads to homogeneous M^{2+}/M^{3+} mixed oxides exhibiting high specific surface areas and strong basic character; moreover, HT containing reducible cations lead to metal particles exhibiting strong metal-support interaction after appropriate reduction treatment. These properties which can be tailored by varying the chemical composition and the activation conditions [1], make the HT-derived mixed oxides very attractive as supports or precursors of supported metal catalysts. In particular, Pd multifunctional catalysts supported on Mg/Al mixed oxides can be interestingly employed for the "one-pot" synthesis of methyl isobutyl ketone (MIBK) from acetone [3], which represents an economically viable and by far less pollutant alternative to the classical three-step catalytic process. Pd is chosen as a metal function due to its ability to promote selectively the C=C bond hydrogenation in mesityl oxide at the expense of the C=O bond hydrogenation of ketones to alcohols. The catalytic performances depend on the balance between the metallic and basic functions, therefore a careful characterisation of these two aspects appears of vital importance to highlight the catalytic behaviour of these solids. In this work, CO and CO_2 were used as probe molecules for this purpose, followed by FT-IR spectroscopy.

2. EXPERIMENTAL.

2.1. Sample preparation.

The hydrotalcite was prepared by coprecipitation at constant pH of suitable amounts of $Mg(NO_3)_2 \cdot 6H_2O$ and $Al(NO_3)_3 \cdot 6H_2O$ ($Mg^{2+}/Al^{3+} = 3$) with NaOH + Na_2CO_3. The suspension was aged at 353 K for 15 h under stirring, washed thoroughly with distilled water (Na < 100 ppm) and dried overnight at 353 K. The hydrotalcite was heat-activated in air flow at 673 K for 4 h to yield the Mg/Al mixed oxide support, which was then contacted with either water-free toluene solutions containing dedicate amounts of Pd acetylacetonate ($Pd(acac)_2$) or $PdCl_2$ aqueous solutions. The solids were finally calcined in air flow at 673 K for 4 h. The samples were named PdHTx(a or c), where x is the Pd content (wt-%), ranging from 0.05 to 1.5 wt-%, and a or c indicate the acetylacetonate or chloride precursor.

2.2. Catalyst characterisation.

Chemical analysis of the samples was performed at the Service Central d'Analyse du CNRS (Solaise, France).

XRD patterns were recorded on a CGR Theta 60 instrument using Cu $K_{\alpha 1}$ radiation.

N_2 sorption experiments at 77 K were carried out with a Micromeritics ASAP2000 instrument and specific surface areas calculated using the BET method.

Integral heats of CO_2 adsorption were determined with a SETARAM TG-DSC-111 apparatus on samples previously calcined in air flow at 673 K. The thermal desorption of CO_2 was started in He from 373 to 673 at 5 K min^{-1}.

Absorption/transmission IR spectra were run at RT on a Perkin-Elmer FT-IR 1760-X spectrophotometer equipped with a Hg-Cd-Te cryodetector. For IR analysis powders were pelletted in self-supporting discs (10-15 mg cm^{-2}), activated in vacuo at increasing temperature up to 723 K, heated in dry O_2 at the same temperature and cooled down in oxygen or subsequently reduced in H_2 at increasing temperature up to 673 K and outgassed at the same temperature. CO_2 and CO (Matheson C.P.) were used as probes.

2.3. Catalytic tests.

Catalytic tests were performed in a microflow fixed-bed reactor, using 50 mg of catalyst, previously activated *in situ* in H_2 at 473 K for 12 h. Acetone was fed by bubbling H_2/He mixture through a saturator at 273 K (acetone/H_2 = 10, flow = 60 cm^3 min^{-1}) and the reaction carried out at 423 K. The effluent mixture was analyzed at regular time intervals with an on-line gas chromatograph (Carlo Erba) equipped with a capillary column (Carbowax 20M bonded phase) and a flame ionization detector.

3. RESULTS AND DISCUSSION.

3.1. Catalyst characterisation.

All the PdHTx calcined samples showed XRD patterns characteristic of Mg/Al mixed oxides with rock salt structure, while PdO was not detected. The specific surface areas of the Mg/Al support and of PdHTx(a) samples with Pd loading up to 0.5 wt-% were in the range 220-230 m^2 g^{-1}, while they declined to 175 m^2 g^{-1} for Pd contents around 1.5 wt-%. A lower value (150 m^2 g^{-1}) was determined for PdHT1.2(c) sample.

3.2. Investigation of the surface acid-base properties.

The nature of the acid-base surface sites was investigated by CO_2 adsorption, followed by IR spectroscopy (Figure 1). The main species detected on the oxidised samples were: i) bidentate carbonates formed on acid-base pairs (1680-1570 cm^{-1}, $\nu_{C=O}$, 1350-1250, 1060-1040 cm^{-1}, ν_{OCO} asym and sym, respectively); ii) monodentate carbonates formed on strongly basic oxygens (1550-1500 and 1450-1400 cm^{-1}, ν_{OCO} asym and sym, respectively; the $\nu_{C\text{-}O}$ mode expected at about 850 cm^{-1} is obscured by the metal-oxygen bulk vibration modes); iii) hydrogen carbonates formed through basic hydroxyls (3616 cm^{-1}, ν_{OH}; 1670, 1470 cm^{-1}, ν_{OCO} asym and sym, respectively; 1230 cm^{-1}, δ_{COH}); iv) CO_2 linearly coordinated on Lewis acid sites, mainly Al^{3+} (2357-38 cm^{-1}, ν_{OCO} asym; 2297 cm^{-1}, ν_{OCO} asym of $^{13}CO_2$ with natural abundance). The relative amounts and thermal stability of the different surface species formed upon CO_2 admission depended mainly on the nature of the Pd precursor used and on the Pd loading.

The effect of Pd loading was examined on PdHTx(a) samples. It was noted that on increasing the metal loading from 0.05 to 1.5 wt-%, the amount of carbonates decreased by up to 15%, while that of CO_2 linearly coordinated on Lewis acid sites slightly increased (see curves 1 and 2 in Fig. 1). This accounted for a slight decrease of the basicity on increasing the Pd content. On PdHTx(a) samples, monodentate carbonates showed the highest resistance to heating under vacuum, being completely removed only at 673 K, whereas bidentate carbonates desorbed at 623 K, bicarbonates at 473 K, and linearly coordinated CO_2 at RT.

The sample pre-treatment, oxidation or reduction in H_2, did not modify the relative and total amounts of the different surface species formed upon CO_2 admission, thus evidencing the negligible role of Pd ions as Lewis acid sites.

Conversely, a marked influence of the Pd precursor used was noted. Indeed, on PdHT1.2(c) sample prepared using the chloride precursor, monodentate carbonates and bicarbonates were absent, while bidentate carbonates were formed in lower amounts than on catalysts of similar Pd loading prepared using Pd(acac) (compare curves 2 and 3 in Figure 1).

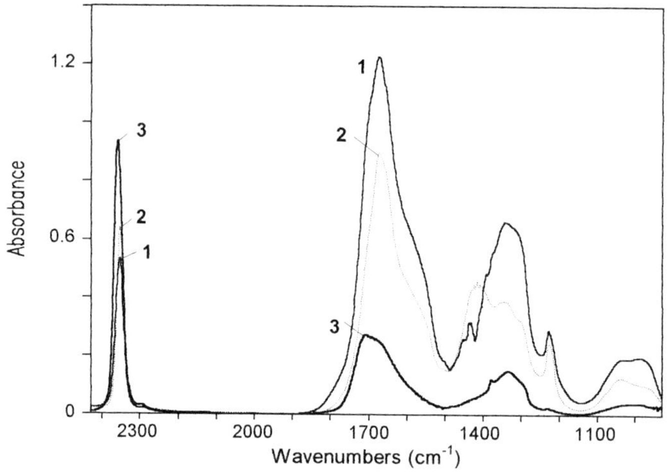

Figure 1. IR difference spectra upon CO_2 adsorption (4 mbar) on PdHT0.2(a) (curve 1), PdHT1.5(a) (curve 2) and PdHT1.2(c) (curve 3) catalysts oxidised at 673 K.

These carbonates also showed a lower stability than on PdHTx(a) samples, about 40% being already removed upon evacuation at RT. Furthermore, the peak due to the linearly coordinated CO_2 showed higher frequency (2359 cm^{-1}) and intensity (+30-40 %) in comparison to the Cl-free samples. These findings evidenced on one hand a lower amount and strength of basic oxygens as well as the absence of basic OH^- and of the strongest basic O^{2-} sites and on the other hand a higher strength and amount of Lewis acid sites due to the presence of chlorine.

The Mg/Al support and PdHTx(a) samples exhibited similar integral heats of CO_2 adsorption ($\Delta H_{m,\,ads}$ CO_2 = - 82-83 kJ mol^{-1}) and number of basic sites (3.1-3.2 x 10^{-6} mol CO_2 adsorbed m^{-2}). Conversely, the values found for PdHT1.2(c) sample containing chlorine were markedly lower ($\Delta H_{m,\,ads}$ CO_2 = - 25 kJ mol^{-1}, 0.9 x 10^{-6} mol CO_2 adsorbed m^{-2}), evidencing the poor basicity, in agreement with IR results.

3.3. Characterisation of metal phase.

The nature of Pd sites was investigated through CO adsorption at RT, followed by IR spectroscopy. On pure Mg/Al mixed oxide only negligible amounts of ketenic and $(CO)_n^{x-}$ species (very weak bands at 2100-1500 cm^{-1}) and carbonates were formed (spectra not reported). Conversely, on oxidised PdHTx(a) catalysts, intense bands due to mono- and bidentate carbonates were formed, similar to those detected upon CO_2 admission (Fig. 2a). Carbonate formation arose from the reduction of PdO species by CO at RT and the subsequent interaction of CO_2 thus formed with the basic oxygens of the support. The formation of reduced Pd atoms was revealed by the bands at 2080-2098 cm^{-1} (the highest frequency corresponds to the highest CO coverage), 1950-1960 and 1900 cm^{-1}, assigned to linear, 2-fold and 3-fold bridged Pd^0 carbonyls, respectively [4]. Weak components at 2130 and 2185 cm^{-1} were ascribed to $Pd^+(CO)$ and $Al^{3+}(CO)$ species.

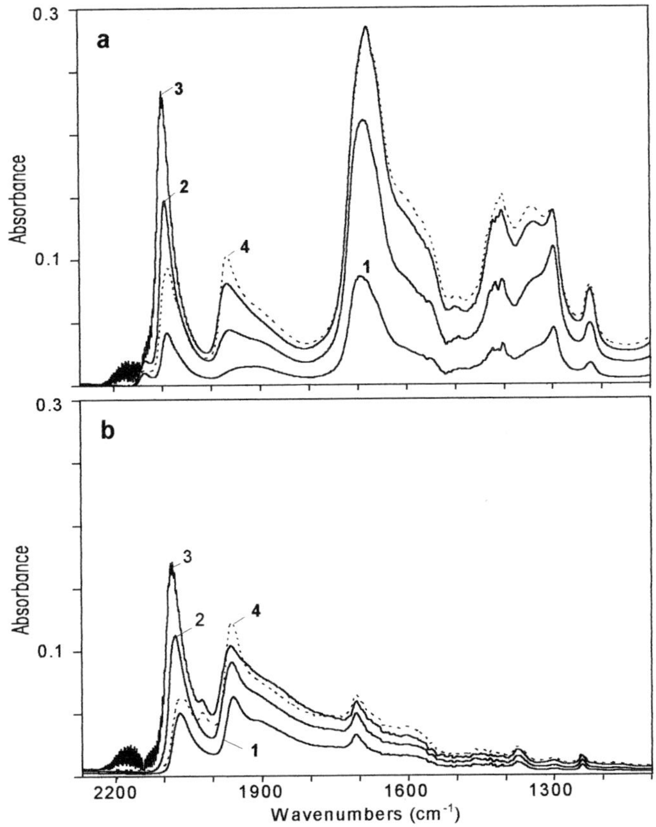

Figure 2. IR difference spectra upon CO adsorption at RT at increasing doses (0.01, 0.5, 50 mbar, curves 1-3) and subsequent evacuation at RT (curves 4, dotted lines) on PdHT1.5(a) sample oxidised at 673 K (section a) or reduced in H_2 at 673 K (section b).

The red shifts of the various absorptions observed by performing the experiments with ^{13}CO were consistent with the isotopic shift expected for monocarbonyl species upon ^{12}C to ^{13}C substitution. Linear Pd carbonyls were partially removed (about 60%) by evacuation at RT, while bridged species were stable up to 573 K. CO admission on samples reduced in H_2 at temperatures in the range 623-673 K gave linear Pd^0 carbonyls (2070-2085 cm^{-1}), bridged (1965 cm^{-1}) and multibridged CO species (1900, 1700 cm^{-1}), while carbonates were not formed, according to what previously proposed (Fig. 2b). A minor component at 2022 cm^{-1} was also detected (*vide infra*).

It is noted that the intensity ratio between linear and bridged carbonyl bands decreased on going from the oxidised to the reduced sample, evidencing some aggregation of the Pd particles caused by the thermal treatment in H_2. Furthermore, a red shift (~10 cm^{-1}) of the band due to linear carbonyls can be observed on going from the oxidised to the reduced sample. This can be explained by a deep reduction of the PdO particles occurring during the thermal treatment in H_2. At variance, the reduction operated by CO at RT most probably concerns only the surface layers of PdO particles, so that electron donation from the surface Pd^0 to $Pd^{\delta+}$ atoms of the sublayers can occur, decreasing the π back-donation from surface Pd sites to CO molecules with a consequent increase of the CO stretching frequency.

The effect of Pd loading was examined in the case of PdHTx(a) samples (Figure 3). It can be noted that the intensity ratio between linear and bridged carbonyl bands slightly increased on catalysts with Pd loading ≤ 0.2 wt-%, evidencing a higher dispersion of the metal phase over the support. Interestingly, on samples with Pd loading in the range 0.2-0.5 wt-% the contribution of the component at 2022 cm^{-1} became important. This absorption showed unusual features, such as a frequency particularly low but still in the range characteristic of linear carbonyls, a low rate of formation and a high stability upon evacuation (Figure 4). Such a behaviour, never observed for Pd single crystals nor for Pd supported on alumina, silica or titania, is most probably due to the interaction with the basic support. Accordingly, similar findings have been reported for Pt dispersed on hydrotalcite-derived mixed oxides [5, 6]. Therefore, the band at 2022 cm^{-1} is assigned to CO adsorbed on Pd atoms strongly interacting with the basic O^{2-} of the support, thus characterised by a particularly high electron density.

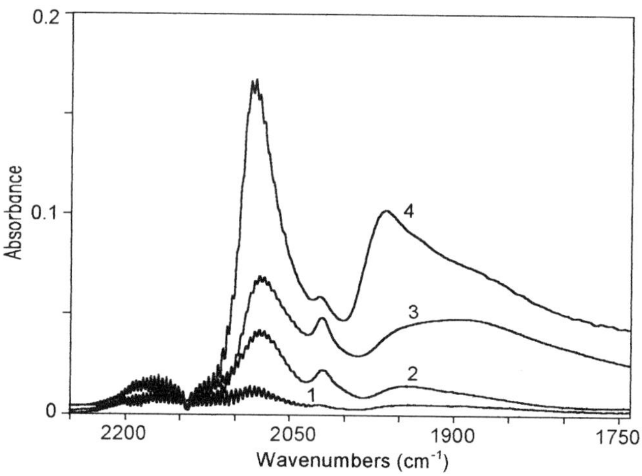

Figure 3. IR difference spectra upon adsorption of CO (50 mbar) at RT on PdHTx(a) samples at different Pd loading reduced in H_2 at 673 K: PdHT0.05(a) (curve 1); PdHT0.2(a) (curve 2); PdHT0.5(a) (curve 3); PdHT1.5(a) (curve 4).

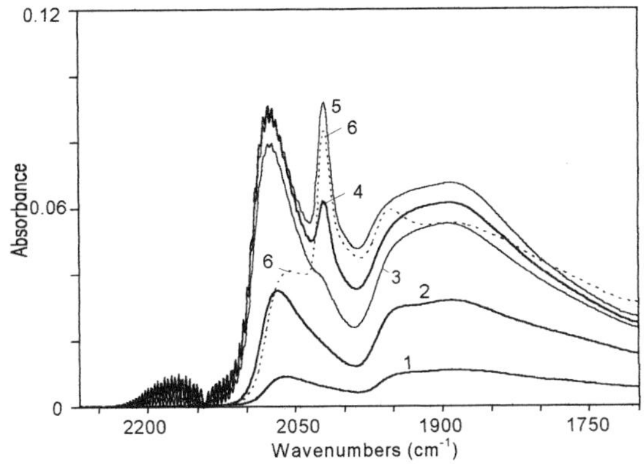

Figure 4. Adsorption of CO at RT on PdHT0.5(a) catalyst reduced in H_2 at 673 K: IR difference spectra upon admission of 0.01, 0.1, 5, 20 mbar (curves 1-4); recorded after 20' of contact with 20 mbar of CO (curve 5); after the subsequent evacuation at RT (curve 6).

The Pd precursor used deeply influenced the nature and amount of the species formed upon CO adsorption, as shown comparing the spectra of samples obtained using the chloride or the acetylacetonate precursors. On oxidised Cl-containing samples, a unique band at 2155 cm^{-1} was formed upon CO admission and assigned to Pd^{2+} linear carbonyls [7], while bridged CO and carbonates were absent (Fig. 5a). These findings suggest the presence of $PdCl_x$ species, where no Pd-O pairs are available for the oxidation of CO to CO_2 and the reduction of Pd ions. After reduction in H_2 at 473-673 K, linear Pd^0 carbonyls were observed at 2100-2090 cm^{-1}, *i.e.* at higher frequencies than on PdHTx(a) reduced samples (Fig. 5b), due to the electron withdrawing effects of chlorine [8].

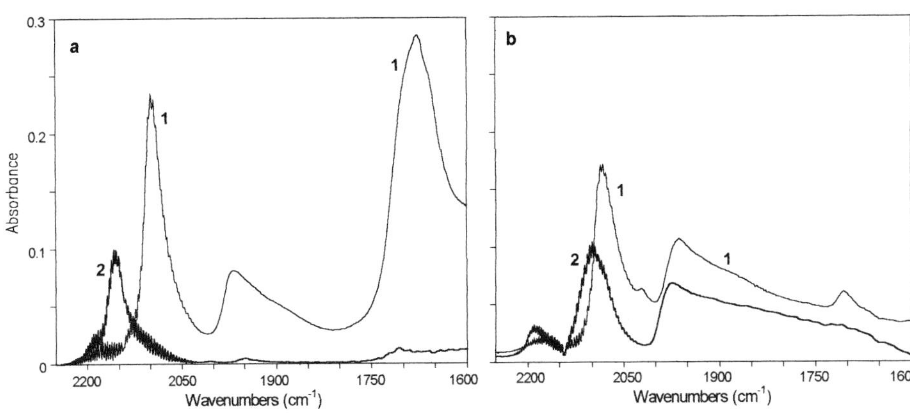

Figure 5. IR difference spectra upon CO adsorption (50 mbar) on PdHT1.5(a) (curves 1), and PdHT1.2(c) (curves 2) oxidised at 673 K (section a) or reduced in H_2 at 673 K (section b).

Furthermore, the intensity ratio between linear and bridged carbonyl bands was lower on PdHT1.2(c) in comparison to Cl-free samples with similar Pd loading, evidencing a lower dispersion of Pd over the support. The absence of the band at 2022 cm^{-1} confirmed that it was related to Pd atoms interacting with strongly basic oxygens of the support, not present on the chlorinated samples, as evidenced both by IR and microcalorimetric measurements upon CO_2 adsorption.

3.4. Catalytic activity in the synthesis of MIBK from acetone.

All PdHTx(a) catalysts showed activity and selectivity for the synthesis reaction of MIBK from acetone (Table 1). The conversion increased with the Pd content, while the selectivity toward MIBK reached a maximum for catalysts with Pd loading in the range 0.2-0.5 wt-%. Indeed, at lower Pd loading, when the basic function of the support prevailed on the hydrogenating ability, condensation reactions took place: condensation of acetone with mesityl oxide to phorone, isophorone and trimethyl-cyclohexanone, and condensation of MIBK with acetone or its self condensation to diisobutyl ketone (DIBK) or trimethyl nonanone (NONA), respectively (Figure 6).

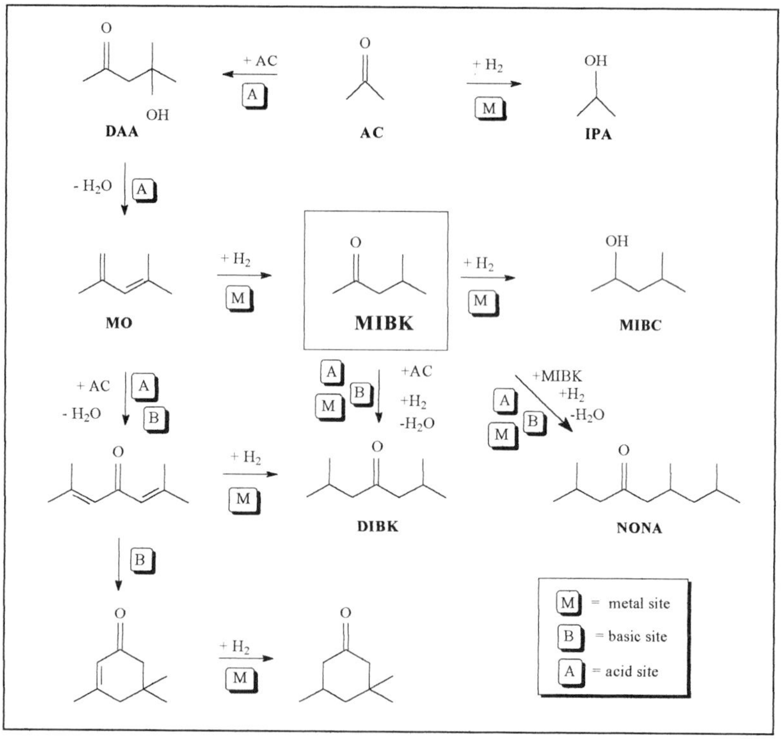

Figure 6. Simplified scheme of the reaction of acetone with hydrogen on a solid catalyst containing basic, acid and metallic sites.

Conversely, at high Pd contents, when the metal function was predominant, the main secondary products were isopropanol (IPA) and methylisobutylcarbinol (MIBC) resulting from the hydrogenation of acetone and MIBK, respectively. Other products were formed in trace amounts. Therefore, the best compromise between basic and hydrogenating properties was achieved with Pd loadings in the range 0.2-0.5 wt-%. We remember that these catalysts exhibited a high dispersion of the metal phase and strong interaction between the metal particles and the strongly basic O^{2-} sites of the support, as shown by IR measurements. Noticeably, PdHTx(a) samples showed activity comparable to that of MgO-supported Pd catalysts with similar Pd loading (see Table 1), but higher selectivity toward MIBK, resulting from the simultaneous presence of acid and basic sites of medium-high strength, and particularly of Al^{3+}-O^{2-} pairs, able to dehydrate the diacetonealcohol (DAA) into mesityl oxide (MO). As a consequence the condensation rate increases at the expense of the hydrogenation rate of acetone to IPA.

PdHTx(c) catalysts obtained using the chloride precursor showed very low activity and selectivity due to the poor basicity and hydrogenating ability induced by the presence of Cl, in agreement with IR results (Table 1).

Table 1.
Activities and selectivities of Pd catalysts in the synthesis of MIBK from acetone at 423 K.

Sample	**Convers. (%)**	**Selectivity (%)**					
		IPA	**MIBK**	**MIBC**	**DIBK**	**MO**	**NONA**
Pd/MgO 0.2 [9]	26	21.05	65.07	1.09	8.73	1.00	2.87
PdHT0.2(a)	27	10.22	73.76	0.76	9.09	1.10	1.38
PdHT1.5(a)	70	34.78	35.87	5.17	18.46	0.93	3.57
PdHT1.2(c)	8	74.7	23.66		1.08		0.54

4. CONCLUSIONS.

Mixed oxides obtained by thermal decomposition of hydrotalcite compounds have shown to be very attractive as supports for multifunctional Pd dispersed catalysts, possessing simultaneously hydrogenating ability and acid-base properties which can be tailored by varying both the Pd loading and the nature of the Pd precursor used for the impregnation. Catalysts obtained using the acetylacetonate precursor, easily decomposed during the catalyst activation, are active and selective in the "one-pot" synthesis of methyl isobutyl ketone from acetone. IR and catalytic measurements evidenced that the best compromise between the metallic and the acid-base functions resulted from the deposition of Pd amounts in the range 0.2-0.5 wt-%. Indeed, these catalysts showed the highest selectivity toward MIBK, due on one hand to the high number and strength of basic sites, and on the other hand to the high dispersion of Pd and specific interaction between the metal particles and the strongly basic O^{2-} sites. At variance, the chlorine introduced using $PdCl_2$ precursor and not eliminated during the catalyst activation had a detrimental influence both on the activity and selectivity of the catalysts, inducing a low basicity and a poor dispersion of the metal.

REFERENCES.

1. A. Vaccari, Appl. Clay. Sc. 14 (1999) 161.
2. F. Cavani, F. Trifirò, A. Vaccari, Catal. Today 1991, Vol. 11, N.2.
3. K. D. Olson, U.S. Patent 4,704,478 (1987).
4. K. Wolter, O. Seiferth, H. Kuhlenbeck, M. Bäumer, H.J. Freund, Surf. Sci. 399 (1998) 190.
5. V. B. Kazanski, V. Yu. Borovkov, A. I. Serykh, F. Figueras, Catal. Lett. 49 (1997) 35.
6. Z. Gandao, B. Coq, L.C. de Ménorval, D. Tichit, Appl. Catal. A, General 147 (1996) 395.
7. Yu.A. Lokhov, A.A. Davydov, Kinet. Katal. 21 N.6 (1980) 1523.
8. M. Kappers, C. Dossi, R. Psaro, S. Recchia, A. Fusi, Catal. Lett. 39 (1996) 183.
9. N. Das, D. Tichit, R. Durand, P. Graffin, B. Coq, Catal. Lett. (submitted).

Studies in Surface Science and Catalysis 140
A. Gamba, C. Colella and S. Coluccia (Editors)

Silica-aluminas: sol-gel synthesis and characterization

C. Rizzo, A. Carati, C. Barabino, C. Perego and G. Bellussi

EniTecnologie S.p.A., Via Maritano 26, I-20097 San Donato Milanese, Italy

Amorphous silica-aluminas with controlled pore size distribution in the micro and mesoporous region (i.e. ERS-8 and MSA respectively) have been synthesised by sol-gel route, changing the SiO_2/Al_2O_3 and tetrapropylammonium hydroxide/SiO_2 molar ratios and the solvent (e.g. methanol, ethanol and 1-propanol). By a proper control of such parameters it is possible to tailor the desired textural properties.

1. INTRODUCTION

Over the last fifteen years, a strong research effort has been devoted to the synthesis of ordered large porous materials both crystalline and amorphous. The largest intracrystalline space materials are microporous metal-phosphates (AET [1], CLO [2], VFI [3]) with a pore diameter in the range 0.8-1.3 nm and ≥14-member rings. Recently aluminosilicates (CFI [4] and DON [5]) having 14-member rings and a pore diameter in the range 0.7-0.8 nm have been discovered. Among amorphous materials several families of aluminosilicates with a narrow pore size distribution in the range 1-10 nm have been described (M41-S [6], FSM-16 [7], HMS [8], SBA [9], MSU [10], KIT-1 [11], MSA [12], ERS-8 [13]). As a general rule, amorphous ordered materials have been obtained by using in the synthesis a templating agent, which can give rise to the formation of micelles (e.g. for MCM-41, FSM-16 and HMS syntheses) or clusters (e.g. for MSA and ERS-8 syntheses).

They are of great interest in catalysis because of their uniform and large pores, which allow sterically hindered molecules (i.e., unable to diffuse effectively through the smaller microporous channels of zeolites) to diffuse easily to internal active sites. These properties could also be interesting for application in gas sorption processes.

Our work is devoted to the investigation of several parameters in sol-gel syntheses from alkoxide reagents. A better understanding of the chemical processes that take place in the earliest stages of sol-gel preparation can provide a potential better control of the microstructural evolution of a catalyst, in terms of chemical homogeneity and textural properties.

2. EXPERIMENTAL

Samples were prepared via sol-gel in alkali-free medium using $Si(OC_2H_5)_4$ (Dynasil-A, Nobel), $Al(sec\text{-}OC_4H_9)_3$ (Fluka), tetrapropylammonium hydroxide (TPAOH, Sachem), alcohol (ROH) selected among CH_3OH, C_2H_5OH, n-C_3H_7OH (Fluka). The preparations were performed at molar ratio: SiO_2/Al_2O_3 = 100, 300, ∞ and TPAOH/SiO_2 = 0.09 and 0.2. A typical synthesis preparation is following described.

Al(sec-$OC_4H_9)_3$ was mixed with $Si(OC_2H_5)_4$ at 60 °C. The obtained homogeneous solution was cooled to room temperature, then added to the required alcohol and TPA-OH in aqueous solution. Monophasic clear solutions were obtained, then transformed in homogeneous compact gel without separation of phases.

After 15 hour ageing at room temperature, the gels were dried at 100 °C and calcined 8 hour in air at 550 °C.

The textural properties of all calcined samples were determined by nitrogen isotherms at liquid N_2 temperature, using a Micromeritics ASAP 2010 apparatus (static volumetric technique). Before determination of adsorption-desorption isotherms the samples (~ 0.2 g) were outgassed for 16 h at 350 °C under vacuum.

The specific surface area (S_{BET}) was evaluated by 2-parameters linear BET plot in the range p/p° 0.01-0.2. The total pore volume (V_T) was evaluated by Gurvitsch rule. Mean pore size (d_{DFT}) and pore size distributions were calculated using DFT method, based on molecular statistical approach. It was applied over the complete range of the isotherm and was not restricted to a confined range of relative pressure or pore sizes. Pore size distribution was calculated by fitting the theoretical set of adsorption isotherms, evaluated for different pore sizes, to the experimental results. DFT model was particularly effective for evaluation in the borderline range between micro and mesopores [14].

3. RESULTS

A fast gelation was observed for all the materials prepared at TPAOH/SiO_2 = 0.09 producing opalescent gels. After calcination, they were characterised by irreversible Type IV + (I) isotherms, characteristic of mesoporous MSA-type materials

By contrast a slow gelation was observed for the materials prepared at TPAOH/SiO_2 = 0.2 and transparent gels were obtained. Besides, the gelation rate increased with the aluminum content. The materials at SiO_2/Al_2O_3 = 300 and ∞ showed reversible Type I isotherms, characteristic of microporous ERS-8-type materials. Instead a border line situation was observed at SiO_2/Al_2O_3 = 100, showing an irreversible Type I + IV isotherm.

The samples synthesised with higher TPAOH/SiO_2 were characterised by a higher specific surface area but a lower pore volume and pore size with respect to the samples synthesised with lower TPAOH/SiO_2. Fig. 1 shows all the isotherms of samples prepared with ethanol both at TPAOH/SiO_2 = 0.09 and 0.2 and Fig. 2 the pore size distributions.

The isotherms and the pore size distributions of the samples synthesised with different alcohols at different SiO_2/Al_2O_3 molar ratio are reported in Figs 3-8. As a general trend, as the aliphatic chain of the alcohol increased, the specific surface area, the pore volume and the mean pore size increased for all SiO_2/Al_2O_3. The results are summarised in Tables 1 and 2.

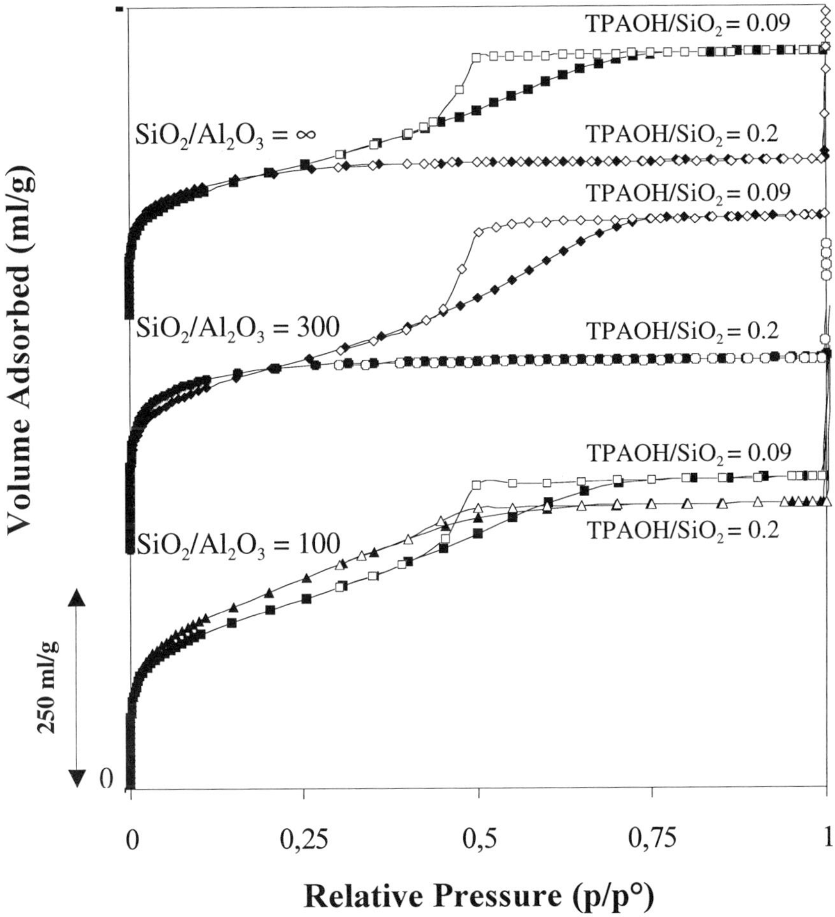

Fig. 1. N_2 adsorption-desorption isotherms of all samples synthesised with ethanol both at $TPAOH/SiO_2 = 0.09$ and 0.2.

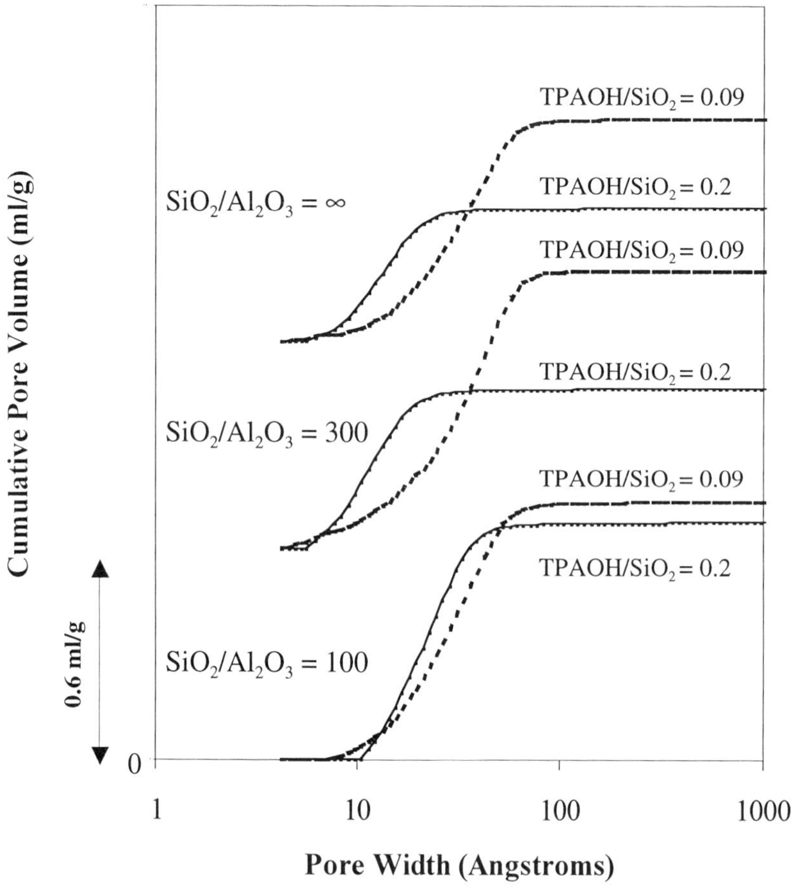

Fig. 2. DFT cumulative pore volume of samples synthesised with ethanol both at $TPAOH/SiO_2 = 0.09$ and 0.2.

Table 1
Influence of the SiO_2/Al_2O_3 molar ratio and of the alcohol on textural properties at $TPAOH/SiO_2 = 0.09$

SiO_2/Al_2O_3	Alcohol	SSA_{BET} (m^2/g)	V_T (ml/g)	d_{DFT} (nm)
100	MeOH	802	0.59	3.0
100	EtOH	798	0.60	2.7
100	PrOH	993	1.01	4.3
300	MeOH	739	0.47	2.2
300	EtOH	835	0.67	3.3
300	PrOH	1040	1.04	4.1
∞	MeOH	519	0.44	3.9
∞	EtOH	668	0.53	3.1
∞	PrOH	667	1.00	7.2

Table 2
Influence of the $TPAOH/SiO_2$ molar ratio on textural properties for samples synthesised with ethanol

SiO_2/Al_2O_3	$TPAOH/SiO_2$	SSA_{BET} (m^2/g)	V_p (ml/g)	d_{DFT} (nm)
100	0.09	798	0.60	2.7
100	0.2	889	0.57	2.2
300	0.09	835	0.67	3.3
300	0.2	875	0.39	1.2
∞	0.09	668	0.53	3.1
∞	0.2	683	0.32	1.3

4. DISCUSSION

As described by Handy *et al.* [15] the gelation time can be related to ability to cross-link polymeric chains. High gelation rate favours the highly-branched cluster aggregates of a colloidal nature producing an opalescent gel, that after calcination gives rise to a highly cross-linked mesoporous material. Low gelation rate favours the formation of linear polymeric chains giving rise to a transparent gel, that after calcination produces weakly cross-linked microporous material. The first situation is characteristic of gel formed in basic medium, instead the second one is observed in acidic medium.

In the present work both transparent and opalescent gels are formed in basic medium, due to the presence of TPAOH.

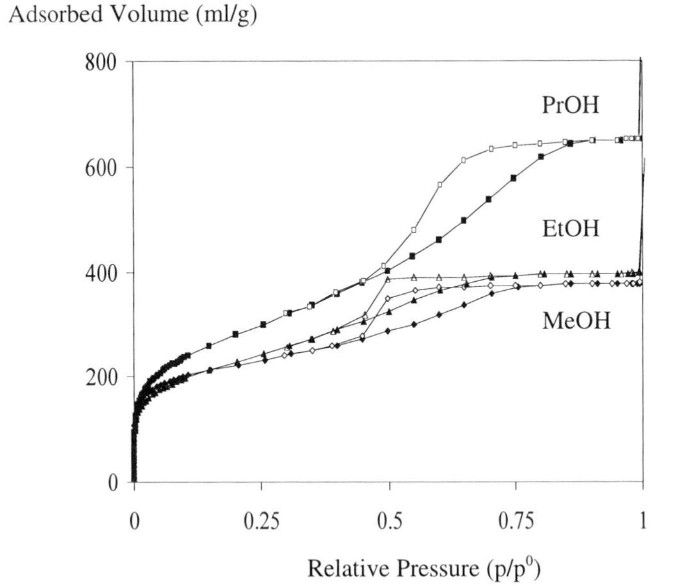

Fig. 3. N_2 adsorption-desorption isotherms of samples synthesised at $SiO_2/Al_2O_3 = 100$.

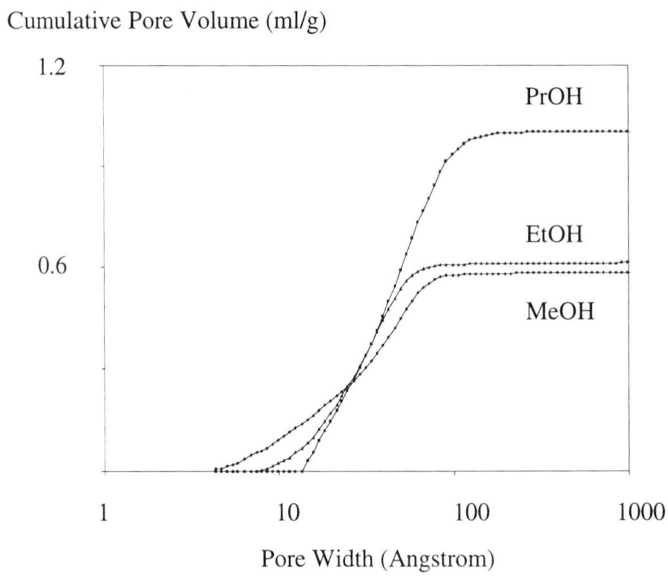

Fig. 4. DFT cumulative pore volume of samples synthesised at $SiO_2/Al_2O_3 = 100$.

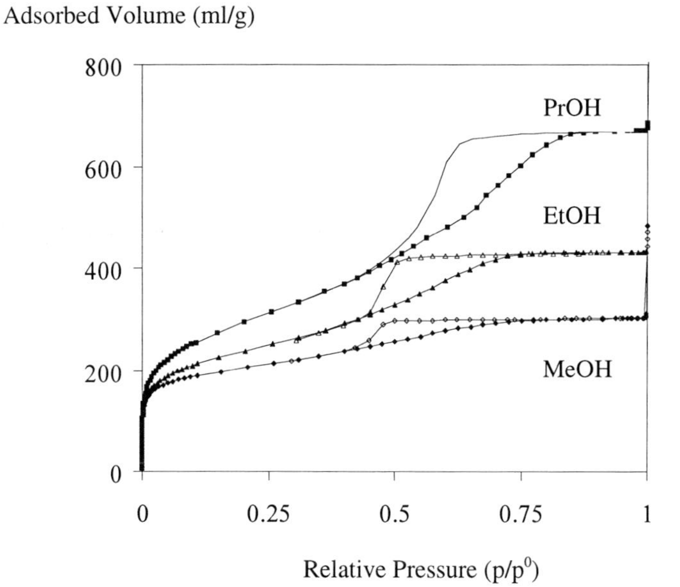

Fig. 5. N_2 adsorption-desorption isotherms of samples synthesised at SiO_2/Al_2O_3 = 300.

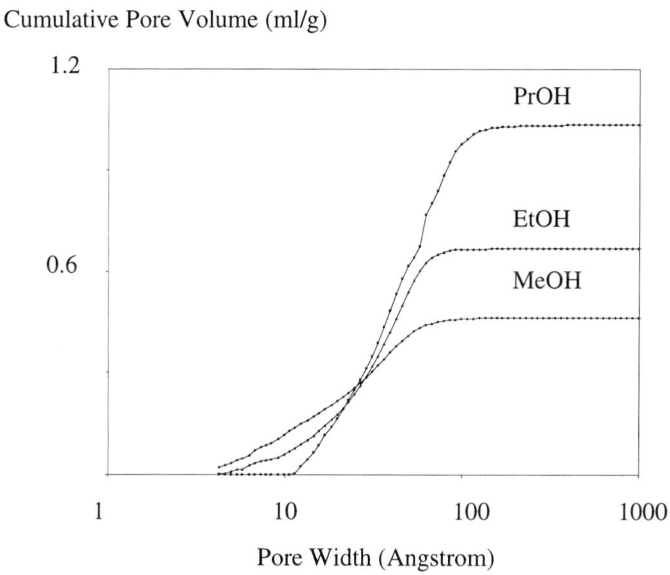

Fig. 6. DFT cumulative pore volume of samples synthesised at SiO_2/Al_2O_3 = 300.

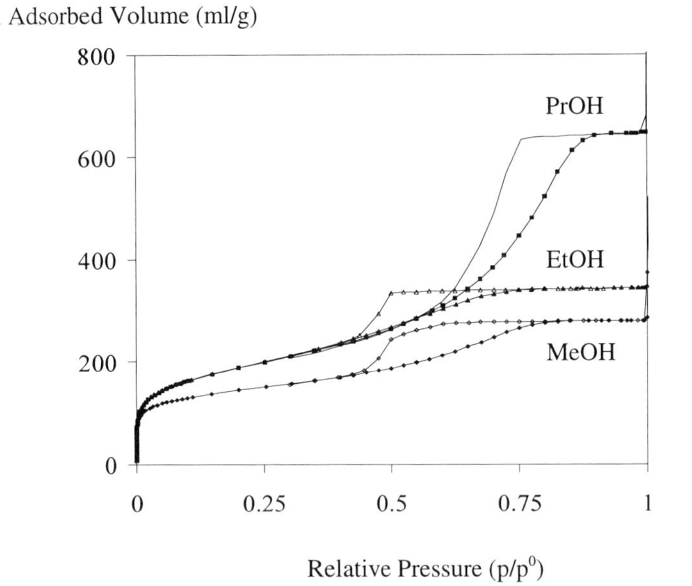

Fig. 7. N_2 adsorption-desorption isotherms of samples synthesised at $SiO_2/Al_2O_3 = \infty$.

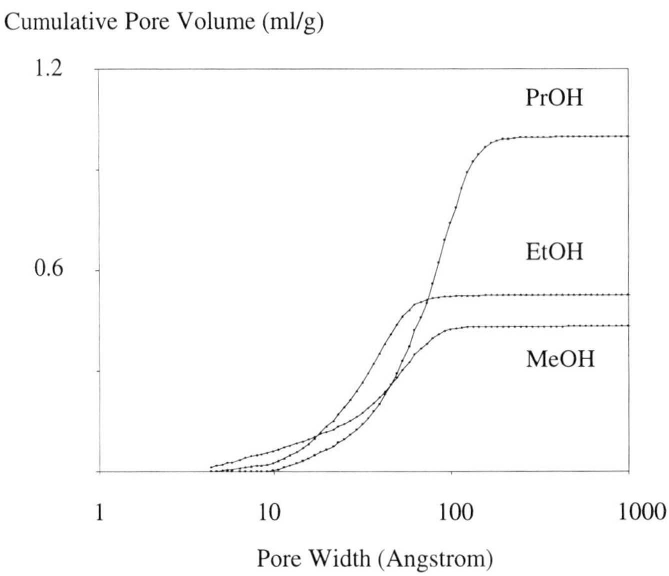

Fig. 8. DFT cumulative pore volume of samples synthesised at $SiO_2/Al_2O_3 = \infty$.

When $TPAOH/SiO_2 = 0.09$ has been used a high gelation rate is observed and mesoporous MSA-type materials have been obtained. On the contrary, when $TPAOH/SiO_2 = 0.2$ has been used a low gelation rate is observed and microporous ERS-8-type materials have been obtained with molar ratios $SiO_2/Al_2O_3 = 300$ and ∞.

At $SiO_2/Al_2O_3 = 100$ a borderline situation (micro-mesoporous material) is observed, underlining an effect of aluminum content.

Perhaps, the presence of a heteroatom, as aluminum, disturbs the chain propagation during polymerisation so the cross-linking is favoured and this explains the borderline situation observed at $TPAOH/SiO_2 = 0.2$ and $SiO_2/Al_2O_3 = 100$.

Therefore, TPAOH plays several roles: it is the catalyst of alkoxide reactions (e.g. hydrolysis, alcoholysis and polymerisation/depolymerisation) and the gelling agent.

Besides, it can act as a templating agent, organising SiO_2 and AlO_2^- units in polymeric structure and addressing the pore formation through the TPA^+ clusters. As reported [13] the relative amount of TPA^+ with respect to silico-aluminate oligomers affects the formation of MSA or ERS-8, as shown in Fig. 9. The three-dimensional MSA is obtained when silico-aluminate oligomers exceed the TPA^+ solvated clusters that are completely surrounded. ERS-8 is obtained when TPA^+ solvated clusters exceed silico-aluminate oligomers that grow as sheets, indeed they can not surround the clusters.

In the sol-gel synthesis water and alcohol play the role of reactants and solvents. As reactants they are able to react with alkoxides by hydrolysis and alcoholysis. Alkoxides with groups having different hydrolysis rate are generated by alcoholysis and this gives rise to a very complex system. As solvents they constitute the solvation sphere of TPA^+ cluster affecting its dimension. That explains as the increase of aliphatic alcohol chain length gives rise to an increase in the pore volume and mean pore size. However, as the alcohol was varied only at low $TPAOH/SiO_2$ molar ratio, MSA-type materials was always obtained, even when an alcohol larger than ethanol was used (e.g. propanol).

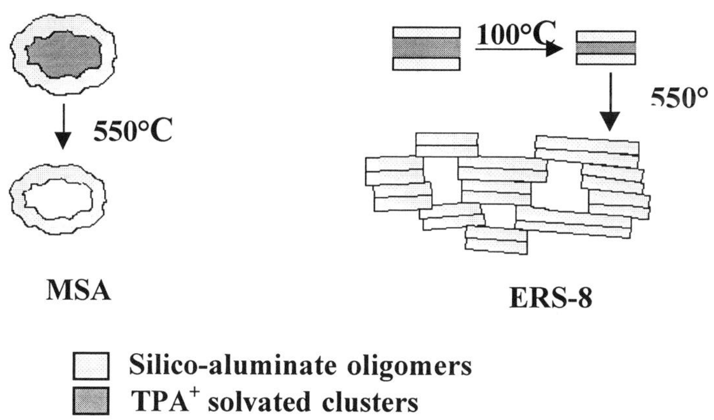

Fig. 9. Proposed formation mechanism for MSA and ERS-8 families.

This seems to indicate that also when propanol is used for all SiO_2/Al_2O_3 molar ratio the silico-aluminate oligomers are able to completely surround the TPA^+clusters according to the mechanism of MSA formation. As already reported [16] a further increasing of the alcoholic aliphatic chain ($n-C_5H_{11}OH$ or $n-C_6H_{13}OH$), do not permit a complete surrounding by aluminosilicates, due to the steric hindrance of the solvated clusters. That gives rise to ERS-8 formation with a decrease of pore size and pore volume.

5. CONCLUSIONS

The textural properties are strongly affected by the $TPAOH/SiO_2$ molar ratio due to the different gelation mechanisms.

The threshold between the range of existence of MSA and ERS-8 is affected not only by TPAOH content but also by SiO_2/Al_2O_3 molar ratio.

The alcoholic solvent strongly influences the textural properties of products: as the aliphatic chain of the alcohol increases the specific surface area, the pore volume and the mean pore size increase for all SiO_2/Al_2O_3.

REFERENCES

1. R.M. Dessau, J.L. Schlenker and J.B. Higgins, Zeolites, 10, (1990), 522.
2. M. Estermann, L. B. Mcusker, Ch. Baerlocher, A. Merrouche and H. Kessler, Nature, 352 (1991) 320.
3. M.E. Davis, C. Saldarriaga, C. Montes, J. Garces and C. Croweder, Nature, 331 (1988) 698.
4. P. Wagner, M. Yoshikawa, M. Lovallo, K. Tsuji, M. Taspatsis and M. Davis, Chem. Comm., (1997) 2179.
5. T. Wessels, C. Baerlocher, L.B. McCusker and E.J. Creyghton, J. Am. Chem. Soc., 121 (1999) 6242.
6. J. S. Beck, J.C. Vartuli, W.J. Roth, M.E. Leonoswicz, C.T. Kresge, K.D. Schmitt, C.T-W. Chu, D.H. Olson, E.W. Sheppard, S.B. McCullen, J.B. Higgins and J.L. Schlenker, J. Am. Chem. Soc., 114 (1992) 10834.
7. S. Inagaki, Y. Fukushima and K. Kuroda, J. Chem. Soc., Chem. Comm., (1993) 680.
8. A. Tuel and S. Gontier, Chem. Mater., 8 (1996) 114.
9. Q. Huo, D.I. Margolese and G.D. Stucky, Chem. Mater., 8 (1996) 1147.
10. S.A. Bagshaw, E. Prouzet and T.J. Pinnavaia, Science, 269 (1995) 1242.
11. R. Ryoo, J.M. Kim, C.H. Shin and J.Y. Lee, Studies in Surface Science and Catalysis, 105 (1997) 45.
12. G. Bellussi, C. Perego, A. Carati, S. Peratello, E. Previde Massara and G. Perego, Studies in Surface Science and Catalysis, 84 (1994) 85.
13. G. Perego, R. Millini, C. Perego, A. Carati, G. Pazzuconi and G. Bellussi, Studies in Surface Science and Catalysis, 105 (1997) 205.
14. C. Rizzo, A. Carati, M. Tagliabue and C. Perego, Studies in Surface Science and Catalysis, 128 (2000) 613.

15. B. Handy, K.L. Walther, A. Wokaun and A. Baiker, Studies in Surface Science and Catalysis, 63 (1991) 239.
16. A. Carati, C. Rizzo, M. Tagliabue, C. Perego, Studies in Surface Science and Catalysis, 130 B (2000) 1085.

Studies in Surface Science and Catalysis 140
A. Gamba, C. Colella and S. Coluccia (Editors)

The Reductive Activation of Molecular Nitrogen on "electron-rich" MgO: Details on the Structure of the Adsorption Site via the N_2^--OH Superhyperfine Interaction.

Z. Sojka [(1)], M. Chiesa [(2)], M. C. Paganini [(2)], E. Giamello [(2)]

[(1)] *Department of Chemistry, Jagellonian University, ul. Ingardena, 3, Cracow, Poland*
[(2)] *Dipartimento di Chimica IFM, Università di Torino, e Unità INFM di Torino, Via Giuria, 7 - 10125 Torino, Italia;*

1. Introduction

The work illustrated in the present paper provides a comprehensive experimental description on the recently discovered reductive activation of the N_2 molecule on basic oxides. The solid systems involved, MgO or CaO containing surface trapped electrons are completely different form the promoted iron catalysts or other heterogeneous systems examined until now in the field of nitrogen fixation [1-4] and lead to the *reversible* reduction of the nitrogen molecule forming an N_2^- radical anion. This radical anion, until now only observed in the bulk of irradiated azides [5-7] or as a temporary negative ion (*resonance*)[8], is stable at the surface of the oxide systems in a limited range of pressure and temperature and is identified on the basis of its EPR spectrum and of the related spin-Hamiltonian parameters. A thorough description of the N_2^- radical ion adsorbed on the *electron-rich* surface of MgO and CaO has been recently reported by us [9,10]. In the present contribution the features of the adsorption site are discussed on the basis of the superhyperfine interaction of the adsorbed radical with a nearby hydroxyl group.

2. Experimental

High surface area (HSA) MgO (*ex Johnson Matthey*) was produced by thermal decomposition of $Mg(OH)_2$ under dynamic vacuum at 523K for 16 h. The HSA MgO was then activated under a vacuum of 10^{-5} Torr at 1073 K for 1h, to obtain a completely dehydrated material. The surface area of the activated oxide was ≈170-200 m^2g^{-1}. The $F_S^+(H)$ or $F_S^+(D)$ colour centres were generated on the surface of the activated oxides as described in detail elsewhere.[11-13] Briefly, hydrogen or deuterium (~100 Torr, 1 Torr = 133 Pa) was added to the activated oxide at 298 K and the powder was subsequently cooled to 77 K. The sample was then irradiated using a low pressure UV mercury vapour lamp for about 1 h. The excess H_2 or D_2 was then slowly evacuated at 298 K from the pale blue coloured sample before recording the EPR spectrum. High purity $^{14}N_2$ and $^{15}N_2$ gases (*ex Cambridge Isotopes)* were used without further purification.

X-band CW-EPR spectra were recorded at 298 K and 77 K on a Varian E-109 and on an a Bruker EMX spectrometer both equipped with a cylindrical cavity operating at a 100 KHz field modulation. Varian pitch (g=2.0028) was used for g value calibration.

3. Results and discussion

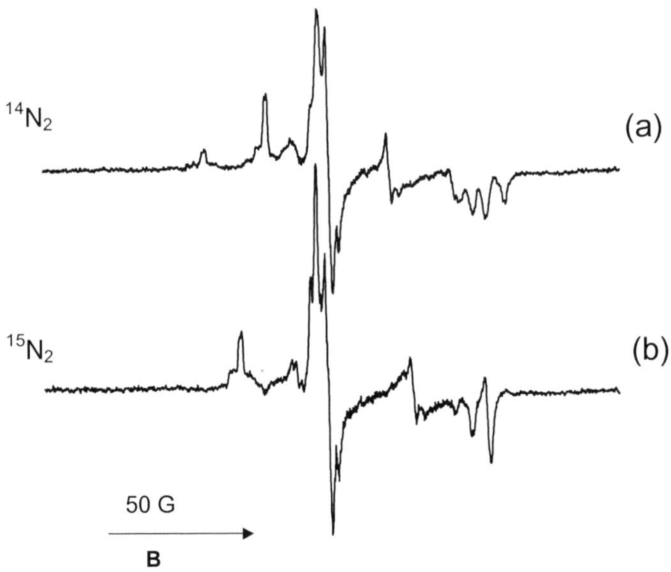

Figure 1: EPR spectra of the (a) $^{14}N_2^-$ and (b) $^{15}N_2^-$ radical anion adsorbed on polycrystalline MgO.

Surface-trapped electrons can be formed at the surface of alkali-earth oxides by different methods, among these methods the UV irradiation of the solid in the presence of hydrogen has been found the most reliable and reproducible and leads to the formation of a particular type of surface colour centre named $F_s^+(H)$ centre[11,13]. The mechanism leading to the formation of these centres implies the heterolytic chemisorption of hydrogen at the surface of activated MgO and the formation of H^+ and H^- ions stabilised onto a couple of low-coordinated O^{2-}-Mg^{2+} ions. Upon UV irradiation a fraction of the hydride H^- ions are ionised and the released electrons stabilised into suitable surface anion vacancies close to the OH^- formed by reaction of H^+ with surface O^{2-} ions. At the end of this process the sample develops a blue colour and exhibits an EPR signal with g values slightly lower than the free spin value

(g_e=2.0023). The $F_s^+(H)$ EPR spectrum is characterised by a hyperfine splitting due to the magnetic interaction between the trapped electron and the nearby hydroxyl protons.

Adsorption of N_2 (100 Torr) at low temperature onto the MgO electron rich surface gives rise to a complex EPR spectrum which has been assigned to a surface N_2^- radical ion [9,10]. The spectrum in Fig. 1a has been interpreted using the following spin-Hamiltonian;

$$\mathcal{H} = \mu_B \boldsymbol{B}\boldsymbol{g}\boldsymbol{S} + \mathrm{S}\boldsymbol{A}\mathrm{I}_a + \mathrm{S}\boldsymbol{A}\mathrm{I}_b$$

The spin Hamiltonian parameters derived by computer simulation are listed in Table 1 and were obtained using a single S = 1/2 species having rhombic $\boldsymbol{g}$ and $\boldsymbol{A}$ tensors, with two g values close to the free spin and one g value lower (1.9677) as expected for the $^2\Pi_{1/2}$ state of an 11-electron radical like N_2^-. The electronic configuration of such a species (isoelectronic with NO, CO^-, O_2^+) is characterised by the presence of a single unpaired electron in the π antibonding orbitals. The EPR spectrum can only be observed for such a radical if the degeneracy of the two antibonding orbitals is lifted by an asymmetric electric field, such as

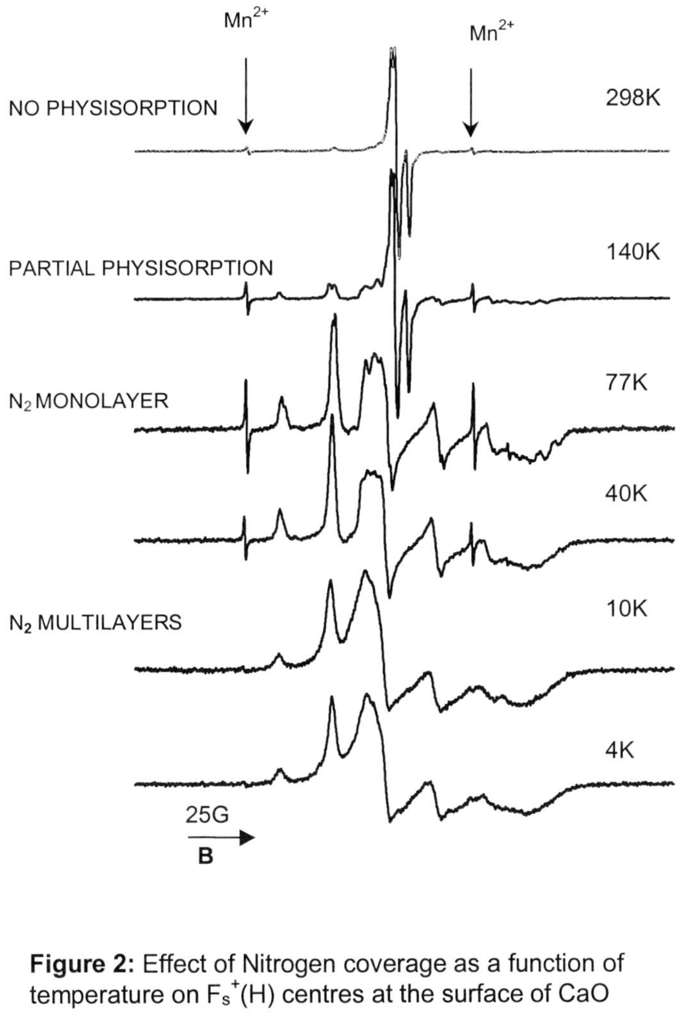

Figure 2: Effect of Nitrogen coverage as a function of temperature on $F_s^+(H)$ centres at the surface of CaO

that exerted by a surface cationic site.

Table 1 Spin-Hamiltonian parameters for N_2^- in different systems

	g_{xx}	g_{yy}	g_{zz}	A_{xx} (MHz)	A_{yy} (MHz)	A_{zz} (MHz)	Ref.
$^{14}N_2^-$ on MgO	2.0018	2.0042	1.9719	8.13	60.31	11.59	(10)
$^{15}N_2^-$ on MgO	2.0018	2.0042	1.9719	11.37	84.43	16.23	(10)
$^{14}N_2^-$ on CaO	2.0006	2.0048	1.9677	7.00	56.12	11.84	(10)
$^{14}N_2^-$ in KN_3	2.0008	2.0027	1.9832	10.64	33.63	11.10	(5)
$^{14}N_2^-$ in NaN_3	≈2.00	≈2.00	1.75	30.82	65.86	9.55	(6)
$^{14}N_2^-$ in $Ba(N_3)_2$	≈1.99	≈1.99	1.979	10.02	57.65	11.36	(7)

The observed hyperfine structure is typical of two equivalent ^{14}N nuclei (I = 1) producing a quintuplets pattern with the intensity ratio of 1:2:3:2:1. The validity of this assignment was confirmed using $^{15}N_2$ (I = 1/2) which produced the expected triplet pattern with a 1:2:1 ratio as shown in Fig.1b. The ^{14}N hyperfine tensor has been carefully analysed in a previous paper [10]. The results indicates that the electron transfer towards the N_2 molecule (mainly in the π_y^* orbital) is about 92%. Formation of N_2^- was found to depend on the nitrogen surface coverage, since the EPR spectrum varied as a function of the N_2 equilibrium pressure. On reducing the N_2 coverage, by either reducing the equilibrium pressure or by raising the temperature, the spectral features of the N_2^- radical vanish and the original spectrum of the $F_s^+(H)$ centres is restored. This effect can be followed in Fig. 2 where the EPR spectra of $F_s^+(H)$ centres under 100 Torr of molecular nitrogen are recorded in the range of temperature between 298K and 4K. A progressive appearance of the N_2^- signal can be observed at the expense of the $F_S^+(H)$ signal as the temperature is lowered. The spectral pattern of the $F_s^+(H)$ centres dominates up to 140 K when nitrogen physisorption starts to occur in appreciable amount. At 77K the N_2^- signal reaches the maximum intensity while at lower temperature only a broadening of the spectral lines is observed.

3.1. Nature of the adsorption site; Analysis of the ^1H superhyperfine (shf) tensor

The surface site responsible for adsorption of N_2^- is the same anionic vacancy or trap from which the electron is initially transferred to the adsorbed N_2 molecule. This statement can be easily and confidently made by considering the complete reversibility of the process during which the unpaired electron is transferred back into the vacancy after N_2 desorption occurs, as the original signal from the electron trapped in the vacancy ($F_S^+(H)$) is restored (Fig. 2). This process could not be reconciled with the idea of a radical spill-over process occurring after electron transfer, since in that case reversed electron transfer from the separated reaction centres on such an ionic surface is difficult if not impossible to explain. This argument is strongly supported also by theoretical calculations at the DFT level which indicates that a small energy barrier separates the unbound state F_s^+/N_2 from the bound one F_s^{2+}/N_2^- [10].The nature of the proposed adsorption site is also supported by a number of extra features which are resolved in the spectrum obtained by N_2 adsorption on $F_S^+(H)$ but absent for the $F_S^+(D)$ centres. These features will be commented in the following.

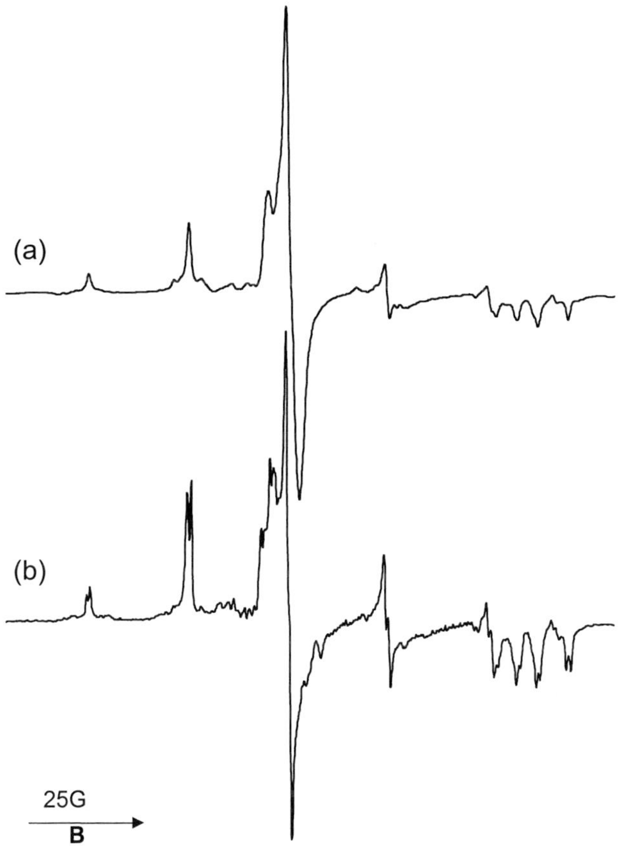

Figure 3: Comparison of the spectra of $^{14}N_2^-$ obtained on (a) MgO containing $F_s^+(D)$ and (b) $F_s^+(H)$ centres respectively

$F_S^+(D)$ centres are analogues to $F_S^+(H)$ centres but obtained by irradiation in D_2 atmosphere. The hyperfine interaction of the trapped electron with the deuteroxyl D nucleus (I=1) leads to an unresolved triplet due to the small magnetic moment of deuterium. The comparison between the N_2^- signal obtained by adsorption of N_2 on $F_S^+(D)$ centres reveals that each of the main lines in the N_2^- spectrum (Fig 3a) are split into doublets in the case of N_2^- formed over the $F_S^+(H)$ centres (Fig 3b). This weak superhyperfine splitting of less than 1 Gauss indicates that a single proton (I = 1/2) interacts with the N_2^- radical, and this proton must originate from the original surface OH group associated with the $F_S^+(H)$ centre. This indicates that the adsorption site for N_2^- is the same assembly of ions constituting the electron trap for the surface electron. The hypothesis is further justified by considering (as mentioned before) the facile and reversible electron transfer from N_2^- back to the surface trap and this seems possible only if the adsorbed molecule and electron trap are close to each other.

The ^1H coupling constants, related to the spectrum shown in Fig. 3b, are $^HA_x = 2$ MHz, $^HA_y = 2.5$ MHz and $^HA_z = 2.8$ MHz and are determined along the g-tensor principal directions (x,y,z) These values, by the way, may not correspond to the principal values of the shf tensor (along the x', y' and z' axes). The small values of the H coupling constants may be accounted for by a weak dipolar through space interaction. Taking into account that the electronic spin is largely localised on the N_2^- radical and that for simple geometrical reasons the proton must be at distance larger than at least 0.2 nm, the ^1H shfs can be adequately rationalised by a point dipole approximation.

Because the actual position of the proton is *a priori* unknown, following an earlier paper[14] at the beginning we have considered two highly symmetrical structures with a collinear arrangement of the N_2^- radical and the proton and a T-shaped structure of isosceles symmetry. Initially the principal axes of the *g* (x,y,z) and ^HA (x',y',z') tensors are assumed to be coincident. For the collinear structure the components of the ^HA tensor are $^HA_i = g_H\beta_n(3\cos^2\theta_i - 1)[1/r^3 + 1/(r+d)^3]$, where *r* is the distance between the N_2^- radical and the proton, *d* is the N-N bond length, while θ_i is the angle of the *r* vector with the corresponding axes. Simple calculations indicate that in such a case the shfs tensor should have the form (-*a*, -*a*, 2*a*), which contradicts the experimental values. For isosceles geometry one has $^HA_{x'} = g_H\beta_n(3\cos^2\theta_{x'} - 1)/r^3$, $^HA_{y'} = = -g_H\beta_n/r^3$ and $^HA_{z'} = = g_H\beta_n(3\cos^2\theta_{z'} - 1)/r^3$, where $\theta_{z'} = \cos^{-1}(2r/d)$ and $\theta_{x'} = \pi/2 - \theta_z$. The calculation has shown that there is no single *r* distance which can reproduce the observed values of the ^HA tensor so also this model has to be rejected. This suggests that a more advanced approach allowing for the principal axes non-coincidence is required. For an arbitrary orientation of the shfs principal axes with respect to the reference frame the Euler transformation matrix is required. However, in the case of a quasi-axial symmetry a single rotation angle relates the two sets of the coordinates. In our calculations we have followed the approach of Attanasio used for the analysis of the frozen solution proton ENDOR spectra[15]. Because the plane of the non coincident axes is *a priori* unknown, we have tested two possibilities corresponding to rotation about z' (non coincidence in x'y' plane) or x' direction (non coincidence in z'y' plane, Fig 4). The appropriate rotation formulae have been proposed by Hoffman[16]. For rotation about x' axis we may write

$$\begin{aligned} {}^HA_{y'} &= [({}^HA^2_y - ({}^HA^2_y + {}^HA^2_z)\sin^2\theta)/(1-2\sin^2\theta)]^{1/2} \\ {}^HA_{z'} &= [({}^HA^2_z - ({}^HA^2_y + {}^HA^2_z)\sin^2\theta)/(1-2\sin^2\theta)]^{1/2} \qquad (1) \\ {}^HA_{x'} &= {}^HA_x \end{aligned}$$

The rotation angle θ may be then calculated as

$$\theta = \arcsin[({}^HA^2_z - {}^HA^2_x)/[({}^HA^2_y + {}^HA^2_z - 2{}^HA^2_x)]^{1/2} \qquad (2)$$

The above method allows derivation of the principal ^1H shf values of the axial tensor $^HA_\parallel = {}^HA_{y'}$ and $^HA_\perp = {}^HA_{z'} = {}^HA_{x'}$. Assumption of the non coincidence in the x'y' plane led us to unacceptable results with $^HA_\parallel < {}^HA_\perp$ and an unrealistically large $a_{iso} = -1.3$ MHz, while following the model for distant proton the latter value should be rather small. In contrast, consistent data were obtained when x' was chosen as the rotation axis. We found in fact the following values: $\theta = 39^\circ$, ${}^{H}A_{\|} = 3.2$ MHz, ${}^{H}A_{\perp} = -1.98$ MHz. The average of the latter values gives the isotropic component $a_{iso} = -0.25$ MHz, whereas the anisotropic part can be used to assess the proton N_2^- radical distance r according to the dipole formula

$$A_D = {}^{H}A_{\|} - a_{iso} = g_e\beta_e g_H\beta_n(3\cos^2\phi - 1)/r^3 \qquad (3)$$

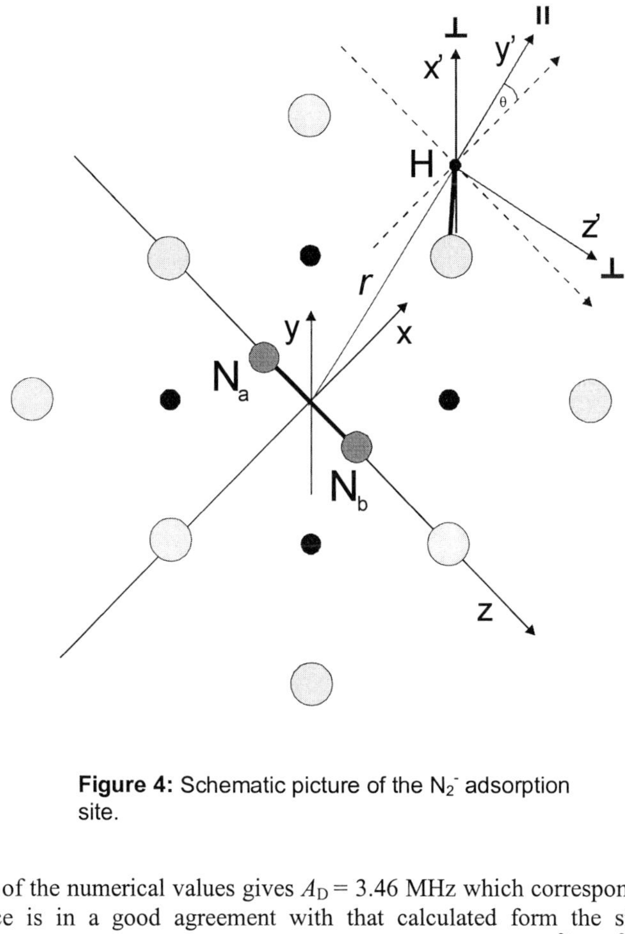

Figure 4: Schematic picture of the N_2^- adsorption site.

Substitution of the numerical values gives $A_D = 3.46$ MHz which corresponds to $r = 0.36$ nm. This distance is in a good agreement with that calculated form the simple geometrical consideration based on the model shown in Fig. 4. Taking the Mg^{2+}—O^{2-} distance equal to 0.212 nm, the OH bond length of 0.096 nm and the N_2^--O-H angle $(\pi-\theta)=140^\circ$, the 'geometrical' dinitrogen radical - proton distance was evaluated to be 0.38 nm. The value of the angle θ indicates that hydroxyl group is located in the closest possible position with respect to the surface cavity where the N_2^- radical is trapped. A good agreement between both values encourages us to assume the structure shown in Fig. 4 as representative for N_2^- - ${}^{1}H$ species.

Acknowledgements

The present work is part of a project coordinated by G. Granozzi and cofinanced by the Italian MURST (Cofin00). The support of INFM through a PRA project (ISADORA) is also gratefully acknowledged.

References

1. Leigh, G. J. *Science* **1998**, 279, 506.
2. Pelikan, P.; Boca, R. *Coord. Chem. Rev.* **1984**, 55, 55.
3. Orme-Johnson, W. H. *Science* **1992**, 257, 1639.
4. Bazhenova, T. A.; Shilov, A. E. *Coord. Chem. Rev.* **1995**, 144, 64.
5. Horst, R. B.; Anderson, J. H.; D. Milligan,E. *J. Phys. Chem. Solids* **1962**, 26, 157.
6. Gelerinter, E.; Silsbee, R. H. *J. Chem. Phys.* **1966**, 45, 1703.
7. Marinkas, P. L.; Bartram, R. H. *J. Chem. Phys.* **1968**, 48, 927.
8. G. J. Schultz, *Rev. Mod. Phys.,* **1973**, 45, 423.
9. Giamello E., Paganini M.C., Chiesa M., Murphy D.M., Pacchioni G., Soave R., Rockembauer A., *J. Phys. Chem. B,* **2000,** 104, 1887.
10. Chiesa M., Giamello E., Paganini M.C., Murphy D.M., Pacchioni G., Sojka Z., *J. Phys. Chem B,* **2001**, 105, 497.
11. Giamello, E.; Paganini, M. C.; Murphy, D.; Ferrari,A. M.; Pacchioni,G. *J. Phys. Chem. B* **1997**, 101, 971.
12. Chiesa, M.; Paganini, M. C.; Giamello, E.; Murphy,D. M. *Langmuir* **1997**, 13, 5306.
13. Paganini, M. C.; Chiesa, M.; Giamello, E.; Martra, G. M.; Coluccia, S.; Murphy, D. M. Pacchioni G., *Surf. Sci.* **1999**, 421, 246.
14. E.Giamello P. Ugliengo, E. Garrone, *J. Chem. Soc., Faraday Trans. I,* **1989**, 85, 1373.
15. D. Attanasio, *J. Phys. Chem.*, **1986**, 90, 4952.
16. B.M. Hoffman, J. Martinsen, R.A. Venters, *J. Mag. Res.*, **1984**, 59, 110.

Studies in Surface Science and Catalysis 140
A. Gamba, C. Colella and S. Coluccia (Editors)

Kinetics of NO Photocatalytic Reduction by CO over MoO_3/SiO_2 Catalysts

I.R. Subbotina[a], B.N. Shelimov[a], M. Che[b♣], and S. Coluccia[c]

[a]Zelinsky Institute of Organic Chemistry, Academy of Sciences, 117913 Moscow, Russia;

[b]Laboratoire de Réactivité de Surface, UMR 7609, CNRS, Université Pierre et Marie Curie, 4, Place Jussieu, 75252 Paris Cedex 05, France;

[c]Dipartimento di Chimica Inorganica, Chimica Fisica e Chimica dei Materiali, Università di Torino, Via Pietro Giuria 7, 10125 Torino, Italy

The kinetics of the photocatalytic reduction of NO by CO to CO_2, N_2O, and N_2 over MoO_3/SiO_2 catalysts at ambient temperature has been studied mass-spectroscopically using ^{13}C labeled carbon monoxide. The kinetic data obtained for CO–NO mixtures of different compositions fit well the proposed redox mechanism, which suggest a paramagnetic complex ($Mo^{6+}...NO^{2-}$) as reaction intermediate. The formation of this complex is proven by EPR experiments.

1. INTRODUCTION

Catalytic elimination of nitric oxides (NOx) admixtures have received much attention because of its importance for the environmental protection [1,2]. UV-irradiation was also used for photocatalytic decomposition of NOx over oxide and zeolite catalysts [3-5].

In our previous publication [6], we have reported for the first time on the photocatalytic reduction of NO by carbon monoxide over a MoO_3/SiO_2 catalyst (2.5 wt % of Mo). The reduction proceeds under UV-irradiation ($\lambda < 360$ nm) at ambient temperature, N_2, N_2O, and CO_2 being the reaction products. A redox reaction mechanism was proposed in [6] consisting of two stages: (*a*) photoinduced reduction of the initial surface Mo^{6+} to Mo^{4+} by CO and (*b*) dark oxidation of Mo^{4+} by NO to yield N_2O and N_2 and to restore Mo^{6+}. Both stages can be accomplished separately or combined in one photocatalytic process. A Mo–nitrosyl paramagnetic complex was suggested to be an intermediate in the photocatalytic NO reduction by CO.

Recently Tsumura et al. [7] have shown that the photocatalytic decomposition of NO to N_2 and CO_2 can proceed in the presence of CO on a related Mo–MCM-41 system. The same reaction scheme as that proposed by us in [6] was adopted for the photocatalytic process on Mo–MCM-41.

The objective of this work was a kinetic study of the photocatalytic NO reduction by CO over MoO_3/SiO_2 catalysts with different Mo loadings to get some quantitative estimates of the process combined with EPR measurements aimed to detect the intermediate Mo–nytrosyl paramagnetic complex.

♣ Institut Universitaire de France

2. EXPERIMENTAL

MoO_3/SiO_2 samples were prepared by two different methods: (1) by impregnation of a commercial KSK-2-5 silica gel (S_{sp} = 300 m^2/g) with $(NH_4)_6Mo_7O_{24}$ aqueous solutions (Mo contents 1, 2.5 and 5 wt. %) and (2) by the grafting method, i.e., by air- and water-free reaction between $MoCl_5$ and hydroxyl groups of the silica support (Mo loading 0.9 wt. %) as described in [8].

Catalyst samples (~0.3 g) were pretreated in a quartz cell connected to an evacuable gas-circulation loop (V = 880 cm^3). The pretreatment was carried out at 750–800° C in a flow oxygen and under vacuum by the procedure described in detail in [6].

Nitric oxide was obtained by solid-state thermal decomposition of a KNO_2–Fe_2O_3–Cr_2O_3 mixture and freed from admixtures as described in [6]. The admixture of N_2O in NO was found mass-spectroscopically to be 2%. ^{13}C-labeled carbon monoxide contained 70 at. % of ^{13}C and was used as received.

As an UV-light source, a high-pressure mercury lamp supplied with a water filter and quartz light collimator was employed. The catalyst samples were irradiated in the quartz cell at ambient temperature in a circulating flow of NO–CO mixtures of different compositions. The composition of the gas phase above the catalyst was continuously monitored during UV-irradiation with a monopole-type MKh-7304 mass-spectrometer. Pressures of gaseous reactants were measured with a capacity-type gauge VDG-1 with an accuracy of about 3%.

EPR spectra were recorded on a Bruker ER200D spectrometer in the X-band range at 77 and 300 K. Magnetic field intensities were measured by a built-in NMR-type magnetometer. g-Values were calculated using a DPPH sample (g = 2.0036) as a reference.

3. RESULTS AND DISCUSSION

3.1. Specific rates of the photocatalytic NO reduction by CO over MoO_3/SiO_2 catalysts

Table 1 summarises the data on specific rates of NO photoreduction by CO over MoO_3/SiO_2 catalysts with different molybdenum loadings prepared by the impregnation or grafting methods. The specific rates were calculated by extrapolating the kinetic curves of NO consumption to a zero irradiation time and are referred either to 1 g of the catalyst or to 1 Mo atom.

Table 1.
Specific rates of NO reduction by CO over MoO_3/SiO_2 catalysts

Mo content, %	CO/NO ratio	Specific rate	
		μmol NO • g(Cat)$^{-1}$ • min^{-1}	molec NO • Mo atom^{-1} • min^{-1} x 10^2
0.9 (*G*)	1 : 1	2.2	2.4
1.0 (*I*)	1 : 1	3.8	3.8
2.5 (*I*)	1 : 1	8.2	3.3
2.5 (*I*)	2 : 1	17.7	7.1
5.0 (*I*)	1 : 1	3.4	0.7

(*G*) is the catalyst prepared by grafting method; (*I*) are the catalysts prepared by impregnation method.

The following conclusions can be drawn from Table 1:

(*a*) The main tendency is that, for a CO : NO = 1 : 1 mixture, the rate of NO reduction referred to 1 Mo atom decreases on going to higher Mo loadings. This implies that the fraction of surface Mo atoms of the catalyst participating in the reaction decreases with increasing Mo loading.

(*b*) The rates of NO decomposition referred to a catalyst weight unit is highest for 2.5% MoO_3/SiO_2.

(*c*) For the same Mo loading, the rate of NO photoreduction is approximately proportional to CO partial pressure in the reaction mixture at a constant NO pressure.

(*d*) Molybdenum loading being equal, the impregnated catalysts are more active than the grafted catalyst.

Based on these findings, 2.5% MoO_3/SiO_2 was selected for a thorough kinetics examination.

3.2. Kinetics of NO photocatalytic reduction by CO over 2.5% MoO_3/SiO_2 for CO-NO mixtures of different compositions

The kinetics of the photocatalytic reaction was investigated using isotopically labeled ^{13}CO. This made it possible to perform full mass-spectroscopic analyses of the complex reaction mixtures consisting of CO, NO, N_2O, N_2, and CO_2. The experiments were carried out in the pressure range of 7.5–15 Torr with CO–NO mixtures of different compositions.

Figure 1 shows the gas-phase composition during UV-irradiation of 2.5% MoO_3/SiO_2 in a circulating CO : NO = 1 : 1 mixture as a function of time. This plot clearly demonstrates the occurrence of photocatalytic NO reduction by CO. At irradiation time shorter than approximately 40 min, the main reaction products are CO_2 and N_2O which are produced in equal amounts. No or very little N_2 is formed until N_2O partial pressure reaches a maximum. N_2O approaches a maximum when most of NO in the gas phase has been already consumed.

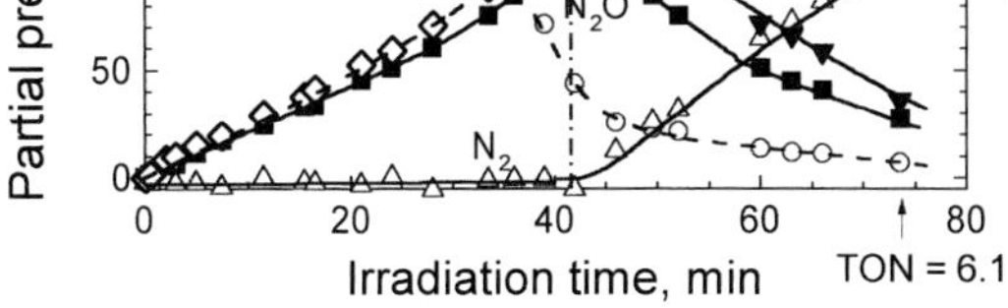

Fig. 1. Kinetics of the photocatalytic reaction over 2.5% MoO_3/SiO_2 in a ^{13}CO : NO = 1 : 1 mixture. 5 Torr ^{13}CO + 5 Torr NO; 0.3 g of the catalyst; $(NO + CO)_{gas}/Mo_{cat}$ = 6.5.

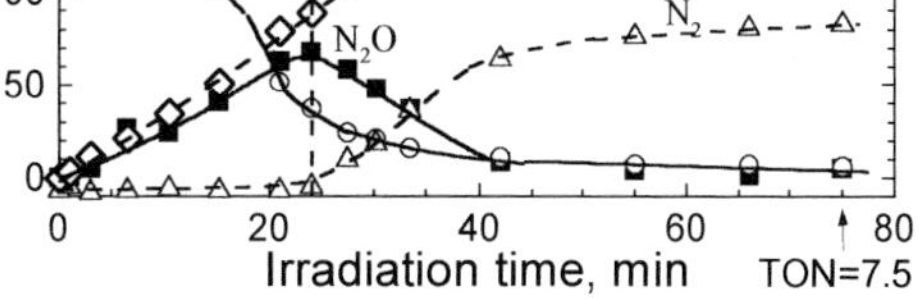

Fig. 2. Kinetics of the photocatalytic reaction over 2.5% MoO_3/SiO_2 in a ^{13}CO : NO = 2 : 1 mixture. 10 Torr ^{13}CO + 5 Torr NO; 0.3 g of the catalyst; $(NO + CO)_{gas}/Mo_{cat}$ = 9.7.

At a longer irradiation time, the N_2O pressure starts to decrease, N_2 appears and is accumulated. The amount of N_2 formed at the end of this run is approximately equal to the amount of N_2O decomposed. This means that the nitrogen is produced via N_2O decomposition. It should be noted that all the kinetic curves are nonlinear.

Thus, the reaction process can be conditionally divided into two parts: the left-hand side of Fig. 1 corresponds to $CO + 2\ NO \rightarrow CO_2 + N_2O$ reaction, whereas the right-hand side corresponds to $CO + N_2O \rightarrow CO_2 + N_2$ reaction.

From the experiment of Fig. 1 we can calculate that, at the end of the run, turnover number (TON), i.e., the number of CO and NO molecules converted over 1 Mo atom of the catalyst sample, is 6.1. Certainly, TON = 6.1 is not a limit value. After removal of the gaseous mixture by brief evacuation at room temperature and admission of a new portion of a CO : NO = 1 : 1 mixture onto the catalyst, the photocatalytic process proceeds with the same rate as during the first irradiation. The same procedure repeated several times did not result in a marked decrease in the photocatalytic activity.

Similar kinetic results were obtained with a CO : NO = 2 : 1 mixture (Fig. 2). The major difference is that the maximum N_2O partial pressure is reached at a shorter irradiation time than in the case of a CO : NO = 1 : 1 mixture (*cf.* Fig. 1). The TON at the end of this run is 7.5.

A different kinetic picture emerged with a CO : NO = 1 : 2 mixture which is a stoichiometric mixture for $CO + 2\ NO \rightarrow CO_2 + N_2O$ reaction (Fig. 3). In contrast to other CO–NO mixtures, all the kinetic curves are linear in this case. No N_2 is formed even under prolonged UV-irradiation. The TON at the end of the run is smaller than those in previous cases, but it is still greater than unity (TON = 2.3).

The kinetic data can be rationalized by the following reaction scheme.

$$\sim Mo^{6+}{=}O^{2-} + h\nu \xrightarrow{(\alpha I)} (\sim Mo^{5+}{-}O^{-})^* \quad (1)$$

$$(\sim Mo^{5+}{-}O^{-})^* \xrightarrow{(k_D)} \sim Mo^{6+}{=}O^{2-} \quad (2)$$

$$(\sim Mo^{5+}{-}O^{-})^* + CO \xrightarrow{(k_R)} Mo^{4+} + CO_2 \quad (3)$$

$$(\sim Mo^{5+}{-}O^{-})^* + NO \xrightarrow{(k_Q)} \sim Mo^{6+}{=}O^{2-} + NO^* \quad (4)$$

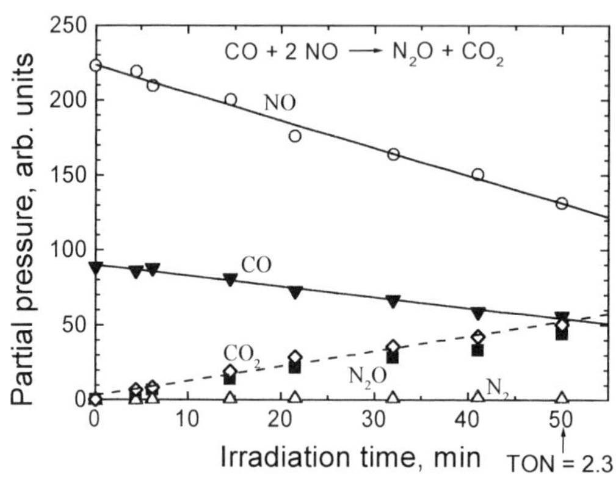

Fig. 3. Kinetics of the photocatalytic reaction over 2.5% MoO_3/SiO_2 in a ^{13}CO : NO = 1 : 2 mixture. 2.6 Torr ^{13}CO + 5.2 Torr NO; 0.3 g of the catalyst; $(NO + CO)_{gas}/Mo_{cat}$ = 4.9.

Here, $\sim Mo^{6+}=O^{2-}$ denotes one of the two molybdenyl bonds of a tetrahedrally coordinated Mo^{6+} ion on the silica surface. $(\sim Mo^{5+}-O^{-})^*$ stands for a short-lived charge-transfer excited state which is formed upon absorption of UV-light quantum ($h\nu$). The excited state was earlier detected in photoluminescence experiments. I is the light intensity, and k_D is the rate constant for deactivation of the excited state. Reaction (3) is the interaction of CO with the excited state to yield Mo^{4+} and CO_2. Reaction (4) is quenching of the excited state $(\sim Mo^{5+}-O^{-})^*$ by NO molecules without formation of reaction products. Reaction (4) is of importance for the kinetic consideration.

Further reactions of the scheme suggest interaction of the photoreduced Mo^{4+} ions with NO which does not require UV-irradiation:

$$\sim Mo^{4+} + NO \xrightarrow{(k_1)} (Mo^{6+}...NO^{2-}) \quad (5)$$

$$(Mo^{6+}...NO^{2-}) + NO \xrightarrow{(k_2)} \sim Mo^{6+}=O^{2-} + N_2O \quad (6)$$

Reaction (5) results in the formation of an intermediate paramagnetic complex $(Mo^{6+}...NO^{2-})$ which arises from a two-electron transfer from Mo^{4+} ion to a NO molecule. This complex has been detected by EPR (see below). Reaction (6) is the interaction of the complex with a second NO molecule to yield N_2O and to restore initial Mo^{6+} species.

Reactions (1)–(6) describe the first part of the kinetic curves. Most likely, the formation of N_2 proceeds via N_2O decomposition on Mo^{4+} produced by reaction (3) (the second part of the kinetic curves):

$$\sim Mo^{4+} + N_2O \xrightarrow{(k_3)} \sim Mo^{6+}=O^{2-} + N_2 \quad (7)$$

A kinetic description of the full reaction scheme including all seven reactions is more difficult, because the system of corresponding differential equations cannot be analytically solved. However, as a first approximation, we can restrict ourselves by considering only reactions (1) to (6) which describe N_2O and CO_2 formation by $CO + 2\ NO \rightarrow CO_2 + N_2O$ reaction. In this case, corresponding kinetic description is applicable either to the full photocatalytic process or to the first part of it (until the N_2O maximum), depending on the composition of initial CO–NO mixtures.

The following set of differential equations emerges from reaction scheme (1)–(6):

$$-d[CO]/dt = d[CO_2]/dt = d[N_2O]/dt = \alpha I k_R[\sim Mo^{6+}=O^{2-}][CO]/(k_R[CO] + k_Q[NO]) \quad (8)$$

$$-d[NO]/dt = 2\ \alpha I k_R[\sim Mo^{6+}=O^{2-}][CO]/ (k_R[CO] + k_Q[NO]) \quad (9)$$

under the assumption that the deactivation rate constant k_D is much smaller than k_R and k_Q and that $(\sim Mo^{5+}-O^{-})^*$, Mo^{4+}, and $(Mo^{6+}...NO^{2-})$ being intermediate species are at steady-state concentrations under UV-irradiation.

It is worth noting that: (*a*) the rate of NO consumption is as twice as high as those of CO consumption and CO_2 and N_2O formation; (*b*) neither k_1 nor k_2 enter equations (8) and (9).

Integration of equations (8) and (9) gives a set of interrelated equations (10)–(12):

$$(1 + 2\ K)(1 - x) - K\ (\gamma - 2)\ \ln x = \alpha I\ [\sim Mo^{6+}=O^{2-}]\ t/\ [CO]_0 \quad (10)$$

$$(0.5 + K)(\gamma - x_1) - K\ (\gamma - 2)\ \ln\ [0.5(x_1 - \gamma) + 1] = \alpha I\ [\sim Mo^{6+}=O^{2-}]\ t\ /\ [CO]_0 \quad (11)$$

$$(1 + 2\ K)\ x_2 - K\ (\gamma - 2)\ \ln\ (1 - x_2) = \alpha I\ [\sim Mo^{6+}=O^{2-}]\ t\ /\ [CO]_0 \quad (12)$$

where $\gamma = [NO]_0/[CO]_0$, $x = [CO]/[CO]_0$, $x_1 = [NO]/[CO]_0$, and $x_2 = [CO_2]/[CO]_0 = [N_2O]/[CO]_0$ are dimensionless concentrations and $K = k_Q\ /\ k_R$. Index '0' refers to the initial concentrations.

Equations (10)–(12) can be further reduced to equations (13)–(15) assuming that $2K >> 1$, i.e., that NO is a stronger quencher than CO. This assumption was found to be valid from the results of photoluminescence experiments which will be reported elsewhere.

$$2(1 - x) - (\gamma - 2)\ \ln x = \alpha I\ [\sim Mo^{6+}=O^{2-}]\ t\ /\ K\ [CO]_0 \quad (13)$$

$$\gamma - x_1 - (\gamma - 2)\ \ln[0.5(x_1 - \gamma) + 1] = \alpha I\ [\sim Mo^{6+}=O^{2-}]\ t\ /\ K\ [CO]_0 \quad (14)$$

$$2x_2 - (\gamma-2)\ln(1-x_2) = \alpha I\,[\sim Mo^{6+}=O^{2-}]\;t\,/\,K\,[CO]_0 \qquad (15)$$

These equations allow us to find the dependence of dimensionless concentrations of the reactants and products on irradiation time t.

Some examples of the fitting of equations (13)–(15) to the experimental data of Figs. 1–3 are demonstrated in Figs. 4–6. The points are experimental data, and the curves are calculated by equations (13)–(15) using a computer best fit procedure. Satisfactorily good agreement between the experimental kinetic data and the calculations have been obtained for different compositions of NO–CO mixtures.

Equations (13)–(15) correctly describe the following characteristic features of the kinetic data obtained:

(*i*) For a *stoichiometric* CO : NO = 1 : 2 mixture ($\gamma = 2$), equations (13)–(15) predict a *linear* dependence of CO, NO, N_2O, and CO_2 partial pressures vs. time of UV-irradiation (Fig. 3). For *nonstoichiometric* CO–NO mixtures ($\gamma \neq 2$), the corresponding kinetic curves are *nonlinear* (Figs. 1 and 2).

(*ii*) Taking into account the above assumption that $K = k_Q / k_R >> 1$, equation (8) can be reduced to the following equation:

$$-d[CO]/dt = d[CO_2]/dt = d[N_2O]/dt = \alpha I k_R[\sim Mo^{6+}=O^{2-}][CO]/\,k_Q[NO] \qquad (16)$$

A linear dependence of the initial rates of CO consumption on the initial CO partial pressure at a constant initial NO partial pressure as follows from equation (16) is shown in Fig. 7.

Thus, we can conclude that the proposed reaction mechanism describes adequately the kinetics of NO photoreduction by CO over Mo/SiO_2 catalysts.

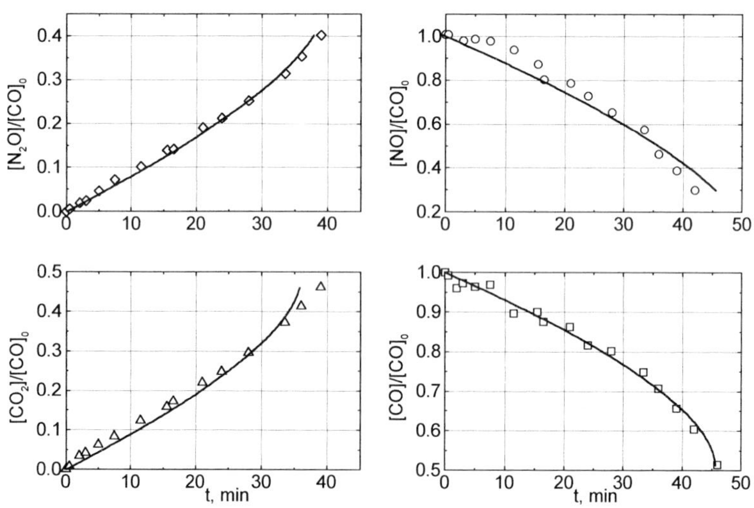

Fig. 4. Analysis of the experimental data for a ^{13}CO : NO = 1 : 1 mixture. Points are experimental data, curves are calculated by equations (13)–(15) at $\gamma = 1$.

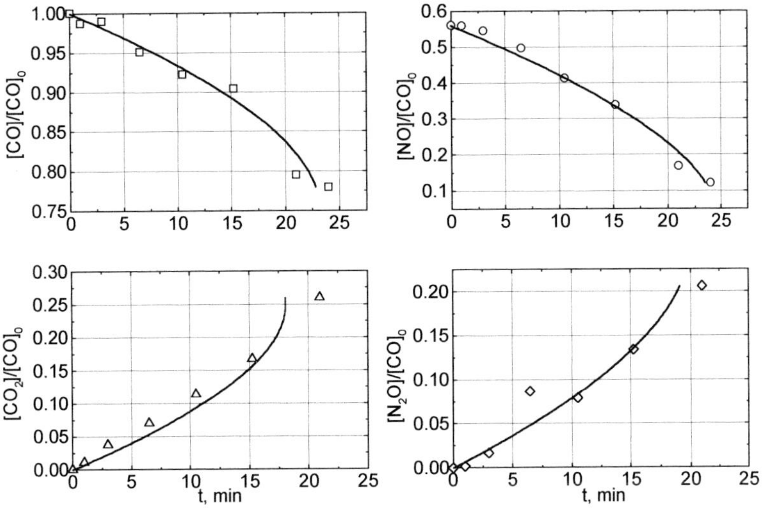

Fig. 5. Analysis of the experimental data for a ^{13}CO : NO = 2 : 1 mixture. Points are experimental data, curves are calculated by equations (13)–(15) at $\gamma = 0.5$.

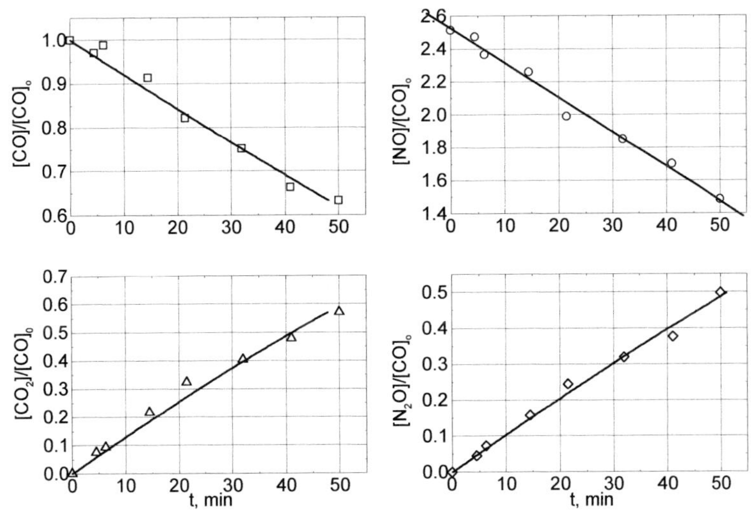

Fig. 6. Analysis of the experimental data for a ^{13}CO : NO = 1 : 2 mixture. Points are experimental data, curves are calculated by equations (13)–(15) at $\gamma = 2$.

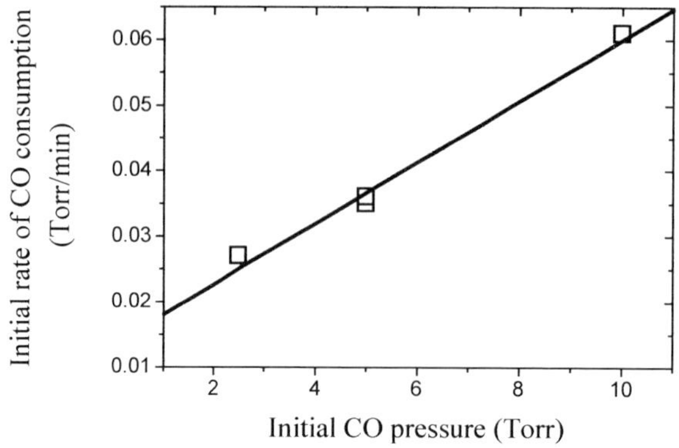

Fig. 7. Initial rates of CO consumption vs. initial CO partial pressure under UV-irradiation of 2.5% MoO_3/SiO_2 in NO–CO mixtures. The initial NO pressure was 5 Torr in all runs.

3.3. EPR study of intermediate (Mo^{6+}...NO^{2-}) complex

The reaction mechanism of the NO photocatalytic reduction by CO discussed above suggests that N_2O is formed via intermediate paramagnetic complexes (Mo^{6+}...NO^{2-}) by reactions (5) and (6).

We attempted to directly detect complexes (Mo^{6+}...NO^{2-}) by EPR. When a 2.5% MoO_3/SiO_2 catalyst sample (~0.1 g) was UV-irradiated in the presence of a CO : NO = 5 : 1 mixture in an EPR sample tube for 10 min at 300 K, EPR spectrum shown in Fig. 8a was recorded.

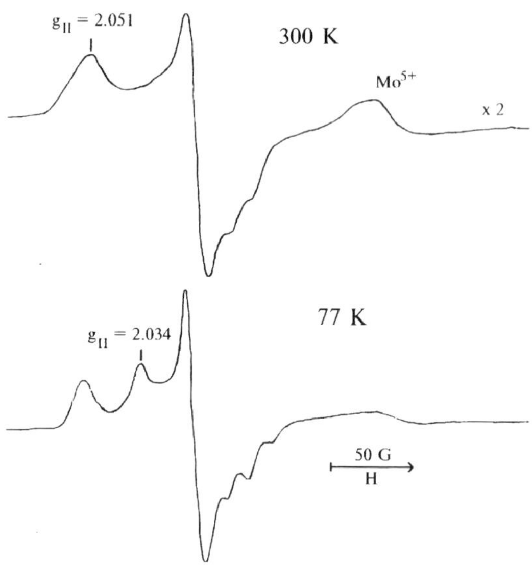

Fig. 8. EPR spectrum of (Mo^{6+}...NO^{2-}) complex. UV-irradiation of 2.5% MoO_3/SiO_2 in a CO : NO = 5 : 1 mixture for 10 min at 300 K.

At 300 K, in addition to a broad Mo^{5+} signal with g < 2, a signal with $g_{\parallel} = 2.051$, $g_{\perp} = 1.996$, and a hyperfine splitting of $g_{\perp}$ with $A^{N}_{\perp} = 15$ G due to interaction of unpaired electron with a nitrogen nucleus appears. The latter signal was earlier assigned to (Mo^{6+}...NO^{2-}) [9].

The parameters of the signal depend on the recording temperature: a new $g_{\parallel} = 2.034$ line appears at 77 K (Fig. 8b). Probably, one or several CO molecules enter the molybdenum ion coordination sphere thus affecting the symmetry and strength of the crystal field. This change in the line shape is completely reversible: when going back to 300 K, the original signal is restored.

The signal is relatively stable at 300K and does not change considerably after switching off the UV-irradiation (at least, for several hours), but it rapidly disappears when more NO was admitted onto the sample (reaction (6)). This supports our suggestion that (Mo^{6+}...NO^{2-}) is an intermediate.

Another evidence that (Mo^{6+}...NO^{2-}) is an intermediate in the photocatalytic reaction is the dependence of the EPR signal intensity on the amount of NO admitted to a preliminary photoreduced catalyst sample (Fig. 9). In these experiments, 2.5% MoO_3/SiO_2 was first photoreduced in CO and evacuated at 423 K and then exposed to successive small portions of NO. Absolute concentrations of (Mo^{6+}...NO^{2-}), which result from NO interaction with the photoreduced Mo^{4+} (reaction (5)), were measured by EPR after admitting each portion of NO at 300 K ('NO titration'). As seen from Fig. 9, the intensity of (Mo^{6+}...NO^{2-}) signal passes through a maximum. Such a behavior is quite typical of a consecutive process described by reactions (5) and (6). Note that the maximum (Mo^{6+}...NO^{2-}) concentration ($\sim 2.5 \cdot 10^{18}$ g^{-1}) is only about 1/10th of the number of NO molecules consumed by the photoreduced sample ($\sim 20 \cdot 10^{18}$ g^{-1}). This implies that the contribution of (Mo^{6+}...NO^{2-}) to the overall mass balance is negligibly small.

In conclusion, it appears that the proposed redox reaction mechanism agrees well the kinetic and EPR data obtained in this work. The results of a kinetic study of $N_2O + CO \rightarrow N_2 + CO_2$ photocatalytic reaction over MoO_3/SiO_2 as well as those of photoluminescence experiments in the presence of different quenchers will be reported in forthcoming publications.

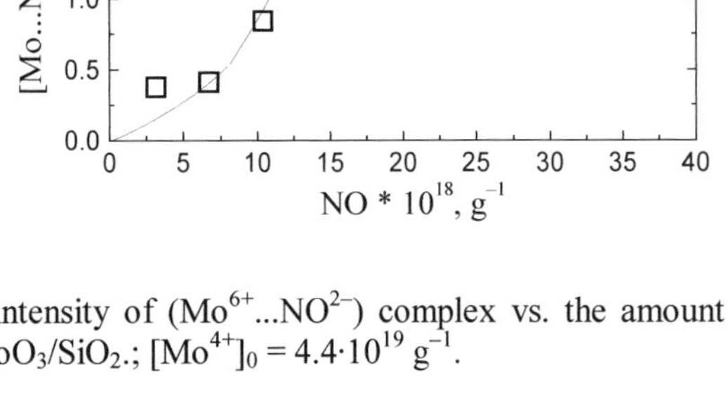

Fig. 9. EPR signal intensity of (Mo^{6+}...NO^{2-}) complex vs. the amount of NO adsorbed on photoreduced 1% MoO_3/SiO_2.; $[Mo^{4+}]_0 = 4.4 \cdot 10^{19}$ g^{-1}.

ACKNOWLEDGMENT
The authors wish to thank INTAS for supporting this work under grant 96-1408.

REFERENCES

1. V.I. Pârvulescu, P. Grange, and B. Delmon, Catalysis Today, 46 (1998) 233.
2. G. Busca, L. Lietti, G. Ramis, and F. Berti, Applied Catal. B: Environmental, 18 (1998) 1.
3. M. Anpo, E. Matsuda, and H. Yamashita, Catalysis Today, 35, (1997) 177.
4. A. Kudo and T. Sakata, Che. Lett. (1992) 2381.
5. N.W. Cant and J.R. Cole, J. Catal., 134 (1992) 317.
6. I.R. Subbotina, B.N. Shelimov, V.B. Kazansky, A.A. Lisachenko, M. Che, and S. Coluccia, J. Cat., 184 (1999) 390.
7. R. Tsumura, S. Higashimoto, M. Matsuoka, H. Yamashita, M. Che, and M. Anpo, Catal. Lett., 68 (2000) 101.
8. M. Anpo, M. Kondo, Y. Kubokawa, C. Louis, and M. Che, J. Chem. Soc. Faraday Trans. 1, 84 (1988) 2771.
9. A.N. Pershin, B.N. Shelimov, and V.B. Kazansky, Kinet. Katal. 21 (1980) 753.

Author Index

STUDIES IN SURFACE SCIENCE AND CATALYSIS

Volume 1 **Preparation of Catalysts I.** Scientific Bases for the Preparation of Heterogeneous Catalysts. Proceedings of the First International Symposium, Brussels, October 14–17, 1975
edited by **B. Delmon, P.A. Jacobs and G. Poncelet**

Volume 2 **The Control of the Reactivity of Solids.** A Critical Survey of the Factors that Influence the Reactivity of Solids, with Special Emphasis on the Control of the Chemical Processes in Relation to Practical Applications
by **V.V. Boldyrev, M. Bulens and B. Delmon**

Volume 3 **Preparation of Catalysts II.** Scientific Bases for the Preparation of Heterogeneous Catalysts. Proceedings of the Second International Symposium, Louvain-la-Neuve, September 4–7, 1978
edited by **B. Delmon, P. Grange, P. Jacobs and G. Poncelet**

Volume 4 **Growth and Properties of Metal Clusters.** Applications to Catalysis and the Photographic Process. Proceedings of the 32nd International Meeting of the Société de Chimie Physique, Villeurbanne, September 24–28, 1979
edited by **J. Bourdon**

Volume 5 **Catalysis by Zeolites.** Proceedings of an International Symposium, Ecully (Lyon), September 9–11, 1980
edited by **B. Imelik, C. Naccache, Y. Ben Taarit, J.C. Vedrine, G. Coudurier and H. Praliaud**

Volume 6 **Catalyst Deactivation.** Proceedings of an International Symposium, Antwerp, October 13–15, 1980
edited by **B. Delmon and G.F. Froment**

Volume 7 **New Horizons in Catalysis.** Proceedings of the 7th International Congress on Catalysis, Tokyo, June 30–July 4, 1980. Parts A and B
edited by **T. Seiyama and K. Tanabe**

Volume 8 **Catalysis by Supported Complexes**
by **Yu.I. Yermakov, B.N. Kuznetsov and V.A. Zakharov**

Volume 9 **Physics of Solid Surfaces.** Proceedings of a Symposium, Bechyňe, September 29–October 3, 1980
edited by **M. Láznička**

Volume 10 **Adsorption at the Gas–Solid and Liquid–Solid Interface.** Proceedings of an International Symposium, Aix-en-Provence, September 21–23, 1981
edited by **J. Rouquerol and K.S.W. Sing**

Volume 11 **Metal-Support and Metal-Additive Effects in Catalysis.** Proceedings of an International Symposium, Ecully (Lyon), September 14–16, 1982
edited by **B. Imelik, C. Naccache, G. Coudurier, H. Praliaud, P. Meriaudeau, P. Gallezot, G.A. Martin and J.C. Vedrine**

Volume 12 **Metal Microstructures in Zeolites.** Preparation - Properties – Applications. Proceedings of a Workshop, Bremen, September 22–24, 1982
edited by **P.A. Jacobs, N.I. Jaeger, P. Jirů and G. Schulz-Ekloff**

Volume 13 **Adsorption on Metal Surfaces.** An Integrated Approach
edited by **J. Bénard**

Volume 14 **Vibrations at Surfaces.** Proceedings of the Third International Conference, Asilomar, CA, September 1–4, 1982
edited by **C.R. Brundle and H. Morawitz**

Volume 15 **Heterogeneous Catalytic Reactions Involving Molecular Oxygen**
by **G.I. Golodets**

Volume 71 **Catalysis and Automotive Pollution Control II.** Proceedings of the 2nd International Symposium (CAPoC 2), Brussels, Belgium, September 10–13, 1990
edited by **A. Crucq**

Volume 72 **New Developments in Selective Oxidation by Heterogeneous Catalysis.** Proceedings of the 3rd European Workshop Meeting on New Developments in Selective Oxidation by Heterogeneous Catalysis, Louvain-la-Neuve, Belgium, April 8–10, 1991
edited by **P. Ruiz and B. Delmon**

Volume 73 **Progress in Catalysis.** Proceedings of the 12th Canadian Symposium on Catalysis, Banff, Alberta, Canada, May 25–28, 1992
edited by **K.J. Smith and E.C. Sanford**

Volume 74 **Angle-Resolved Photoemission. Theory and Current Applications**
edited by **S.D. Kevan**

Volume 75 **New Frontiers in Catalysis, Parts A-C.** Proceedings of the 10th International Congress on Catalysis, Budapest, Hungary, 19–24 July, 1992
edited by **L. Guczi, F. Solymosi and P. Tétényi**

Volume 76 **Fluid Catalytic Cracking: Science and Technology**
edited by **J.S. Magee and M.M. Mitchell, Jr.**

Volume 77 **New Aspects of Spillover Effect in Catalysis. For Development of Highly Active Catalysts.** Proceedings of the Third International Conference on Spillover, Kyoto, Japan, August 17–20, 1993
edited by **T. Inui, K. Fujimoto, T. Uchijima and M. Masai**

Volume 78 **Heterogeneous Catalysis and Fine Chemicals III.** Proceedings of the 3rd International Symposium, Poitiers, April 5 - 8, 1993
edited by **M. Guisnet, J. Barbier, J. Barrault, C. Bouchoule, D. Duprez, G. Pérot and C. Montassier**

Volume 79 **Catalysis: An Integrated Approach to Homogeneous, Heterogeneous and Industrial Catalysis**
edited by **J.A. Moulijn, P.W.N.M. van Leeuwen and R.A. van Santen**

Volume 80 **Fundamentals of Adsorption.** Proceedings of the Fourth International Conference on Fundamentals of Adsorption, Kyoto, Japan, May 17–22, 1992
edited by **M. Suzuki**

Volume 81 **Natural Gas Conversion II.** Proceedings of the Third Natural Gas Conversion Symposium, Sydney, July 4–9, 1993
edited by **H.E. Curry-Hyde and R.F. Howe**

Volume 82 **New Developments in Selective Oxidation II.** Proceedings of the Second World Congress and Fourth European Workshop Meeting, Benalmádena, Spain, September 20–24, 1993
edited by **V. Cortés Corberán and S. Vic Bellón**

Volume 83 **Zeolites and Microporous Crystals.** Proceedings of the International Symposium on Zeolites and Microporous Crystals, Nagoya, Japan, August 22–25, 1993
edited by **T. Hattori and T. Yashima**

Volume 84 **Zeolites and Related Microporous Materials: State of the Art 1994.** Proceedings of the 10th International Zeolite Conference, Garmisch-Partenkirchen, Germany, July 17–22, 1994
edited by **J. Weitkamp, H.G. Karge, H. Pfeifer and W. Hölderich**

Volume 85 **Advanced Zeolite Science and Applications**
edited by **J.C. Jansen, M. Stöcker, H.G. Karge and J. Weitkamp**

Volume 86 **Oscillating Heterogeneous Catalytic Systems**
by **M.M. Slin'ko and N.I. Jaeger**

Volume 87 **Characterization of Porous Solids III.** Proceedings of the IUPAC Symposium (COPS III), Marseille, France, May 9–12, 1993
edited by **J. Rouquerol, F. Rodriguez-Reinoso, K.S.W. Sing and K.K. Unger**

Volume 88 **Catalyst Deactivation 1994.** Proceedings of the 6th International Symposium, Ostend, Belgium, October 3–5, 1994
edited by **B. Delmon and G.F. Froment**

Volume 89 **Catalyst Design for Tailor-made Polyolefins.** Proceedings of the International Symposium on Catalyst Design for Tailor-made Polyolefins, Kanazawa, Japan, March 10–12, 1994
edited by **K. Soga and M. Terano**

Volume 90 **Acid-Base Catalysis II.** Proceedings of the International Symposium on Acid-Base Catalysis II, Sapporo, Japan, December 2–4, 1993
edited by **H. Hattori, M. Misono and Y. Ono**

Volume 91 **Preparation of Catalysts VI.** Scientific Bases for the Preparation of Heterogeneous Catalysts. Proceedings of the Sixth International Symposium, Louvain-La-Neuve, September 5–8, 1994
edited by **G. Poncelet, J. Martens, B. Delmon, P.A. Jacobs and P. Grange**

Volume 92 **Science and Technology in Catalysis 1994.** Proceedings of the Second Tokyo Conference on Advanced Catalytic Science and Technology, Tokyo, August 21–26, 1994
edited by **Y. Izumi, H. Arai and M. Iwamoto**

Volume 93 **Characterization and Chemical Modification of the Silica Surface**
by **E.F. Vansant, P. Van Der Voort and K.C. Vrancken**

Volume 94 **Catalysis by Microporous Materials.** Proceedings of ZEOCAT'95, Szombathely, Hungary, July 9–13, 1995
edited by **H.K. Beyer, H.G. Karge, I. Kiricsi and J.B. Nagy**

Volume 95 **Catalysis by Metals and Alloys**
by **V. Ponec and G.C. Bond**

Volume 96 **Catalysis and Automotive Pollution Control III.** Proceedings of the Third International Symposium (CAPoC3), Brussels, Belgium, April 20–22, 1994
edited by **A. Frennet and J.-M. Bastin**

Volume 97 **Zeolites: A Refined Tool for Designing Catalytic Sites.** Proceedings of the International Symposium, Québec, Canada, October 15–20, 1995
edited by **L. Bonneviot and S. Kaliaguine**

Volume 98 **Zeolite Science 1994: Recent Progress and Discussions.** Supplementary Materials to the 10th International Zeolite Conference, Garmisch-Partenkirchen, Germany, July 17–22, 1994
edited by **H.G. Karge and J. Weitkamp**

Volume 99 **Adsorption on New and Modified Inorganic Sorbents**
edited by **A. Dąbrowski and V.A. Tertykh**

Volume 100 **Catalysts in Petroleum Refining and Petrochemical Industries 1995.**
Proceedings of the 2nd International Conference on Catalysts in Petroleum Refining and Petrochemical Industries, Kuwait, April 22-26, 1995
edited by **M. Absi-Halabi, J. Beshara, H. Qabazard and A. Stanislaus**

Volume 101 **11th International Congress on Catalysis - 40th Anniversary.**
Proceedings of the 11th ICC, Baltimore, MD, USA, June 30-July 5, 1996
edited by **J.W. Hightower, W.N. Delgass, E. Iglesia and A.T. Bell**

Volume 102 **Recent Advances and New Horizons in Zeolite Science and Technology**
edited by **H. Chon, S.I. Woo and S.-E. Park**

Volume 103 **Semiconductor Nanoclusters - Physical, Chemical, and Catalytic Aspects**
edited by **P.V. Kamat and D. Meisel**

Volume 104 **Equilibria and Dynamics of Gas Adsorption on Heterogeneous Solid Surfaces**
edited by **W. Rudziński, W.A. Steele and G. Zgrablich**

Volume 105 **Progress in Zeolite and Microporous Materials**
Proceedings of the 11th International Zeolite Conference, Seoul, Korea, August 12-17, 1996
edited by **H. Chon, S.-K. Ihm and Y.S. Uh**

Volume 106 **Hydrotreatment and Hydrocracking of Oil Fractions**
Proceedings of the 1st International Symposium / 6th European Workshop, Oostende, Belgium, February 17-19, 1997
edited by **G.F. Froment, B. Delmon and P. Grange**

Volume 107 **Natural Gas Conversion IV**
Proceedings of the 4th International Natural Gas Conversion Symposium, Kruger Park, South Africa, November 19-23, 1995
edited by **M. de Pontes, R.L. Espinoza, C.P. Nicolaides, J.H. Scholtz and M.S. Scurrell**

Volume 108 **Heterogeneous Catalysis and Fine Chemicals IV**
Proceedings of the 4th International Symposium on Heterogeneous Catalysis and Fine Chemicals, Basel, Switzerland, September 8-12, 1996
edited by **H.U. Blaser, A. Baiker and R. Prins**

Volume 109 **Dynamics of Surfaces and Reaction Kinetics in Heterogeneous Catalysis.**
Proceedings of the International Symposium, Antwerp, Belgium, September 15-17, 1997
edited by **G.F. Froment and K.C. Waugh**

Volume 110 **Third World Congress on Oxidation Catalysis.**
Proceedings of the Third World Congress on Oxidation Catalysis, San Diego, CA, U.S.A., 21-26 September 1997
edited by **R.K. Grasselli, S.T. Oyama, A.M. Gaffney and J.E. Lyons**

Volume 111 **Catalyst Deactivation 1997.**
Proceedings of the 7th International Symposium, Cancun, Mexico, October 5-8, 1997
edited by **C.H. Bartholomew and G.A. Fuentes**

Volume 112 **Spillover and Migration of Surface Species on Catalysts.**
Proceedings of the 4th International Conference on Spillover, Dalian, China, September 15-18, 1997
edited by **Can Li and Qin Xin**

Volume 113 **Recent Advances in Basic and Applied Aspects of Industrial Catalysis.**
Proceedings of the 13th National Symposium and Silver Jubilee Symposium of Catalysis of India, Dehradun, India, April 2-4, 1997
edited by **T.S.R. Prasada Rao and G. Murali Dhar**

Volume 114 **Advances in Chemical Conversions for Mitigating Carbon Dioxide.**
Proceedings of the 4th International Conference on Carbon Dioxide Utilization, Kyoto, Japan, September 7-11, 1997
edited by **T. Inui, M. Anpo, K. Izui, S. Yanagida and T. Yamaguchi**

Volume 115 **Methods for Monitoring and Diagnosing the Efficiency of Catalytic Converters.**
A patent-oriented survey
by **M. Sideris**

Volume 116 **Catalysis and Automotive Pollution Control IV.**
Proceedings of the 4th International Symposium (CAPoC4), Brussels, Belgium, April 9-11, 1997
edited by **N. Kruse, A. Frennet and J.-M. Bastin**

Volume 117 **Mesoporous Molecular Sieves 1998**
Proceedings of the 1st International Symposium, Baltimore, MD, U.S.A., July 10-12, 1998
edited by **L. Bonneviot, F. Béland, C. Danumah, S. Giasson and S. Kaliaguine**

Volume 118 **Preparation of Catalysts VII**
Proceedings of the 7th International Symposium on Scientific Bases for the Preparation of Heterogeneous Catalysts, Louvain-la-Neuve, Belgium, September 1-4, 1998
edited by **B. Delmon, P.A. Jacobs, R. Maggi, J.A. Martens, P. Grange and G. Poncelet**

Volume 119 **Natural Gas Conversion V**
Proceedings of the 5th International Gas Conversion Symposium, Giardini-Naxos, Taormina, Italy, September 20-25, 1998
edited by **A. Parmaliana, D. Sanfilippo, F. Frusteri, A. Vaccari and F. Arena**

Volume 120A **Adsorption and its Applications in Industry and Environmental Protection.**
Vol I: Applications in Industry
edited by **A. Dąbrowski**

Volume 120B **Adsorption and its Applications in Industry and Environmental Protection.**
Vol II: Applications in Environmental Protection
edited by **A. Dąbrowski**

Volume 121 **Science and Technology in Catalysis 1998**
Proceedings of the Third Tokyo Conference in Advanced Catalytic Science and Technology, Tokyo, July 19-24, 1998
edited by **H. Hattori and K. Otsuka**

Volume 122 **Reaction Kinetics and the Development of Catalytic Processes**
Proceedings of the International Symposium, Brugge, Belgium, April 19-21, 1999
edited by **G.F. Froment and K.C. Waugh**

Volume 123 **Catalysis: An Integrated Approach**
Second, Revised and Enlarged Edition
edited by **R.A. van Santen, P.W.N.M. van Leeuwen, J.A. Moulijn and B.A. Averill**

Volume 124 **Experiments in Catalytic Reaction Engineering**
by **J.M. Berty**

Volume 125 **Porous Materials in Environmentally Friendly Processes**
Proceedings of the 1st International FEZA Conference, Eger, Hungary, September 1-4, 1999
edited by **I. Kiricsi, G. Pál-Borbély, J.B. Nagy and H.G. Karge**

Volume 126 **Catalyst Deactivation 1999**
Proceedings of the 8th International Symposium, Brugge, Belgium, October 10-13, 1999
edited by **B. Delmon and G.F. Froment**

Volume 127 **Hydrotreatment and Hydrocracking of Oil Fractions**
Proceedings of the 2nd International Symposium/7th European Workshop, Antwerpen, Belgium, November 14-17, 1999
edited by **B. Delmon, G.F. Froment and P. Grange**

Volume 128 **Characterisation of Porous Solids V**
Proceedings of the 5th International Symposium on the Characterisation of Porous Solids (COPS-V), Heidelberg, Germany, May 30- June 2, 1999
edited by **K.K. Unger, G. Kreysa and J.P. Baselt**

Volume 129 **Nanoporous Materials II**
Proceedings of the 2nd Conference on Access in Nanoporous Materials, Banff, Alberta, Canada, May 25-30, 2000
edited by **A. Sayari, M. Jaroniec and T.J. Pinnavaia**

Volume 130 **12th International Congress on Catalysis**
Proceedings of the 12th ICC, Granada, Spain, July 9-14, 2000
edited by **A. Corma, F.V. Melo, S. Mendioroz and J.L.G. Fierro**

Volume 131 **Catalytic Polymerization of Cycloolefins**
Ionic, Ziegler-Natta and Ring-Opening Metathesis Polymerization
by **V. Dragutan and R. Streck**

Volume 132 **Proceedings of the International Conference on Colloid and Surface Science, Tokyo, Japan, November 5-8, 2000**
25th Anniversary of the Division of Colloid and Surface Chemistry, The Chemical Society of Japan
edited by **Y. Iwasawa, N. Oyama and H. Kunieda**

Volume 133 **Reaction Kinetics and the Development and Operation of Catalytic Processes**
Proceedings of the 3rd International Symposium, Oostende, Belgium, April 22-25, 2001
edited by **G.F. Froment and K.C. Waugh**

Volume 134 **Fluid Catalytic Cracking V**
Materials and Technological Innovations
edited by **M.L. Occelli and P. O'Connor**